T0348582

THE PHYSIOLOGY OF POLAR FISHES

This is Volume 22 in the

FISH PHYSIOLOGY series

Edited by Anthony P. Farrell and David J. Randall
Honorary Editor: William S. Hoar

A complete list of books in this series appears at the end of the volume

THE PHYSIOLOGY OF POLAR FISHES

Edited by

ANTHONY P. FARRELL
Biological Sciences Department
Simon Fraser University
Burnaby, Canada

JOHN F. STEFFENSEN
Marine Biological Department
University of Copenhagen
Helsingør, Denmark

ELSEVIER
ACADEMIC
PRESS

AMSTERDAM • BOSTON • HEIDELBERG • LONDON
NEW YORK • OXFORD • PARIS • SAN DIEGO
SAN FRANCISCO • SINGAPORE • SYDNEY • TOKYO

Cover photo credit: "Polar cod (Boreogadus saida) in ice" photo by B. Gulliksen

Elsevier Academic Press
525 B Street, Suite 1900, San Diego, California 92101-4495, USA
84 Theobald's Road, London WC1X 8RR, UK

This book is printed on acid-free paper.

For all information on all Elsevier Academic Press publications visit our Web site at www.books.elsevier.com

ISBN-13: 978-0-12-350446-3
ISBN-10: 0-12-350446-5

Printed and bound by CPI Group (UK) Ltd, Croydon, CR0 4YY
Transferred to Digital Print 2012

CONTENTS

CONTRIBUTORS — ix

PREFACE — xi

1. The Arctic and Antarctic Polar Marine Environments
   Arthur L. DeVries and John F. Steffensen

   I. Introduction — 1
   II. The Polar Marine System — 2
   III. Physical and Chemical Properties of Polar Seawater — 5
   IV. Primary Production and Seasonal Light Availability — 7
   V. The Arctic Marine Region — 8
   VI. The Antarctic Marine Region — 12
   VII. Glacial and Geological History of the Polar Regions and its Relation to
       Polar Fish Faunas — 19
       References — 22

2. Systematics of Polar Fishes
   Peter Rask Møller, Jørgen G. Nielsen, and M. Eric Anderson

   I. Introduction — 25
   II. Biogeography — 30
   III. A Classification of Fishes Occurring in Arctic and Antarctic Regions,
       with an Annotated List of Fish Families and Notes on
       Revisions and Phylogenetic Hypotheses — 33
       References — 69

3. Metabolic Biochemistry: Its Role in Thermal Tolerance and in
   the Capacities of Physiological and Ecological Function
   H. O. Pörtner, M. Lucassen, and D. Storch

   I. Introduction — 79
   II. Cold Adaptation: Performance Levels and Mode of Metabolism — 88

  III. Membrane Functions and Capacities: Constraints in Ion and
       Acid–Base Regulation   113
  IV. Molecular Physiology of Metabolic Functions   120
   V. Trade-Offs in Energy Budgets and Functional Capacities:
      Ecological Implications   128
  VI. Summary   138
      References   139

4.   Antifreeze Proteins and Organismal Freezing Avoidance in Polar Fishes
Arthur L. DeVries and C.-H. Christina Cheng

    I. Introduction   155
   II. Freezing Challenge to Hyposmotic Teleost Fish in Ice-Laden Freezing
      Marine Environments   157
  III. Freezing Avoidance Strategies   158
  IV. Organismal Freezing Points   161
   V. Types of Antifreeze Proteins   163
  VI. Noncolligative Lowering of the Freezing Point by Antifreeze Protein   168
  VII. Environmental Ice and Exogenous/Endogenous Ice in Polar Fish   174
 VIII. The Integument as a Physical Barrier to Ice Propagation   179
  IX. Synthesis and Distribution of Antifreeze Proteins in Body Fluids   181
   X. Stability of Undercooled Fish Fluids Lacking Antifreeze Proteins   185
  XI. Serum Hysteresis Levels and Environmental Severity   187
  XII. Mechanism of Organismal Freeze Avoidance   191
      References   193

5.   Respiratory Systems and Metabolic Rates
John F. Steffensen

    I. The Respiratory System   203
   II. Metabolic Rates   214
  III. Conclusion   231
      References   233

6.   The Circulatory System and its Control
Michael Axelsson

    I. The Cardiovascular System in the Cold   239
   II. Anatomy and Control of the Fish Heart   251
  III. The Branchial and Systemic Vasculature   262
  IV. Integrated Cardiovascular Responses   268
   V. Summary   272
      References   273

7.   Blood-Gas Transport and Hemoglobin Function in Polar Fishes: Does Low Temperature Explain Physiological Characters?
     R. M. G. Wells

       I. Introduction   281
      II. Mechanism of Evolution in Physiological Systems   283
     III. Expression and Significance of Multiple Hemoglobin Components   284
      IV. Functional Properties of Hemoglobins   291
       V. Respiratory Functions of Blood: Role of Blood in Maintaining Homeostasis 296
      VI. Reflections and Perspectives   302
          References   306

8.   Antarctic Fish Skeletal Muscle and Locomotion
     William Davison

       I. Introduction   317
      II. Basic Fish Muscle Anatomy   319
     III. Mechanical and Physiological Properties of Isolated Muscle   327
      IV. Protecting the Cell   329
       V. Exercise and Energy Supply   330
      VI. Swimming   334
     VII. Exercise and Temperature   337
          Reference   341

9.   The Nervous System
     John MacDonald and John Montgomery

       I. Introduction   351
      II. Nervous System: Structure and Function   353
     III. Sensory Physiology of Polar Fishes   366
      IV. Concluding Remarks   375
          References   378

INDEX   385

OTHER VOLUMES IN THE SERIES   395

CONTRIBUTORS

The numbers in parentheses indicate the pages on which the authors' contributions begin.

M. ERIC ANDERSON *(25)*, *The South African Institute for Aquatic Biodiversity, J.L.B. Smith Institute of Ichthyology, South Africa*

MICHAEL AXELSSON *(239)*, *Department of Zoology/Zoophysiology, University of Göteborg, Göteborg, Sweden*

C.-H. CHRISTINA CHENG *(155)*, *Department of Animal Biology, University of Illinois, Urbana-Champaign, Illinois*

WILLIAM DAVISON *(317)*, *School of Biological Sciences, University of Canterbury, Christchurch, New Zealand*

ARTHUR L. DE VRIES *(1, 155)*, *Department of Animal Biology, University of Illinois, Urbana-Champaign, Illinois*

M. LUCASSEN *(79)*, *Alfred-Wegener-Institut, für Polar- und Meeresforschung, Bremerhaven, Germany*

JOHN MCDONALD *(351)*, *School of Biological Sciences, The University of Auckland, Auckland, New Zealand*

JOHN MONTGOMERY *(351)*, *Leigh Marine Laboratory, School of Biological Sciences, The University of Auckland, Auckland, New Zealand*

PETER RASK MØLLER *(25)*, *Zoological Museum, University of Copenhagen, Copenhagen, Denmark*

JØRGEN G. NEILSEN *(25)*, *Zoological Museum, University of Copenhagen, Copenhagen, Denmark*

H. O. PÖRTNER *(79)*, *Alfred-Wegener-Institut, für Polar- und Meeresforschung, Bremerhaven, Germany*

JOHN F. STEFFENSEN *(1, 203)*, *Marine Biological Laboratory, University of Copenhagen, Helsingør, Denmark*

D. STORCH *(79)*, *Alfred-Wegener-Institut, für Polar- und Meeresforschung, Bremerhaven, Germany*

R. M. G. WELLS *(281)*, *School of Biological Sciences, The University of Auckland, Auckland, New Zealand*

PREFACE

Compared with the temperature extremes that prevail in terrestrial environments of the Arctic and Antarctic, the polar aquatic environment provides a relatively stable temperature for animals. Nevertheless, the temperatures experienced by polar fishes in the Arctic and Antarctic environments can be sub-zero and these temperatures are certainly frigid in terms of physiological processes, especially when compared with tropical fishes (see volume 21 of the *Fish Physiology* series). Despite these frigid temperatures, polar aquatic environments are remarkably productive and support large populations of birds and mammals that feed extensively upon fish. This tells us that polar fishes exploit these environments either seasonally through acclimatization processes or are year round residents with special adaptations for frigid water temperatures, which are especially challenging during the winter: water temperature reaches a nadir; gas exchange through the ice is limited; and the lack of sunlight limits photosynthetic activity at the bottom of the food chain.

The overarching focus of this book is on the physiological adaptations that evolved to allow certain fishes to exploit the Arctic and Antarctic. Therefore, the primary question answered by each author is: What is special about the physiology of fish from the stenothermal Arctic and Antarctic environments? It turns out that certain aspects of the physiology of polar fishes can show some remarkable and unique features. For example, some polar fish have no red blood cells; the blood is straw-colored rather than the usual red color of vertebrate blood. Similarly, some polar fishes have pale-colored heart muscle in which the red intracellular respiratory pigment myoglobin is not expressed. In other ways the physiology of polar fishes is normal compared with temperate species. For example, extensive and somewhat controversial earlier literature had suggested that polar fishes had an elevated metabolic rate – termed metabolic compensation. However, more careful measurements of metabolic rate in a variety of polar fishes have shown that the metabolic compensation theory should be abandoned; the earlier studies appear to have reported elevated metabolic rates from stressed fish. An unwritten message here is that research on the physiology of polar fishes presents difficult technical challenges. Indeed, some aspects of their physiology are more difficult to work on than others. Therefore, although an attempt was made to provide broad coverage of physiological processes, certain aspects of physiology are more amply represented than others in the chapters that follow. This uneven coverage is a true reflection of the knowledge base.

A more difficult central question that the authors tried to answer was: are there common themes to the physiology for these fishes that live poles apart but in frigid water? It is evident that the majority of the knowledge base on the physiology of polar fishes is for a limited number of Antarctic species, in particular the notothenioids. The physiology of Arctic species is woefully under-represented. While a comprehensive answer to the possibility of a convergent evolution of polar survival strategies must await future research, it is clear that the physiogeographies of the Arctic and Antarctic show fundamental differences. Since it is the physical environment that shapes the physiology of animals, at least from the perspective of evolutionary adaptation, the greater geographic isolation of Antarctic waters may possibly explain the more extreme physiological adaptations seen for certain Antarctic fishes, which are, as yet, undiscovered in Arctic fishes.

We know that certain eurythermal fishes can acclimate to temperatures experienced in the polar environments. Also, fish in deep lakes and deep seas (see volume 16 of the *Fish Physiology* series) face stenothermal environments close to the upper temperature range of the polar environments. Therefore, this observation raises a third question that each author addressed: how do polar fishes differ from those fishes that are more eurythermal and can acclimatize to severe cold? To address this question, authors included brief comparisons with the eurythermal fishes.

As this book reveals, our state of knowledge on the physiology of polar fishes is limited and fragmentary, and greatly favours ocean over freshwater species despite the existence of numerous deep polar lakes. Also, human nature (and often research funding) is such that we tend to study fascinating but extreme forms, and if information is limited, a bias may appear in our view of the general physiology of animals, such as polar fishes, if indeed common ground can be found. Consequently, because much remains to be discovered, many of the authors' conclusions are cautious ones.

One thing is certain, however. Polar temperatures are warming, a phenomenon that is especially evident in the Canadian Arctic. To what degree this will challenge and change Arctic fishes is unknown, but the changes have already started. I recently participated in a research expedition to Greenland, hoping to study Arctic cod at the southern end of their known geographic distribution. However, we were unable to capture even one Arctic cod at the Danish Arctic Marine Station. Apparently, the seawater temperature in this location is now too warm for Arctic cod. Their southern distribution is now further north. Clearly, we had arrived a few years too late.

Anthony P. Farrell
Vancouver, May 2005

1

THE ARCTIC AND ANTARCTIC POLAR MARINE ENVIRONMENTS

ARTHUR L. DEVRIES
JOHN F. STEFFENSEN

I. Introduction
II. The Polar Marine System
III. Physical and Chemical Properties of Polar Seawater
   A. Effect of Temperature on the Density and Viscosity of Seawater
   B. Salinity and Freezing of Polar Waters
   C. Oxygen Solubility and Temperature
IV. Primary Production and Seasonal Light Availability
V. The Arctic Marine Region
VI. The Antarctic Marine Region
   A. The Antarctic Continental Shelf
   B. McMurdo Sound
VII. Glacial and Geological History of the Polar Regions and its Relation to Polar Fish Faunas

I. INTRODUCTION

The overarching focus of this book on polar fishes is on the physiological adaptations that evolved to allow certain fishes to exploit the frigid and yet biologically productive Arctic and Antarctic marine regions. Although the available information may limit the scope of any given chapter, the intent is that three central themes run through each chapter and the volume as a whole:

- What is special about the physiology of fish from the stenothermal Arctic and Antarctic environments?
- Are there common themes to the physiology of these fishes that live poles apart but in frigid water? (If not, what might be the basis for the differences?)

The Physiology of Polar Fishes: Volume 22
FISH PHYSIOLOGY

- How do these fishes differ from those fishes that are more eurythermal and can acclimatize to severe cold?

With respect to the second of these questions, and as will become apparent in each chapter, consideration must be given to the important fact that the Arctic and Antarctic Oceans differ in their physiogeographic characteristics, despite that both share many features beyond being just frigid. Consideration of physiogeographic characteristics is important for two reasons. Foremost, it is the physical environment that shapes the physiology of animals, from the perspectives of both evolutionary adaptation and acclimatory responses to short-term environmental changes. Second, the information base available to authors is limited and fragmentary. Therefore, comparisons both between the Arctic and Antarctic and within each of these environments cannot always be made evenly, so authors, out of necessity, may caution the reader about some of the generalities they make. In view of this, we open this volume with a succinct comparison of the physiogeographic characteristics of the Arctic and Antarctic marine systems to illustrate similarities and differences.

II. THE POLAR MARINE SYSTEM

The northern and southern polar marine regions have some striking physiogeographic differences but share a number of climatic and oceanographic features. The Arctic Ocean is mostly surrounded by continental land masses and connects to the Atlantic and Pacific Oceans through a number of cold shallow seas and narrow passageways separating the continental land masses and Arctic islands (Figure 1.1). The Arctic Ocean is an oceanic basin with restricted circulation and a large input of Atlantic Ocean water that lacks oceanic fronts. The outflow of cold saline water is largely through the Fram Strait into the shallow Greenland Sea where it flows southwards as a distinct water mass. In contrast, Antarctica is an ice-covered continent surrounded by a vast Southern Ocean that is separated from the Atlantic, Pacific, and Indian Oceans by a distinct oceanographic front but nevertheless exchanges large amounts of heat and water with the surrounding oceans (Smith and Sakshaug, 1990) (Figure 1.2).

Both marine regions mostly freeze during the winter and their surface waters are at their freezing point (-1.8 to $-1.9\,°C$); however, much of the ice melts during the summer. The freezing conditions of the marine waters are largely an expression of the cold climates of the regions. Also, during their respective winter season, both regions experience long periods of darkness. In the high latitudes, the sun may be below the horizon for as much as

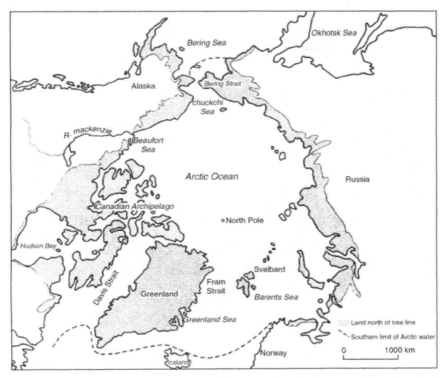

Fig. 1.1. Arctic marine region showing Arctic Ocean, major Arctic seas, and northern land masses. (Adapted from Sugden, 1982.)

4 months of the year, while in the summer there may be 4 months of 24 hours of daylight. Because biological production is highly dependent on light availability, there is a substantial pulse of primary production but only during the summer, necessitating carbon storage in primary consumers in the form of lipids to survive the winter. Indeed, diurnal fluctuations of activity, if they exist, are unlikely to be dominated by photoperiod effects, as is the case in temperate and tropical regions.

These characteristics, namely, freezing seawater, thick ice cover, and pulses of primary production, are the important environmental extremes that members of the fish fauna of the southern and northern polar regions have adapted to over evolutionary time during the colonization of these waters. These environmental extremes are most relevant in the shallow waters. We, therefore, focus in detail on the shallow-water marine environments as they pertain to the aforementioned environmental extremes.

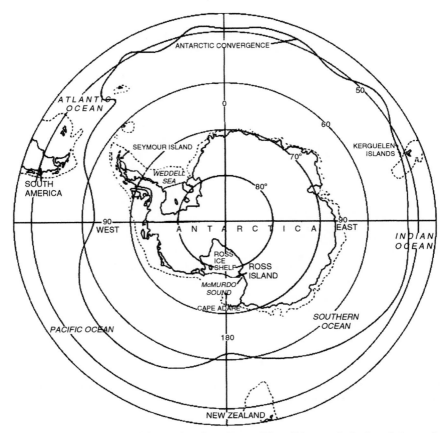

Fig. 1.2. Antarctica centered in the Southern Ocean. The 1000-meter isobath and Antarctic Front or Convergence are indicated by the dashed and solid line, respectively. (Adapted from Eastman, 1993.)

Although these conditions may indirectly affect the little-known deepwater fish fauna of the Southern Ocean too, there are fewer deepwater (>1000 m) species, they are difficult to capture and keep alive, and few physiological studies have been done with them.

Before describing the two polar environments in detail, a brief review of the physical and chemical properties of seawater is in order because at these low temperatures, some of the properties of seawater differ substantially from those at higher temperatures and have implications for the physiology of the fishes.

III. PHYSICAL AND CHEMICAL PROPERTIES OF POLAR SEAWATER

A. Effect of Temperature on the Density and Viscosity of Seawater

The surface waters of the polar marine environments may warm to 0°C and above during the summer, but during the winter, they are at their freezing point of −1.8 to −1.9°C. Although the density of freshwater is at its maximum at 4°C, the density of seawater increases as temperature decreases, even though the relationship is not linear. The thermal expansion coefficient at 20–30°C is much greater than it is around 0°C (Kennett, 1982), so the small temperature changes that occur in the polar waters will have less of an effect on density than small changes in salinity. The freezing point of seawater is also influenced by pressure, and although it appears to be minor (0.005°C/101.3 kilopascals), it does have significant implications for the Antarctic fishes living in the vicinity of ice shelves where, because of the effect of pressure (Fujino *et al.*, 1974), formation of very cold water (−2.5°C) can occur at the underside of the ice shelves (Foldvik and Kvinge, 1977). Advection of this super-cold water upwards reduces the hydrostatic pressure and results in a rise in its freezing point, leading to formation of many small ice crystals in the upper part of the water column through which pelagic fishes swim. These minute ice crystals are potential ice-nucleating agents for the fish that make contact with them.

The viscosity of water increases with decreasing temperature and is approximately twice as high at 0°C as at 25°C (Vogel, 1981). The relatively high viscosity at low temperatures has significant implications for locomotion and fluid convection of all marine organisms. In the case of fishes, more energy will be required for circulation of blood, by the "opercular pumps" for ventilation of the gills with oxygenated seawater and muscular performance for swimming. There are no studies other than the reports of the relatively high viscosity of Antarctic fish blood and endolymph (Macdonald and Wells, 1987). Knowledge of the energetic costs of circulating viscous blood, movement of water over the gills, and movement through it at subzero temperatures are of interest.

B. Salinity and Freezing of Polar Waters

Salinity was originally expressed as the quantity of dissolved salts in parts per thousand (‰). It was originally measured by titration of the chloride ion concentration, [Cl], and converted to salinity by multiplying the [Cl] by 1.80655 (Sverdrup *et al.*, 1942). Presently, salinity determination is based on the measurement of conductivity using a conductivity cell and is expressed as practical salinity units (PSUs). The density of seawater is influenced by temperature and pressure, so all three parameters need to be measured simultaneously to determine density, which is important for understanding the stability and potential movement of water masses. A conductivity, temperature, and depth (CTD) logging instrument is commonly used to make high-resolution measurements from the surface to depths of a few thousand meters. From such measurements, accurate salinities and densities can be calculated, which in turn are used to identify discrete water masses and ascertain their stability characteristics. The salinity in the polar oceans varies from near zero ‰ in surface riverine plumes to approximately 35 ‰ in the deep water, so the freezing point can vary from near zero to $-1.9\,°C$.

During the initial stages of the freezing of seawater, about 70% of the salts is excluded because the ice crystal lattice cannot accommodate salt ions (Wadhams, 2000). The residual salt concentrates in the spaces between adjacent ice crystals and melts small channels through the congelation ice. As the very dense cold brine sinks, it partially mixes with the less dense water beneath but eventually reaches the bottom and accumulates on the shelf as dense cold water. Upon further mixing with the less dense underlying water, it flows off the continental shelf into the deep ocean basins and is called *bottom water*. In the case of the Antarctic Ocean, its movement toward the equator is very important in the dynamics of global ocean circulation.

The melting of salt-depleted ice generates enough freshwater that when mixed with surface water produces a reduced salinity layer that may be between 2 and 30 m deep. This surface layer is stable because it is separated from unlying water by a pycnocline and, therefore, does not readily mix with it (Smith and Sakshaug, 1990). Such stable surface layers are readily warmed by solar radiation, and most importantly they maintain the phytoplankton in the euphotic zone. Riverine input in the Arctic is also important because it not only supplies nutrients but also freshens the saline surface water imparting stability to the mixed layer, again keeping phytoplankton in the surface water. These stable water masses are often sites of intensive primary production, and they tend to concentrate zooplankton at their interfaces with more saline water masses (Ainley and DeMaster, 1990). The Antarctic Ocean lacks riverine input, and its surface is only slightly diluted by melting of the pack ice, icebergs, and precipitation. Thus, one would expect to find euryhaline species in the Arctic shallow waters in regions of riverine input, and indeed this is the case. In fact many of the rivers are inhabited by anadromous salmonids that feed in the productive marine waters during the summer and over winter in the rivers and lakes to avoid freezing. Marine smelts, herring, and two species of cod are known to inhabit brackish or nearly freshwater during part of their life cycle (Christiansen *et al.*, 1995). The polar cod,

Boreogadus, is known to congregate at freshwater saline interfaces during the summer and thus is exposed to low salinities (Ainley and DeMaster, 1990). The Antarctic notothenioid fishes on the other hand are stenohaline (O'Grady and DeVries, 1982) most likely because during their evolution they experienced only full-strength seawater.

C. Oxygen Solubility and Temperature

Oxygen solubility is inversely related to temperature and salinity (Green and Carrit, 1967). Seawater with a salinity of 35 ‰ at 0°C saturated with oxygen will contain 50% more oxygen than seawater at 20°C. The surface waters of the polar oceans are saturated with oxygen, so Arctic and Antarctic fishes probably never experience environmental hypoxic conditions (Eastman, 1993), although cessation of breathing, say, when swallowing prey, may interrupt gill ventilation. The impact of human activities on marine environments (e.g., sewage disposal and the local oxygen depletions that may occur) is poorly documented. A few deepwater Arctic fishes might, particularly in deep fjords where shallow entrance thresholds limit water exchange with the surrounding sea. The mixing of the Antarctic surface waters by wave action results in most of the continental shelf water being saturated with oxygen. In the ice-covered McMurdo Sound, Antarctica (77° 50′S, 166° 30′E), oxygen saturation has been reported to vary between 74% and 100% (Littlepage, 1965), so oxygen is never a limiting factor at all depths in McMurdo Sound.

IV. PRIMARY PRODUCTION AND SEASONAL LIGHT AVAILABILITY

Both polar regions experience periods of total darkness and 24 hours of light during part of the year. The extended daylight during the summer results in a pulse of primary production followed by an increase in phytoplankton grazers. With the return of the light, primary production is initially associated with the ice algae at the underside of the thin ice (Sullivan *et al.*, 1984). Despite snow and ice cover reducing light to low levels, ice algae still make a significant contribution to the summer primary production because they are adapted to low light levels (Holm-Hansen and Mitchell, 1991). In addition, they are the source of many species of phytoplankton that bloom in the water at the receding ice edge. In both the Arctic and the Antarctic as the season progresses, a strong pulse occurs in the stabilized surface layer associated with the receding pack ice and then spreads to the surrounding open water. Melt water from the pack ice and river runoff are important because when mixed with the upper part of the water column, they create a stable layer that ensures that phytoplankton bloom does not quickly sink below the euphotic zone, and because they are sources of essential nutrients (Smith and Sakshaug, 1990). Only in the semi-permanent thick pack ice of the Arctic Ocean is primary production reduced because of the lack of light penetration (Buch, 2001).

The grazers or primary consumers associated with the summer polar blooms often store carbon in the form of wax esters or triglycerides, which are used as energy sources in the absence of winter primary production (Smith and Schnack-Schiel, 1990). These grazers, which are mostly long-lived copepods and euphausiids are year-round prey for pelagic consumers including a few pelagic fishes in both polar regions (Ainley and DeMaster, 1990). Much of the phytoplankton in many of the shallow shelf regions sinks to the bottom and is used by a rich benthos that forms the base of an extensive food web, which includes a variety of benthic fishes in both polar regions.

V. THE ARCTIC MARINE REGION

The Arctic marine region is composed of a "land-locked" ice-covered Arctic Ocean, mostly surrounded by continental land masses. It connects to the Pacific Ocean via the shallow-water Bering Sea and to the Atlantic Ocean via the shallow Greenland and Norwegian Seas. There are also a number of small connecting channels to the north Atlantic through the Canadian Archipelago. For the purpose of this volume on polar fishes, probably the most meaningful boundary between the Arctic marine region and the Atlantic and Pacific Oceans is the mean maximum extent of winter sea ice along with its freezing seawater (Figures 1.1 and 1.3). In the Pacific sector, this would include the Bering and Okhotsh Seas (68°N) (Sugden, 1982). In the eastern Atlantic sector, the winter ice limit is around Svalbard (78°N), the edge of the continental shelves of the east and west coasts of Greenland. In the western Atlantic, it includes the water south from the Davis Strait down to 47°S along the eastern coast of Canada (Lewis, 1982). In comparison to the Antarctic marine system, that of the Arctic is only about one-third the size of the Southern Ocean marine system and the continental shelves are shallow (50–100 m).

The thermal characteristics of the Arctic Ocean itself are largely controlled by the climate of the surrounding continents and from the influx of relatively warm Atlantic surface water that flows beneath the less saline cold Arctic surface water. Approximately 80% of the water that flows into this region is Atlantic surface water, while the remaining 20% is Pacific Ocean

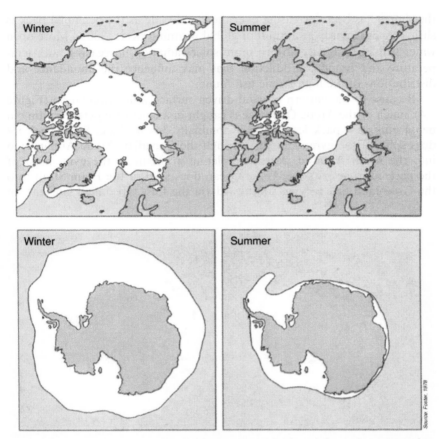

Fig. 1.3. Ice cover during the winter and summer in the Arctic and Antarctic marine regions. (Modified from Foster, 1978.)

water that enters from the Bering Sea, by way of the Bering Straits. Only 2% of the inflow is from primarily Siberian rivers, and though small on the grand scale, it is large relative to the size of the Arctic Ocean (Treshnikov and Baranov, 1973). The river water freshens the Eurasian sector of the surface water, reducing its salinity to 27‰, but it increases to 34.5‰ toward the Atlantic Ocean. The temperature of the surface water is generally close to the freezing point of saline water in the winter and varies from -0.5 to $-1.9°C$ depending on the salinity at that location. In contrast to the Southern Ocean, the ice-free surface waters of the continental shelves of the Arctic Ocean and Arctic seas may be several degrees above zero during the summer (Sugden, 1982). The ice-free water area between the perennial pack ice and the shores has been increasing over the last decade as the perennial pack is shrinking apparently as a result of global warming (Kerr, 2005). More open water will undoubtedly absorb more solar radiation, increasing both temperature and primary production that may influence the abundance and distributions of the high Arctic fish fauna.

Because of the pattern of wind-driven surface water circulation (Figure 1.4), much of the Arctic Ocean ice is caught in a clockwise gyre, resulting in dense multiyear pack ice, which is nominally 3–4 m thick. Some pack ice does split off the gyre and exits through the Canadian Arctic Archipelago into the north Atlantic. After completing a circuit in the gyre, most of the pack ice that exits the Arctic Ocean does so thru the Fram Strait into the Greenland Sea moving southward via the East Greenland Current. In

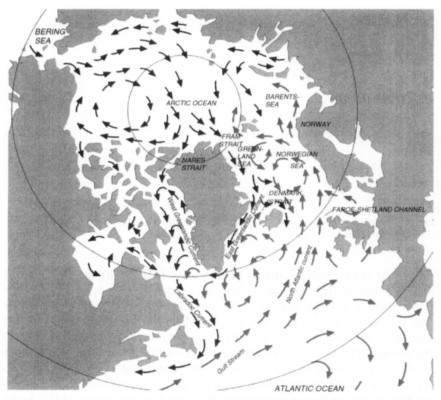

Fig. 1.4. Major ocean currents in the Arctic marine region. Dark arrows represent cold Arctic water, while the lighter arrow represents warm Atlantic water fed by the Gulf Stream. (Modified from Dietrick, 1957.)

the Arctic Ocean, because the pack is constrained by the surrounding land mass, it remains consolidated with open water only in the small leads within and with wide bands of open water between the pack and the shore. Although the shallow waters near the shore may warm considerably above zero (6°C) during the summer season, the leads within the shifting pack ice are still near freezing. The only fish that frequents the freezing leads is the polar cod, *Boreogadus saida*, and thus this species is in need of year-round freeze protection. Many of the adjoining connecting seas become ice free during the summer because of melting, wind, and current-driven ice breakup. The southerly direction of the near-shore currents moves the ice south, and the 24 hours of solar radiation of ice-free waters warms them well above the freezing point. The major outflow of cold water and ice from the Arctic Ocean is through the Fram Strait, although a small amount does flow out through the straits of the Canadian Arctic and the Nares Strait between northwestern Greenland and Ellesmere Island (Figure 1.4). Along the east coast of Greenland, the flow of the ice-laden current is through the Greenland Sea and becomes the East Greenland Current. This continues along the southern coast and around the tip of Greenland, becoming the West Greenland Current running north on the eastern side of the Davis Strait. In the northern part of the Baffin Bay, the water leaving the Arctic Ocean via the Nares Strait meets the Greenland Current. The cold mixed waters flow south and become the cold Labrador Current that contributes to the freezing conditions that exist during the winter in shallow water along much of eastern coast of Canada and the New England states during the winter (Figure 1.4).

The cold water that exits the Arctic Ocean is mostly associated with the shelf regions of Greenland and Baffin Islands. Although the Bering Sea receives little Arctic water, it has a very broad shallow shelf and associated cold water (Dayton, 1990). The shelves of the Arctic region are relatively shallow (50–200 m) and broad compared to the Antarctic regions where they are often 500–600 m deep and narrow (Anderson, 1991). High levels of primary production are associated with the shelf waters in both polar regions (Smith and Sakshaug, 1990). Luxuriant phytoplankton blooms occur over Arctic shelves because of increased surface temperatures, abundant nutrients from mixing of fronts, and from upwelling along with high levels of incident radiation. The primary production in the waters over these shallow shelves supports a complex food web. Sustainable shrimp, crab, and fin fisheries exist in many of these shelf areas (Andersen, 2001), as well as summer populations of various marine mammals and birds (Ainley and DeMaster, 1990; Dayton, 1990). Likewise in the Southern Ocean, especially in areas such as the Ross and Weddell Seas, the receding ice edge over the shelves is also the area of highest primary production with abundant primary and secondary consumers and apex predators (Ainley and DeMaster, 1990; Smith and Sakshaug, 1990).

VI. THE ANTARCTIC MARINE REGION

The most obvious extreme physical characteristics of the Southern Ocean are the year-round frigid waters and, in the southern extremes, the abundance of ice. These two environmental factors are probably the main selective forces that drove the adaptive radiation of the ancestral Antarctic fish stock that survived the cooling and freezing of the high-latitude Antarctic Ocean 10–15 million years ago (MYA).

The Antarctic marine region is an ice-covered continent surrounded by an unrestricted Southern Ocean. Geographically, the Southern Ocean is loosely defined as that body of water that encompasses the southern extremities of the Pacific, Atlantic, and Indian Oceans including the Subtropical Convergence and the waters south to the Antarctic continent. The water adjacent to the Antarctic continent is sometimes referred to as the *Antarctic Ocean*. That part of the Southern Ocean between Antarctica and the Antarctic Polar Front (50–60°S) is characterized by near-freezing surface waters for much of the year (Knox, 1970). The sub-Antarctic is that region between the Polar Front and the Subtropical Convergence (40°S), and though still cold, it is ice free throughout the year. The waters of the South Island of New Zealand and the southern tip of South America are included in this region, and as might be expected, a few notothenioid fishes (the predominant suborder of Antarctic fishes) inhabit their coastal waters (Eastman, 1993). The characteristics of these environments are more like those of northern hemisphere cold-temperate water environments in that the light regimen is diurnal and water temperatures are well above freezing throughout the year (4–12°C).

Like the Arctic Ocean and its adjoining seas, the Southern Ocean has extensive ice cover during the winter. Of importance for the fish faunas is that these waters are at their freezing point ($-1.90\,°C$) immediately beneath the ice, and in some areas of the Antarctic Ocean, unlike the Arctic, the entire water column to a depth of 600 m or more is close to its freezing point.

In contrast to the Arctic pack ice, the Southern Ocean pack ice is thinner and less dense, with much of it melting during the summer. The Southern Ocean south of 60°S is ice covered during the winter maximum, which is the month of September, and by March the coverage is reduced by 75% (Foster, 1984). In the winter, fast sea ice (ice attached to the shore) can break away and be dispersed because of the strong katabatic winds blowing off the Antarctic continent. They often cause large leads to open and the formation of near-shore polynyas (Kurtz and Bromwich, 1985; Zwally and Comiso, 1985), the latter of which play an important role in the formation of cold dense seawater because of the freezing out of mostly salt-free water. These ice-free areas are also one of the initial sites of the primary production when the light returns in the spring. Most of the fast ice adjacent to the continent that forms each winter breaks away early in the summer because of storms and becomes floating pack ice, with much of it melting as the summer season progresses. Less than 15% of the ice is carried over to the next year as multiyear pack ice. Some embayments, like McMurdo Sound, remain ice covered for all but a few weeks to a few months during the warmest part of the summer, and there are years when the ice does not break up at all and forms multiyear fast ice. Eventually with storms and associated sea swells, the multiyear ice will break up and then the freezing and breakup reverts to an annual cycle.

During the winter, there is a dramatic increase in sea ice cover in both the Arctic and the Antarctic region, and in the case of the Southern Ocean, the ice cover during the period of maximum coverage is nearly 20 million km^2, about three times that observed during the summer minimum. For the fish fauna, this wide expanse of sea ice and freezing temperatures have implications for their freezing avoidance and activity because of the effects of low temperature on energy production and activity.

One of the more important features of the Southern Ocean is the Antarctic Circumpolar Current (ACC), which flows around the continent in a clockwise direction driven by persistent westerly winds (Figure 1.5). This broad current is 200–1200 km wide in various parts of the Southern Ocean (Foster, 1984), extends to the bottom, and has surface velocities between 25 and 30 cm/s (Gordon, 1971, 1988) depending on its location. Its center is located between 47°S and 60°S depending on longitude. This current is thought to have developed about 38 MYA, thermally isolating the waters surrounding Antarctica by preventing the intrusion of northerly warm currents into the high latitudes. The thermal isolation of Antarctica is one of the contributing factors that lead to the cooling and glaciations of the continent, followed by the extensive cooling and freezing of the seawater surrounding it. This major oceanic cooling leads to extinction of the Eocene fish fauna (Eastman, 1993), leaving an underused environment that was then colonized by a fauna that was able to adapt to the cold. This current, which extends from the surface to the bottom, not only forms a physical barrier to the northerly dispersion of eggs and larvae of the Antarctic marine organisms but also serves as a mode of transport around the continent for them (Knox, 1970). Some Southern Ocean organisms have, however, escaped the ACC and colonized the more temperate shallow-water habitats of many of the sub-Antarctic islands and the cold waters around the tip of South America.

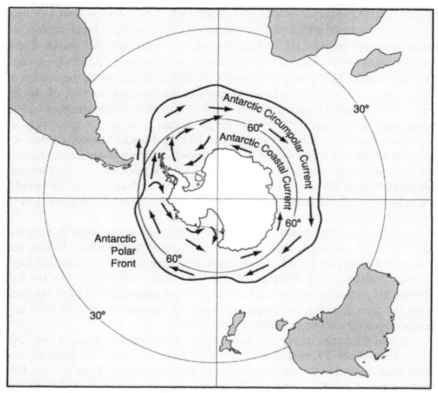

Fig. 1.5. The Southern Ocean Circumpolar Current and Antarctic Coastal Current. The heavy line is the approximate position of the Antarctic Front or Convergence. (Modified from Knox, 1985.)

The East-Wind Drift, or Antarctic Current, is a westward-flowing current along much of the coastline of Antarctica, though not circumpolar (Figure 1.5). The coastal flow into the large embayments of the Ross, Weddell, and Bellingshausen Seas creates clockwise gyres that feed into the ACC. At the interface of the ACC and the Antarctic Current, a region of divergence exists and causes upwelling of the intrusive southward-moving Circumpolar Deep Water (CDW) (Figure 1.6) near the edge of the continental shelf. This relatively warm (1–2 °C), oxygen-poor (4 mg/l), nutrient-rich water replaces the northward flow (Ekman flow) of the Antarctic surface water and the Antarctic bottom water (ABW) that flows off the continental slopes (Gordon, 1971) and is important for primary production in the continental shelf break region (Knox, 1994). North of the ACC is a zone

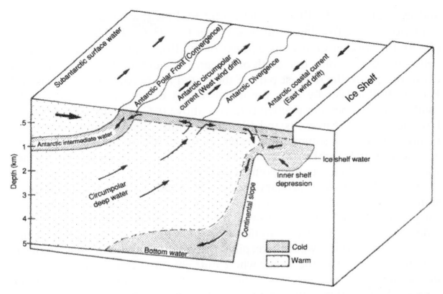

Fig. 1.6. Cross section of the Southern Ocean showing the major water masses, currents, and the intrusion of the relatively warm, nutrient-rich Circumpolar Deep Water that mixes with the cold surface water and the cold shelf water. (Modified from Gordon and Goldberg, 1970.)

of convergence, called the *Antarctic Polar Front*, also referred to as the *Antarctic Convergence*, where downwelling occurs as the denser Antarctic surface water slides beneath the less dense sub-Antarctic surface water, creating Antarctic intermediate water. The approximate position of this front is 50°S and 60°S, depending on which sector of the Southern Ocean is involved, and its position is not stationary, but it meanders with loops as far as 150 km north or south of the mean position (Foster, 1984). Abrupt temperature changes occur across the front, with the change being 4–8°C during the summer and 1–3°C during winter where the front is relatively narrow (Knox, 1970). This abrupt temperature change is most likely an important thermal barrier for the northerly distribution of Antarctic pelagic fish eggs and larvae and probably restricts their distribution primarily to the colder Antarctic surface waters (Knox, 1970). In some sectors of the Southern Ocean, this front appears to be a relatively biologically rich area in that zooplankton is concentrated there and fed on by sea birds and marine mammals (Ainley and DeMaster, 1990).

The water temperature of large portions of the Antarctic surface waters, south of 60°S is generally less than 0°C (Deacon, 1984). Temperature differences of the various water masses are only a few degrees with depth and season (Knox, 1970). Near the continental shelf, water temperatures are at or close to the freezing point of seawater ($-1.93\,°C$). In McMurdo Sound (78°S), the southern most embayment of the Ross Sea, water temperatures below 100 m vary only one-tenth of a degree throughout the year down to a depth of 750 m, the deepest part of the sound (Littlepage, 1965; Lewis and Perkin, 1985). Usually only for a few weeks during the summer in the shallow water does the temperature rise above 0°C (Figure 1.7). At lower latitudes such as Anvers Island (Palmer Station), Signy Island, and South Georgia Island, the summertime water temperatures are considerably warmer than those in McMurdo Sound; however, during the winter, the surface waters cool to their freezing points (Figure 1.7).

During the summer, primary production in the Southern Ocean follows a similar pattern as described for the Arctic region. Ice algae contribute to the early part of the bloom, followed by the receding ice edge (Smith and Sakshaug, 1990). As in the Arctic, the melt water from the pack ice helps stabilize the surface waters, keeping the phytoplankton in the euphotic region. In contrast to the Arctic where nutrients are depleted in a few weeks in the stabilized surface layer, they are rarely limiting in the Antarctic regions. In part this is due to more vertical mixing but mostly due to the high initial levels that result from the intrusion of the CDW (Lutjeharms, 1990). The mixing of the CDW with the surface water results in its conversion into fresher, colder, highly oxygenated, nutrient-rich water. The high levels of nutrients and intense summer irradiation result in high levels of primary production in the surface waters over the continental shelf and northward beyond the shelf break. In these waters, the predominant

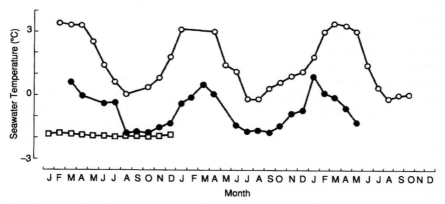

Fig. 1.7. Annual shallow-water temperature record for high-latitude Southern Ocean water (□, McMurdo Sound, 78°S), mid-latitude water (●, Signy Island, 60°S), and low-latitude water (○, South Georgia, 54°S). (Modified from Eastman, 1993; Everson, 1977.)

zooplanktors are copepods and the Antarctic krill *Euphausia superba*, as well as *Euphausia crystallorophias* (Smith and Schnack-Schiel, 1990). These zooplanktors are important prey for many of the fishes, and in some cases, they are the exclusive prey of notothenioid fishes inhabiting the continental shelf of the Antarctic Peninsula region (Kock, 1985). The krill, as well as the fish that feed on them, are important prey for the marine birds and mammals that inhabit the area during the summer season (Ainley and DeMaster, 1990).

A. The Antarctic Continental Shelf

Much of the Southern Ocean is between 3000 and 5000 m in depth, with shallow areas found along the coast, islands, and sub-sea ridges. There are also some plateaus where depths are between 500 and 1000 m, mostly in the sub-Antarctic regions. In contrast to broad shallow (100 m) continental shelves in the Arctic Ocean and adjacent seas, the Antarctic continental shelf depth averages around 500 m, and with the exception of the Ross and Weddell Seas, it is relatively narrow. It has a rugged topography in most places because of extensive past glaciations that resulted from their extension to the shelf break (Anderson, 1991). The shelf also has inner depressions that in some places are as deep as 1200 m, which were gouged out by large glaciers. In many places, the shelf break is shallower than the inner shelf, because during the last glacial maximum, some ice shelves pushed up moraines at their leading edge (Anderson, 1991). The weight of the Antarctic Ice Cap also depresses the continent, which is also a contributing factor to the depth of the shelf regions (Anderson, 1991). The lack of continental erosion and absence of riverine input also may be contributing factors to the rugged topography of the shelves, which is in sharp contrast to the Arctic region. Unlike in the Arctic, there is an absence of freshwater rivers and estuaries and organisms associated with the brackish water habitats. No fish are present in the isolated freshwater lakes in ice-free coastal areas. The relatively deep water of the continental shelves with its inner shelf depression and absence of freshwater input has probably had a significant effect on evolution of the fish fauna of the Antarctic. The deep depressions most likely form unique habitats, and some of the fish species are found only in these depressions, suggesting that the isolation of deep habitats may have played a role in the speciation of some members of the notothenioid plunder fishes.

In addition, the thick ice shelves such as the Ross Ice Shelf, the Filchner, and the Ronne cover a considerable portion of the continental shelf in the Ross and Weddell Seas. These shelves average 500 m in thickness. The absence of light beneath them precludes primary production, but nevertheless, there is a relatively abundant benthos in some areas, as well as fishes

(Littlepage and Pearce, 1964; Bruchhausen *et al.*, 1979). Presumably plankton advected in from the Ross Sea is the necessary food source for the benthos.

B. McMurdo Sound

Since a great deal is known about the physiology of the notothenioid fishes of McMurdo Sound, it is worthwhile to describe the McMurdo Sound environment in some detail. As pointed out above, its temperature varies little with season (Hunt *et al.*, 2003). Most of the sound is covered by ice for 10–12 months of the year, and it reaches a thickness of 2–3 m in the southern-most reaches. The ice generally breaks out each year to near the Ross Ice Shelf because of northerly swells and southerly winds. Occasionally, the sound may remain ice covered through the year, and in some cases, it may not break out for several years when huge icebergs dampen the northerly storm swells that usually enter the sound from the Ross Sea. In general, ice reforms in the sound in April, and by late August, it may be more than 1.5 m thick, at which time a sub-ice platelet layer begins to form beneath the congelation ice (Hunt *et al.*, 2003). The sub-ice platelet layer is composed of ice platelets about 0.2 mm in thickness by 10–20 cm in diameter, and the ice platelets freeze to each other randomly, forming a network of large crystals. Platelet ice also forms on the bottom to depths of about 30 m in large masses called *anchor ice*. Anchor ice prevents settling of sessile benthic invertebrates in many areas (Dayton *et al.*, 1969). Although the exact mechanism of ice platelet formation is unclear, we do know that it is associated with the formation of undercooled water beneath the Ross Ice Shelf and its northward flow into the sound. The platelet ice layer and bottom anchor ice appear in mid to late August and generally melt away by mid December when the current flow is predominately from the Ross Sea into the sound. The platelet layer and anchor ice are a habitat for many motile invertebrates such as starfish, isopods, and amphipods, as well as the adult notothenioid *Trematomus* fishes and *Pagothenia borchgrevinki*, the latter that forages for invertebrates and juvenile fish among the sub-ice platelets. The latter also hides in the platelet layer to avoid its predators, the Weddell seal and the emperor penguin (Fuiman *et al.*, 2002). Thus, for many of the McMurdo Sound notothenioid fishes, the platelet and anchor ice formations represent an important habitat that is the coldest and iciest environment in Antarctica. There, they have had to cope with the challenge of avoiding freezing.

The formation of undercooled water at the underside of ice shelves deserves further comment in that this freezing water pervades much more of the water column than that represented by the platelet layer and anchor ice formations. At the underside of thick ice shelves, the ice can either be forming or melting. In either case, because of the effect of hydrostatic pressure, the freezing point of seawater is depressed (Fujino *et al.*, 1974; Lewis and Perkin, 1985). Often, this water, which is at its *in situ* freezing point, flows from the underside of the ice shelf and is advected upwards either because of a density difference or because it flows over a shoal area. With the decrease of hydrostatic pressure, it then becomes undercooled water that readily nucleates, forming minute ice crystals. The small ice crystals are thought to grow into large ice platelets by the addition of water molecules mostly on planes parallel to the c-axis. At times, in the upper 30 m of the column of McMurdo Sound, one can see these minute ice crystals in a light beam as reflective particles in the water column and their presence is always associated with growth of the sub-ice platelet layer and anchor ice formations. The "capture" of several kilograms of 10–15 cm diameter ice platelets at a depth of 225 m in a self-closing net near the Filchner Ice Shelf in the Weddell Sea indicates that nucleation of undercooled water and free growth of ice crystals can occur at considerable depths in the water column (Dieckmann *et al.*, 1986). Thus, fish inhabiting the water column within 30–50 km of the leading edge of the ice shelves and near floating glacial ice tongues can encounter temperatures a few tenths of a degree below the surface water freezing point and have contact with ice. The depth of ice formation in the water column near ice shelves often corresponds to a depth slightly shallower than the underside of the ice shelf.

In contrast to environmental extremes that exist in the Antarctic waters, there are no substantial ice shelves in the Arctic region. Theoretically, the underside of floating glacier tongues off the coast of Greenland could generate similar freezing conditions, but there are no reports of undercooled water or platelet ice formation beneath that sea ice. Thus, the freezing conditions in marine environments of the Arctic region are apparently somewhat less severe from the point of view of temperature and abundance of ice, so fishes inhabiting these waters are expected to show a slightly lower resistance to freezing than their Antarctic counterparts, which indeed is the case.

VII. GLACIAL AND GEOLOGICAL HISTORY OF THE POLAR REGIONS AND ITS RELATION TO POLAR FISH FAUNAS

An understanding of the glacial and geological events that led to the present-day geography and physiogeographic characteristics of the polar regions is necessary for understanding the origin and evolution of their extant fish faunas. The Antarctic fish fauna is largely confined to the narrow continental shelves of Antarctica and is represented by the monophyletic perciform suborder Notothenioidae, composed of five closely related families. These circumpolar stenothermal fishes are separated from other southern hemisphere land masses by a deep ocean with a well-defined thermal front (Polar Front). Both of these represent barriers to the northerly dispersal of their eggs and larvae, as well as substantial barriers to southerly immigration of other fish taxa.

The breakup of Gondwanaland is thought to have been complete about 38 MYA, and the opening of the deepwater Drake Passage followed at about 25 MYA. At this time, the development of the ACC isolated the Southern Ocean from warm intrusive currents, and it began to cool. As it cooled, the Eocene fish fauna disappeared, most likely because they failed to adapt to the low water temperatures. This fauna is thought to have been replaced by a benthic blennioid ancestor (Eastman, 1993) that was able to adapt. Whatever the ancestral form, it adapted to the cold, and when the Antarctic Ocean cooled to its freezing point around 10–15 MYA, a novel biological antifreeze system evolved in the ancestral notothenioids (Cheng, 1998). This evolutionary innovation provided the essential freeze avoidance mechanism and appears to have occurred about the time the notothenioid ancestor radiated into many diverse body forms. Some evolved mouth structures specialized for a particular food source, whereas others evolved specializations for pelagic (neutral buoyancy) and semi-pelagic niches in the resource-rich unexploited water column. Although the notothenioids are not the most specious group, they are the most obvious and represent the largest biomass of the Antarctic fauna. Thus, most of the physiological and biochemical studies have been done with members of this taxa, whereas only a few deal with the zoarcid and liparid fishes whose taxonomic affinities reside mostly in the deep temperate and shallow Arctic waters.

The notothenioids have been widely used in physiological studies for a number of reasons. First, most are hardy fishes that can be captured at depths of 1000 m and brought to the surface without problem because they lack swim bladders. Many are thick-skinned benthic forms with firmly attached scales, whereas others are scaleless and are good models for studying gas transport through the integument. The ice fishes, family Channichthyidae, lack hemoglobin and have become a model for examining hemopoiesis, as well as the various circulatory adaptations necessary for a life without red blood cells. Finally, in contrast to the Arctic fauna, the Antarctic fauna appears to be stenohaline (O'Grady and DeVries, 1982) and is stenothermal (Somero and DeVries, 1967), most dying at a temperature as low as 6°C. The stenothermal and stenohaline condition is most likely a result of the long residence time in a constant low temperature and salinity environment. The glacial record of Antarctica indicates that the continent has been ice covered and the surrounding waters near their freezing points for several million years (Kennett, 1982). During glacial maxima, the ice shelves even extended to the continental shelf breaks (Anderson, 1991), leaving only the narrow continental margin as a habitat for the fish fauna. Thus, it is not unreasonable to suggest that because of their prolonged isolation in a cold stable environment, the high latitude notothenioids lost their genetic plasticity, which would allow acclimation to higher temperatures. There are some indications that those notothenioids experiencing more variable thermal environments at the higher latitudes have higher upper lethal temperatures than the strictly high-latitude species and, therefore, may have retained more genetic plasticity.

In contrast to the Antarctic, the Arctic glaciations and winter ice cover is a much more recent event having begun only 2 or 3 MYA. The Arctic Ocean is an open system in that it is connected to the Atlantic and Pacific Oceans, largely through shallow coastal shelves, and thus, a migratory route exists between the cold and warm environments. Given the taxonomic diversity of the Arctic fish fauna, it is clear that many taxa possessed the genetic plasticity to evolve adaptations to the cold.

The present-day Arctic region fish fauna is represented by diverse families that are widely separated in their taxonomic affinities (Leim and Scott, 1966). These include pleuronectids, cottids, salmonids, gadids, liparids, zoarcids, and others. These families also have wide latitudinal distributions. For example, the genus *Myoxocephalus*, a benthic sculpin in the family Cottidae ranges from Europe across the Arctic shallows down the coast of the Pacific Ocean and into the warm estuaries of Long Island on the Atlantic Coast. One pleuronectid species ranges from the freezing Bering Sea to central California (Clemens and Wilby, 1961).

As with the Antarctic fishes, colonization of the freezing Arctic waters was made possible by the evolution of a biological antifreeze system and in contrast to the Antarctic notothenioids structurally unrelated antifreezes arose several times by various molecular mechanisms in unrelated taxa (Cheng, 1998). In one case, however, the antifreeze glycopeptide that evolved in the gadids is identical to the one in the notothenioids and is now a classic example of convergent evolution at the protein level (Chen *et al.*, 1997).

When the Arctic waters cooled, the shallow-water habitats provided a path for the migration of fishes either into or out of the freezing seawater environment. There is evidence that during the last glaciations, the Arctic Ocean was completely frozen over and its exits in the Atlantic may have been blocked (Ewing and Donn, 1958). Even during extreme glaciations, the shallow water shelves at the glacial interfaces most likely provided an avenue for escape to warmer environments. It is possible that there were multiple migrations into and out of the high Arctic region during glacial minima and maxima.

The Arctic fish fauna includes species that are euryhaline such as the pleuronectids and the anadromous salmonids (Dempson and Krlstofferson, 1987). The fauna is also much more eurythermal than the Antarctic fauna, with some living at summer temperatures as high as 15°C. The responses to ranges of salinities and temperature are not unexpected because the fauna originated in more temperate waters and has not been isolated in a constant temperature/salinity environment like the Antarctic fauna. Even the Arctic polar cod, whose range is largely restricted to the high Arctic, is much more eurythermal and euryhaline (Andersen, 2001) than members of the Antarctic fauna; however, it appears to be more stenothermal than some of its gadid relatives that inhabit the freezing waters of the lower latitudes of the Arctic (Enevoldsen *et al.*, 2003).

As indicated in the introduction, it is largely the physical environment that shapes the animals physiology, and the differences in the two polar fish faunas is a relevant example of this. The glacial geological changes that led to the present physiogeographic differences in the two polar regions must also be considered when comparing and interpreting the physiological similarities and differences in the fish faunas of the two regions.

REFERENCES

Ainley, D. G., and DeMaster, D. P. (1990). The upper trophic levels in polar marine ecosystems. *In* "Polar Oceanography, Part B: Chemistry, Biology, and Geology" (Smith, W. O., Ed.), pp. 599–630. Academic Press, San Diego.

Andersen, O. N. (2001). The marine Ecosystem. *In* "The Ecology of Greenland" (Born, E., and Bocher, J., Eds.), pp. 123–135. Ministry of Environment and Natural Resources, Nuuk, Greenland.

Anderson, J. B. (1991). The Antarctic continental shelf: Results from marine geological and geophysical investigations. *In* "The Geology of Antarctica" (Tingey, R. J., Ed.), pp. 285–334. Oxford University Press, Oxford.

Bruchhausen, P. M., Raymond, J. A., Jacobs, S. S., DeVries, A. L., Thorndike, E. M., and De Witt, H. H. (1979). Fish, crustaceans and the sea floor under the Ross Ice Shelf. *Science* **203**, 449–451.

Buch, H. (2001). The ocean environment. *In* "The Ecology of Greenland" (Born, E. W., and Bocher, J., Eds.), pp. 111–122. Ministry of Environments and Natural Resources, Nuuk, Greenland.

Chen, L., DeVries, A. L., and Cheng, C. C.-H. (1997). Convergent evolution of antifreeze glycoproteins in Antarctic notothenioid fish and Arctic cod. *Proc. Natl. Acad. Sci. USA* **94**, 3817–3822.

Cheng, C.-H. C. (1998). Evolution of the diverse antifreeze proteins. *Curr. Opin. Genet. Dev.* **8**, 715–720.

Christiansen, J. S., Chernitsky, A. G., and Karamushko, O. V. (1995). An Arctic teleost fish with a noticeably high body fluid osmolality: A note on the navaga, *Eleginus navaga* (Pallas 1811), from the White Sea. *Polar Biol.* **15**, 303–306.

Clemens, W. A., and Wilby, G. V. (1961). "Fishes of the Pacific Coast of Canada." Fisheries Research Board of Canada, Ottawa.

Dayton, P. K. (1990). Polar benthos. *In* "Polar Oceanography, Part B: Chemistry, Biology, and Geology" (Smith, W. O., Ed.), pp. 631–685. Academic Press, San Diego.

Dayton, P. K., Robilliard, G. A., and DeVries, A. L. (1969). Anchor ice formation in McMurdo Sound, Antarctica, and its biological effects. *Science* **163**, 273–274.

Deacon, G. (1984). "The Antarctic Circumpolar Ocean." Cambridge University Press, Cambridge.

Dempson, J. B., and Kristofferson, A. H. (1987). Spatial and temporal aspects of the ocean migration of anadromous Arctic char. *Am. Fish. Soc. Symp.* **1**, 340–357.

Dieckmann, G., Rohardt, G., Hellmer, H., and Kipfstuhl, J. (1986). The occurrence of ice platelets at 250m depth near the Filchner Ice Shelf and its significance for sea ice biology. *Deep-Sea Res.* **33**, 141–148.

Eastman, J. T. (1993). "Antarctic Fish Biology: Evolution in a Unique Environment." Academic Press, San Diego.

Enevoldsen, L. T., Heiner, I., DeVries, A. L., and Steffensen, J. F. (2003). Does fish from the Disko Bay area of Greenland possess antifreeze proteins during the summer? *Polar Biol.* **26**, 365–370.

Ewing, M., and Donn, W. L. (1958). A theory of ice ages I. *Science* **123**, 1061–1066.

Foldvik, A., and Kvinge, T. (1977). Thermohaline convection in the vicinity of an ice shelf. *In* "Polar Oceans" (Dunbar, M. J., Ed.), pp. 247–255. Arctic Institute of North America, Calgary.

Foster, T. D. (1984). The marine environment. *In* "Antarctic Ecology" (Laws, R. M., Ed.), Vol. 2, pp. 345–371. Academic Press, London.

Fuiman, L., Davis, R., and Williams, T. (2002). Behavior of midwater fishes under the Antarctic ice: Observations by a predator. *Mar. Biol.* **140**, 815–822.

Fujino, K., Lewis, E. L., and Perkin, R. G. (1974). The freezing point of seawater at pressures up to 100 Bars. *J. Geophys. Res.* **79**, 1792–1797.

Gordon, A. L. (1971). Recent physical oceanographic studies of Antarctic waters. *In* "Research in the Antarctic" (Quam, L. O., Ed.), pp. 609–629. American Association for Advancement of Science, Washington, DC.

Gordon, A. L. (1988). Spatial and temporal variability within the Southern Ocean. *In* "Antarctic Ocean and Resources Variability" (Sahrhage, D., Ed.), pp. 41–56. Springer-Verlag, Berlin, Heidelberg.

Green, E. J., and Carrit, D. E. (1967). New tables for oxygen saturation of seawater. *J. Mar. Biol.* **25**, 140–147.

Holm-Hansen, O., and Mitchell, B. G. (1991). Spatial and temporal distribution of phytoplankton and primary production in the western Bransfield Strait. *Deep Sea Res.* **38**, 961–980.

Hunt, B. M., Hoefling, K., and Cheng, C.-H. C. (2003). Annual warming episodes in seawater temperatures in McMurdo Sound in relationship to endogenous ice in notothenioid fish. *Antarct. Sci.* **15**, 333–338.

Kennett, J. P. (1982). "Marine Geology." Prentice-Hall, New Jersey.

Kerr, R. A. (2005). Scary Arctic ice loss? Blame the wind. *Science* **307**, 203.

Knox, G. A. (1970). Antarctic marine ecosystems. *In* "Antarctic Ecology" (Holdgate, M. W., Ed.), Vol. 1, pp. 69–96. Academic Press, London.

Knox, G. A. (1994). "The Biology of the Southern Ocean." Cambridge University Press, Cambridge.

Kock, K.-H. (1985). Krill consumption by Antarctic notothenioid fish. *In* "Antarctic Nutrient Cycles and Food Webs" (Siegfried, W. R., Condy, P. R., and Laws, R. M., Eds.), pp. 437–451. Springer-Verlag, Berlin, Heidelberg.

Kurtz, D. D., and Bromwich, D. H. (1985). A recurring, atmospherically forced polynya in Terra Nova Bay. *In* "Oceanology of the Antarctic Continental Shelf" (Jacobs, S. S., Ed.), Vol. 43, pp. 177–201. American Geophysical Union, Washington, DC.

Leim, A. H., and Scott, W. B. (1966). "Fishes of the Atlantic Coast of Canada." Fish Research Board Canada, Ottawa.

Lewis, E. L. (1982). The Arctic Ocean: Water masses and energy exchanges. *In* "The Arctic Ocean" (Rey, L., Ed.), pp. 43–68. John Wiley & Sons, New York.

Lewis, E. L., and Perkin, R. G. (1985). The winter oceanography of McMurdo Sound, Antarctica. *In* "Oceanology of the Antarctic Continental Shelf" (Jacobs, S. S., Ed.), Vol. 43, pp. 145–165. American Geophysical Union, Washington, DC.

Littlepage, J. L. (1965). Oceanographic investigations in McMurdo Sound, Antarctica. *In* "Biology of the Antarctic Seas II" (Llano, G. A., Ed.), Vol. 5, pp. 1–37. American Geophysical Union, Washington, DC.

Littlepage, J. L., and Pearce, J. C. (1964). Biological and oceanographic observations under an Antarctic ice shelf. *Science* **137**, 679–680.

Lutjeharms, J. R. E. (1990). The oceanography and fish distribution of the Southern Ocean. *In* "Fishes of the Southern Ocean" (Gon, O., and Heemstra, P. C., Eds.), pp. 6–27. J. L. B. Smith Institute of Ichthyology, Grahamstown.

Macdonald, J. A., and Wells, R. M. G. (1987). Viscosity of body fluids from Antarctic notothenioid fish. *In* "Biology of Antarctic Fish" (Prisco, G. D., Maresca, B., and Tota, T., Eds.). Springer-Verlag, Berlin.

O'Grady, S. M., and DeVries, A. L. (1982). Osmotic and ionic regulation in polar fishes. *J. Exp. Mar. Biol. Ecol.* **57**, 219–228.

Smith, S. L., and Schnack-Schiel, S. B. (1990). Polar zooplankton. *In* "Polar Oceanography" (Smith, W. O., Ed.). Academic Press, San Diego.

Smith, W. O., and Sakshaug, E. (1990). Polar phytoplankton. *In* "Polar Oceanography" (Smith, W. O., Ed.), Vol. B, pp. 477–525. Academic Press, San Diego.

Somero, G. N., and DeVries, A. L. (1967). Temperature tolerance of some Antarctic fishes. *Science* **156**, 257–258.

Sugden, D. (1982). "Arctic and Antarctic." Basil Blackwell, Oxford.

Sullivan, C. W., Palmisano, A. C., and Soo Hoo, J. B. (1984). Influence of sea ice and sea ice biota on downwelling irradiance and spectral composition of light in McMurdo Sound. *In* "Proceeding of the SPIE-The International Society for Optical Engineering" (Blizarrd, M. A., Ed.), Vol. 489, pp. 159–165. SPIE-The International Society for Optical Engineering, Bellingham, Washington.

Sverdrup, H. E., Johnson, M. W., and Flemming, R. H. (1942). "The Oceans: Their Physics, Chemistry, and General Biology." Prentice-Hall, Englewood Cliffs, New Jersey.

Treshnikov, A. F., and Baranov, G. I. (1973). "Water Circulation in the Arctic Basin." Israel Program for Scientific Translations, Jerusalem.

Vogel, S. (1981). "Life in Moving Fluids: The Physical Biology of Flow." Willard Grant, Boston.

Wadhams, P. (2000). "Ice in the Ocean." Overseas Publisher Association, Amsterdam.

Zwally, J. H., and Comiso, J. C. (1985). Antarctic offshore leads and polynyas and oceanographic effects. *In* "Oceanology of the Antarctic Continental Shelf" (Jacobs, S. S., Ed.), Vol. 43, pp. 203–226. American Geophysical Union, Washington, DC.

2

SYSTEMATICS OF POLAR FISHES

PETER RASK MØLLER
JØRGEN G. NIELSEN
M. ERIC ANDERSON

I. Introduction
   A. Definition
   B. Historical View
II. Biogeography
   A. Distribution Patterns
   B. Origin of Polar Fishes
III. A Classification of Fishes Occurring in Arctic and Antarctic Regions, with an Annotated List of Fish Families and Notes on Revisions and Phylogenetic Hypotheses

I. INTRODUCTION

Description and ordering of taxa form the basic fundament of all biological disciplines, including physiology. The systematics of polar fishes is an active research field: New species are still being described from both Arctic and Antarctic areas, because of taxonomic revisions and new collections from continuing fisheries and scientific expeditions. The phylogenetic relationships of families, genera, and species are now being studied more intensively than ever, mainly because of the development of new molecular methods, in which the sequences of various DNA nucleotides are used. Phylogenetic analyses based on morphological characters are still equally relevant, but because they are usually based on time-consuming examination of large samples, the molecular-based phylogenies have become more common. Ideally, both morphological and molecular data should be employed in the reconstructions of evolution. In the family accounts below, we mention as many phylogenetic studies as possible, because an understanding of phylogenetic relationships is often essential for the interpretation of results in comparative physiological studies. Mapping of physiological "characters"

The Physiology of Polar Fishes: Volume 22
FISH PHYSIOLOGY

on phylogenetic trees provides an excellent overview of how the physiological characters evolved (e.g., did the antifreeze proteins evolve more than once in codfishes?). From Figure 2.1, it is evident that it happened either two independent times (in the *Gadus-Arctogadus* clade [with a secondary loss in *Theragra*] and in the *Eleginus-Microgadus* clade) or only once in the *Gadus-Microgadus* clade with secondary losses in *Theragra*, *Melanogrammus-Merlangius*, and *Pollachius* clades. The first-mentioned hypothesis is the most parsimonious, because it involves only three evolutionary steps in contrast to four in the last mentioned. Some precautions to the above example should be taken because six species (*M. proximus*, *P. pollachius*, *T. chalcogramma*, *T. minutes*, *T. luscus*, and *G. argenteus*) have not been examined for antifreeze proteins (C. Cheng, personal communication, April 2003). The mutual benefit of the two disciplines is seen when physiological characteristics in conflict with phylogenies inspire evolutionary biologists to reexamine the phylogenetic reconstruction of a group. Many families have, however, not been studied phylogenetically, and the available work should be viewed as a first stage in a far from finished process.

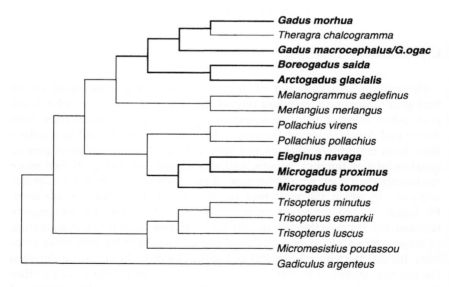

Fig. 2.1. Presence of antifreeze proteins (in bold) mapped on a cytochrome *b*–based phylogenetic reconstruction of all recent codfishes, Gadinae, except *Eleginus gracilis*, *Micromesistius australis*, and *Theragra finnmarchica*. (Modified from Møller *et al.*, 2002.)

A. Definition

In this chapter, we define polar fishes from a biogeographical point of view, employing areas with major shifts in species composition as borders for the Arctic and Antarctic areas. These are often defined by oceanic frontal systems and topographical structures such as submerged ridges. Sub-Arctic and sub-Antarctic areas are not included because these have a degree of endemism and are often visited by summer migrants from temperate areas.

In the North Atlantic, we follow Andriashev and Chernova (1995), with a few exceptions (Figure 2.2a). In the west, the border goes from Belle Isle Strait (*c.* 52 °N) up the coast of Canada (depths <50 m) to Durban Island (*c.* 68 °30′N), along the Canada-Greenland submerged ridge (maximum depth *c.* 700 m) to near Assiat (*c.* 68 °40′N), along the coast of southern Greenland (depths <50 m), along the Greenland-Iceland submerged ridge (maximum depth *c.* 630 m), northern Iceland, along the Iceland-Faroe Island-Scotland ridge (maximum depth 657 m), including the Faroese Trench, in the Norwegian Sea at depths more than 500 m, to the Svalbard Archipelago, in the Barents Sea north of *c.* 74 °N to Novaya Zemlya, but including the White Sea east to Murmansk. Outside this area, many relict populations of Arctic marine fishes are found (e.g., in the Baltic Sea, Gulf of St. Lawrence, and Norwegian fjords) (Ekman, 1953; Scott and Scott, 1988; Christiansen and Fevolden, 2000).

In the North Pacific, Arctic waters are defined as shallow (<90 m) Bering Sea water, south to near Cape Navarin, St. Lawrence Island, and the mouth of the Yukon River (Andriashev and Chernova, 1995).

For the Arctic freshwater fishes, we generally follow the 10 °C July isotherm (Stonehouse, 1989) as the definition for the southern limit of the polar area (Figure 2.2A). The species included generally follow Berg (1932) for the Eurasian area and Mecklenburg *et al.* (2002) for the American area.

Cold-adapted fishes are also found in lower latitudes, for instance, in the northern Sea of Okhotsk, and Sea of Japan, but are not included here. Cold-adapted freshwater fishes are also found in high-altitude alpine areas in low latitudes, but these are not included here.

In the Antarctic, we follow the regional classification synthesized by Anderson (1990) (Figure 2.2B). Antarctic polar waters were divided into two provinces, the South Polar (continental areas) and the South Georgian (eastern Scotia Sea). These areas are well defined by being south of the temperature-driven Antarctic Polar Front and do not include the sub-Antarctic Burdwood Bank (an extension of the Magellan Province), the Prince Edward, and Crozet Islands, Kerguelen Plateau, and the sub-Antarctic islands of New Zealand. Deepwater fishes known from Banzare Bank (*ca.* 60 °S, 80 °E) are included as Antarctic as they have, or most likely

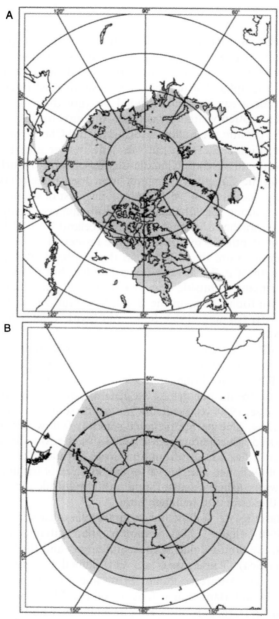

Fig. 2.2. Map of the Arctic (A) and Antarctic (B) regions. See the text for definitions.

have, widespread distributions around the continent. Other species found on Banzare Bank are excluded if they have an otherwise subantarctic distribution.

B. Historical View

The description of fishes occurring in the Arctic area began with Linnaeus (1758), who included 22 Atlantic boreal fishes in his *Systema Naturae*. The first endemic Arctic species were described in the late eighteenth century, mainly from shallow waters. During the eighteenth and nineteenth centuries, the rate of description of new species from the Arctic zone was relatively stable, with a peak toward the end of the nineteenth century, when many expeditions took place. Toward the end of the twentieth century, the addition of new species declined, but new species are still being described (Figure 2.3).

In the Antarctic area, a few widespread species were known from the late eighteenth century, and the first endemic Antarctic species was described as late as 1844 by Richardson. The description of the Antarctic fish fauna had a peak around 1900, also due to several expeditions. After a stable addition of species until the middle of the twentieth century, a large number of species

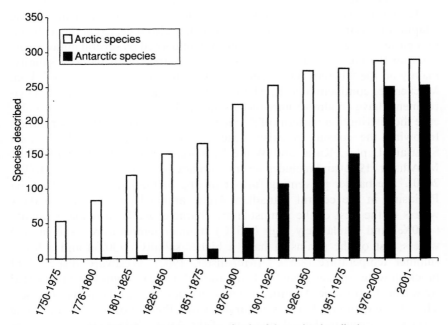

Fig. 2.3. Cumulative number of polar fish species described.

were described toward the end of the century. Judging from the number of species known from only the holotype, there are reasons to expect a continuous growth in the fauna identified in the twenty-first century.

II. BIOGEOGRAPHY

A. Distribution Patterns

The two polar areas of the world contain 538 species of fish, 289 in the Arctic, and 252 in Antarctica (Figure 2.3). Many families contain antitropical species, but only three widespread pelagic species (*Lamna nasus*, *Cyclothone microdon*, and *Arctozenus risso*) are found within both the Arctic and Antarctic zones. Only 12 (*Lamna*, *Amblyraja*, *Bathyraja*, *Cyclothone*, *Arctozenus*, *Protomyctophum*, *Coryphaenoides*, *Macrourus*, *Micromesistius*, *Paraliparis*, *Careproctus*, and *Lycenchelys*) of 214 polar fish genera and 10 (Lamnidae, Rajidae, Gonostomatidae, Bathylagidae, Myctophidae, Paralepididae, Gadidae, Macrouridae, Liparidae, Zoarcidae) of the 72 polar fish families are found in both areas (see Appendix 2.1). The following section describes the rather different evolutionary forces acting on the two polar areas and discusses the biogeographic links between the two fish faunas.

Despite its youth, the Arctic fish fauna displays a number of interesting adaptations to the harsh environment, which is also reflected in the distribution patterns (see below). Arctic waters are influenced by several major rivers entering the coastal waters, resulting in huge areas with reduced salinity, mainly on the Siberian (Ob, Yenisei, Lena), and eastern Canadian (Yukon, Mackenzie) continental shelf, whereas the Atlantic part and deep-sea environments have an almost unreduced salinity. These differences clearly influence the distribution patterns of Arctic fishes. The species with a distribution following the coastal brackish waters of the Beaufort, Chukchi, East Siberian, Laptev, Kara, and White Seas were termed the *Inuit fauna* by McAllister (1962) and includes several families (see also Craig, 1984; Møller, 2000). The barrier preventing further mixing of the faunas is the Boothia Peninsula in the central Canadian Arctic archipelago since the shallow waters to the west of the peninsula are characterized by a reduced salinity (<30‰) and a relatively high (>5 °C) summer temperature, whereas the waters east of the peninsula have a higher salinity (>30‰) and a lower summer temperature (<5 °C) (McAllister, 1962). Westward dispersal of brackish water species from the Kara and White Seas seems to be prevented by warm, high saline Atlantic currents entering the southern Barents Sea. The aforementioned barriers naturally work the opposite way for saline species occurring in the Atlantic part of the Arctic. Briggs (1974) mentions the deep Fram Strait between Svalbard and Greenland as an important barrier for Arctic fishes, but this does not seem to be the case because examples of Arctic species occurring in Svalbard and not in northeast Greenland are lacking. Barents Sea fishes not found in East Greenland are all confined to the easternmost brackish part of the sea and are not found in the western saline part, including Svalbard. Several species occurring in West Greenland (e.g., *Lycodes mucosus*, *Lycodes polaris*, and *Gadus ogac*) are not found in Spitsbergen/Barents Sea, and they are also lacking in East Greenland waters. This means that the barrier is not the Fram Strait, but at the northernmost part and southern tip (Cape Farewell) of Greenland. They are probably related to ocean currents and temperatures, but the topic needs further investigation.

Some Arctic species tend to be found at greater depth in the southern part than in the northern part of their ranges (e.g., *Somniosus microcephalus* [Compagno, 1984] and *Lycodes pallidus* [Møller, 2001a]). This is also known for some Arctic invertebrates and can be explained by their avoidance of warm Atlantic surface water in the southern areas.

Brackish habitats are absent in the Antarctic and ice cover plays a more important role in fish distributions than salinity. Zoogeographic schemes for the Antarctic region date from the early twentieth century, but recent divisions are based on the northward limit of the Antarctic Polar Front as a barrier and eastern and western faunal subdivisions (Andriashev, 1965; Anderson, 1990; Eastman, 1993). Although collections tend to corroborate eastern, western, and circumpolar distributional patterns, much of coastal Antarctica is still poorly known, particularly the Amundsen and Bellingshausen seas in the west and most far-eastern seas.

Deepwater sampling is still inadequate anywhere in Antarctica, and most deep-sea species are known from very few specimens. Outer bathyal (slope) areas are dominated by liparids, macrourids, and zoarcids, fishes whose origins lie outside Antarctica. Many areas of the inner shelf are characterized by deep basins, termed the *pseudo*-bathyal zone by Andriashev (1965). These trench-like depressions, from 1000 to 1600 m deep, are separated from the outer true bathyal zone, by sills around 500 m deep. Dominant in this habitat are typically nearshore notothenioids, and outer slope species or relatives are much less abundant. The age and isolation of pseudo-bathyal habitats seems to have served as an evolutionary theater for notothenioids.

Development and behavior of water masses and thermal fronts are the chief physical factors influencing fish distributions in Antarctica. The seasonal variability of physical environmental factors is extreme in both polar regions, and oceanographic sampling in the Southern Ocean is still inadequate to fully document these changes (Lutjeharms, 1990). Andriashev (1965) suggested that Southern Ocean fronts might act as biogeographic barriers because of their flow regimens, but more research is needed to test this. Small-scale changes may be important in dispersing or concentrating organisms and thus may have aided the creation of faunal provinces over time or at least unfolding geographic patterns (Kock, 1985).

B. Origin of Polar Fishes

The present composition of any fish fauna is formed by the biogeographic history of the taxa and evolution of the present climatic conditions. The Arctic fish fauna is dominated by phylogenetically young families, with few representatives of old families, particularly in deep waters (Andriashev, 1954; Dunbar, 1968). It is thought that old groups were eliminated during the rapid cooling of the Middle Miocene (Savin, 1977), and that younger families invaded mainly from the Pacific (e.g., Scorpaenidae, Stichaeidae, and Zoarcidae), when the Bering Strait opened 3–3.5 million years ago (Einarsson *et al.*, 1967; Briggs, 1974, 2003). Continued cooling during and after the Pliocene (Herman and Hubkins, 1980) eventually forced Arctic Ocean species south into warmer Atlantic and Pacific waters. This vicariant event created some amphiboreal sister species (e.g., *Hippoglossus hippoglossus* and *Hippoglossus stenolepis*), an observation first made by Berg (1918, 1934) and since widely accepted. Other species seem to have avoided subzero temperatures by adapting to brackish habitats (e.g., *Clupea pallasii, Lycodes jugoricus*, and *Pleuronectes glacialis*) or by adapting to deep ice-free waters (e.g., *Lycodes frigidus, Paraliparis bathybius*, and *Triglops nybelini*)—the so-called secondary deep-sea fishes (Andriashev, 1953). A few families appear to have invaded the Arctic from the Atlantic (e.g., Lotidae, Gadidae, and Anarhichadidae), but the creation of the present Arctic fish fauna has probably had about an equal number of Pacific and Atlantic components. An alternative hypothesis to the origins of deep-sea Arctic fishes with Pacific ancestry was suggested by Andriashev (1990): Since the Bering Strait is rather shallow, it seems unlikely that this barrier was penetrated by certain groups (e.g., *Bathyraja*, psychrolutids, zoarcids, and liparids). Instead, Andriashev (1990) suggested a pathway of dispersal along the western coast of South America, through Drakes Passage into the south Atlantic and to the north by means of trans-equatorial migration in deep cold waters through the tropics. This theory has long been used to explain the antitropical distribution pattern of many marine organisms (Ekman, 1953), and the catch of a large Patagonian toothfish, *Dissostichus eleginoides*, off Greenland indicates that this type of deepwater migration does take place (Møller *et al.*, 2003). More phylogenetic research is required to further assess this theory.

Perhaps the most important ecological process limiting polar fish diversity is the extreme seasonality of trophic structures (Dunbar, 1968). Thomson

(1977) suggested that the basal limiting factor to species diversity in both polar areas may be imposed by food resources and primary productivity, essentially bound within the summer months. Still, the biogeographic and climatic history of Antarctica has shaped a fish fauna with origins much older than the present Arctic fauna.

Physical processes leading to the isolation of Antarctica since the Cretaceous were summarized by Anderson (1990), Eastman (1993), Miller (1993), and Briggs (2003). As the ancient continent Gondwana broke up, geographic isolation began with the separation of shallow connecting ridges to other land masses and the formation of Antarctica's characteristic inner shelf basins, which would become evolutionary theaters during the Cenozoic. The continent's isolation was further ensured oceanographically by rapid cooling after the Eocene and the formation of the circumpolar Antarctic Convergence by the late Oligocene. Cold deepwater continental boundary currents and the northward flow of Antarctic bottom water from the Weddell Sea then formed, probably hindering colonization of continental areas by temperate species.

This isolation and exclusion provided niches for the radiation of the dominant Notothenioidei. The monophyly of this group has been questioned on the basis of insufficient knowledge of their morphology, genetics, and fossil record (Eastman, 1993; Lecointre *et al.*, 1997). However, Hennigian definitions of higher groups (based on shared derived characters) do not always work in taxonomy (lack of data does not mean lack of corroboration), and Lecointre *et al.* (1997) did not know about Balushkin's (1994) identification of an Eocene notothenioid. Thus, with few similar groups to compare, the non-bovichtid notothenioids are considered here as a monophyletic group and the whole assemblage must be related to the zoarcids and pinguipedids, all at least as old as the Eocene (Anderson, 1990, 1994). Skates seem to have invaded the South Atlantic and then Antarctica in the Eocene as well (Long, 1994), but the high diversity of zoarcids and liparids today suggests several invasions by these fishes, particularly from bathyal areas of the southeastern Pacific (Andriashev, 1965, 1986; Anderson, 1990, 1994).

III. A CLASSIFICATION OF FISHES OCCURRING IN ARCTIC AND ANTARCTIC REGIONS, WITH AN ANNOTATED LIST OF FISH FAMILIES AND NOTES ON REVISIONS AND PHYLOGENETIC HYPOTHESES

The classification of families follows Eschmeyer (1998), except for the family status of Liparidae (snailfishes). The main source of information about the number of species and genera in each family is Nelson (1994), which is used if nothing else is cited. The family accounts below are very short and include mainly a status of taxa known to date and their distribution. References to major revisions and phylogenetic studies are included, whereas data for identification should be found in regional books (e.g., Bering Sea: Allen and Smith, 1988; Alaska: Mecklenburg *et al.*, 2002; Canada: Coad *et al.*, 1995; Greenland: Nielsen and Bertelsen, 1992, Okamura *et al.*, 1995; Iceland: Jónsson, 1992; Faroe Islands: Joensen and Tåning, 1970; northeastern Atlantic/Arctic: Whitehead *et al.*, 1984–86; Russia: Berg, 1932, Andriashev, 1954; Antarctica: Gon and Heemstra, 1999, Miller, 1993, and original papers).

   Myxinidae (hagfishes): Five genera with 61 species. One species, *Myxine glutinosa*, in Arctic waters, but more common in Atlantic waters. No Antarctic species, because records of *Myxine australis* (Fernholm, 1990) are probably not correct (B. Fernholm, personal communication, 19 December 2003). For a systematic update, see Fernholm (1998), Mincarone (2001), Mok and Kuo (2001), and Mok (2002).

   Petromyzontidae (lampreys): About six to eight genera, with approximately 40 species. Two genera and five species in polar areas. *Geotria australis* is known from Antarctic South Georgia, three species are known from North Pacific Arctic waters, and one, *Lampetra marinus*, is known from the Atlantic Arctic. *Lampetra camtschatica* (*Lethenteron japonica* in literature.) is by far the most common and widespread of the Arctic species (Andriashev, 1954). The number of subfamilies, genera, and species are still a matter of debate (Mecklenburg *et al.*, 2002). The molecular phylogeny of *Lampetra* was studied by Docker *et al.* (1999).

   Lamnidae (mackerel sharks): Three genera with five species. *Carcharodon carcharias* and two species of *Lamna* enter Arctic waters: *Lamna ditropis* and *C. carcharias* from the North Pacific side and *Lamna nasus* from the North Atlantic side. The latter is also found in a circumglobal band throughout the southern hemisphere, where it enters Antarctic waters (Compagno, 2001). The species are migratory, mainly seen in polar areas in summer on feeding migrations. They are able to elevate their body temperature more than 10 °C above that of the surrounding water (Smith and Rhodes, 1983). The molecular phylogeny, the family, and the genetic identification of species were studied by Naylor *et al.* (1997) and Shivji *et al.* (2002).

   Cetorhinidae (basking sharks): The single species, *Cetorhinus maximus*, has an antitropical distribution in temperate waters (Compagno, 2001). The species seems to prefer relatively warm waters (8–24 °C) but has been recorded from considerably colder waters (e.g., in the White Sea and in a West Greenland fjord [Muus, 1981]. The phylogenetic position of the basking shark through molecular methods was studied by Naylor *et al.* (1997).

Dalatiidae (sleeper sharks): Formerly included in Squalidae but given family status by Nelson (1994) and Eschmeyer (1998), partly based on Shirai (1992). The 18 genera hold approximately 50 species, two in polar waters. The Pacific sleeper shark *Somniosus pacificus* is one of the most abundant sharks in the Pacific Arctic region and is found in the southwest Atlantic (de Astarloa *et al.*, 1999). The Greenland shark, *Somniosus microcephalus*, is the most common shark in the Atlantic Arctic region, where it often occurs in dense aggregations. It is also known from sub-Antarctic areas of the southern Indian and Pacific Oceans (Compagno, 1990). A molecular study on the phylogeny and complicated distribution patterns of the four species in the genus *Somniosus* is much needed.

Squalidae (dogfish sharks): Two genera with 10 species, with one abundant representative, *Squalus acanthias*, in Pacific Arctic waters. It has a wide antitropical distribution in the Atlantic and Pacific Oceans, with a broad temperature and depth range. It enters sub-arctic waters off West Greenland and northern Norway, and was recently found in sub-Antarctic waters (near Kerguelen Island) as well (Pshenichnov, 1997).

Rajidae (skates) (Figure 2.4): The largest cartilaginous fish family with about 26 genera and 230 known and a large number of undescribed species (Last and Stevens, 1994; McEachran and Dunn, 1998; Long and McCosker, 1999; Gomes and Paragó, 2001). Skates are benthic fishes mainly found in deep waters. Four species are endemic to the Antarctic, four live in Atlantic Arctic and nine in Pacific Arctic waters. Species of "soft-nosed skates," *Bathyraja*, are represented in all areas and appear to be the most diverse in polar waters. Also, the genus *Amblyraja* has species in both Arctic and Antarctic waters. The most northerly recorded species is *Amblyraja hyperborea*, which is one of the dominant large predatory fishes in the Arctic Ocean, whereas *Amblyraja radiata* is widespread and abundant in Atlantic sub-arctic areas. Last and Stevens (1994) mentioned that the identity of *A. hyperborea* off Tasmania and New Zealand should be further studied. The phylogenetic relationships of genera based on morphology were studied by McEachran and Dunn (1998).

Acipenseridae (sturgeons): The four genera include 24 anadromous and freshwater species, all from the northern hemisphere. Two species of *Acipenser* are Siberian endemics and *A. sturio* enters the White Sea (Sokolov and Berdichevskii, 1989). Molecular phylogenetics was given by Birstein and DeSalle (1998) and Birstein *et al.* (2002).

Notacanthidae (spiny eels): Ten benthic or benthopelagic deepwater species in three genera. *Notacanthus chemnitzii* occurs worldwide in temperate parts of the Pacific, Atlantic, and Indian Oceans and enters polar waters in Baffin Bay and Barents Sea (Sulak, 1986).

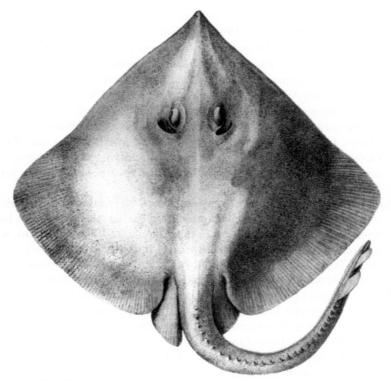

Fig. 2.4. Rajidae: *Bathyraja spinicauda*, 740-mm TL, female, Davis Strait. (From Jensen, 1914.)

Anguillidae (freshwater eels): One genus with 15 catadromous species from tropical and temperate seas. *Anguilla anguilla* enters the White Sea, and a few specimens of *Anguilla rostrata* are known from southern Greenland. Bastrop *et al.* (2000) and Lin *et al.* (2001) presented phylogenetic analyses inferred from mitochondrial genes.

Clupeidae (herrings): Fifty-six genera with about 180 species. Only two occur in polar waters (Whitehead, 1985). Atlantic herring, *Clupea harengus*, are found in southwestern Greenland fjords, and Pacific herring, *Clupea pallasi*, along the Alaskan and western Canadian Arctic coastline and in the White Sea. The Pacific herring was once thought to be a subspecies of Atlantic herring, but it is now clear that it is a separate species (Eschmeyer and Herald, 1983; Uyeno and Sato, 1984). The nomenclature of the White Sea population has also been discussed for many years, but it is now clear that it belongs to

2. SYSTEMATICS OF POLAR FISHES

C. pallasi and not the geographically much closer *C. harengus* (Rass and Wheeler, 1991). The disjunct Arctic distribution of *C. pallasi* appears to be related to a preference for brackish water (Craig, 1984).

Cyprinidae (minnows and carps): The largest freshwater fish family with more than 210 genera and 2100 species. Of the approximate 270 species native to North America, only the lake chub *Couesius plumbeus* lives in the Arctic area. The systematics and morphology-based phylogeny of American cyprinids was revised by Mayden (1991) and Coburn and Cavender (1992). Ten species occur in the Eurasian Arctic (Berg, 1932). A molecular phylogeny of European and North American taxa (not *Couesius*) was presented by Briolay *et al.* (1998).

Catostomidae (suckers): A freshwater family of 14 genera and 76 species, almost endemic to North America (Harris and Mayden, 2001). One species *Catostomus catostomus* is found in the Arctic parts of North America and Siberia. A systematic revision and a morphological phylogeny were given by Smith (1992). Molecular phylogenetic relationships of major clades of Catostomidae were analyzed by Harris and Mayden (2001).

Cobitidae (loaches): The 18 genera include about 110 species from Eurasian freshwaters. The only Arctic species, *Cobitis taenia*, has a wide distribution in Eurasia (Berg, 1932).

Balitoridae (river loaches): The approximate 40 genera include about 470 species from Eurasian freshwaters. Some specialists recognize about 60 genera. The only Arctic species, *Nemacheilus barbatulus*, has a wide Eurasian distribution. Sawada (1982) presented a phylogeny.

Esocidae (pikes and mudminnows): The systematics of this family and the interrelationships with the family Umbridae has been controversial for many years (Johnson and Patterson, 1996). Following the molecular phylogeny and classifications by López *et al.* (2000), Esocidae consists of three genera and eight species found in fresh and brackish waters. Two species occur in the Arctic area: The northern pike, *Esox lucius*, which has a circumarctic distribution and enters brackish waters in the Baltic Sea. *Dallia pectoralis* occurs in Alaska and eastern Siberia (Gudkov, 1998; Mecklenburg *et al.*, 2002).

Microstomatidae (microstomatids): Three genera with about 17 species, in all oceans from the sub-arctic to Antarctica. One mesopelagic species, *Nansenia antarctica*, in Antarctic and sub-Antarctic waters.

Bathylagidae (deepsea smelts): Nelson (1994) recognized one genus (*Bathylagus*) and about 15 species, whereas Kobyliansky (1990) includes eight genera and 19 species. In either case, the five species in polar waters are included in *Bathylagus*. All are relatively widely distributed oceanic species, but their exact distributional patterns need study. Two species enter Arctic waters from the North Atlantic, but the identity of the small-eyed specimens identified as *Bathylagus bericoides* needs to be confirmed (P. R. M., personal observations). Of the three species in Antarctic waters, only *Bathylagus antarcticus* occurs near the continent.

Osmeridae (smelts): Seven genera with 15–16 species in the northern hemisphere (Saruwatari *et al.*, 1997), of which three genera and four species are found in the Arctic area. Marine, freshwater, brackish, and anadromous species and populations exist. The capelin, *Mallotus villosus*, is one of the most important fishes in the Arctic ecosystem as a major food item for seals, birds, and fishes (Sakshaug *et al.*, 1994). The relationships of osmerid genera were reviewed by Johnson and Patterson (1996). Saruwatari *et al.* (1997) revised the genus *Hypomesus*, including the pond smelt, *H. olidus*, which occurs in freshwaters of the American Arctic. A molecular analysis of relationships of *Osmerus* was published by Taylor and Dodson (1994). Finally, Waters *et al.* (2002) analyzed the phylogenetic position of Osmeridae, based on mitochondrial sequence data.

Salmonidae (chars, graylings, salmons, trouts, whitefishes, and allies): Nelson (1994) included 11 genera and 66 species originally native to the northern hemisphere but now introduced to all continents, except Antarctica. In the Arctic area, nine genera and 28 species are found. Based on the morphological phylogenetic work of Sanford (1990), most authors (e.g., Johnson and Patterson, 1996; Kottelat, 1997) have suggested family status for Coregonidae and Salmonidae (Thymallinae + Salmonidae), reflecting the relationship of the three groups. However, following Eschmeyer (1998), they are here treated as one family.

The whitefishes (Coregoninae) are represented by 12 species, 10 of which belong to the genus *Coregonus*. The taxonomy of this group is particularly difficult, and many local forms and varieties exist. Most of these were treated as separate species by Kottelat (1997), but unfortunately, many are already extinct or mixed with other forms or species. Bernatchez and Dodson (1994) analyzed the phylogeny of 63 species and populations of *Coregonus* based on mitochondrial sequence data. The results support the idea that subalpine forms are not the same species as Scandinavian and Arctic ones (Kottelat, 1997).

Graylings (Thymallinae) are represented by two species in the Arctic area, one in the Eurasian part and one circumpolar. Weiss *et al.* (2002) analyzed the phylogeny and historical biogeography of the European grayling *Thymallus thymallus*.

Chars, salmons, and trouts (Salmoninae) are represented by five genera and 14 species in the Arctic area. The arctic char, *Salvelinus alpinus*, is the most northerly distributed of freshwater fishes, whereas the Atlantic salmon is mainly boreal and present only in the Arctic by some stocks in White Sea rivers and one in West Greenland. The many species of Pacific salmons dominate Pacific Arctic rivers and streams. Stearly (1992) presented a phylogeny of Salmoninae, arguing that the anadromous nature of both Pacific and Atlantic genera (*Oncorhynchus* and *Salmo*) represents a possible convergent apomorphic adaptation from non-anadromic ancestors, and that present nonmigratory populations are a result of secondary land-locking. The anadromous behavior enables the Arctic species to feed in rich marine environments during summer and to stay in the often warmer freshwater streams and lakes during winter, when near-shore marine waters drop to $-1.8\,°C$.

> *Gonostomatidae* (bristlemouths) (Figure 2.5): Four genera with 21 mesopelagic and bathypelagic species from all oceans. Four species of the cosmopolitan genus *Cyclothone* are found in the Antarctic and one species, *C. microdon*, is known from both Arctic and Antarctic waters. All species with photophores. Ahlstrom *et al.* (1984) presented a phylogeny.
>
> *Stomiidae* (barbeled dragonfish): The 26 genera hold 230 mesopelagic and bathypelagic species from all oceans. *Stomias boa* is known from sub-Arctic to Antarctic. No Arctic species. All species have photophores. Fink (1985) presented a phylogeny of the family.
>
> *Scopelarchidae* (pearleyes): Four genera with 18 mesopelagic species from all oceans. Two species of *Benthalbella* are found in the Antarctic as well. No Arctic species. All have tubular eyes. A phylogeny was presented by Johnson (1984).
>
> *Notosudidae* (waryfishes): Three genera with 19 mesopelagic and bathypelagic species from all oceans. *Scopelosaurus hamiltoni* is known from Antarctic and sub-Antarctic waters. No Arctic species. Revised by Bertelsen *et al.* (1976).
>
> *Paralepididae* (barracudinas): The 12 genera hold 56 epipelagic to bathypelagic and a few demersal species from all oceans. Four species present in Antarctic waters, one endemic species and three species

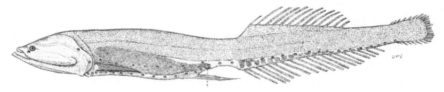

Fig. 2.5. Gonostomatidae: *Cyclothone kobayashii*, 66.6-mm SL, Drake Passage. (From Gon and Heemstra, 1990.)

entering from surrounding oceans. One of the latter, *Arctozenus rissoi*, is also found in Arctic waters. One species is Atlantic Arctic. The subfamily Paralepidinae was revised by Post (1987).

Anotopteridae (daggertooth): One genus with one species, *Anotopterus pharao*, occurring in all oceans including Antarctic waters. Also known from sub-Arctic waters, but not within the Arctic zone (Templeman, 1970).

Myctophidae (lanternfishes): The 32 genera include about 235 mesopelagic to bathypelagic species from all oceans. Two species are known from Arctic and 22 from Antarctic waters. Only *Electrona antarctica* is strictly polar. Most of the species make diurnal vertical migrations, often several hundred meters. All species with photophores. A phylogeny of the family was presented by Paxton *et al.* (1984), and the evolution of the tribe Electronini, with many Antarctic species, was presented by Hulley (1998).

Lampridae (opahs): One genus with two pelagic species. Both are found in Antarctic waters. *Lampris guttatus* is furthermore present in all oceans and *Lampris immaculatus* is found in the southern hemisphere south of 34 °S. Not represented in the Arctic zone. Olney (1984) discussed the relationships.

Percopsidae (trout-perches): One genus with two species, both from North American freshwaters. *Percopsis omiscomaycus* is found in a few localities in the Arctic zone.

Muraenolepididae (eel cods) (Figure 2.6): One genus, *Muraenolepis*, with four bottom-dwelling Antarctic species, two of which also occur in the sub-Antarctic zone. They are found along continental shelves and slopes. Once considered rather primitive in the order, some works suggest relationships to "higher" gadiforms (Markle, 1989; Howes, 1991). Markle (1989) and Endo (2002) suggested a sister relationship to bregmacerotids.

Macrouridae (grenadiers or rattails) (Figure 2.7): The 36 genera include about 300 species from the slopes to the abyss of all oceans. A few species are bathypelagic. Only *Macrurus berglax* and rarely *Coryphaenoides rupestris* (southern Baffin Bay) occur in the Atlantic portion of the Arctic. Ten species in four genera occur in the Antarctic, including the, bathypelagic *Cynomacrurus piriei*. Most systematic works on grenadiers are regional reviews, but an overview was provided by Cohen *et al.* (1990). Molecular phylogenetic analyses of the deep-sea genus *Coryphaenoides* were provided by Morita (1999) and Wilson and Attia (2003).

Moridae (deep-sea cods): The 18 genera include about 100 species, mostly living on continental slopes. They are found worldwide and a few occur inshore, even entering estuaries; others occur on abyssal plains. None enters the Arctic region, but three are found in the South

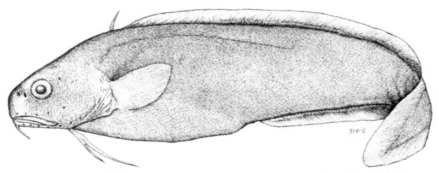

Fig. 2.6. Muraenolepididae: *Muraenolepis marmoratus*, 28-cm TL, (53°50′S, 37°25′W). (From Gon and Heemstra, 1990.)

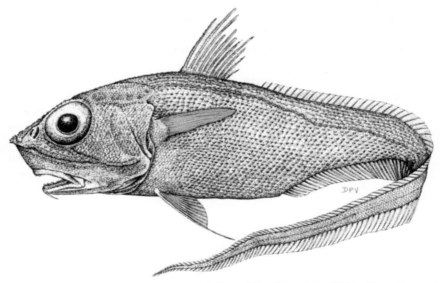

Fig. 2.7. Macrouridae: *Macrourus whitsoni*, 33-cm TL, 59°34′S, 27°18′W. (from Gon and Heemstra 1990).

Georgian Province of the Antarctic (Chiu *et al.*, 1990). Morid systematics is in a state of flux, but overviews are found in Paulin (1989) and Cohen *et al.* (1990).

Melanonidae (pelagic cods) (Figure 2.8): One genus, *Melanonus*, of uncertain affinity includes two mesopelagic and bathypelagic species. Both species are widespread, and one, *Melanonus gracilis*, occurs throughout the Antarctic region and the sub-Antarctic.

Gadidae (cods and haddocks): Contains 12 genera and 20–21 benthopelagic and pelagic species (Cohen *et al.*, 1990). All except *Micromesistius
australis*, which occurs in the southern hemisphere including Antarctic
waters, are confined to temperate to Arctic waters. Two species are
Arctic endemics, five occur in both Atlantic and Arctic waters, three
occur in both northern Pacific and Arctic waters, and *Boreogadus saida*
is found in all three oceans. The Arctic species, *Arctogadus borisovi*,
was placed in the synonymy of *Arctogadus glacialis* (Møller *et al.*,
2002; Jordan *et al.*, 2003). *Gadus ogac* has been suggested as a junior
synonym of *G. macrocephalus* (Carr *et al.*, 1999), a view not held by all
researchers (see Mecklenburg *et al.*, 2002). An isolated population of
G. macrocephalus/ogac exists in the White Sea, and land-locked populations of *Gadus morhua* are found in several places (Cohen *et al.*,
1990). Molecular phylogenies were provided by Carr *et al.* (1999) and
Møller *et al.* (2002), suggesting three independent invasions of the
Arctic Ocean from the Atlantic side (Figure 2.1). Morphological phylogenies were given by Dunn (1989) and Endo (2002), showing rather
different results.

Lotidae (burbots): Six genera and approximately 22 bottom-dwelling
species, sometimes included in Gadidae, subfamily Lotinae (Nelson,
1994). Most species occur in marine habitats, a few enter brackish
waters, and *Lota lota* is the only freshwater species in the family. Three
species occur in Atlantic and Arctic waters (Cohen *et al.*, 1990). Two
species of *Gaidropsarus* are among the few species found in great
depth (>1000 m) in the Arctic Ocean, where they can be very abundant

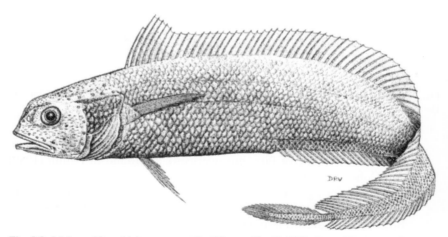

Fig. 2.8. Melanonidae: *Melanonus gracilis* 187-mm SL, 59°37′S, 88°54′W. (From Gon and
Heemstra, 1990.)

(P. R. Møller, personal observations, Baffin Bay). In the classification provided by Nelson (1994), Lotinae and Gadinae are regarded as subfamilies in Gadidae.

Ophidiidae (cusk-eels and brotulas): The 48 genera include about 240 species, the great majority of which are found in tropical and subtropical areas (Nielsen *et al.*, 1999). One Antarctic species, *Holcomycteronus brucei*, known from one specimen caught at 4500 m. No Arctic species.

Carapidae (pearlfishes): Seven genera with 32 circumglobal, marine species. *Echiodon cryomargarites* occurs from the subtropics to the Antarctic zone. None in the Arctic zone. Many species live inside bivalves and holothurians. A phylogeny was presented by Markle and Olney (1990).

Oneirodidae (dreamers) (Figure 2.9): The 16 genera include about 60 species. They are all mesopelagic and bathypelagic and found worldwide. The genus *Oneirodes* contains about half the species. None is known to penetrate into the Arctic. In the Antarctic, *Oneirodes notius* is known from continental to sub-Antarctic regions. A review of the group was given by Pietsch (1974).

Ceratiidae (seadevils) (Figure 2.10): Two genera hold four mesopelagic and bathypelagic species found worldwide except in the Arctic. In the Southern Ocean, *Ceratias tentaculatus* is known from continental Antarctica, the sub-Antarctic, and temperate seas. A review of the group was given by Pietsch (1986).

Melanocetidae (blackdevils): The single genus, *Melanocetus*, of five species is found worldwide mainly in tropical and subtropical seas. None is known from Arctic areas, but *Melanocetus rossi* was described from the Ross Sea, Antarctica (Balushkin and Fedorov, 1981). The remaining species were treated by Pietsch and van Duzer (1980).

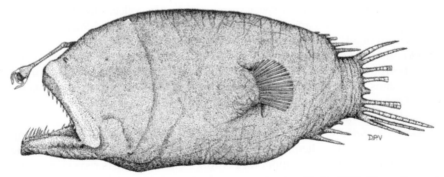

Fig. 2.9. Oneirodidae: *Oneirodes notius*, 60-mm SL, female, 62°05′S, 89°56′W. (From Gon and Heemstra, 1990.)

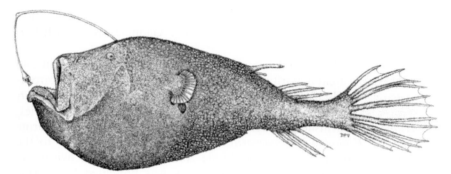

Fig. 2.10. Ceratiidae: *Ceratias tentaculatus*, 163-mm SL, female, 58 °10′S, 59 °13′W. (From Gon and Heemstra, 1990.)

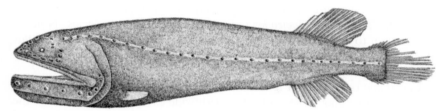

Fig. 2.11. Cetomimidae: *Gyrinomimus andriashevi*, 25-cm SL, Lazarev Sea. (From Gon and Heemstra, 1990.)

Melamphaidae (bigscale fishes): Five genera with 33 circumglobal meso-pelagic to abyssopelagic species. Two species occur, besides in the main oceans, also in the Antarctic zone. None in the Arctic zone. A phylogeny was presented by Keene and Tighe (1984).

Cetomimidae (flabby whalefishes) (Figure 2.11): Nine genera with about 35 marine species known from all oceans, except the Arctic mainly at depths of 1000–4000 m. Most species are very rare in collections. Two strictly Antarctic species and one species also in the Atlantic and Pacific Oceans south of 34 °S. Colgan *et al.* (2000) presented a phylogeny.

Oreosomatidae (oreos): Four genera with nine species all living near the bottom down to almost 2000 m. Occurs in temperate waters of all oceans except the Arctic. Most common in the southern hemisphere. One species, *Pseudocyttus maculatus*, is partly Antarctic. James *et al.* (1988) revised the family.

Gasterosteidae (sticklebacks): Five genera with seven species occurring on the northern hemisphere in marine, brackish, and freshwaters. The number of species is under discussion. Some specialists recognize many more species. Two species are partly Arctic and partly in the northern temperate region. Fritzsche (1984) discussed relationships to other families. A comprehensive literature exists on every aspect of the biology of this family, especially on the three-spine stickleback *Gasterosteus aculeatus*. Other phylogenetic works include Ortí *et al.* (1994), Peichel *et al.* (2001), Reusch *et al.* (2001), and Takahashi and Goto (2001).

Sebastidae (rockfishes): Often treated as two subfamilies (Sebastinae and Sebastolobinae) of the family Scorpaenidae. Seven genera with 130–135 marine species, of which many are viviparous. Occurs in northern temperate waters with the majority in the northern Pacific. Eight species in the Arctic, of which five also occur in the northern Pacific and three in the northern Atlantic. No species in Antarctic waters. Washington *et al.* (1984) discussed the scorpaeniform families. The molecular evolution was studied by Rocha-Olivares *et al.* (1999), but because only two of the Arctic species were included, little can be said about their evolution.

Hexagrammidae (greenlings): Five genera with 11 marine, mainly littoral species. All are found in the northern Pacific, from Japan to northern Mexico, and three species also in Arctic waters.

Bathylutichthyidae (no name): One genus with one species, *Bathylutichthys taranetzi*, known from a depth of 1650 m off the Antarctic island of South Georgia.

Cottidae (sculpins) (Figure 2.12): About 70 genera and 300 mainly benthic species (Nelson, 1994). Represented in marine, brackish, and freshwaters mainly in the northern hemisphere, but four species live in the southwestern Pacific. None in Antarctica. Seven species are Arctic endemics, 3 are Atlantic Arctic, and 25 Pacific Arctic. Five species all in the genus *Cottus* live fully or partly in Arctic freshwaters. Baker *et al.* (2001) used mitochondrial sequences in the identification of *Cottus* species, including two Arctic species *Cottus bairdii* and *Cottus cognatus*. An osteology-based phylogeny of the superfamily

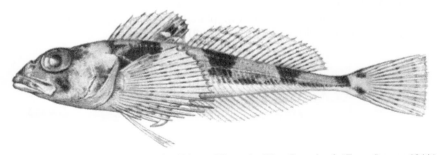

Fig. 2.12. Cottidae: *Triglops pingelii*, 129-mm TL, male, West Greenland. (From Jensen, 1944.)

Cottoidea was presented by Yabe (1985). Other, mainly regional, revisions of Arctic genera include: *Artediellus*: Neyelov (1979), Van Guelpen (1986); *Icelus*: Jensen and Volsøe (1949), Nelson (1984); *Myoxocephalus*: Cowan (1971), Neyelov (1979); *Triglops*: Andriashev (1949), Pietsch (1993).

Hemitripteridae (sailfin sculpins): Three genera with eight marine species known from shallow waters in the northwestern Atlantic and northern Pacific. A single Atlantic species also occurs in the Arctic zone (Hudson Bay), and of the seven Pacific species, three are Arctic as well. No Antarctic species. Yabe (1985) presented a phylogeny.

Psychrolutidae (fathead sculpins): Seven genera with about 30 marine, benthic species known from all oceans down to almost 2200 m. Seven species occur in Arctic waters, of which four are found also in the northern Pacific, one also in the northern Atlantic, and two are restricted to the Arctic. No species in Antarctica. Yabe (1985) presented a phylogeny.

Agonidae (poachers) (Figure 2.13): The 20 genera include 44 marine, benthic species found from shallow water to about 1600 m. Most are from the northern Pacific, four from the North Atlantic, and one species occurs off southern South America. *Ulcina olrikii* is strictly Arctic, seven are found in the northern Pacific and Arctic waters, and two species both in the North Pacific and Atlantic as well as in Arctic waters. No species in Antarctica. Kanayama (1991) presented a phylogeny.

Cyclopteridae (lumpsuckers) (Figure 2.14): Consists of seven genera and 28 marine species in temperate to Arctic parts of the northern hemisphere. None in Antarctica. They are mainly benthic, but some species (e.g., *Cyclopterus lumpus*) make long feeding migrations from coastal shallow waters to the pelagic parts of the open ocean (Davenport, 1985). Four species are Arctic endemics, two are Atlantic Arctic, and four are Pacific Arctic. A comprehensive revision was provided by Ueno (1970), but many taxonomic problems remain to be solved, because of the variability and paucity of some species in museum collections (Mecklenburg *et al.*, 2002).

Fig. 2.13. Agonidae: *Leptagonus decagonus*, 171-mm TL, Davis Strait. (From Jensen, 1942.)

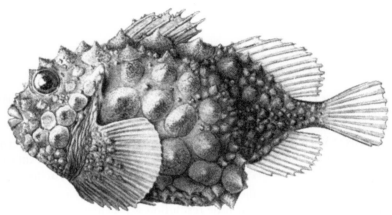

Fig. 2.14. Cyclopteridae: *Eumicrotremus spinosus*, 100-mm TL, Davis Strait. (From Jensen, 1944.)

Liparidae (snailfishes): About 20 genera and more than 200 species known; many undescribed in collections. Snailfishes are found in temperate to polar seas from tide pools to the abyssal plains. Sixteen species in four genera occur in the Arctic. In Antarctic waters, 61 species occur, including three genera endemic to the Southern Ocean. The snailfishes are included as subfamily Liparinae in the family Cyclopteridae by many authors (e.g., Eschmeyer, 1998), but most specialists have them in a separate family. Morphological phylogenies were provided by Kido (1988) and Balushkin (1996a), and these differ from one another substantially. Revisions of polar genera include Andriashev and Stein (1998, *Careproctus*), Able (1990, *Liparis*), Andriashev (1986, *Paraliparis*), and Andriashev (2003, Southern Ocean Liparidae). Many taxonomic problems remain at the generic level (e.g., if the Arctic endemic monotypic genus *Rhodichthys* is valid or a junior synonym of *Paraliparis*).

Percidae (perches): Ten genera with about 160 species from the northern hemisphere. With a few brackish exceptions, all from freshwater. Two Arctic species, *Perca fluviatilis* and *Acerina cernua*, have a wide Eurasian distribution. Wiley (1992) and Song *et al.* (1998) presented phylogenies.

Bathymasteridae (ronquils): Three genera with seven marine bottom-dwelling species found near shore in the northern Pacific. One species in Arctic waters as well.

Zoarcidae (eelpouts) (Figure 2.15): The 47 genera include approximately 250 species, with new species being described yearly (Anderson, 1994). Two genera, *Lycodapus* and *Melanostigma*, are mesopelagic, the rest benthic. Eelpouts occur worldwide from littoral habitats to trench

Fig. 2.15. Zoarcidae: *Lycodes mcallisteri* 230-mm SL, female, Baffin Bay. (From Møller, 2001b.)

depths (5320 m) and are one of the few families found at hydrothermal vents. Forty-two species are found in Arctic and 19 in Antarctic waters. Only the genus *Lycenchelys* occurs in both polar areas.

Anderson (1994) studied the osteology and phylogeny of the family and supported four subfamilies, but little resolution was achieved in Lycodinae, which contains most of the species. Møller and Gravlund (2003) studied the systematics and molecular phylogeny of the diverse genus *Lycodes*, but shallow-water Arctic and western Pacific species were not included.

In some polar habitats, eelpouts are dominant, especially the genera *Gymnelus* and *Lycodes* in the Arctic. Though diverse, Antarctic eelpouts are a minor component of the fish fauna there. Several Arctic species tolerate brackish water. Revisions and other important studies of genera with Polar species include: *Gymnelus*: Anderson (1982), Chernova (2000); *Lycodes*: Toyoshima (1985), Møller and Jørgensen (2000), Møller and Gravlund (2003); *Lycenchelys*: Andriashev (1985), Toyoshima (1955), Anderson (1995).

Stichaeidae (pricklebacks) (Figure 2.16): The 36 genera include about 65 marine species from northern Pacific and northern Atlantic waters, with most in the former area. Twelve species occur in Arctic waters; one of these, *Eumesogrammus praecius*, is strictly Arctic. The Pacific species *Lumpenella longirostris* was surprisingly reported and illustrated from Greenland waters by Miki (1995), but the specimen cannot be traced in the Hokkaido University collection. (H. Imamura, personal communication, 19 Nov. 2002).

Pholidae (gunnels) (Figure 2.17): The three to four genera include about 15 bottom-dwelling species in the northern hemisphere. Three species occur in the Arctic area: The Pacific *Rhodymenichthys dolichogaster* is caught as far north as St. Lawrence Island; *Pholis fasciatus* occurs from the northern Pacific to western Greenland and Canada, where it overlaps with the Atlantic Arctic *P. gunnelus*. Gunnels are common shallow-water fishes and are important food items for birds and fishes. The family was revised by Yatsu (1981) and later supplemented by a phylogenetic analysis including the Arctic species (Yatsu, 1985). The genus *Allopholis* was not recognized by Mecklenburg *et al.* (2002).

Fig. 2.16. Stichaeidae: *Stichaeus punctatus*, 152-mm SL, West Greenland. (From Jensen, 1944.)

Fig. 2.17. Pholidae: *Pholis fasciata*, 225-mm TL, West Greenland. (From Jensen, 1942.)

Anarhichadidae (wolffishes): Two genera and five species of bottom-dwelling fishes. The genus *Anarhichas* includes three species with a northern Atlantic and Arctic distribution, and one species with a northern Pacific and Arctic distribution.

Zaproridae (prowfish): One genus with one species occurring in the northern Pacific just entering the Arctic zone. Systematic notes in McAllister and Krejsa (1961).

Bovichtidae (thornfishes): This family of southern hemisphere fishes contains three genera: *Bovichtus* with eight species, the monotypic *Cottoperca* and *Halaphritis* (Last *et al.*, 2002). Formerly included was an Australian freshwater species, *Pseudaphritis urvillii*, now removed to its own family (Balushkin, 1992; Lecointre *et al.*, 1997; Eastman and Eakin, 2000), with the bovichtids the stem notothenioids. Only one species, *Bovichtus elongatus*, penetrates into continental Antarctica (Hureau and Tomo, 1977). Balushkin (2000) presented a phylogeny.

Nototheniidae (notothens, Antarctic "cods") (Figure 2.18): The 12 genera and 48 species make Nototheniidae the largest of the families in the southern hemisphere suborder Notothenioidei (16 genera and 58 species: Balushkin, 2000). Twenty-nine species (22 endemics) are present in Antarctic coastal and bathyal areas where they are numerically dominant. Several species, such as the abundant *Pleuragramma antarcticum*, are free swimming under sea ice (cryopelagic). Eastman (1993) provides a review of the origins, relationships, systematics, and biology of Antarctic fishes, with the main focus on notothenioids. On the basis of molecular evidence, Lecointre *et al.* (1997) suggested the whole group is not monophyletic, but Bargelloni *et al.* (1994, 2000)

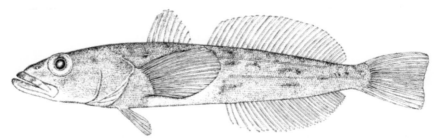

Fig. 2.18. Nototheniidae: *Dissostichus eleginoides*, about 58-cm SL. (From Gon and Heemstra, 1990.)

used combinations of mitochondrial and nuclear DNA sequences to support a phylogeny for the suborder, corroborating the erection of families Pseudaphritidae and Eleginopidae for two sub-Antarctic species (Balushkin, 1992). An Eocene fossil, *Proeleginops grandeastmanorum*, was described in the *Eleginopsidae* (Balushkin, 1994). Bargelloni *et al.* (2000) removed *Dissostichus* from Balushkin's (1992) subfamily Eleginopinae, resulting in the Nototheniidae to now contain three subfamilies, Nototheniinae, Pleuragramminae, and Trematominae. These three were also recognized by Balushkin (2000), who placed *Dissostichus* in his Pleuragrammatinae and recognized 16 genera of notothens. The molecular systematics and phylogenies of some trematomines were studied by McDonald *et al.* (1992) and Ritchie *et al.* (1996). Earlier classifications and the limits of genera of the notothens given by Andersen (1984) and Balushkin (1992) were not followed entirely by DeWitt *et al.* (1990) and Eastman and Eakin (2000). Balushkin (2000) discusses the origin, evolution and phylogeny of the notothens, as well as the other members of the suborder.

Artedidraconidae (barbeled plunderfishes) (Figure 2.19): Placed in four genera with 25 Antarctic-endemic species, 68% of which are in the genus *Pogonophryne* Regan; these fishes were classified with the spiny plunderfishes (below) in the Harpagiferidae by Eakin (1981). This was followed by Nelson (1994) and Eschmeyer *et al.* (1998) in their classifications. They were placed in their own family by Hureau (1986), which was followed by Balushkin (1992, 2000), Eastman (1993), and Eastman and Eakin (2000). These fishes are benthic shelf to upper slope-dwelling species except for *Pogonophryne immaculata*, taken at 2473–2542 m. The evolution and phylogeny was discussed by Balushkin (2000).

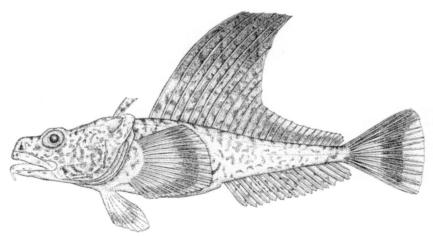

Fig. 2.19. Artedidraconidae: *Pogonophryne barsukovi*, 185-mm SL, female, 60°52′S, 55°36′W. (From Gon and Heemstra, 1990.)

Harpagiferidae (spiny plunderfishes): One genus with 11 species. This is a Southern Ocean morphological and ecological equivalent of the mainly northern Cottidae. They are all benthic, inshore fishes occurring in tidal areas out to 320 m depth (Hureau, 1990; Prirodina, 2000). Two species occur in Antarctic waters.

Bathydraconidae (Antarctic dragonfishes) (Figure 2.20): The 11 genera hold 15–16 deepwater benthic Antarctic species. There is doubt as to the validity of *Bathydraco joannae* (Ofer Gon's, unpublished original research). One species is also found on the subantarctic Kerguelen Plateau. Some have been collected inshore, others to depths down to 2950 m. The systematics, osteology, and morphological phylogeny have been studied by Voskoboinikova (1988), Andriashev *et al.* (1989), and Balushkin and Voskoboinikova (1995). A phylogeny based on both anatomical and molecular data was given by Derome *et al.* (2002).

Channichthyidae (icefishes) (Figure 2.21): The 11 genera include 20 Southern Ocean species. The popular name refers to the lack of hemoglobin. Sixteen species are found in Antarctic waters (Balushkin, 1996b, 2000; La Mesa *et al.*, 2002). All are benthic, shelf, and slope-dwelling fishes, but *Chionobathyscus dewitti* reaches 2000 m. Iwami (1985), Balushkin (2000), and Chen *et al.* (1998) studied the anatomy and phylogeny of the group.

Ammodytidae (sand lances): Five genera with about 19 marine, benthic species. *Ammodytes dubius* is restricted to Arctic waters and two species are partly Arctic. None in the Antarctic zone. Relationships are discussed by Pietsch and Zabetian (1990).

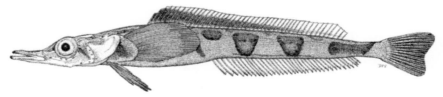

Fig. 2.20. Bathydraconidae: *Gerlachea australis*, 167-mm SL, Elephant Island. (From Gon and Heemstra, 1990.)

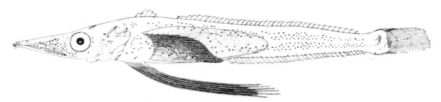

Fig. 2.21. Channichthyidae: *Chionodraco myersi*, 31-cm SL, Ross Sea. (From Gon and Heemstra, 1990.)

Tripterygiidae (threefin blennies): Mainly tropical and temperate cryptic bottom fishes in 20 genera with 115 species (Nelson, 1994). One species, *Helcogrammoides antarcticus*, known from only five specimens is endemic to Paradise Bay, Western Antarctica (Tomo, 1981; Gon, 1990). Stepien *et al.* (1997) analyzed the position of the family, but without including *Helcogrammoides*.

Gempylidae (snake mackerels): The 16 genera include 23 mainly pelagic, marine species found in all oceans from the surface to 2–3000 m. Adults of *Paradiplospinus gracilis* are restricted to Antarctic waters, larvae, and juveniles in the Subantarctic zone. No Arctic species. Relationships are discussed by Collette *et al.* (1984).

Centrolophidae (medusafishes): Seven genera with 27 marine, pelagic species known from all oceans except the Arctic. One species, *Icichthys australis*, partly occurring in Antarctic waters. A phylogeny was presented by Horn (1984).

Achiropsettidae (armless, or southern flounders) (Figure 2.22): Four genera with four species, endemic to the Southern Ocean, but only a single species, *Mancopsetta maculata*, enters the Antarctic zone. Evseenko (2000) reviewed the anatomy, phylogeny, and taxonomic history of the family.

Pleuronectidae (righteye flounders): The 39 genera include about 95 species, with most found in boreal waters. Seventeen species occur in Arctic and cold-temperate waters. No Arctic endemics and no

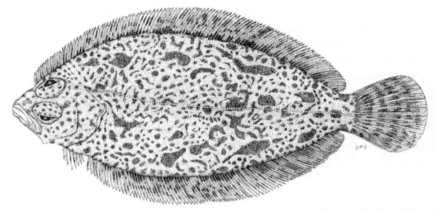

Fig. 2.22. Achiropsettidae: *Mancopsetta maculata*, 22-cm SL, South Georgia. (From Gon and Heemstra, 1990.)

Antarctic species. Cooper and Chapleau (1998) presented a cladistic analysis of the family (excluding Rhombosoleinae) based on morphological and osteological characters and Berendzen and Dimmick (2002) discussed the phylogenetic relationships of the Pleuronectiformes based on molecular evidence.

APPENDIX 2.1

Checklist of Polar Fishes in Alphabetical Order by Family

Amr = Northamerica, **An** = Antarctic, **Ar** = Arctic, At =Atlantic, Euras = Eurasien, Pa = Pacific, Samr = South America, Nz = New Zealand, Aust. = Australia

Achiropsettidae (armless, or southern flounders, p. 52)
Mancopsetta maculata (Günther, 1880) At, Pa, **An**

Acipenseridae (sturgeons, p. 35)
Acipenser baeri; Brandt, 1869; **Ar**
Acipenser ruthenus; Linnaeus, 1758; **Ar**
Acipenser sturio; Linnaeus, 1758; At, **Ar**

Agonidae (poachers, p. 46)
Aspidophoroides monopterygius (Bloch, 1786) At, Pa, **Ar**
Hypsagonus quadricornis (Valenciennes, 1829) Pa, **Ar**

Leptagonus decagonus (Bloch and Schneider, 1801) At, Pa, **Ar**
Leptagonus frenatus (Gilbert, 1896) Pa, **Ar**
Occella dodecahedron (Tilesius, 1813) Pa, **Ar**
Pallasina barbata (Steindachner, 1876) Pa, **Ar**
Percis japonica (Pallas, 1769) Pa, **Ar**
Podothecus veternus; Jordan and Starks, 1895; Pa, **Ar**
Podothecus acipenserinus (Tilesius, 1803) Pa, **Ar**
Ulcina olrikii (Lütken, 1876) **Ar**

Ammodytidae (sand lances, p. 51)
Ammodytes hexapterus; Pallas, 1814; Pa, **Ar**
Ammodytes dubius; Reinhardt, 1838; **Ar**
Ammodytes marinus; Raitt, 1934; At, **Ar**

Anarhichadidae (wolffishes, p. 49)
Anarhichas denticulatus; Krøyer, 1845; At, **Ar**
Anarhichas orientalis; Pallas, 1814; Pa, **Ar**
Anarhichas minor; Olafsen and Poulsen, 1772; At, **Ar**
Anarhichas lupus; Linnaeus, 1758; At, **Ar**

Anguillidae (freshwater eels, p. 36)
Anguilla anguilla; Linnaeus, 1758; At, **Ar**
Anguilla rostrata (Lesueur, 1817) At, **Ar**

Anotopteridae (daggertooth, p. 40)
Anotopterus pharao; Zugmayer, 1911; At, In, Pa, **An**

Artedidraconidae (barbeled plunderfishes, p. 50)
Artedidraco glareobarbatus; Eastman and Eakin, 1999; **An**
Artedidraco loennbergi; Roule, 1913; **An**
Artedidraco mirus; Lönnberg, 1905; **An**
Artedidraco orianae; Regan, 1914; **An**
Artedidraco shackletoni; Waite, 1911; **An**
Artedidraco skottsbergi; Lönnberg, 1905; **An**
Dolloidraco longedorsalis; Roule, 1913; **An**
Histiodraco velifer (Regan, 1914) **An**
Pogonophryne albipinna; Eakin, 1981; **An**
Pogonophryne barsukovi; Andriashev, 1967; **An**
Pogonophryne cerebropogon; Eakin and Eastman, 1998; **An**
Pogonophryne dewitti; Eakin, 1988; **An**
Pogonophryne eakini; Balushkin, 1999; **An**
Pogonophryne fusca; Balushkin and Eakin, 1998; **An**
Pogonophryne immaculate; Eakin, 1981; **An**
Pogonophryne lanceobarbata; Eakin, 1987; **An**

Pogonophryne macropogon; Eakin, 1981; **An**
Pogonophryne marmorata; Norman, 1938; **An**
Pogonophryne mentella; Andriashev, 1967; **An**
Pogonophryne orangiensis; Eakin and Balushkin, 1998; **An**
Pogonophryne permitini; Andriashev, 1967; **An**
Pogonophryne platypogon; Eakin, 1988; **An**
Pogonophryne scotti; Regan, 1914; **An**
Pogonophryne squamibarbata; Eakin and Balushkin, 2000; **An**
Pogonophryne ventrimaculata; Eakin, 1987; **An**

Balitoridae (river loaches, p. 37)
Nemachilus barbatulus; Linnaeus, 1758; Euras, **Ar**

Bathydraconidae (Antarctic dragonfishes, p. 51)
Acanthodraco dewitti; Skora, 1995; **An**
Akarotaxis nudiceps (Waite, 1916) **An**
Bathydraco antarcticus; Günther, 1878; **An**
Bathydraco macrolepis; Boulenger, 1907; **An**
Bathydraco marri; Norman, 1938; **An**
Bathydraco scotiae; Dollo, 1906; **An**
Cygnodraco mawsoni; Waite, 1916; **An**
Gerlachea australis; Dollo, 1900; **An**
Gymnodraco acuticeps; Boulenger, 1902; **An**
Parachaenichthys charcoti (Vaillant, 1906) **An**
Parachaenichthys georgianus (Fischer, 1885) **An**
Prionodraco evansii; Regan, 1914; **An**
Psilodraco breviceps; Norman, 1937; **An**
Racovitzia glacialis; Dollo, 1900; **An**
Vomeridens infuscipinnis (DeWitt, 1964) **An**

Bathylagidae (deepsea smelts, p. 37)
Bathylagus antarcticus; Günther, 1878; At, In, Pa, **An**
Bathylagus gracilis; Lönnberg, 1905; At, Pa, **An**
Bathylagus tenuis; Kobyliansky, 1986; At, In, Pa, **An**
Bathylagus bericoides (Borodin, 1929) At, In, Pa, **Ar**
Bathylagus euryops; Goode and Bean 1896; At, **Ar**

Bathylutichthyidae (no English name, p. 45)
Bathylutichthys taranetzi; Balushkin and Voskoboinikova, 1990; **An**

Bathymasteridae (ronquils, p. 47)
Bathymaster signatus; Cope, 1873; Pa, **Ar**

Bovichtidae (thornfishes, p. 49)
Bovichtus elongatus (Hureau and Tomo, 1977) **An**

Carapidae (pearlfishes, p. 43)
Echiodon cryomargarites; Markle, Williams, and Olney, 1983; At, In, Pa, **An**

Catostomidae (suckers, p. 37)
Catostomus catostomus (Foster, 1773) Amr, Euras, **Ar**

Centrolophidae (medusafishes, p. 52)
Icichthys australis; Haedrich, 1966; At, In, Pa, **An**

Ceratiidae (seadevils, 43)
Ceratias tentaculatus; Norman, 1930; At, In, Pa, **An**

Cetomimidae (flabby whalefishes, p. 44)
Gyrinomimus andriashevi; Fedorov, Balushkin, and Trunov, 1987; **An**
Gyrinomimus grahami; Richardson and Garrick, 1964; At, In, Pa, **An**
Notocetichthys trunovi; Fedorov, Balushkin, and Paxton, 1989; **An**

Cetorhinidae (basking sharks, p. 34)
Cetorhinus maximus (Gunnerus, 1765) At, In, Pa, **Ar**

Channichthyidae (icefishes, p. 51)
Chaenocephalus aceratus (Lönnberg, 1906) **An**
Chaenodraco wilsoni; Regan, 1914; **An**
Champsocephalus esox (Günther, 1861) At, Pa, **An**
Champsocephalus gunnari; Lönnberg, 1905; **An**
Chionobathyscus dewitti; Andriashev and Neyelov, 1978; **An**
Chionodraco hamatus; Lönnberg, 1905; **An**
Chionodraco kathleenae; Regan, 1914; **An**
Chionodraco myersi; DeWitt and Tyler, 1960; **An**
Chionodraco rastrospinosus; DeWitt and Hureau, 1979; **An**
Cryodraco antarcticus; Dollo, 1900; **An**
Cryodraco atkinsoni; Regan, 1914; **An**
Dacodraco hunteri; Waite, 1916; **An**
Neopagetopsis ionah; Nybelin, 1947; **An**
Pagetopsis macropterus; Boulenger, 1907; **An**
Pagetopsis maculatus; Barsukov and Permitin, 1958; **An**
Pseudochaenichthys georgianus; Norman, 1937; **An**

Clupeidae (herrings, p. 36)
Clupea harengus; Linnaeus, 1758; At, **Ar**
Clupea pallasii; Valenciennes, 1847; Pa, **Ar**

Cobitidae (loaches, p. 37)
Cobitis taenia; Linnaeus, 1758; Euras, **Ar**

Cottidae (sculpins, P. 45)

Artediellus camtschaticus; Gilbert and Burke, 1912; Pa, **Ar**

Artediellus gomojunovi; Taranets, 1933; Pa, **Ar**

Artediellus miacanthus; Gilbert and Burke, 1912; Pa, **Ar**

Artediellus ochotensis; Gilbert and Burke, 1912; Pa, **Ar**

Artediellus pacificus; Gilbert, 1896; Pa, **Ar**

Artediellus scaber; Knipowitsch, 1907; Pa, **Ar**

Artediellus atlanticus; Jordan and Evermann, 1898; At, **Ar**

Artediellus uncinatus (Reinhardt, 1835) **Ar**

Cottus aleuticus; Gilbert, 1896; Amr, Pa, **Ar**

Cottus bairdii (Girard, 1855) Amr, **Ar**

Cottus cognatus; Richardson, 1836; Amr, Euras, **Ar**

Cottus ricei (Nelson, 1876) Amr, **Ar**

Cottus sibiricus; Kessler, 1899; Euras, **Ar**

Enophrys diceraus (Pallas, 1788) Pa, **Ar**

Enophrys lucasi (Jordan and Gilbert, 1898) Pa, **Ar**

Gymnocanthus galeatus; Bean, 1881; Pa, **Ar**

Gymnocanthus pistilliger (Pallas, 1814) Pa, **Ar**

Gymnocanthus tricuspis (Reinhardt, 1830) **Ar**

Hemilepidotus jordani; Bean, 1881; Pa, **Ar**

Hemilepidotus papilio (Bean, 1830) Pa, **Ar**

Icelus bicornis (Reinhardt, 1840) **Ar**

Icelus canaliculatus; Gilbert, 1896; Pa, **Ar**

Icelus eryops; Bean, 1890; Pa, **Ar**

Icelus spatula; Gilbert and Burke, 1912; Pa, **Ar**

Icelus spiniger; Gilbert, 1896; Pa, **Ar**

Megalocottus platycephalus (Pallas, 1814) Pa, **Ar**

Microcottus sellaris (Gilbert, 1896) Pa, **Ar**

Myoxocephalus aenaeus (Mitchill, 1814) At, **Ar**

Myoxocephalus jaok (Cuvier, 1829) Pa, **Ar**

Myoxocephalus polyacanthocephalus (Pallas, 1814) Pa, **Ar**

Myoxocephalus quadricornis (Linnaeus, 1758) **Ar**

Myoxocephalus scorpioides (Fabricius, 1780) **Ar**

Myoxocephalus scorpius (Linnaeus, 1758) Pa, **Ar**

Porocottus quadrifilis; Gill, 1859; **Ar**

Trichocottus brashnikovi; Soldatov and Pavlenko, 1915; Pa, **Ar**

Triglops forficatus (Gilbert, 1896) Pa, **Ar**

Triglops nybelini; Jensen, 1944; **Ar**

Triglops pingelii; Reinhardt, 1837; Pa, **Ar**

Triglops scepticus; Gilbert, 1896; Pa, **Ar**

Triglops murrayi; Günther, 1888; At, **Ar**

Cyclopteridae (lumpsuckers, p. 46)
Aptocyclus ventricosus (Pallas, 1769) Pa, **Ar**
Eumicrotremus andriashevi; Perminov, 1936; **Ar**
Eumicrotremus birulai; Popov, 1928; Pa, **Ar**
Eumicrotremus derjugini; Popov, 1926; **Ar**
Eumicrotremus lindbergi (Soldatov, 1930) Pa, **Ar**
Eumicrotremus orbis (Günther, 1861) Pa, **Ar**
Eumicrotremus spinosus (Fabricius, 1776) At, **Ar**
Cyclopteropsis jordani; Soldatov, 1929; **Ar**
Cyclopteropsis macalpini (Fowler, 1914) **Ar**
Cyclopterus lumpus; Linnaeus, 1758; At, **Ar**

Cyprinidae (minnows and carps, p. 37)
Abramis brama; Linnaeus, 1758; Euras, **Ar**
Alburnus alburnus; Linnaeus, 1758; Euras, **Ar**
Carassius carassius; Linnaeus, 1758; Euras, **Ar**
Gobio gobio; Linnaeus, 1758; Euras, **Ar**
Couesius plumbeus (Agassiz, 1850) Amr, **Ar**
Leuciscus idus; Linnaeus, 1758; Euras, **Ar**
Leuciscus leuciscus; Linnaeus, 1758; Euras, **Ar**
Phoxinus percnurus (Pallas, 1811) Euras, **Ar**
Phoxinus phoxinus; Linnaeus, 1758; Euras, **Ar**
Rutilus rutilus; Linnaeus, 1758; Euras, **Ar**
Tinca tinca; Linnaeus, 1758; Euras, **Ar**

Dalatiidae (sleeper sharks, p. 35)
Somniosus microcephalus (Bloch and Schneider, 1801) At, In, Pa, **Ar**
Somniosus pacificus; Bigelow and Schroder, 1944; At, Pa, **Ar**

Esocidae (pikes and mudminnows, p. 37)
Esox lucius; Linnaeus, 1758; Amr, Euras, **Ar**
Dallia pectoralis; Bean, 1880; Amr, **Ar**

Gadidae (cods and haddocks, p. 42)
Arctogadus glacialis (Peters, 1874) **Ar**
Boreogadus saida (Lepechin, 1774) At, Pa, **Ar**
Eleginus gracilis (Tilesius, 1810) Pa, **Ar**
Eleginus navaga (Pallas, 1811) **Ar**
Gadus macrocephalus; Tilesius, 1810; Pa, **Ar**
Gadus morhua; Linnaeus, 1758; At, **Ar**
Gadus ogac; Richardson, 1836; At, **Ar**
Melanogrammus aeglefinus (Linnaeus, 1758) At, **Ar**
Micromesistius australis; Norman, 1937; At, Pa, **An**
Micromesistius poutassou (Risso, 1826) At, **Ar**

Pollachius virens (Linnaeus, 1758) At, **Ar**
Theragra chalcogramma (Pallas, 1814) Pa, **Ar**

Gasterosteidae (sticklebacks, p. 44)
Gasterosteus aculeatus; Linnaeus, 1758; Amr, Euras, **Ar**
Pungitius pungitius (Linnaeus, 1758) Amr, Euras, **Ar**

Gempylidae (snake mackerels, p. 52)
Paradiplospinus gracilis (Brauer, 1906) In, **An**

Gonostomatidae (bristlemouths, p. 39)
Cyclothone acclinidens; Garman, 1889; At, In, Pa, **An**
Cyclothone microdon (Günther, 1878) At, In, Pa, **An**, **Ar**
Cyclothone pallida; Brauer, 1902; At, In, Pa, **An**
Cyclothone kobayashii; Miya, 1994; At, In, Pa, **An**

Harpagiferidae (spiny plunderfishes, p. 51)
Harpagifer antarcticus; Nybelin, 1947; **An**
Harpagifer georgianus; Nybelin, 1947; **An**

Hemitripteridae (sailfin sculpins, p. 46)
Blepsias bilobus; Cuvier, 1829; Pa, **Ar**
Hemitripterus bolini (Myers, 1934) Pa, **Ar**
Hemitripterus americanus (Gmelin, 1789) At, **Ar**
Nautichthys pribilovius (Jordan and Gilbert, 1898) Pa, **Ar**

Hexagrammidae (greenlings, p. 45)
Hexagrammos lagocephalus (Pallas, 1810) Pa, **Ar**
Hexagrammos octogrammus (Pallas, 1814) Pa, **Ar**
Hexagrammos stelleri; Tilesius, 1810; Pa, **Ar**

Lamnidae (mackerel sharks, p. 34)
Lamna ditropis; Hubbs and Follett, 1947; Pa, **Ar**
Lamna nasus (Bonnaterre, 1788) At, In, Pa, **An**, **Ar**
Carcharodon carcharias (Linnaeus, 1758) At, In, Pa, **Ar**

Lampridae (opahs, p. 40)
Lampris guttatus; Brünnich, 1788; At, In, Pa, **An**
Lampris immaculatus; Gilchrist, 1904; At, In, Pa, **An**

Liparidae (snailfishes, p. 47)
Careproctus acifer; Andriashev and Stein, 1998; **An**
Careproctus ampliceps; Andriashev and Stein, 1998; **An**
Careproctus catherinae; Andriashev and Stein, 1998; **An**
Careproctus continentalis; Andriashev and Prirodina, 1990; **An**
Careproctus credispinulosus; Andriashev and Prirodina, 1990; **An**

Careproctus dubius; Zugmayer, 1911; At, **Ar**
Careproctus eltaninae; Andriashev and Stein, 1998; **An**
Careproctus fedorovi; Andriashev and Stein, 1998; **An**
Careproctus georgianus; Lönnberg, 1905; **An**
Careproctus guillemi; Matallanas, 1998; **An**
Careproctus improvisus; Andriashev and Stein, 1998; **An**
Careproctus inflexidens; Andriashev and Stein, 1998; **An**
Careproctus lacmi; Andriashev and Stein, 1998; **An**
Careproctus leptorhinus; Andriashev and Stein, 1998; **An**
Careproctus longipectoralis; Duhamel, 1992; **An**
Careproctus longipinnis; Burke, 1912; **Ar**
Careproctus micropus (Günther, 1887) At, **Ar**
Careproctus novaezelandiae; Andriashev, 1990; Pa, **An**
Careproctus parini; Andriashev and Prirodina, 1990; **An**
Careproctus parviporatus; Andriashev and Stein, 1998; **An**
Careproctus polarsterni; Duhamel, 1992; **An**
Careproctus profundicola; Duhamel, 1992; **An**
Careproctus pseudoprofundicola; Andriashev and Stein, 1998; **An**
Careproctus reinhardti; Krøyer, 1862; **Ar**
Careproctus rimiventris; Andriashev and Stein, 1998; **An**
Careproctus sandwichensis; Andriashev and Stein, 1998; **An**
Careproctus scaphopterus; Andriashev and Stein, 1998; **An**
Careproctus steini; Andriashev and Prirodina, 1990; **An**
Careproctus tricapitidens; Andriashev and Stein, 1998; **An**
Careproctus vladibeckeri; Andriashev and Stein, 1998; **An**
Careproctus zispi; Andriashev and Stein, 1998; **An**
Edentoliparis terraenovae (Regan, 1916) **An**
Elassodiscus tremebundus; Gilbert and Burke, 1912; Pa, **Ar**
Genioliparis lindbergi; Andriashev and Neyelov, 1976; **An**
Liparis atlanticus (Jordan and Evermann, 1898) At, **Ar**
Liparis bristolensis (Burke, 1912) Pa, **Ar**
Liparis callyodion (Pallas, 1814) Pa, **Ar**
Liparis fabricii; Krøyer, 1847; At, **Ar**
Liparis gibbus; Bean, 1881; Pa, **Ar**
Liparis marmoratus; Schmidt, 1950; Pa, **Ar**
Liparis ochotensis; Schmidt, 1904; Pa, **Ar**
Liparis tunicatus; Reinhardt, 1837; **Ar**
Notoliparis kurchatovi; Andriashev, 1975; **An**
Notoliparis macquariensis; Andriashev, 1978; Pa, **An**
Paraliparis andriashevi; Stein and Tompkins, 1989; **An**
Paraliparis antarcticus; Regan, 1914; **An**
Paraliparis balgueriasi; Matallanas, 1999; **An**

Paraliparis bathybius (Collett, 1879) **Ar**
Paraliparis ceracinus; Andriashev, 1986; **An**
Paraliparis charcoti; Duhamel, 1992; **An**
Paraliparis copei; Goode and Bean, 1896; At, **An**
Paraliparis devriesi; Andriashev, 1980; **An**
Paraliparis diploprora; Andriashev, 1986; **An**
Paraliparis fuscolingua; Stein and Tompkins, 1989; **An**
Paraliparis gracilis; Norman, 1930; **An**
Paraliparis hureaui; Matallanas, 1999; **An**
Paraliparis incognita; Stein and Tompkins, 1989; **An**
Paraliparis kreffti; Andriashev, 1986; **An**
Paraliparis leobergi; Andriashev, 1982; **An**
Paraliparis leucogaster; Andriashev, 1986; **An**
Paraliparis leucoglossus; Andriashev, 1986; **An**
Paraliparis macrocephalus; Chernova and Eastman, 2001; **An**
Paraliparis mawsoni; Andriashev, 1986; **An**
Paraliparis meganchus; Andriashev, 1982; **An**
Paraliparis monoporus; Andriashev and Neyelov, 1979; **An**
Paraliparis neelovi; Andriashev, 1982; **An**
Paraliparis orcadensis; Matallanas and Pequeno, 2000; **An**
Paraliparis operculosus; Andriashev, 1979; In, **An**
Paraliparis rossi; Chernova and Eastman, 2001; **An**
Paraliparis somovi; Andriashev and Neyelov, 1979; **An**
Paraliparis stehmanni; Andriashev, 1986; **An**
Paraliparis tetrapteryx; Andriashev and Neyelov, 1979; **An**
Paraliparis thalassobathyalis; Andriashev, 1982; At, **An**
Paraliparis trilobodon; Andriashev and Neyelov, 1979; **An**
Paraliparis valentinae; Andriashev and Neyelov, 1984; **An**
Paraliparis violaceus; Chernova, 1991; **Ar**
Rhodichthys regina; Collett, 1879; **Ar**

Lotidae (burbots, p. 42)
Brosme brosme (Müller, 1776) At, **Ar**
Gaidropsarus ensis (Reinhardt, 1838) At, **Ar**
Gaidropsarus argentatus (Reinhardt, 1838) At, **Ar**
Lota lota (Linnaeus, 1758) Amr, Euras, **Ar**

Macrouridae (grenadiers or rattails, p. 40)
Caelorinchus fasciatus (Günther, 1878) At, In, Pa, **An**
Caelorinchus marinii; Hubbs, 1934; At, **An**
Coryphaenoides armatus (Hector, 1875) At, In, Pa, **An**
Coryphaenoides ferrieri (Regan, 1913) At, In, Pa, **An**
Coryphaenoides filicauda; Günther, 1878; At, In, Pa, **An**

Coryphaenoides lecointei; Dollo, 1900; At, **An**
Coryphaenoides leptolepis; Günther, 1877; At, In, Pa, **An**
Coryphaenoides rupestris; Gunnerus, 1765; At, **Ar**
Cynomacrurus piriei; Dollo, 1909; At, In, Pa, **An**
Macrourus holotrachys; Günther, 1878; At, Pa, **An**
Macrourus whitsoni (Regan, 1913) At, In, Pa, **An**
Macrourus berglax; Lapépede, 1801; At, **Ar**

Melamphaidae (bigscale fishes, p. 44)
Poromitra crassiceps (Günther, 1878) At, In, Pa, **An**
Sio nordenskjoldii; Lönnberg, 1905; At, In, Pa, **An**

Melanocetidae (blackdevils, p. 43)
Melanocetus rossi; Balushkin and Fedorov, 1981; **An**

Melanonidae (pelagic cods, p. 42)
Melanonus gracilis; Günther, 1878; At, In, Pa, **An**

Microstomatidae (microstomatids, p. 37).
Nansenia antarctica; Kawaguchi and Butler, 1984; At, In, Pa, **An**

Moridae (deepsea cods, p. 40).
Antimora rostrata (Günther, 1878) At, In, Pa, **An**
Halargyreus johnsonii; Günther, 1862; At, In, Pa, **An**
Lepidion ensiferus? (Günther, 1887) At, In, Pa, **An**

Muraenolepididae (eel cods, p. 40)
Muraenolepis marmoratus; Günther, 1880; In, Pa, **An**
Muraenolepis microcephalus; Norman, 1937; **An**
Muraenolepis microps; Lönnberg, 1905; **An**
Muraenolepis orangiensis; Vaillaint, 1888; At, In, **An**

Myctophidae (lanternfishes, p. 40)
Benthosema glaciale (Reinhardt, 1837) At, Pa, **Ar**
Ceratoscopelus warmingii (Lütken, 1892) At, In, Pa, **An**
Electrona antarctica (Günther, 1878) **An**
Electrona carlsbergi; Tåning, 1933; At, In, Pa, **An**
Electrona subasper (Günther, 1864) At, In, Pa, **An**
Gymnoscopelus bolini; Andriashev, 1962; At, In, Pa, **An**
Gymnoscopelus braueri; Lönnberg, 1905; At, In, Pa, **An**
Gymnoscopelus fraseri (Fraser-Brunner, 1931) At, **An**
Gymnoscopelus hintonoides; Hulley, 1981; At, In, Pa, **An**
Gymnoscopelus microlampas; Hulley, 1981; At, In, Pa, **An**
Gymnoscopelus nicholsi (Gilbert, 1911) At, In, Pa, **An**
Gymnoscopelus opisthopterus; Fraser-Brunner, 1949; At, In, Pa, **An**

Gymnoscopelus piabilis (Whitley, 1931) At, In, Pa, **An**
Krefftichthys anderssoni (Lönnberg, 1905) At, In, Pa, **An**
Lampanyctus achirus; Andriashev, 1962; At, In, Pa, **An**
Lampanyctus macdonaldi; Goode and Bean, 1896; At, In, Pa, **An**
Notoscopelus resplendens (Richardson, 1845) At, In, Pa, **An**
Protomyctophum andriashevi (Bekker, 1963) At, In, Pa, **An**
Protomyctophum arcticum (Lütken, 1892) At, **Ar**
Protomyctophum bolini (Fraser-Brunner, 1949) At, In, Pa, **An**
Protomyctophum choriodon; Hulley, 1981; At, In, Pa, **An**
Protomyctophum parallelum (Lönnberg, 1905) At, In, Pa, **An**
Protomyctophum tenisoni (Norman, 1930) At, In, Pa, **An**
Taaningichthys bathyphilus (Tåning, 1928) At, In, Pa, **An**

Myxinidae (hagfishes, p. 34)
Myxine glutinosa; Linnaeus, 1758; At, **Ar**

Notacanthidae (spiny eels, p. 35)
Notacanthus chemnitzii; Block, 1788; At, In, Pa, **Ar**

Notosudidae (waryfishes, p. 39)
Scopelosaurus hamiltoni (Waite, 1916) At, In, Pa, **An**

Nototheniidae (notothens or Antarctic "cods," p. 49)
Aethotaxis mitopteryx; DeWitt, 1962; **An**
Cryothenia peninsulae; Daniels, 1981; **An**
Dissostichus eleginoides; Smitt, 1888; At, In, Pa, **An**
Dissostichus mawsoni; Norman, 1937; Pa, **An**
Gobionotothen angustifrons (Fischer, 1885) **An**
Gobionotothen gibberifrons (Lönnberg, 1905) **An**
Gvozdarus svetovidovi; Balushkin, 1989; **An**
Lepidonotothen larseni (Lönnberg, 1905) In, **An**
Lepidonotothen nudifrons (Lönnberg, 1905) **An**
Lepidonotothen squamifrons (Günther, 1880) At, In, **An**
Notothenia coriiceps; Richardson, 1844; **An**
Notothenia rossii; Richardson, 1844; At, In, Pa, **An**
Pagothenia borchgrevinki; Boulenger, 1902; **An**
Pagothenia brachysoma (Pappenheim, 1912) **An**
Paranotothenia dewitti; Balushkin, 1990; **An**
Paranotothenia magellanica (Forster, 1801) At, Pa, **An**
Patagonotothen guntheri (Norman, 1937) At, **An**
Pleuragramma antarcticum; Boulenger, 1902; **An**
Trematomus bernacchii; Boulenger, 1902; **An**
Trematomus eulepidotus; Regan, 1914; **An**
Trematomus hansoni; Boulenger, 1902; **An**

Trematomus lepidorhinus (Pappenheim, 1911) **An**
Trematomus loennbergii; Regan, 1913; **An**
Trematomus newnesi; Boulenger, 1902; **An**
Trematomus nicolai; Boulenger, 1902; **An**
Trematomus pennellii; Regan, 1914; **An**
Trematomus scotti (Boulenger, 1907) **An**
Trematomus tokarevi; Andriashev, 1988; **An**
Trematomus vicarius; Lönnberg, 1905; **An**

Oneirodidae (dreamers, p. 43)
Oneirodes notius (Pietsch, 1974) At, Pa, **An**

Ophidiidae (cusk-eels and brotulas, p. 43)
Holcomycteronus brucei (Dollo, 1906) **An**

Oreosomatidae (oreos, p. 44)
Pseudocyttus maculatus; Gilchrist, 1906; At, In, Pa, **An**

Osmeridae (smelts, p. 38)
Hypomesus olidus (Pallas, 1814) Pa, **Ar**
Mallotus villosus (Müller, 1776) At, Pa, **Ar**
Osmerus eperlanus; Linnaeus, 1758; At, **Ar**
Osmerus mordax (Mitchill, 1814) Pa, **Ar**

Paralepididae (barracudinas, p. 39)
Arctozenus risso (Bonnaparte, 1840) At, In, Pa, **An, Ar**
Magnisudis prionosa (Rofen, 1963) At, Pa, **An**
Notolepis annulata; Post, 1978; At, **An**
Notolepis coatsi; Dollo, 1908; **An**
Paralepis coregonoides; Risso, 1820; At, **Ar**

Percidae (perches, p. 47)
Perca fluviatilis; Linnaeus, 1758; Euras, **Ar**
Acerina cernua (Linnaeus, 1758) Euras, **Ar**

Percopsidae (trout-perches, p. 40)
Percopsis omiscomaycus (Walbaum, 1792) Amr, **Ar**

Petromyzontidae (lampreys, p. 34)
Geotria australis; Gray, 1851; Aust, Nz, Samr, **An**
Lampetra alaskense (Vladykov and Kott, 1978) Amr, **Ar**
Lampetra camtschatica (Tilesius, 1811) Amr, Euras, **Ar**
Lampetra marinus (Linnaeus, 1758) Amr, Euras, **Ar**
Lampetra tridentatus (Richardson, 1836) Amr, Euras, **Ar**

Pholidae (gunnels, p. 48)
Pholis gunnellus (Linnaeus, 1758) At, **Ar**

Pholis fasciata (Bloch and Schneider, 1801) Pa, **Ar**
Rhodymenichthys dolichogaster (Pallas, 1814) Pa, **Ar**

Pleuronectidae (righteye flounders, p. 52)
Acanthopsetta nadeshnyi; Schmidt, 1904; Pa, **Ar**
Hippoglossoides platessoides (Fabricius, 1780) At, **Ar**
Hippoglossoides elassodon; Jordan and Gilbert, 1880; Pa, **Ar**
Hippoglossoides robustus; Gill and Townsend, 1897; Pa, **Ar**
Hippoglossus stenolepis; Schmidt, 1904; Pa, **Ar**
Hippoglossus hippoglossus; Linnaeus, 1758; At, **Ar**
Lepidopsetta polyxystra; Orr and Mataresse, 2000; Pa, **Ar**
Limanda aspera (Pallas, 1814) Pa, **Ar**
Limanda proboscidea; Gilbert, 1896; Pa, **Ar**
Limanda sakhalinensis; Hubbs, 1915; Pa, **Ar**
Platichthys stellatus (Pallas, 1787) Pa, **Ar**
Liopsetta glacialis (Pallas, 1776) Pa, **Ar**
Pleuronectes platessa; Linnaeus, 1758; At, **Ar**
Pleuronectes quadrituberculatus (Pallas, 1814) Pa, **Ar**
Platichthys flesus; Linnaeus, 1758; At, **Ar**
Reinhardtius hippoglossoides (Walbaum, 1792) Pa, **Ar**
Atheresthes stomias (Jordan and Gilbert, 1880) Pa, **Ar**

Psychrolutidae (fathead sculpins, p. 46)
Cottunculus sadko; Essipov, 1937; **Ar**
Cottunculus microps; Collett, 1875; At, **Ar**
Dasycottus setiger; Bean, 1890; Pa, **Ar**
Eurymen gyrinus; Gilbert and Burke, 1912; Pa, **Ar**
Malacocottus zonurus; Bean, 1890; Pa, **Ar**
Psychrolutes paradoxus; Günther, 1861; Pa, **Ar**
Psychrolutes subspinosus (Jensen, 1902) **Ar**

Rajidae (skates, p. 35)
Amblyraja georgiana; Norman, 1938; **An**
Amblyraja hyperborea; Collett, 1879; At, Pa, **Ar**
Amblyraja radiata; Donovan, 1808; At, **Ar**
Bathyraja eatonii (Günther, 1876) **An**
Bathyraja maccaini; Springer, 1971; **An**
Bathyraja meridionalis; Stehmann, 1987; **An**
Bathyraja spinicauda (Jensen, 1914) At, **Ar**
Rajella fyllae; Lütken, 1888; At, **Ar**
Bathyraja abyssicola (Gilbert, 1896) Pa, **Ar**
Bathyraja aleutica (Gilbert, 1896) Pa, **Ar**
Bathyraja interrupta (Gill and Townsend, 1897) Pa, **Ar**

Bathyraja maculata; Ishiyama and Ishihara, 1977; Pa, **Ar**
Bathyraja minispinosa; Ishiyama and Ishihara, 1977; Pa, **Ar**
Bathyraja parmifera (Bean, 1881) Pa, **Ar**
Bathyraja taranetzi (Dolganov, 1983) Pa, **Ar**
Bathyraja violacea (Suvorov, 1935) Pa, **Ar**
Raja binoculata; Girard, 1854; Pa, **Ar**

Salmonidae (chars, graylings, salmons, trouts, whitefishes and allies, p. 38)
Brachymystax lenok (Pallas, 1773) Euras, **Ar**
Hucho taimen (Pallas, 1773) Euras, **Ar**
Salmo salar; Linnaeus, 1758; Amr, Euras, **Ar**
Salmo trutta; Linnaeus, 1758; Euras, **Ar**
Oncorhynchus gorbuscha (Walbaum, 1792) Amr, Euras, **Ar**
Oncorhynchus keta (Walbaum, 1792) Amr, Euras, **Ar**
Oncorhynchus kisutch (Walbaum, 1792) Amr, Euras, **Ar**
Oncorhynchus nerka (Walbaum, 1792) Amr, Euras, **Ar**
Oncorhynchus tshawytscha (Walbaum, 1792) Amr, Euras,'**Ar**
Salvelinus alpinus (Linnaeus, 1758) Amr, Euras, **Ar**
Salvelinus fontinalis (Mitchill, 1814) Amr, **Ar**
Salvelinus leucomaenis (Pallas, 1814) Euras, **Ar**
Salvelinus malma (Walbaum, 1792) Amr, Euras, **Ar**
Salvelinus namaycush (Walbaum, 1792) Amr, **Ar**
Coregonus artedi; Lesueur, 1818; Euras, **Ar**
Coregonus autumnalis (Pallas, 1776) Amr, Euras, **Ar**
Coregonus clupeaformis (Mitchill, 1818) Amr, **Ar**
Coregonus laurettae; Bean, 1881; Amr, Euras, **Ar**
Coregonus muksun (Pallas, 1814; Euras, **Ar**
Coregonus nasus (Pallas, 1776) Amr, Euras, **Ar**
Coregonus nelsonii; Bean, 1884; Euras, **Ar**
Coregonus peled; Gmelin, 1789; Euras, **Ar**
Coregonus pidschian (Gmelin, 1789) Amr, Euras, **Ar**
Coregonus sardinella; Valenciennes, 1848; Amr, Euras, **Ar**
Prosopium cylindraceum (Pallas, 1784) Amr, Euras, **Ar**
Stenodus leucichthys (Güldenstadt, 1772) Amr, Euras, **Ar**
Thymallus arcticus (Pallas, 1776) Amr, Euras, **Ar**
Thymallus thymallus (Linnaeus, 1758) Euras, **Ar**

Scopelarchidae (pearleyes, p. 39)
Benthalbella elongata (Norman, 1937) At, In, Pa, **An**
Benthalbella macropinna; Bussing and Bussing, 1966; At, In, Pa, **An**

Sebastidae (rockfishes, p. 45)
Sebastes aleutianus (Jordan and Evermann, 1898) Pa, **Ar**

Sebastes alutus (Gilbert, 1890) Pa, **Ar**
Sebastes borealis; Barsukov, 1970; Pa, **Ar**
Sebastes mentella (Travin, 1951) At, **Ar**
Sebastes polyspinis (Taranetz and Moiseev, 1933) Pa, **Ar**
Sebastes marinus (Linnaeus, 1758) At, **Ar**
Sebastes viviparus; Krøyer, 1845; At, **Ar**
Sebastolobus alascanus; Bean, 1890; Pa, **Ar**

Squalidae (dogfish sharks, p. 35)
Squalus acanthias; Linnaeus, 1758; At, Pa, **Ar**

Stichaeidae (pricklebacks, p. 48)
Acantholumpenus mackayi (Gilbert, 1896) Pa, **Ar**
Alectrias alectrolophus (Pallas, 1814) Pa, **Ar**
Anisarchus medius (Reinhardt, 1837) Pa, **Ar**
Chirolophis decoratus (Jordan and Snyder, 1902) Pa, **Ar**
Chirolophis snyderi (Taranetz, 1938) Pa, **Ar**
Eumesogrammus praecius (Krøyer, 1837) **Ar**
Leptoclinus maculatus (Fries, 1837) Pa, **Ar**
Lumpenella longirostris (Evermann and Goldsborough, 1907) Pa, **Ar**
Lumpenus fabricii Reinhardt (Valenciennes, 1836) Pa, **Ar**
Lumpenus lampretaeformis (Walbaum, 1792) At, **Ar**
Lumpenus sagitta; Wilimovsky, 1956; Pa, **Ar**
Stichaeus punctatus (Fabricius, 1780) Pa, **Ar**

Stomiidae (barbeled dragonfish, p. 39).
Stomias boa (Risso, 1810) At, **An**

Tripterygiidae (threefin blennies, p. 52)
Helcogrammoides antarcticus (Tomo, 1981) **An**

Zaproridae (prowfish, p. 49)
Zaprora silenus; Jordan, 1896; Pa, **Ar**

Zoarcidae (eelpouts, p. 47)
Dieidolycus leptodermatus; Anderson, 1988; **An**
Gymnelus hemifasciatus; Andriashev, 1937; Pa, **Ar**
Gymnelus viridis (Fabricius, 1780) **Ar**
Gymnelus andersoni; Chernova, 1998; **Ar**
Gymnelus barsukovi; Chernova, 1999; **Ar**
Gymnelus bilabrus; Andriashev, 1937; Pa, **Ar**
Gymnelus knipowitschi; Chernova, 1999; **Ar**
Gymnelus retrodorsalis; Le Danois, 1913; **Ar**
Lycenchelys antarctica; Regan, 1913; Pa, **An**
Lycenchelys aratrirostris; Andriashev and Permitin, 1968; **An**

Lycenchelys argentina; Marschoff, Torno and Tomo, 1977; **An**
Lycenchelys bellingshauseni; Andriashev and Permitin, 1968; **An**
Lycenchelys kolthoffi; Jensen, 1904; **Ar**
Lycenchelys muraena (Collett, 1878) **Ar**
Lycenchelys nanospinata; Anderson, 1988; **An**
Lycenchelys nigripalatum; DeWitt and Hureau, 1979; **An**
Lycenchelys paxillus (Goode and Bean, 1879) At, **Ar**
Lycenchelys platyrhina (Jensen, 1902) **Ar**
Lycenchelys sarsii (Collett, 1871) At, **Ar**
Lycenchelys tristichodon; DeWitt and Hureau, 1979; **An**
Lycenchelys wilkesi; Anderson, 1988; **An**
Lycodapus antarcticus; Tomo, 1981; At, **An**
Lycodapus pachysoma; Peden and Anderson, 1978; Pa, **An**
Lycodes adolfi; Nielsen and Fosså, 1993; **Ar**
Lycodes brevipes; Bean, 1890; Pa, **Ar**
Lycodes diapterus; Gilbert, 1892; Pa, **Ar**
Lycodes esmarkii; Collett, 1875; At, **Ar**
Lycodes eudipleurostictus; Jensen, 1902; **Ar**
Lycodes frigidus; Collett, 1879; **Ar**, **Ar**
Lycodes gracilis; Sars, 1867; At, **Ar**
Lycodes jugoricus; Knipowitsch, 1906; **Ar**
Lycodes lavalei; Vladykov and Tremblay, 1936; At, **Ar**
Lycodes luetkenii; Collett , 1880; **Ar**
Lycodes mcallisteri; Møller, 2001; **Ar**
Lycodes marisalbi; Knipowitsch, 1906; **Ar**
Lycodes mucosus; Richardson, 1855; **Ar**
Lycodes palearis; Gilbert, 1896; Pa, **Ar**
Lycodes paamiuti; Møller, 2001; **Ar**
Lycodes pallidus; Collett, 1879; At, **Ar**
Lycodes polaris (Sabine, 1824) **Ar**
Lycodes raridens; Taranets and Andriashev, 1937; Pa, **Ar**
Lycodes reticulatus; Reinhardt, 1835; **Ar**
Lycodes rossi; Malmgreen, 1865; **Ar**
Lycodes sagittarius; McAllister, 1976; **Ar**
Lycodes seminudus; Reinhardt, 1837; **Ar**
Lycodes squamiventer; Jensen, 1904; **Ar**
Lycodes turneri; Bean, 1879; Pa, **Ar**
Lycodes vahlii; Reinhardt, 1831; At, **Ar**
Lycodichthys antarcticus; Pappenheim, 1911; **An**
Lycodichthys dearborni (DeWitt, 1962) **An**
Lycodonus flagellicauda (Jensen, 1902) **Ar**
Lycodonus mirabilis; Goode and Bean, 1883; At, **Ar**

Melanostigma bathium; Bussing, 1965; Pa, **An**
Melanostigma gelatinosum; Günther, 1881; At, In, Pa, **An**
Oidiphorus mcallisteri; Anderson, 1988; **An**
Ophthalmolycus amberensis; Tomo, Marschoff, and Torno, 1977; **An**
Ophthalmolycus bothriocephalus (Pappenheim, 1912) **An**
Pachycara brachycephalum (Pappenheim, 1912) **An**
Pachycara goni; Anderson, 1991; **An**
Seleniolycus laevifasciatus (Torno, Tomo and Marschoff, 1977) **An**
Zoarces viviparus (Linnaeus, 1758) At, **Ar**

REFERENCES

Able, K. W. (1990). A revision of Arctic snailfishes of the genus *Liparis* (Scorpaeniformes: Cyclopteridae). *Copeia* **2**, 476–492.

Ahlstrom, E. H., Richards, W. J., and Weitzman, S. H. (1984). Families Gonostomatidae, Sternoptychidae and associated stomiiform groups: Development and relationships. *In* "Ontogeny and Systematics in Fishes" (Moser, H. G., Richards, W. J., Cohen, D. M., Fahay, M. P., Kendall, A. W., Jr., and Richardson, S. L., Eds.), pp. 184–198. *ASIH, Special publication no. 1.*

Allen, M. J., and Smith, G. B. (1988). Atlas and zoogeography of common fishes in the Bering Sea and northeastern Pacific. *NOAA Tech. Rep. NMFS* **66**, 1–151.

Andersen, N. C. (1984). Genera and subfamilies of the family Nototheniidae (Pisces, Perciformes) from the Antarctic and Subantarctic. *Steenstrupia* **10**(1), 1–34.

Anderson, M. E. (1982). Revision of the fish genera *Gymnelus* Reinhardt and *Gymnelopsis* Soldatov (Zoarcidae), with two new species and comparative osteology of *Gymnelus viridis*. *Natl. Mus. Can. Publ. Zool.* **17**, 1–76.

Anderson, M. E. (1990). The origin and evolution of the Antarctic Ichthyofauna. *In* "Fishes of the Southern Ocean" (Gon, O., and Heemstra, P. C., Eds.), pp. 28–33. J. L. B. Smith Institute of Ichthyology, Grahamstown, South Africa.

Anderson, M. E. (1994). Systematics and osteology of the Zoarcidae (Teleostei: Perciformes). *Ichthyol. Bull.* **60**, 1–120.

Anderson, M. E. (1995). The eelpout genera *Lycenchelys* Gill and *Taranetzella* Andriashev (Teleostei: Zoarcidae) in the eastern Pacific, with descriptions of nine new species. *Proc. Calif. Acad. Sci.* **49**, 55–113.

Andriashev, A. P. (1949). On the species composition and distribution of sculpins of the genus *Triglops* Reinh. in the northern seas. *Trudy Vsesoyuznogo Gidrobiologicheskogo Obshchestva Akademii Nauk SSSR* **1**, 194–209. (Translation by Ichthyological Laboratory, Bureau of Commercial Fisheries, US National Museum, Washington, DC.).

Andriashev, A. P. (1953). Ancient deep-water and secondary deep-water fishes and their importance in a zoogeographical analysis. *In* "Notes on General Problems in Ichthyology" (Lindberg, G. U., Ed.), pp. 58–64. Akademii Nauk SSSR Ikhtiol. Kom., Moscow-Leningrad.

Andriashev, A. P. (1954). "Fishes of the Northern Seas of the USSR" Academy of Sciences of the Union of Soviet Socialist Republics, Moscow-Leningrad. (Translated from Russian by Israel program for scientific translations Jerusalem, 1964).

Andriashev, A. P. (1955). A review of the genus *Lycenchelys* Gill (Pisces, Zoarcidae) and related forms in the seas of the USSR and adjacent waters. *Trudy Zool. Inst. Akad. Nauk SSSR* **18**, 349–384.

Andriashev, A. P. (1965). A general review of the Antarctic fish fauna. *In* "Biogeography and Ecology in Antarctica" (van Oye, P., and van Mieghen, J., Eds.), Vol. 15, pp. 491–550.

Andriashev, A. P. (1986). Review of the snailfish genus *Paraliparis* (Scorpaeniformes: Liparididae) of the Southern Ocean. *Theses Zool.* **7**, 1–204.

Andriashev, A. P. (1990). Possible pathways of *Paraliparis* (Pisces: Liparididae) and some other North Pacific secondary deep-sea fishes into the North Atlantic and Arctic depths. *Polar Biol.* **11**, 213–218.

Andriashev, A. P. (2003). Liparid fishes (Liparidae, Scorpaeniformes) of the southern ocean and adjacent waters [in Russian]. *Biol. Results Russian Antarctic Expeditions* **9**, 1–474.

Andriashev, A. P., Balushkin, A. V., and Voskoboinikova, O. S. (1989). Morphological substantiation of the subfamily Gymnodraconinae (Bathydraconidae). *Vopr. Ikhtiol.* **29**(4), 515–523.

Andriashev, A. P., and Chernova, N. V. (1995). Annotated list of fishlike vertebrates and fish of the Arctic Seas and Adjacent waters. *J. Ichthyol.* **35**(1), 81–123.

Andriashev, A. P., and Stein, D. L. (1998). Review of the snailfish genus *Careproctus* (Liparidae, Scorpaeniformes) in Antarctic and adjacent waters. *Contrib. Sci. (Los Ang.)* 1–63.

Astarloa, J. M. D., Figueroa, D. E., Lucifora, L., Menni, R. C., Prenski, B. L., and Chiaramonte, G. (1999). New records of the Pacific sleeper shark; *Somniosus pacificus* (Chondrichthyes: Squalidae), from the south-west Atlantic. *Ichthyol. Res.* **46**(3), 303–308.

Baker, R. L., Chandler, M. L., and Eckdahl, T. T. (2001). Identification of *Cottus* species in Montana using mitochondrial RFLP analysis. *Bios* **72**, 83–87.

Balushkin, A. V. (1992). Classification, phylogenetic relationships, and origins of the families of the suborder Notothenioidei (Perciformes). *J. Ichthyol.* **32**(7), 90–110.

Balushkin, A. V. (1994). *Proeleginops grandeastmanorum* gen et sp. nov. (Perciformes, Notothenioidei, Eleginopsidae) from the Late Eocene of Seymour Island (Antarctica) is a fossil notothenioid, not a gadiform. *Vopr. Ikhtiol.* **34**(3), 298–307.

Balushkin, A. V. (1996a). A new genus and species of liparid fish *Palmoliparis beckeri* from the northern Kurile Islands (Scorpaeniformes, Liparidae) with consideration of phylogeny of the family. *J. Ichthyol.* **36**(4), 281–287.

Balushkin, A. V. (1996b). Similarity of fish of family Channichthyidae (Notothenioidei, Perciformes), with remarks on the species composition of the family and the description of a new species from the Kerguelen Archipelago. *J. Ichthyol.* **36**(1), 1–10.

Balushkin, A. V. (2000). Morphology, classification, and evolution of notothenioid fishes of the Southern Ocean (Notothenioidei, Perciformes). *J. Ichthyol.* **40**(suppl. 1), 74–109.

Balushkin, A. V., and Fedorov, V. V. (1981). On finding the deepwater anglerfishes (*Melanocetus rossi* sp. n. and *Oneirodes notius*) in the Ross Sea (Antarctica). *Biol. Morya* **2**, 79–82.

Balushkin, A. V., and Voskoboinikova, O. S. (1995). Systematics and phylogeny of Antarctic dragonfishes (Bathydraconidae, Notothenioidei, Perciformes). *Vopr. Ikhtiol.* **35**(2), 147–155.

Bargelloni, L., Ritchie, P. A., Patarnello, T., Battaglia, B., Lambert, D. M., and Meyer, A. (1994). Molecular evolution at subzero temperatures: Mitochondrial and nuclear phylogenies of fishes from Antarctica (suborder Notothenioidei), and the evolution of antifreeze glycopeptides. *Mol. Biol. Evol.* **11**(6), 854–863.

Bargelloni, L., Marcato, S., Zane, L., and Patarnello, T. (2000). Mitochondrial phylogeny of notothenioids, a molecular approach to Antarctic fish evolution and biogeography. *Syst. Biol. Mar.* **49**(1), 114–129.

Bastrop, R., Strehlow, B., Jurss, K., and Sturmbauer, C. (2000). A new molecular phylogenetic hypothesis for the evolution of freshwater eels. *Mol. Phylogenet. Evol.* **14**(2), 250–258.

Berendzen, P. B., and Dimmick, W. W. (2002). Phylogenetic relationships of Pleuronectiformes based on molecular evidence. *Copeia* **3**, 642–652.

Berg, L. S. (1918). On the causes of similarity between the fauna of the North Atlantic and the northern Pacific Ocean [in Russian]. *Izvest. Ross. Akad. Nauk.* **16,** 1835–1842.

Berg, L. S. (1932). Übersicht der Verbreitung der Süßwasserfische Europas. *Zoogeographica* **1**(2), 107–208.

Berg, L. S. (1934). On the amphiboreal (discontinuous) distribution of marine fauna in the northern hemisphere [in Russian]. *Izvest. Gos. Geogr. Obshch.* **66,** 69–78.

Bernatchez, L., and Dodson, J. J. (1994). Phylogenetic relationships among Palearctic and Nearctic whitefish (*Coregonus* sp.) populations as revealed by mitochondrial DNA variation. *Can. J. Fish Aquat. Sci.* **51**(suppl. 1), 240–251.

Bertelsen, E., Krefft, G., and Marshall, N. B. (1976). The fishes of the family Notosudidae. *Dana-report* **86,** 1–114.

Birstein, V. J., and DeSalle, R. (1998). Molecular phylogeny of Acipenserinae. *Mol. Phylogenet. Evol.* **9**(1), 141–155.

Birstein, V.J, Doukakis, P., and DeSalle, R. (2002). Molecular phylogeny of Acipenseridae: Nonmonophyly of Scaphirhynchinae. *Copeia* **2,** 287–301.

Briggs, J. C. (1974). "Marine Zoogeography." McGraw–Hill series in Population Biology, New York.

Briggs, J. C. (2003). Marine centres of origin as evolutionary engines. *J. Biogeogr.* **30,** 1–18.

Briolay, J., Galtier, N., Brito, R. M., and Bouvet, Y. (1998). Molecular phylogeny of Cyprinidae inferred from cytochrome *b* DNA sequences. *Mol. Phylogenet. Evol.* **9**(1), 100–108.

Carr, S. M., Kivlichan, D. S., Pepin, P., and Crutcher, D. C. (1999). Molecular systematics of gadid fishes: Implications for the biogeographic origin of Pacific species. *Can. J. Zool.* **77,** 17–26.

Chen, W. J., Bonillo, C., and Lecointre, G. (1998). Phylogeny of the Channichthyidae (Notothenioidei, Teleostei) based on two mitochondrial genes. *In* "Fishes of Antarctica. A Biological Overview" (di Prisco, G., Pisano, E., and Clarke, A., Eds.), pp. 288–298. Springer-Verlag Publishers, Italia.

Chernova, N. V. (2000). Four new species of *Gymnelus* (family Zoarcidae) from the far eastern seas, with genus diagnosis and key to species. *J. Ichthyol.* **40**(1), 1–12.

Chiu, T. S., Markle, D. F., and Melendez, R. (1990). Moridae. *In* "Fishes of the Southern Ocean" (Gon, O., and Heemstra, P. C., Eds.), pp. 183–187. J. L. B. Smith Institute Ichthyology, Grahamstown.

Christiansen, J. S., and Fevolden, S.-E. (2000). The polar cod of Porsangerfjorden, Norway; revisited. *Sarsia* **85,** 189–193.

Coad, B., Waszczuk, H., and Labignan, I. (1995). "Encyclopedia of Canadian fishes." Canadian Museum of Nature and Canadian Sportfishing Production.

Coburn, M. M., and Cavender, T. M. (1992). Interrelationships of North American Cyprinid Fishes. *In* "Systematics, Historical Ecology and North American Freshwater Fishes" (Mayden, R. L., Ed.), pp. 328–373. Stanford University Press, Stanford, California.

Cohen, D. M., Inada, T., Iwamoto, T., and Scialabba, N. (1990). Gadiform fishes of the world (order Gadiformes): An annotated and illustrated catalogue of cods, hakes, grenadiers and other gadiform fishes known to date. *FAO Species Catalogue* **10,** 1–442.

Colgan, D. J., Zhang, C., and Paxton, J. R. (2000). Phylogenetic investigations of the Stephanoberyciformes and Beryciformes, particularly whalefishes (Euteleostei: Cetomimidae), based on partial 12S rDNA and 16S rDNA sequences. *Mol. Phylogenet. Evol.* **17**(1), 15–25.

Collette, B. B., Potthoff, T., Richards, W. J., Ueyanagi, S., Russo, J. L., and Nishikawa, Y. (1984). Scombroidei: Development and relationships. *In* "Ontogeny and Systematics in Fishes" (Moser, H. G., Richards, W. J., Cohen, D. M., Fahay, M. P., Kendall, A. W., Jr., and Richardson, S. L., Eds.), pp. 591–620. *ASIH. Spec. Publ. No. 1.*

Compagno, L. J. V. (1984). Sharks of the world. *FAO Species Catalogue* **4**(1), 1–249.

Compagno, L. J. V. (1990). Sharks. *In* "Fishes of the Southern Ocean" (Gon, O., and Heemstra, P. C., Eds.), pp. 81–85. J. L. B. Smith Institute of Ichthyology, Grahamstown, South Africa.

Compagno, L. J. V. (2001). Sharks of the world. An annotated and illustrated catalogue of shark species known to date. *FAO Species Catalogue for Fishery Purposes* **1**(2), 1–269.

Cooper, J. A., and Chapleau, F. (1998). Monophyly and interrelationships of the family Pleuronectidae (Pleuronectiformes), with a revised classification. *Fish. Bull.* **96**, 686–726.

Cowan, G. I. M. (1971). Comparative morphology of the cottid genus *Myoxocephalus* based on meristic, morphometric, and other anatomical characters. *Can. J. Zool.* **49**(11), 1479–1496.

Craig, P. C. (1984). Fish use of coastal waters of the Alaskan Beaufort Sea: A Review. *Trans. Am. Fish Soc.* **113**, 265–282.

Davenport, J. (1985). Synopsis of biological data on the lumpsucker *Cyclopterus lumpus* (Linnaeus 1758). *FAO Fisheries Synopsis No.* **147**, 1–31.

Derome, N., Chen, W. J., Dettai, A., Bonillo, C., and Lecointre, G. (2002). Phylogeny of Antarctic dragonfishes (Bathydraconidae, Notothenioidei, Teleostei) and related families based on their anatomy and two mitochondrial genes. *Mol. Phylogenet. Evol.* **24**(1), 139–152.

Dewitt, H. H., Heemstra, P. C., and Gon, O. (1990). Nototheniidae. *In* "Fishes of the Southern Ocean" (Gon, O., and Heemstra, P. C., Eds.), pp. 279–331. J. L. B. Smith Institute of Ichthyology, Grahamstown, South Africa.

Docker, M. F., Youson, J. H., Beamish, R. J., and Devlin, R. H. (1999). Phylogeny of the lamprey genus *Lampetra* inferred from mitochondrial cytochrome b and ND3 gene sequences. *Can. J. Fish. Aquat. Sci.* **56**, 2340–2349.

Dunbar, M. J. (1968). "Ecological development in polar regions: A study of evolution." Prentice-Hall, Englewood Cliff, New Jersey.

Dunn, J. R. (1989). A provisional phylogeny of gadid fishes based on adult and early life-history characters. *In* "Papers on the Systematics of Gadiform Fishes" (Cohen, D. M., Ed.), pp. 209–235. *Nat. Hist. Mus. Los Angeles Count. Sci. Ser.* **32**.

Eakin, R. R. (1981). Osteology and relationships of the fishes of the Antarctic family Harpagiferidae (Pisces, Notothenioidei) from the Ross Sea Antarctica. *Ant. Res. Ser.* **31**(5), 81–147.

Eastman, J. T. (1993). "Antarctic Fish Biology. Evolution in a Unique Environment." Academic Press, San Diego.

Eastman, J. T., and Eakin, R. R. (2000). An updated species list for notothenioid fish (Perciformes; Notothenioidei), with comments on Antarctic species. *Arch. Fish. Mar. Res.* **48**(1), 11–20.

Einarsson, T., Hopkins, D. M., and Doell, R. R. (1967). The stratigraphy of Tjörnes, northern Iceland, and the history of the Bering land bridge. *In* "The Bering Land Bridge" (Hopkins, D. M., Ed.), pp. 312–325. Stanford University Press, Stanford.

Ekman, S. (1953). "Zoogeography of the Sea." Sidgwick and Jackson, London.

Endo, H. (2002). Phylogeny of the order Gadiformes (Teleostei, Paracanthopterygii). *Mem. Grad. Sch. Fish. Sci. Hokkaido University* **49**(2), 75–149.

Eschmeyer, W. N. (Ed.) (1998). Catalog of fishes. Special publication no. 1 of the center for biodiversity research and information California Academy of Sciences, San Francisco.

Eschmeyer, W. N., and Herald, E. S. (1983). "A Field Guide to Pacific Coast Fishes of North America." Peterson Field Guide Series. Field Guide Pacific Coast Fishes i–xii, pp. 1–336.

Evseenko, S. A. (2000). Family Achiropsettidae and its position in the taxonomic and ecological classifications of Pleuronectiformes. *J. Ichthyol.* **40**(suppl. 1), 110–138.

Fernholm, B. (1990). Myxinidae, Hagfishes. *In* "Fishes of the Southern Ocean" (Gon, O., and Heemstra, P. C., Eds.), pp. 77–78. J. L. B. Smith Institute of Ichthyology, Grahamstown, South Africa.

Fernholm, B. (1998). Hagfish systematics. *In* "The Biology of Hagfishes" (Jørgensen, J. M., Lomholt, J. P., Weber, R. E., and Malte, H., Eds.), pp. 33–44. Chapman and Hall, London.

Fink, W. L. (1985). Phylogenetic interrelationships of the stomiid fishes (Teleostomi: Stomiiformes). *Misc. Publ. Mus. Zool. Univ. Mich.* **171**, 1–127.

Fritzsche, R. A. (1984). Gasterosteiformes: Development and relationships. *In* "Ontogeny and Systematics in Fishes" (Moser, H. G., Richards, W. J., Cohen, D. M., Fahay, M. P., Kendall, A. W., Jr., and Richardson, S. L., Eds.), pp. 398–405. *ASIH. Spec. publ. No.* **1**.

Gomes, U. L., and Paragó, C. (2001). Espécie nova de rajídeo (Chondrichthyes, Rajiformes) do Atlântico Sul Ocidental. *Bol. Do Mus. Nac., N.S., Zool., Rio de Janeiro* **448**, 1–10.

Gon, O. (1990). Tripterygiidae. *In* "Fishes of the Southern Ocean" (Gon, O., and Heemstra, P. C., Eds.), pp. 400–401. J. L. B. Smith Institute of Ichthyology, Grahamstown, South Africa.

Gon, O., and Heemstra, P. C. (Eds.), (1990). *In* "Fishes of the Southern Ocean," pp. 1–462. J. L. B. Smith Institute of Ichthyology, Grahamstown, South Africa.

Gudkov, P. K. (1998). Bering Sea *Dallia pectoralis* in the Chukchi Peninsula. *J. Ichthyol.* **38**(2), 199–203.

Harris, P. M., and Mayden, R. L. (2001). Phylogenetic relationships of major clades of Catostomidae (Teleostei: Cypriniformes) as inferred from mitochondrial SSU and LSU rDNA sequences. *Mol. Phylogenet. Evol.* **20**(2), 225–237.

Herman, Y., and Hubkins, D. M. (1980). Arctic oceanic climate in late Cenozoic time. *Science* **209**, 557–562.

Horn, M. H. (1984). Stromateoidei: Development and relationships. *In* "Ontogeny and Systematics in Fishes" (Moser, H. G., Richards, W. J., Cohen, D. M., Fahay, M. P., Kendall, A. W., Jr., and Richardson, S. L., Eds.), pp. 620–628. *ASIH. Spec. Publ. No.* **1**.

Howes, G. J. (1991). Biogeography of gadoid fishes. *J. Biogeogr.* **18**, 595–622.

Hulley, P. A. (1998). Preliminary investigations on the evolution of the tribe Electronini (Myctophiformes, Myctophidae). *In* "Fishes of Antarctica. A Biological Overview" (di Prisco, G., Pisano, E., and Clarke, A., Eds.), pp. 75–85. Springer-Verlag, Italia.

Hureau, J.-C. (1986). Relations phylogénétiques au sein des Notothenioidei. *Oceanis* **12**(5), 367–376.

Hureau, J.-C. (1990). Harpagiferidae. *In* "Fishes of the Southern Ocean" (Gon, O., and Heemstra, P. C., Eds.), pp. 357–363. J. L. B. Smith Institute of Ichthyology, Grahamstown, South Africa.

Hureau, J.-C., and Tomo, A. P. (1977). *Bovichthys elongatus* n. sp., poisson Bovichthyidae, famille novelle pour l'Antarctique. *Cybium* **3**(1), 67–74.

Iwami, T. (1985). Osteology and relationships of the family Channichthyidae. *Mem. Natl. Inst. Polar Res., Series E.* **36**, 1–69.

James, G. D., Inada, T., and Nakamura, I. (1988). Revision of the oreosomatid fishes (family Oreosomatidae) from the southern oceans, with description of a new species. *New Zealand J. Zool.* **15**, 291–326.

Jensen, A. S., and Volsøe, H. (1949). A revision of the genus *Icelus* (Cottidae) with remarks on the structure of its urogenital papilla. *Kongelige Danske Videnskabernes Selskab Biologiske Meddelelser* **21**(6), 1–26.

Joensen, J. S., and Tåning, Å. V. (1970). Marine and freshwater fishes. *In* "Zoology of the Faroes" (Jensen, A. S., Lundbeck, W., and Mortensen, T., Eds.), pp. 1–241. Copenhagen.

Johnson, R. K. (1984). Scopelarchidae: Development and relationships. *In* "Ontogeny and Systematics in Fishes" (Moser, H. G., Richards, W. J., Cohen, D. M., Fahay, M. P., Kendall, A. W., Jr., and Richardson, S. L., Eds.), pp. 245–250. *ASIH. Spec. Publ. No. 1*.

Johnson, G. D., and Patterson, C. (1996). Relationship of lower Euteleostean fishes. *In* "Interrelationships of Fishes" (Stiassny, M. L., Parenti, L. R., and Johnson, G. D., Eds.), pp. 251–332. Academic Press, San Diego, California.

Jónsson, G. (1992). "Islenskir Fiskar." pp. 1–568. Fiolvi, Reykjavik.

Jordan, A. D., Møller, P. R., and Nielsen, J. G. (2003). Revision of the Arctic cod genus *Arctogadus*. *J. Fish Biol.* **62,** 1339–1352.

Kanayama, T. (1991). Taxonomy and phylogeny of the family Agonidae (Pisces: Scorpaeniformes). *Mem. Fac. Fish. Hokkaido Univ.* **38,** 1–199.

Keene, M. J., and Tighe, K. A. (1984). Beryciformes: Development and relationships. *In* "Ontogeny and Systematics of Fishes" (Moser, H. G., Richards, W. J., Cohen, D. M., Fahay, M. P., Kendall, A. W., Jr., and Richardson, S. L., Eds.), pp. 383–392. *ASIH. Spec. Publ. No.* **1.**

Kido, K. (1988). Phylogeny of the family Liparididae, with the taxonomy of the species found around Japan. *Mem. Fac. Fish. Hokkaido Univ. (MFFHU)* **35**(2), 125–256.

Kobyliansky, S. H. (1990). Taxonomic status of microstomatid fishes and problems of classification of suborder Argentinoidei (Salmoniformes, Teleostei). *Trudy Inst. Okeanol. Akad. Nauk. SSSR (TIO),* **125,** 148–177.

Kock, K.-H. (1985). Marine habitats—Antarctic fish. *In* "Key Environments—Antarctica" (Bonner, W. N., and Walton, D. W. H., Eds.), pp. 173–192. Pergamon Press, Oxford.

Kottelat, M. (1997). European freshwater fishes. *Biologia* **52**(suppl. 5), 1–271.

La Mesa, M., Vacchi, M., Iwami, T., and Eastman, J. T. (2002). Taxonomic studies of the Antarctic icefish genus *Cryodraco* Dollo, 1900 (Notothenioidei: Channichthyidae). *Polar Biol.* **25,** 384–390.

Last, P. R., and Stevens, J. D. (1994). "Sharks and Rays of Australia." CSIRO, Australia.

Lecointre, G., Bonillo, C., Ozouf-Costaz, C., and Hureau, J.-C. (1997). Molecular evidence for the origins of Antarctic fishes: paraphyly of the Bovichtidae and no indication for the monophyly of the Notothenioidei (Teleostei). *Polar Biol.* **18,** 193–208.

Lin, Y. S., Poh, Y. P., and Tzeng, C. S. (2001). A phylogeny of freshwater eels inferred from mitochondrial genes. *Mol. Phylogenet. Evol.* **20,** 252–261.

Linnaeus, C. (1758). "Systema Naturae per Regna Tria Naturae secundum Classes, Ordinus, Genera, Species cum Characteribus, Differentiis Synonymis, Locis,". 10th edn., Vol. 1. Holmiae Salvii.

Long, D. J. (1994). Quaternary colonization or Paleogene persistence? Historical biogeography of skates (Chondrichthyes: Rajidae) in the Antarctic ichthyofauna. *Paleobiology* **20**(2), 215–228.

Long, D. J., and McCosker, J. E. (1999). A new species of deep-water skate, *Rajella eisenhardti* (Chondrichthyes, Rajidae) from the Galápagos Islands. *Proc. Biol. Soc. Wash.* **112**(1), 45–51.

López, J. A., Bentzen, P., and Pietsch, T. W. (2000). Phylogenetic relationships of esocoid fishes (Teleostei) based on partial cytochrome b and 16S mitochondrial DNA sequences. *Copeia* **2,** 420–431.

Lutjeharms, J. R. E. (1990). The oceanography and fish distribution of the Southern Ocean. *In* "Fishes of the Southern Ocean" (Gon, O., and Heemstra, P. C., Eds.), pp. 6–27. J. L. B. Smith Institute of Ichthyology, Grahamstown, South Africa.

McDonald, M. A., Smith, M. H., Smith, M. W., Novak, J. M., Johns, P. E., and De Vries, A. L. (1992). Biochemical systematics of notothenioid fishes from Antarctica. *Biochem. Syst. Ecol.* **20**(3), 233–241.

Markle, D. F. (1989). Aspects of character homology and phylogeny of the Gadiformes. *In* "Papers on the Systematics of Gadiform Fishes" (Cohen, D. M., Ed.), pp. 59–88 *Nat. Hist. Mus. Los Angeles Count. Sci. Ser.* **32.**

Markle, D. F., and Olney, J. E. (1990). Systematics of the Pearlfishes (Pisces: Carapidae). *Bull. Mar. Sci.* **47,** 269–410.

Mayden, R. L. (1991). Cyprinids of the New World. *In* "Cyprinid Fishes: Systematics, Biology and Exploitation" (Winfield, I. J., and Nelson, J. S., Eds.), pp. 240–263. Chapman and Hall, London.

McAllister, D. E. (1962). Fishes of the 1960 "*Salvelinus*" program from Western Arctic Canada. *Natl. Museum Can. Bull.* **185**, 17–39.

McAllister, D. E., and Krejsa, R. J. (1961). Placement of the prowfishes, Zaproridae, in the superfamily Stichaeoidae. *Nat. Hist. Pap. Natl. Mus. Can.* **11**, 1–4.

McEachran, J. D., and Dunn, K. A. (1998). Phylogenetic analysis of skates, a morphologically conservative clade of elasmobranchs (Chondrichthyes: Rajidae). *Copeia* **2**, 271–290.

Mecklenburg, C. W., Mecklenburg, T. A., and Thorsteinson, L. K. (2002). "Fishes of Alaska." American Fisheries Society, Bethesda, Maryland.

Miki, T. (1995). Stichaeidae. *In* "Fishes collected by R/V "Shinkai Maru" in the waters around Greenland" (Okamura, O., Amaoka, K., Takeda, M., Yano, K., Okada, K., and Chikuni, S., Eds.), pp. 207–211. Japan Marine Fishery Resources Research Center, Tokyo.

Miller, R. G. (1993). "A History and Atlas of the Fishes of the Antarctic Ocean." Foresta Institute for Ocean and Mountain Studies, Carson City.

Mincarone, M. M. (2001). *Myxine sotoi*, a new species of hagfish (Agnatha, Myxinidae) from Brazil. *Bull. Mar. Sci.* **68**(3), 479–483.

Mok, H.-K. (2002). *Myxine kuoi*, a new species of hagfish from southwestern Taiwanese waters. *Zool. Stud.* **41**(1), 59–62.

Mok, H.-K., and Kuo, C.-H. (2001). *Myxine formosana*, a new species of hagfish (Myxiniformes: Myxinidae) from southwestern waters of Taiwan. *Ichthyol. Res.* **48**(3), 295–297.

Morita, T. (1999). Molecular phylogenetic relationships of the deep-sea fish genus *Coryphaenoides* (Gadiformes: Macrouridae) based on mitochondrial DNA. *Mol. Phylogenet. Evol.* **13** (3), 447–454.

Muus, B. J. (1981). Fisk. *In* "Grønlands fauna Fisk, Fugle, Pattedyr" (Muus, B. J., Salomonsen, F., and Vibe, C., Eds.), pp. 23–157. Gyldendalske Boghandel, Nordisk Forlag A/S, Copenhagen.

Møller, P. R. (2000). Restoration of the taxon, *Lycodes marisalbi*, with notes on its disjunct Arctic distribution. *J. Fish Biol.* **57**, 1404–1415.

Møller, P. R. (2001a). Redescription of the *Lycodes pallidus* species complex (Pisces, Zoarcidae), with a new species from the Arctic/North Atlantic Ocean. *Copeia* **4**, 972–996.

Møller, P. R. (2001b). A new zoarcid, *Lycodes mcallisteri* from eastern Arctic Canada (Teleostei: Perciformes). *Ichthyol. Res.* **48**(2), 111–116.

Møller, P. R., and Jørgensen, O. A. (2000). Distribution and abundance of eelpouts (Pisces, Zoarcidae) off West Greenland. *Sarsia* **85**, 23–48.

Møller, P. R., Jordan, A. D., Gravlund, P., and Steffensen, J. F. (2002). Phylogenetic position of the cryopelagic cod-fish genus *Arctogadus* Drjagin, 1932 based on partial mitochondrial cytochrome b sequences. *Polar Biol.* **25**, 342–349.

Møller, P. R., Nielsen, J. G., and Fossen, I. (2003). Patagonian toothfish found off Greenland. *Nature (London)* **421**, 599.

Møller, P. R., and Gravlund, P. (2003). Phylogeny of the eelpout genus *Lycodes* (Pisces, Zoarcidae) as inferred from mitochondrial cytochrome b and 12S rDNA. *Mol. Phylogenet. Evol.* **26**(3), 369–388.

Naylor, G. J. P., Martin, A. P., Mattison, E. G., and Brown, W. M. (1997). Interrelationships of lamniform sharks: Testing phylogenetic hypotheses with sequence data. *In* "Molecular Systematics of Fishes" (Kocher, T. D., and Stepien, C. A., Eds.), pp. 195–214. Academic Press, San Diego.

Nelson, D. W. (1984). Systematics and distribution of cottid fishes of the genera *Rastrinus* and *Icelus*. *Occasional Papers of the California Academy of Sciences* **138**(iv), 58.

Nelson, J. S. (1994). "Fishes of the World," 3 edn. pp. 1–600.

Neyelov, A. V. (1979). Seismosensory system and classification of cottid fishes (Cottidae: Myoxocephalinae, Artediellinae). *Akad. Nauka, Leningrad.* 1–208.

Nielsen, J. G., and Bertelsen, E. (1992). "Fisk I Grønlandske Farvande [in Danish]," pp. 1–65. Atuakkiorfik, Nuuk, Greenland.

Nielsen, J. G., Cohen, D. M., Markle, D. F., and Robins, C. R. (1999). Ophidiiform fishes of the world (order Ophidiiformes). An annotated and illustrated catalogue of pearlfishes, cuskeels, brotulas and other ophidiiform fishes known to date. FAO species catalogue. *FAO Fisheries Synopsis* **18**, 1–178.

Okamura, O., Amaoka, K., Takeda, M., Yano, K., Okada, K., and Chikuni, S. (1995). "Fishes collected by R/V 'Shinkai Maru' in the Waters around Greenland," pp. 1–304. Japan Marine Fishery Resources Research Center, Tokyo.

Olney, J. E. (1984). Lampridiformes: Development and relationships. *In* "Ontogeny and Systematics in Fishes" (Moser, H. G., Richards, W. J., Cohen, D. M., Fahay, M. P., Kendall, A. W., Jr., and Richardson, S. L., Eds.), pp. 368–379. *ASIH. Spec. Publ. No.* **1**.

Ortí, G., Bell, M. A., Reimchen, T. E., and Meyer, A. (1994). Global survey of mitochondrial DNA sequences in the threespine stickleback: Evidence for recent migrations. *Evolution* **48** (3), 608–622.

Paulin, C. D. (1989). Review of the morid genera *Gadella*, *Physiculus*, and *Salilota* (Teleostei: Gadiformes) with descriptions of seven new species. *N. Z. J. Zool.* **16**, 93–133.

Paxton, J. R., Ahlstrom, E. H., and Moser, H. G. (1984). Myctophidae: Relationships. *In* "Ontogeny and Systematics in Fishes" (Moser, H. G., Richards, W. J., Cohen, D. M., Fahay, M. P., Kendall, A. W., Jr., and Richardson, S. L., Eds.), pp. 239–244. *ASIH. Spec. Publ. No.* **1**.

Pietsch, T. W. (1974). Osteology and relationships of ceratioid anglerfishes of the family Oneirodidae, with a review of the genus *Oneirodes* Lütken. *Sci. Bull. Nat. Hist. Mus., Los Angeles Co.* **18**, 1–113.

Pietsch, T. W. (1986). Systematics and distribution of bathypelagic anglerfishes of the family Ceratiidae (order: Lophiiformes). *Copeia* **2**, 479–493.

Pietsch, T. W. (1993). Systematics and distribution of cottid fishes of the genus *Triglops* Reinhardt (Teleostei: Scorpaeniformes). *Zool. J. Linn. Soc.* **109**, 335–393.

Pietsch, T. W., and Van Duzer, J. P. (1980). Systematics and distribution of ceratioid anglerfishes of the family Melanocetidae with the description of a new species from the eastern North Pacific Ocean. *US Natl. Mar. Fish. Serv. Fish. Bull.* **78**(1), 59–87.

Pietsch, T. W., and Zabetian, C. P. (1990). Osteology and interrelationships of the sand lances (Teleostei: Ammodytidae). *Copeia* **1**, 78–100.

Peichel, C. L., Nereng, K. S., Ohgi, K. A., Cole, B. L., Colosimo, P. F., Buerkle, C. A., Schluter, D., and Kingsley, D. M. (2001). The genetic architecture of divergence between threespine stickleback species. *Nature (London)* **414**, 901–905.

Post, A. (1987). Results of the research cruises of FRV "Walther Herwig" to South America. LXVII. Revision of the subfamily Paralepidinae (Pisces, Aulopiformes, Alepisauroidei, Paralepididae). I. Taxonomy , morphology and geographical distribution. *Arch. Fischereiwiss.* **38**, 75–131.

Prirodina, V. P. (2000). On the systematic position of littoral and deep-water species of the genus *Harpagifer* (Harpagiferidae, Notothenioidei) from Macquarie Island with a description of two new species. *J. Ichthyol.* **40**(7), 488–494.

Pshenichnov, L. K. (1997). A dogfish species new for ichthyofauna of sub-Antarctic regions: *Squalus acanthias* (Squalidae). *J. Ichthyol.* **37**(8), 678–679.

Rass, T., and Wheeler, A. (1991). The nomenclature of the Pacific herring, *Clupea pallasii* Valenciennes, 1847. *J. Fish Biol.* **39**, 137–138.

Reusch, T. B. H., Wegner, K. M., and Kalbe, M. (2001). Rapid genetic divergence in postglacial populations of threespine stickleback (*Gasterosteus aculeatus*): The role of habitat type, drainage and geographical proximity. *Mol. Ecol.* **10**, 2435–2445.

Richardson, J. (1844–48). Ichthyology of the voyage of H. M. S. Erebus and Terror. *In* "The Zoology of the Voyage of H. M. S. 'Erebus and Terror,' under the command of Captain Sir J. C. Ross during 1839–43, Book 2" (Richardson, J., and Gray, J. E., Eds.), London.

Ritchie, P. A., Bargelloni, L., Meyer, A., Taylor, J. A., Mac Donald, J. A., and Lambert, D. M. (1996). Mitochondrial phylogeny of trematomid fishes (Nototheniidae, Perciformes) and the evolution of Antarctic fish. *Mol. Phylogenet. Evol.* **5**(2), 383–390.

Rocha-Olivares, A., Rosenblatt, R. H., and Vetter, R. D. (1999). Molecular evolution, systematics, and zoogeography of the rockfish subgenus *Sebastomus* (*Sebastes*, Scorpaenidae) based on mitochondrial cytochrome *b* and control region sequences. *Mol. Phylogenet. Evol.* **11**(3), 441–458.

Sakshaug, E., Bjørge, A., Gulliksen, B., Loeng, H., and Mehlum, F. (1994). Structure, biomass distribution, and energetics of the pelagic ecosystem in the Barents Sea: A synopsis. *Polar Biol.* **14**, 405–411.

Sanford, C. P. J. (1990). The phylogenetic relationships of salmonid fishes. *Bull. Br. Mus. (Natl. Hist.) Zool.* **56**(2), 145–153.

Saruwatari, T., López, J. A., and Pietsch, T. W. (1997). A revision of the osmerid genus *Hypomesus* Gill (Teleostei: Salmoiformes), with the description of a new species from the southern Kuril Islands. *Species Diversity* **2**(1), 59–82.

Savin, S. M. (1977). The history of the Earth's surface temperature during the past 100 million years. *Ann. Rev. Earth Plan. Sci.* **5**, 319–355.

Sawada, Y. (1982). Phylogeny and zoogeography of the superfamily Cobitoidea (Cyprinoidei, Cypriniformes). *Mem. Fac. Fish. Hokkaido Univ.* **28**, 65–223.

Scott, W. B., and Scott, M. G. (1988). Atlantic fishes of Canada. *Can. Bull. Fish. Aquat. Sci.* **219**, 1–731.

Shirai, S. (1992). "Squalean Phylogeny: A New Framework of 'Squaloid' Sharks and Related Taxa." Hokkaido University Press, Sapporo, 209.

Shivji, M., Clarke, S., Pank, M., Natanson, L., Kohler, N., and Stanhope, M. (2002). Genetic identification of pelagic shark body parts for conservation and trade monitoring. *Conserv. Biol.* **16**(4), 1036–1047.

Smith, R. L., and Rhodes, D. (1983). Body temperature of the salmon shark, *Lamna ditropis.* *J. Mar. Biol. Ass. UK* **63**(1), 243–244.

Smith, G. R. (1992). Phylogeny and biogeography of the Catostomidae, Freshwater fishes of North America and Asia. *In* "Systematics, Historical Ecology and North American Freshwater Fishes" (Mayden, R. L., Ed.), pp. 778–826. Stanford University Press, Stanford, California.

Sokolov, L. I., and Berdicheskii, L. S. (1989). Acipenseridae. *The freshwater fishes of Europe.* **1**(2), 150–443.

Song, C. B., Near, T. J., and Page, L. M. (1998). Phylogenetic relations among percid fishes as inferred from mitochondrial cytochrome b DNA sequence data. *Mol. Phylogenet. Evol.* **10** (3), 343–353.

Stearly, R. F. (1992). Historical ecology of Salmonidae, with special reference to *Oncorhynchus.* *In* "Systematics, Historical Ecology and North American Freshwater Fishes" (Mayden, R. L., Ed.), pp. 622–658. Stanford University Press, Stanford, California.

Stepien, C. A., Dillon, A. K., Brooks, M. J., Chase, K. L., and Hubers, A. N. (1997). The evolution of blennioid fishes based on an analysis of mitochondrial 12S rDNA. *In* "The Molecular Systematics of Fishes" (Kocher, T. D., and Stepien, C. A., Eds.), pp. 245–270. Academic Press, San Diego.

Stonehouse, B. (1989). "Polar Ecology." Blachie, Chapman, and Hall, New York.

Sulak, K. J. (1986). Notacanthidae. *In* "Fishes of the North-Eastern Atlantic and the Mediterranean" (Whitehead, P. J. P., Bauchot, M.-L., Hureau, J.-C., Nielsen, J., and Tortonese, E., Eds.), Vol. 2, pp. 599–603. UNESCO, Paris.

Takahashi, H., and Goto, A. (2001). Evolution of East Asian ninespine sticklebacks as shown by mitochondrial DNA control region sequences. *Mol. Phylogenet. Evol.* **21**(1), 135–155.

Taylor, E. B., and Dodson, J. J. (1994). A molecular analysis of relationships and biogeography within a species complex of Holarctic fish (genus *Osmerus*). *Mol. Ecol.* **3**(3), 235–248.

Templeman, W. (1970). Distribution of *Anotopterus pharao* in the North Atlantic and comparison of specimens of *A. pharao* from the western North Atlantic with those from other areas. *J. Fish. Res. Board Can.* **27**(3), 499–512.

Thomson, K. S. (1977). The pattern of diversification among fishes. *In* "Pattern of Evolution as Illustrated by the Fossil Record" (Hallam, A., Ed.), pp. 377–404. Elsevier, Amsterdam.

Tomo, A. P. (1981). Contribucion al conocimiento de la fauna ictiologica del sector Antarctico Argentine. *Inst. Antarct. Argent.* **14**, 1–242.

Toyoshima, M. (1985). Taxonomy of the subfamily Lycodinae (Family Zoarcidae) in Japan and adjacent waters. *Mem. Fac. Fish. Hokkaido Univ.* **32**(2), 131–243.

Ueno, T. (1970). "Fauna Japonica. Cyclopteridae (Pisces)." Academic Press of Japan, Tokyo. Blackwell Science.

Uyeno, T., and Sato, Y. (1984). *Clupea pallasii* Valenciennes. *In* "The Fishes of the Japanese Archipelago" (Masuda, H., Amaoka, K., Araga, C., Uyeno, T., and Yoshino, T., Eds.), p. 19. Tokai University Press.

Van Guelpen, L. (1986). Hookear sculpins (genus *Artediellus*) of the North American Atlantic: Taxonomy, morphological variability, distribution, and aspects of life history. *Can. J. Zool.* **64**(3), 677–690.

Voskoboinikova, O. S. (1988). Comparative osteology of the Gymnodraconinae (Bathydraconidae). *Trudy Zool. Inst., Akad. Nauk SSSR* **181**, 44–55.

Waters, J. M., Saruwatari, T., Kobayashi, T., Oohara, I., McDowall, R. M., and Wallis, G. P. (2002). Phylogenetic placement of retropinnid fishes: Data set incongruence can be reduced by using asymmetric character state transformation costs. *Syst. Biol.* **51**(3), 432–449.

Washington, B. B., Moser, H. G., Laroche, W. A., and Williams, W. J. (1984). Scorpaeniformes: Development. *In* "Ontogeny and Systematics of Fishes" (Moser, H. G., Richards, W. J., Cohen, D. M., Fahay, M. P., Kendall, A. W., Jr., and Richardson, S. L., Eds.), pp. 405–428. *ASIH. Spec. Publ. No.* **1**.

Weiss, S., Persat, H., Eppe, R., Schloetterer, C., and Uiblein, F. (2002). Complex patterns of colonization and refugia revealed for European grayling *Thymallus thymallus*, based on complete sequencing of the mitochondrial DNA control region. *Mol. Ecol.* **11**(8), 1393–1407.

Whitehead, P. J. P., Bauchot, M.-L., Hureau, J.-C., Nielsen, J., and Tortonese, E. (Eds.) (1984–86). "Fishes of the North-Eastern Atlantic and the Mediterranean," Vol. I/III. UNESCO, Paris.

Whitehead, P. J. P. (1985). FAO species catalog. Clupeoid fishes of the world (suborder Clupeoidei). Part 1—Chirocentridae, Clupeidae and Pristigasteridae. *FAO Fish. Synop.* i–x, 1–303.

Wiley, E. O. (1992). Phylogenetic relationships of the Percidae (Teleostei: Perciformes): A preliminary hypothesis. *In* "Systematics, Historical Ecology, and North American Freshwater Fishes" (Meyden, R. L., Ed.) Stanford University Press, Stanford, California.

Wilson, R. R., and Attia, P. (2003). Interrelationships of the subgenera of *Coryphaenoides* (Teleostei: Gadiformes: Macrouridae): Synthesis of allozyme, peptide mapping, and DNA sequence data. *Mol. Phylogenet. Evol.* **27**(21), 343–347.

Yabe, M. (1985). Comparative osteology and myology of the superfamily Cottoidea (Pisces, Scorpaeniformes), and its phylogenetic classification. *Mem. Fac. Fish. Hokkaido Univ.* **32**, 1–130.

Yatsu, A. (1981). A revision of the gunnel family Pholididae (Pisces, Blennioidei). *Bull. Natl. Sci. Mus. Ser. A (Zool.)* **7**(4), 165–190.

Yatsu, A. (1985). Phylogeny of the family Pholididae (Blennioidei) with a redescription of Pholis Scopoli. *Jpn. J. Ichthyol.* **32**(3), 273–282.

METABOLIC BIOCHEMISTRY: ITS ROLE IN THERMAL TOLERANCE AND IN THE CAPACITIES OF PHYSIOLOGICAL AND ECOLOGICAL FUNCTION

H.O. PÖRTNER
M. LUCASSEN
D. STORCH

I. Introduction
  A. Thermal Specialization and Thermal Niches in Marine Environments
  B. Thermal Tolerance and Metabolic Design: The Conceptual Framework
II. Cold Adaptation: Performance Levels and Mode of Metabolism
  A. Enhancing Oxygen Supply and Aerobic Metabolic Capacities
  B. Reduced Anaerobic Capacity?
  C. Larval Fish
  D. Substrates in the Cold: Lipid Metabolism
  E. Biochemistry and Physiology of Cellular Oxygen Supply
III. Membrane Functions and Capacities: Constraints in Ion and Acid–Base Regulation
IV. Molecular Physiology of Metabolic Functions
  A. Regulation of Gene Expression and Functional Protein Levels
  B. Temperature Effects on Protein Structure and Function
V. Trade-Offs in Energy Budgets and Functional Capacities: Ecological Implications
  A. Protein Synthesis, Growth, and Standard Metabolism
  B. Reproduction and Development
VI. Summary

I. INTRODUCTION

A. Thermal Specialization and Thermal Niches in Marine Environments

Because of their inherently high levels of organizational complexity, animals have specialized on environmental temperatures more than unicellular bacteria and algae (Pörtner, 2002a). Accordingly, the ranges of thermal tolerance differ between ectothermal animal species and their populations

The Physiology of Polar Fishes: Volume 22
FISH PHYSIOLOGY

depending on latitude or seasonal temperature acclimatization and are, therefore, related to geographical distribution. Trying to understand the limits and benefits of thermal specialization requires an understanding of the trade-offs and constraints in thermal adaptation. These become visible when temperature-dependent physiological characters are compared in ectotherms specialized to live in various temperature regimens. On a global scale, the marine realm offers clearly defined thermal niches and, thus, is an ideal source for such comparisons. Marine animals of the high Antarctic, for example, rely on constant water temperatures at the low end of the temperature continuum in marine environments. They are (possibly life's most extreme) permanent stenotherms and are unable to sustain the complete set of life functions required for survival and fitness at temperatures above 3–6 °C. In contrast, eurytherms tolerate wider temperature fluctuations and, in temperate to subpolar zones, are able to dynamically change the range of thermal tolerance between summer and winter.

In this chapter, we investigate the special characteristics of metabolism in the cold by considering their role in tissue and whole-organism functional adjustments to cold, as well as the environmental forces that cause temperature-dependent trade-offs in biochemical and physiological tissue and cellular design. This relates to the key question of which unifying physiological and biochemical mechanisms animals use to shift the upper and lower limits of thermal tolerance during adaptation. Such questions have regained interest in light of global warming (Wood and MacDonald, 1997; Pörtner *et al.*, 2001; Pörtner, 2002a) and associated shifts in the geographical distribution of ectothermic animals and in ecosystem structure and functioning.

A comparison of subpolar to temperate eurytherms with polar stenotherms indicates that energy turnover increases at low, but variable, temperatures (Pörtner *et al.*, 2000). Study of adaptation to cold, therefore, needs to consider the stability of ambient temperatures, which is greater in the Antarctic than in the Arctic. Also, the same amplitude of temperature oscillations may have larger effects in the cold than at the warm end of the temperature scale. The time scales of those changes are also relevant, ranging from periods of diurnal to seasonal to permanent cold exposure. By focusing on polar species and comparing them with patterns detected in species and populations from various latitudes, we evaluate thermal specialization patterns developed on evolutionary time scales (evolutionary cold adaptation). When appropriate, we contrast these patterns with those observed during seasonal acclimatization to cold. The latter may include metabolic depression strategies during hibernation, a process important in survival of the dark food-limited polar winter.

Climate oscillations and temperature-dependent evolution may display largely different patterns between southern and northern hemispheres. In this context, the much younger age and the lower degree of isolation from adjacent seas of Arctic compared with Antarctic oceans require consideration. Accordingly, Antarctic species have developed features of permanent cold adaptation over millions of years, whereas Arctic species or species subpopulations may be found in transition to life in the permanent cold. Especially for the Arctic, clear definitions have to be applied to distinguish cold-stenothermal from cold-eurythermal fauna. Among cold-eurythermal species, separate populations that display clearly and permanently different levels of cold adaptation may exist. This has become apparent from studies of growth rates and mitochondrial enzymes in liver and white muscle of genetically distinct populations of cod (*Gadus morhua*) from Barents, Norwegian, White, and North Seas (Pörtner *et al.*, 2001; Lannig *et al.*, 2003; T. Fischer *et al.*, unpublished observations).

A problem with those large-scale comparisons of Arctic, Antarctic, and temperate to subpolar fish fauna is that phylogenetically correct comparisons are not always possible. Cosmopolitan fish families should, therefore, be included, which may allow for evaluation of the general validity of patterns elaborated in groups with more limited geographical distribution. For example, much work on the Antarctic fish fauna has focused on the dominant and largely endemic fish group Notothenioideae. Several sub-Antarctic notothenioid species are believed to originate from cold-adapted Antarctic fauna, so plesiomorphic characters originating from the cold past may persist and hamper the comparative analysis of cold adaptation (Stankovic *et al.*, 2001). Other fish groups and invertebrates have rarely been considered (for review, see Pörtner *et al.*, 1998, 2000; Peck, 2002). The Zoarcidae as a cosmopolitan fish family have been more widely studied (Hardewig *et al.*, 1999a; van Dijk *et al.*, 1999; Mark *et al.*, 2002). Little is known about the metabolic biochemistry of Arctic fish, whereas some work on whole-animal respiration during rest or, less frequently, exercise has been carried out (Bushnell *et al.*, 1994; Steffensen *et al.*, 1994; Hop and Graham, 1995; Schurmann and Steffensen, 1997; Zimmermann and Hubold, 1998).

The forces and trade-offs in climate-dependent temperature adaptation may be unifying among metazoans (Pörtner, 2004). Therefore, in this chapter, we draw on examples from invertebrates and even mammals, when appropriate, to emphasize the bigger picture and when data for fish allow only limited insight. Also, such comparisons are included for identification of unifying principles of adaptation to polar cold.

B. Thermal Tolerance and Metabolic Design: The Conceptual Framework

The goal of this chapter is to integrate the present knowledge of metabolic biochemistry of polar fishes, including molecular and cellular design, and of individual tissue functioning toward an understanding of the whole

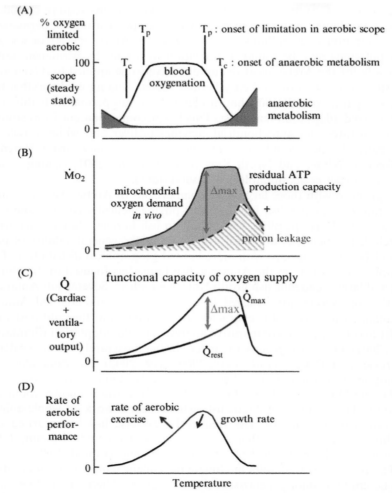

Fig. 3.1. Generalized scheme of the concept of oxygen-limited thermal tolerance in metazoans including fish (modified after Pörtner, 2001, 2002a; Pörtner *et al.*, 2004). Onset of a mismatch between oxygen demand and supply mechanisms set the thermal tolerance windows of metazoans, by limiting (A) aerobic scope, delineated by both upper and lower pejus temperatures (Tp). Critical temperatures (Tc) indicate onset of anaerobic metabolism. In both cold and warm (B), the functional capacities of mitochondria and consecutively, cells, and tissues (including ventilation and circulation) become insufficient. In the cold, this occurs partially through insufficiently low mitochondrial adenosine triphosphate (ATP) formation capacity. In the warm, the rising baseline oxygen demand of organismic (Figure 3.2) and cellular maintenance, caused in part by a rising fraction of mitochondrial proton leakage, elicits a mismatch between oxygen supply and demand. According to capacity limits of ventilatory and circulatory organs, the

animal, its thermal specialization, and its integration into the ecosystem. The hypothesis of an oxygen-limited thermal tolerance as a unifying principle in animals (Pörtner *et al.*, 2000; Pörtner, 2001, 2002a) is important in this context and is briefly summarized and updated here.

According to recent insight, thermal limitation becomes effective first at high hierarchical levels of organization, the intact organism, and then at lower levels in cellular and molecular functions. Similarly, thermal adaptation can be understood only if functional consequences of thermal adjustments at molecular levels have been identified for the whole organism. Molecular to organismic levels are tightly intertwined and influence each other. Following this rationale, work performed on marine invertebrates and fish, including Antarctic species, demonstrates that decreasing whole-animal aerobic scope (i.e., reduction of its flexibility to increase the rate of aerobic metabolism above basal rates) characterizes the onset of thermal limitation at low and high temperature thresholds, called *pejus temperatures* ("pejus" meaning getting worse) (Figure 3.1; e.g., Frederich and Pörtner, 2000; Mark *et al.*, 2002). Toward temperature extremes, the decrease in aerobic scope is indicated by the circulatory and ventilatory systems failing to supply sufficient oxygen to tissues and fully cover maximum oxygen demands. This leads to either constant oxygen levels in the body fluids that are insufficient to enhance oxygen diffusion at rising demands (as seen in the warm in Antarctic fish; Mark *et al.*, 2002) (Figure 3.2) or even to falling oxygen levels in the body fluids in the cold and warm [as seen in temperate crustaceans (Frederich and Pörtner, 2000) and in venous blood of cod *G. morhua* (Lannig *et al.*, 2004)]. Further cooling or warming beyond these limits leads to low or

---

residual capacity of mitochondria to produce ATP is reduced beyond pejus temperatures in the warm, when at the same time energy demand rises exponentially. Accordingly, the temperature-dependent capacity of ventilation and circulation (after Farrell, 1997), especially the difference between maximum and resting output characterizes aerobic scope and oxygen supply limits (C). At the upper pejus temperature, maximum functional reserves in mitochondrial ATP generation (B) and, thus, in oxygen supply (C) support an asymmetric performance curve (D, after Angilletta *et al.*, 2002), with optimal performance (e.g., growth, exercise) close to upper pejus levels. Here, functions are supported by both high temperatures and optimum oxygen supply. Temperature-dependent growth rates depending on aerobic scope are suggested by similarly shaped growth curves found in fish (Jobling, 1997) and in invertebrates (Mitchell and Lampert, 2000; Giebelhausen and Lampert, 2001), as well as by the recent finding that low blood oxygen tension limits protein synthesis rates in feeding crabs (Mente *et al.*, 2003). Aerobic scope and growth rate were found to be related in a population of Atlantic cod (Claireaux *et al.*, 2000). As a trade-off in eurythermal cold adaptation (*arrows*), standard metabolism, and aerobic exercise capacity increase in the cold (D, Pörtner, 2002b), whereas temperature-specific growth performance is reduced. Similar trade-offs are found in pelagic versus benthic Antarctic fish (see V. A., Figure 3.12). As a corollary, mechanisms setting aerobic scope are crucial in thermal adaptation (Figure 3.3). The figure does not consider denaturation temperatures (Td, see Figure 3.3) or mechanisms supporting resistance against freezing.

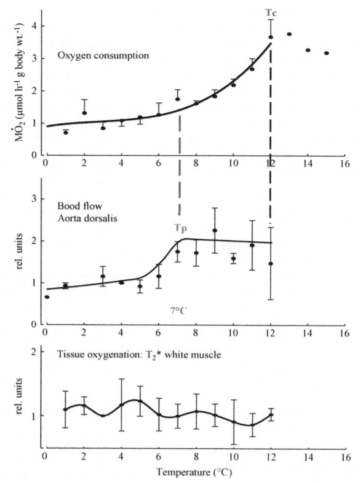

Fig. 3.2. Oxygen-limited thermal tolerance in Antarctic eelpout (*Pachycara brachycephalum*) investigated by use of nuclear magnetic resonance (MR) imaging techniques (after Mark *et al.*, 2002). The exponential rise in oxygen consumption during warming, co-determined by rising costs of ventilation and circulation, is not paralleled by a similar rise in arterial blood flow (derived from flow-weighted MR images) or tissue oxygenation (derived from blood oxygenation level–dependent [BOLD] contrast of T2-weighted MR images, numbers are relative units). Blood flow rather reaches a maximum level, indicating capacity limitation when oxygen demand still rises and tissue oxygenation remains constant. This indicates onset of a fall in aerobic scope and, thus, existence of an upper pejus temperature (Tp) at 7 °C. In accordance with earlier findings (van Dijk *et al.*, 1999), maximum rate of oxygen consumption reached in the warm indicates the upper critical temperature (Tc) at about 12 °C where transition to anaerobic metabolism occurs.

high critical threshold temperatures (Tc). Beyond both Tc's, aerobic scope becomes minimal and transition to an anaerobic mode of mitochondrial metabolism and progressive insufficiency of cellular energy levels occurs. Oxygen deficiency may contribute to oxidative stress and, thereby, favor the thermal denaturation of molecular functions (Prabhakar and Kumar, 2004) (Figure 3.3). In the cold, osmolytes and antifreeze protection or super-cooling extend the tolerance range to below the freezing point of body fluids. These mechanisms are not discussed in this chapter.

Beyond pejus temperatures, only time-limited passive survival is possible and supported by *protective* mechanisms ranging from metabolic depression and the setting of anaerobic capacity at the whole-organism level

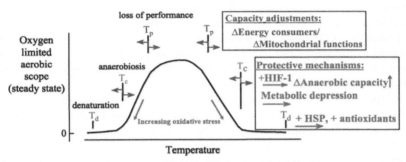

Fig. 3.3. Generalized scheme of processes involved during thermal acclimatization and adaptation, considering the suggested hierarchy of thermal tolerance thresholds and associated mechanisms (updated from Pörtner et al., 2001; Pörtner, 2002a). The modeled depiction considers thermal limitation (Figure 3.1, by onset of a loss in aerobic scope at Tp, of anaerobic metabolism at Tc, of molecular denaturation at Td) and the concomitant shift of low and high thermal tolerance thresholds during temperature adaptation by capacity adjustments, protective mechanisms, and molecular modifications (not shown). The mechanisms causing unidirectional shifts of both high- and low-tolerance thresholds include overall changes of mitochondrial functional capacity at the expense of significant metabolic cold compensation in eurytherms (i.e., a rise in temperature-specific organismic oxygen demand). A drop in oxygen demand and a shift in the onset of anaerobic energy production is expected from a decrease in mitochondrial densities and capacities seen during warming or from metabolic depression in the passive range beyond pejus temperatures. Downward-pointing linear arrows indicate increased oxidative stress during progressive oxygen deficiency at cold or warm temperatures. Hypoxia-inducible factor 1 (HIF-1) may contribute to improvement in anaerobic capacity and oxygen supply through erythropoiesis (not shown) and may, thereby, enhance the capacity for heat and cold endurance (see text). Td likely shifts with molecular modifications (see text), as well as the presence of molecular protection mechanisms like heat shock proteins or antioxidants. The figure is incomplete in that at the low end of the thermal tolerance window, hibernation and associated metabolic depression would extend passive cold tolerance. Antifreeze protection or supercooling extends the tolerance range to below the freezing point of body fluids.

to the effect of heat shock proteins and antioxidants at the molecular level (Figure 3.3). Strictly time-limited tolerance beyond critical temperatures is relevant in those organisms regularly exposed to thermal extremes, like the epibenthic fauna of the intertidal zone. The lengths of passive survival periods are co-determined by capacity adjustments of anaerobic metabolism.

Crucial in thermal adaptation are mitochondria and their capacity to provide metabolic energy. It has been recognized that oxygen and substrate demand by mitochondria is not only linked to adenosine triphosphate (ATP) production capacity and ATP demand. Some baseline oxygen demand is elicited by the dissipation of the mitochondrial proton gradient through proton leakage, which is temperature dependent and is highest in the warm, thereby claiming a large fraction of the oxygen available to the organism (Figure 3.1). At the same time, ATP production capacity (i.e., the summed aerobic capacity of tissue mitochondria) will also increase during warming. In principle, ATP production capacity may even exceed organismic ATP demand in the warm but can only be exploited with enough oxygen being provided. If, however, mitochondrial oxygen demand (co-defined by proton leakage and other dissipative processes) approaches the capacity limits of ventilation and/or circulation, excess ATP production capacity is no longer matched by oxygen supply and, thus, is unavailable to the organism. In consequence, the limiting oxygen supply causes transition to anaerobiosis beyond the Tc, because of insufficient rates of aerobic ATP production (Figure 3.1C). The balance between oxygen demand by proton leakage and ATP production capacity already indicates one key trade-off involved in thermal adaptation.

In the cold, mitochondrial ATP production capacity in itself may become limiting, so functional requirements are again in excess of ATP supply. As with other functions, ventilatory and circulatory capacities are progressively affected by limited aerobic ATP production capacity of mitochondria. For this reason, oxygen supply becomes insufficient again, aerobic scope will fall, and, finally, anaerobic metabolism sets in. As a corollary, a mismatch in oxygen supply and demand will limit thermal tolerance for similar reasons, at both low and high temperatures.

Overall, the functional capacities of ventilation and circulation, co-determined by mitochondrial capacities, appear as key processes involved in thermal adaptation and limitation. On the one hand, aerobic scope can only be made available if PO_2 is kept high in the body fluids, which requires sufficient ventilation and circulation. On the other hand, limited capacity of mitochondria to produce excess energy likely contributes to the loss of function and scope (e.g., in circulatory and ventilatory muscles at thermal extremes).

As a general hypothesis, mechanisms that shift the window of aerobic scope will be key in thermal adaptation (Pörtner, 2002a). Thermal adaptation causes a parallel shift of upper and lower limits of the tolerance window. *Capacity* adjustments to temperature are likely the prime mechanisms restoring full functional aerobic scope of the animals and thereby supporting permanent survival. This involves adjusting tissue mitochondrial capacity, which is defined by mitochondrial density and their respiratory capacity per milligram of protein (Figure 3.3). At the same time, trade-offs and constraints inherent in these processes of thermal adaptation define the limited nature of the thermal window. In the cold, an increase in mitochondrial capacity contributes to eliminate the capacity limitations of ventilation and circulation; however, the same process enhances sensitivity to warm temperatures as a trade-off and for the reasons outlined earlier in this chapter. During adaptation to the warmth, a reduction of overall mitochondrial capacity reduces proton leakage and, thus, baseline oxygen demand of mitochondrial maintenance, thereby allowing upper thermal limits to shift to higher values (Figures 3.1 and 3.3). In this case, the trade-off is reflected in an enhanced sensitivity of the organism to cold. In consequence, aerobic capacity of ventilation and circulation is set to optimize aerobic scope within the thermal envelope of a species.

The signaling mechanisms contributing to the shifts of the various thermal thresholds in Figure 3.1 are incompletely understood (Figure 3.3) (see Section IV). Changes in mitochondrial aerobic capacities may be elicited by factors linked to energy demand. Adjustments in anaerobic capacity may depend on the frequency of hypoxia events in the organism, involving a role for hypoxia-inducible factor 1 (HIF-1) (Treinin *et al.*, 2003). Other protective mechanisms like heat shock protein expression are also sensitive to hypoxia events (for review, see Burdon, 1987; Feder and Hofmann, 1999). The level of antioxidative defense mechanisms may be triggered by a mismatch between thermally enhanced formation of oxygen radical species and the capacity of their removal during or after hypoxia events (Heise *et al.*, 2003) (Figure 3.3). Accordingly, bidirectional mechanistic links likely exist between temperature-induced processes at different levels of organization. Toward thermal extremes, the sequence of events from high to low hierarchical levels includes limited capacity of oxygen supply systems, progressive tissue hypoxia at extreme temperatures, expression of HIF-1, use of anaerobic metabolism, expression of heat shock proteins, and compensatory changes in gene expression of functional proteins involved in capacity adjustments during thermal adaptation. On the one hand, whole-organism functioning would, thus, shape molecular responses. On the other hand, molecular adjustments integrate into a shift of the functional capacity and thermal window of the whole organism. Most of these mechanistic links still need to be qualified and quantified, but the likelihood of their existence suggests that the functional hierarchy seen between high to low levels of organization is also mirrored in a systemic to molecular hierarchy of thermal tolerance (Figures 3.1 and 3.3).

Thermal tolerance windows are narrow, particularly in marine ectotherms from polar areas and most notably in the Southern Ocean. The narrowing of tolerance windows, as found in the Antarctic, has been explained by the secondary reduction of mitochondrial ATP formation and proton leakage capacities (Figures 3.3 and 3.4) and, thus, by low organismic standard metabolic rates (SMRs) in the cold. A high thermal sensitivity of proton leakage in Antarctic species (Hardewig *et al.*, 1999b; Pörtner *et al.*, 1999) will cause an early increase in oxygen demand by this process during moderate warming and, thus, an early reduction of ATP formation capacity and aerobic scope for the whole organism. An early limitation of oxygen supply consequently results in narrow windows of thermal tolerance (Figure 3.1). Nonetheless, despite constant water temperatures between −1.9 and +1 °C, these windows are not the same for all Antarctic species. The capacities of ventilation and circulation are higher in mobile fish and octopods than, for example, in sessile bivalves, and this likely relates to the higher pejus and critical temperatures found in more mobile compared with sessile epifauna species (Pörtner *et al.*, 2000). In contrast to Antarctic stenotherms, elevated SMRs in cold-adapted populations of eurythermal animals are suitable to support even wider tolerance windows.

II. COLD ADAPTATION: PERFORMANCE LEVELS AND MODE OF METABOLISM

Considerable interest has centered on the question of how polar fish compare with temperate and warm-water fish with respect to their capacity for muscular performance. A discussion of aerobic metabolic scopes (Pörtner, 2002b) emphasized that both SMR and factorial scopes of metabolism in Antarctic fish are low (3.9–5.7) compared with those seen in more active temperate and tropical fish, with values between 10 and 12 in salmonids, bass, and mackerel. Among polar fish, only an Arctic cryopelagic species, *Boreogadus saida* reached a factorial scope of up to 8.4, which is close to the latter (Zimmermann and Hubold, 1998). However, the highest performance levels among fish (like tuna) and squid are not reached with high factorial scopes, but with elevated rates of standard metabolism combined with low factorial and large absolute aerobic metabolic scopes (O'Dor and Webber, 1991; Brill, 1996; Korsmeyer *et al.*, 1996; Pörtner and Zielinski, 1998). A high SMR is a requirement for large absolute metabolic scopes and,

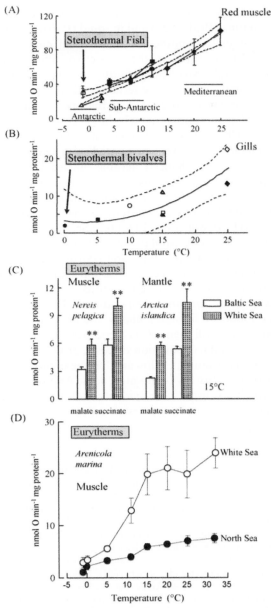

Fig. 3.4. Mitochondrial adenosine triphosphate (ATP) synthesis capacities (state 3 respiration rates) in stenothermal and eurythermal ectotherms (modified from Pörtner *et al.*, 2000). (A) Temperature-dependent state 3 respiration rates of isolated mitochondria from red

at the same time, supports rapid recovery rates from excessive anaerobic exercise, as seen in tuna (Brill, 1996) and in squid (Pörtner et al., 1991, 1993).

The patterns of cold stenotherms versus cold eurytherms developed above are, thus, also mirrored in the respective differences in metabolic scopes and activity levels. Most importantly, metabolic trends seen during seasonal cold acclimation and long-term (particularly eurythermal) cold adaptation appear similar to those visible in high-performance fish and squid with high standard metabolism, namely large absolute but low factorial scopes (Pörtner, 2002b). A lower increment in metabolic rate at high swimming speed characterizes northern compared with temperate cod populations, possibly indicating enhanced aerobic efficiency (data by Bushnell et al., 1994; Schurmann and Steffensen, 1997), similar to that induced by long-term exercise training. At the same time, enhanced recovery rates have been observed in cold-adapted Antarctic compared to cold-acclimated temperate eelpout (Hardewig et al., 1998) (see Section II.B). As a corollary, cold adaptation likely elicits similar changes in whole-animal metabolic physiology as high-performance adaptation (Pörtner, 2002b).

The principle processes of cold compensation are costly and cause elevated SMRs. Particularly sub-Arctic eurytherms require cold-compensated metabolic rates at more unstable temperatures. The low SMRs and low energy turnover mode of lifestyles finally found in the permanent cold, especially the Antarctic, will then reflect a compensation for the cost of cold adaptation (i.e., a secondary reduction of cost), at the expense of enhanced stenothermy in Antarctic fauna. Following this line of thought, because of the relatively young thermal history of this area, some Arctic ectotherms may still be cold eurythermal; however, some evidence indicates that they may be on their way to cold stenothermy in areas with stable cold temperatures.

---

musculature of Antarctic (open symbols), sub-Antarctic (South American, filled circles and squares), and Mediterranean fish species (filled diamonds, redrawn from Johnston et al., 1998). (B) State 3 respiration rates at various temperatures of isolated mitochondria from bivalve tissues (symbols reflect various studies and species, redrawn from Pörtner et al., 2000). (C) Comparison of state 3 respiration rates in White Sea (cold eurythermal) and Baltic Sea populations of Nereis pelagica and Arctica islandica analyzed at 15 °C (redrawn from Tschischka et al., 2000) (** indicates a significant difference between sub-Arctic and temperate population). (D) Temperature-dependent state 3 respiration rates in Arenicola marina from White and North Seas (redrawn from Sommer and Pörtner, 2002). Note the significantly higher state 3 respiration rates in the cold-adapted (sub-Arctic) eurytherms, whereas in southern hemisphere sub-Antarctic and Antarctic stenotherms, mitochondrial capacities remain uncompensated. In general, proton leakage capacities were found to be correlated with ATP synthesis capacities (state 3 respiration) (Hardewig et al., 1999; Pörtner et al., 1999; Tschischka et al., 2000; Sommer and Pörtner, 2002, 2004).

According to this rationale, low SMRs and successively reduced metabolic scopes for activity in polar fish and invertebrates may be possible only in thermally stable environments. At least in the Arctic, they may reflect a secondary situation that starts from a eurythermal ancestor preadapted to cold but variable temperatures in a latitudinal cline. The reasons for these patterns may lie in the level of energy efficiency reached with beneficial consequences for the overall energy budget (see Section V). The question arises how these whole-organism patterns are supported by cellular and tissue metabolic biochemistry.

A. Enhancing Oxygen Supply and Aerobic Metabolic Capacities

Acclimation to seasonal cold in many eurytherms is well known to cause a rise in mitochondrial density or mitochondrial aerobic capacity and is interpreted to maintain aerobic scope in the cold (see Section I.B). Such a key role of mitochondria does not neglect that integrated modifications in lipid saturation, kinetic properties of metabolic enzymes, contractile proteins, and transmembrane transporters contribute to the optimization of higher functions within the window of thermal tolerance (Johnston, 1990; Hazel, 1995; Storelli et al., 1998; Pörtner et al., 1998; see below). Increases in mitochondrial density and/or capacity in the cold are likely capacity and cost determined; however, the detailed mechanistic background for cold-induced mitochondrial proliferation remains obscure (Moyes and Hood, 2003) (see Sections I and III.A).

In fish from temperate waters, these changes have been observed in slow oxidative and fast glycolytic muscle and include a proliferation of aerobic (red) muscle fibers, which is reversed during seasonal warming (Johnston and Maitland, 1980; Sidell, 1980; Tyler and Sidell, 1984; Egginton and Sidell, 1989; Sidell and Moerland, 1989; Guderley and Johnston, 1996; Guderley, 1998; Pörtner et al., 1998, 2000; St.-Pierre et al., 1998). In parallel, a compensatory increase of mitochondrial enzyme activities has been reported in goldfish, carp, green sunfish, striped bass, and flounder during cold acclimation (Shaklee et al., 1977; Sidell, 1980; Wodtke, 1981a; Sidell and Moerland, 1989) but may be absent in some salmonid species or in lake whitefish (Blier and Guderley, 1988; Guderley and Gawlicka, 1992). Similar patterns emerge in invertebrates (Sommer and Pörtner, 2002, 2004). Patterns of seasonal cold acclimatization may differ from those seen in latitudinal cold adaptation (Sommer and Pörtner, 2004). Metabolic depression strategies during the cold season (Guderley and St.-Pierre, 2002; Sommer and Pörtner, 2004) influence the overall picture of cold-adjusted aerobic tissue design in some fish, amphibian, and invertebrate species. Whereas cold compensation leads to enhanced mitochondrial capacities (and associated costs of mitochondrial maintenance through proton leakage) in the winter-active animals, this trend is less or reversed during hibernation periods, where baseline costs and capacities of mitochondria are reduced (Boutilier and St.-Pierre, 2002; Guderley and St.-Pierre 2002).

During eurythermal cold adaptation in general (including cold acclimatization in specimens remaining active during winter), the functional capacity per mitochondrion (or per milligram of mitochondrial protein) rises on top of increasing mitochondrial volume densities. A significant (about two-fold) increase in mitochondrial capacity (per milligram of mitochondrial protein) was observed in cold-acclimated active eurytherms (Guderley and Johnston, 1996; Guderley, 1998) or in cold-adapted populations of eurytherms (invertebrates and fish, including cod) analyzed during summer at (sub-)Arctic latitudes (Tschischka *et al.*, 2000; Sommer and Pörtner, 2002, 2004; T. Fischer *et al.*, unpublished) (Figure 3.4). In rainbow trout, a capacity increment is reflected in higher cristae densities at unchanged mitochondrial volume fractions (St.-Pierre *et al.*, 1998). Mitochondrial proliferation is likely less in cold-adapted eurytherms than in Antarctic stenotherms (see below) (Sommer and Pörtner, 2002). As a consequence, less cellular space is occupied by mitochondria; more space can be allocated to contractile proteins. Exposure to unstable cold temperatures, thus, appears as a major driving force for enhanced metabolic capacities and maximized levels of aerobic motor activity in eurytherms (Pörtner, 2002b, 2004). Nonetheless, eurytherms at cold temperatures still do not reach the same level of aerobic performance in the cold as their warm-acclimated con-specifics at higher temperatures.

Very high mitochondrial densities are found in cold-adapted Antarctic fish, despite low SMRs. Mitochondrial density in Antarctic fish muscle is about 29–33%, whereas in Mediterranean fish at similar activity levels, the value is 8–13% (Johnston *et al.*, 1998). The highest mitochondrial densities found in the cold are 56% for the pelagic notothenioid *Pleuragramma antarcticum* and beyond 50% in icefish, (Johnston, 1987; Dunn *et al.*, 1989; Johnston *et al.*, 1998). Hemoglobin (Hb)-less species tend to have higher mitochondrial densities than red-blooded species (Johnston, 1987; Dunn *et al.*, 1989; O'Brien and Sidell, 2000). In warm waters, even very active fish do not reach far beyond 40%, influenced by body size: 46% was reported for small anchovies and 29% for tuna (*Katsuwonas pelamis*), respectively (Johnston, 1982; Moyes *et al.*, 1992). Hummingbird flight muscle possesses 35% (Suarez *et al.*, 1991). All of these observations clearly indicate that mitochondria are more efficient in the warm, leaving more space for contractile elements. Force generation by contractile fibers is also higher in the warm. Both processes contribute to higher performance levels than in the cold (see Section II.B).

Mitochondrial cristae surface area is traditionally interpreted to correlate with aerobic ATP formation capacity. The surface area of mitochondrial cristae in Antarctic red-blooded fish (36–37 m^2 cm^{-3}) as a measure of membrane folding is similar to that found in temperate and tropical perciform fish with similar lifestyles (Archer and Johnston, 1991; Johnston et al., 1998). This surface area is not especially high compared with the highest values reported, which are for tuna (*K. pelamis*) red muscle (63–70 m^2 cm^{-3}) and hummingbird flight muscle (58 m^2 cm^{-3}) (Suarez et al., 1991; Moyes et al., 1992). The latter values approach more closely a theoretical limit of 83 m^2 cm^{-3}, suggested by Srere (1985) with limited space left for Krebs cycle enzymes.

Evidently, cold-induced mitochondrial proliferation is taken to an extreme in pelagic Antarctic notothenioid fish but does not support the same maximum performance levels as seen in warm-water fish. One reason may be that the space allocated to mitochondria and sarcoplasmic reticulum (SR) (for SR between 3 and 5% in mammalian and up to 10% in hummingbird muscle) (Hochachka et al., 1988) reduces the space available for myofilaments (Pennycuick, 1992) and, thus, limits force development. The question arises why aerobic design is maximized despite reduced rates of aerobic metabolism in Antarctic fish.

Possibly, a trade-off exists between mitochondrial density and cristae surface area during eurythermal versus stenothermal cold. Low standard metabolism in Antarctic fish is reflected in moderate cristae surface areas and reduced mitochondrial aerobic ATP formation (Figure 3.4) capacities at high mitochondrial densities. The excessive mitochondrial densities observed in Antarctic fish, compared to cold-acclimated or possibly even Arctic fish, might indicate that mitochondrial proliferation not only serves the compensation of aerobic ATP synthesis capacity but may also adopt other functions like enhanced oxygen distribution in the cell (see Section II.E). In the cold, this may also loosen the correlation between cristae surface area and aerobic capacity, as indicated by data obtained in icefish (O'Brien and Sidell, 2000). The consequences of low versus high cristae density for the temperature-dependent regulation of metabolic flux are unclear, but the patterns observed might indicate that high cristae density not only goes hand in hand with high ATP formation capacity but also high baseline energy requirements as during adaptation to eurythermal cold.

Similar considerations extend to ventricular muscle of the notothenioid heart (Tota et al., 1991; Axelsson et al., 1998; see Chapter 6 in this volume). Mitochondrial volume densities of 43% in icefish (*Chaenocephalus aceratus*) are the highest reported in any teleost heart. In ventricular more so than in skeletal muscle, density values close to or beyond 50% are considered close to the theoretical limit for muscle function (Tota et al., 1991; Johnston *et al.*, 1998). The volume densities of myofibrils in icefish heart (25–31%) are low compared with other teleosts. This constraint on muscular force may be compensated for to some extent by enhanced size of the myocytes (Zummo *et al.*, 1995) and even the heart, which leads to a relative heart weight in icefish similar to that of small mammals (Tota *et al.*, 1991, 1997). The enhancement of cardiac mass in icefish compared with other notothenioids (Robertson *et al.*, 1998) might appear as a consequence of reduced oxygen carrying capacity in the blood and, therefore, the need to generate a higher cardiac output. In light of a role for membrane lipids in cellular oxygen supply, which is emphasized by the loss of myoglobin (Mb), a large size of the heart in Mb-less icefish may also compensate for the cellular space constraints resulting from the excessive accumulation of membrane lipids. As a result, cardiac tissue size reaches morphometric design limits, indicating the trade-off between enhanced tissue size on the one hand and the benefit of energy savings due to loss of Hb and Mb, as well as to reduced blood viscosity, on the other hand. These trends are made possible with improved oxygen solubilities in water, plasma, and cytosol, as well as low standard metabolism in the permanent cold (see Section II.E).

Although aerobic capacity of individual mitochondria in Antarctic fish and invertebrates evidently is not cold compensated (Johnston *et al.*, 1998; Pörtner *et al.*, 2000) (Figure 3.4), increased mitochondrial densities and associated improvements in cellular oxygen supply (see Section II.E) still suggest some cold-compensated tissue aerobic metabolic capacity in Antarctic ectotherms. Clearly cold-compensated enzyme capacities of aerobic metabolism have been documented in muscle and in brain of Antarctic fish (Kawall *et al.*, 2002). The Antarctic notothenioid *Gobionotothen gibberifrons* displays five times higher cytochrome *c* oxidase (COX) activities in white muscle than a temperate-zone fish with similar lifestyle (Crockett and Sidell, 1990). Citrate synthase (CS) activities were also increased but only 1.4–2.8 times in *newnesi* and *G. gibberifrons* over the levels found in temperate species. Only slightly increased levels of COX were found in white muscle of Antarctic (*Pachycara brachycephalum*) versus temperate eelpout, *Zoarces viviparus* (Hardewig *et al.*, 1999). Otherwise, a larger degree of cold compensation of the activities of membrane-bound enzymes like COX and succinate dehydrogenase (SDH) may be due not only to enhanced accumulation of enzyme protein, but also to cold-induced maintenance of membrane fluidity and stimulation of the enzymes by effects of associated changes in lipid composition (Wodtke, 1981a,b; Hazel, 1995).

Among temperate-zone fish, trends for cold-compensated membrane versus matrix enzyme activities in muscle are not universal. Similar to data available for Antarctic fish, elevated levels of muscle COX and SDH activities in cold-acclimated green sunfish, with unchanged activities of matrix malate dehydrogenase (Shaklee *et al.*, 1977), also indicate lower or no factorial increments for matrix compared with membrane-bound enzymes of mitochondria. In contrast, Battersby and Moyes (1998) found that both COX and CS activities increased in parallel during cold acclimation in white and red muscle of rainbow trout, indicating a fixed ratio of matrix to cristae enzymes for both muscle types at increased mitochondrial volume densities. In red and white muscle of striped bass, an increase in mitochondrial volume density without changes in mitochondrial size and cristae surface density has been observed during cold acclimation (Egginton and Sidell, 1989). The interpretation of variable patterns needs to consider whether fish maintain activity during winter cold or undergo metabolic depression (see above).

Moreover, patterns found in muscle differ from findings in liver. Liver COX activity displayed an increase neither in cold acclimated temperate eelpout *Z. viviparus* nor in cold-adapted Antarctic eelpout *P. brachycephalum* (Hardewig *et al.*, 1999a; Lucassen *et al.*, 2003a). However, liver CS activity was significantly elevated. CS activities in the liver of Antarctic eelpout *P. brachycephalum* were similar to levels found in cold-acclimated eelpout, indicating cold compensation of mass-specific liver mitochondrial functions in the Antarctic species (Lucassen *et al.*, 2003), on top of a cold-induced increase in liver size (Hardewig *et al.*, 1999a; Lannig *et al.*, 2003; Lucassen *et al.*, 2003).

As mentioned earlier, COX activity may not always profit from cold-induced changes in the lipid composition of the inner mitochondrial membrane (homeoviscous adaptation). In fact, mitochondrial membrane lipids in liver and heart of cold-acclimated sea bass (*Dicentrarchus labrax*) did not display homeoviscous adaptation (Trigari *et al.*, 1992). Accordingly, COX activity per milligram of mitochondrial protein was even decreased at lower acclimation temperatures. In liver of *Z. viviparus*, the increased activity of CS, together with a more or less constant level of COX activity, might be due to an increased tissue mitochondrial volume at enhanced levels of membrane viscosity in the cold. A higher copy number of COX might serve to compensate for incomplete homeoviscous adaptation, which causes a lower level of molecular activity (Lucassen *et al.*, 2003a).

In eelpout liver, the observed discrepancies between changes in CS and COX activities imply cold-induced functional changes of mitochondria with a relative increase in matrix over membrane functions. Mitochondria are not only involved in energy metabolism but also in anabolic processes with CS providing excess citrate, for example, for lipid synthesis (see Section II.D). Cold acclimation of *Z. viviparus* led to an increase in the ratio of CS to COX activities by a factor of 2, whereas confamilial Antarctic eelpout demonstrated a CS/COX ratio even slightly higher than temperate eelpout (Lucassen *et al.*, 2003) (Figure 3.5). In the eelpout, cold temperatures cause

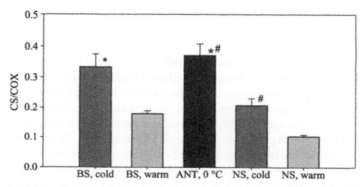

Fig. 3.5. Effect of acclimation and adaptation to cold on the ratio of steady state levels of citrate synthase (CS) to cytochrome *c* oxidase (COX) activities assayed in liver homogenates of eelpouts at 10 °C. *Zoarces viviparus* from Baltic Sea (BS) was acclimated to 3.0 °C (cold acclimated) and 13.5 °C (warm acclimated) for 2 months. *Z. viviparus* from North Sea (NS) was acclimated to 3.0 °C (cold acclimated) and 10.0 °C (warm acclimated) for 4 weeks. The Antarctic eelpout *Pachycara brachycephalum* (filled bar) was adapted to 0 °C. Although remarkable differences in enzyme activities could be observed between *Z. viviparus* populations, the ratio of CS over COX activities rose consistently in the cold by a factor of 2 in both temperate populations, whereas Antarctic eelpout demonstrated the highest CS/COX ratio. Accordingly, cold adaptation and cold acclimation may display similar patterns in that cold temperatures cause similar relative changes in the capacities of metabolic functions. (Data from Lucassen *et al.*, 2003. Values are means ±SEM, n = 4–6; *significant difference from the warm-acclimated Baltic Sea fish. # significant difference from the warm-acclimated North Sea eelpout.)

similar relative changes in the capacities of liver metabolic functions. Low COX and high CS activities in liver would be in line with enhanced anabolic capacities in the cold, thereby possibly supporting the high level of lipid accumulation observed in Antarctic fish (including eelpout). In general, this trend may be supported by the extremely high mitochondrial densities seen in cold stenothermal species (Pörtner, 2002b).

Enhanced densities of mitochondria and aerobic enzymes in the cold may serve to support more rapid recovery from exercise. In Antarctic fish, this conclusion may hold regardless of relatively low SMRs and low absolute aerobic scopes. Restoration of metabolic equilibria will benefit from shortened diffusion distances at elevated mitochondrial densities. Because the largest fraction of lactate in fish muscle is reconverted to glycogen *in situ* (Milligan, 1996), cold compensation of gluconeogenesis may also be required to fully explain the patterns observed.

In temperate eurythermal fish, like herring larvae and rainbow trout, metabolic recovery processes appear at least partially cold compensated (Dalla Via *et al.*, 1989; Kieffer *et al.*, 1994; Franklin *et al.*, 1996; Kieffer, 2000). However, the roach *Rutilus rutilus* not only produces less lactate during burst activity at 4 °C, but despite enhanced aerobic capacity needs about four times longer to metabolize lactate levels than specimens acclimated to 20 °C (Dalla Via *et al.*, 1989). Nonetheless, a Q_{10} below 2 exists for lactate clearance rates (Kieffer, 2000).

In stenothermal notothenioid fish, the time required for recovery of blood composition and rates of oxygen consumption after exercise and capture stress is either somewhat faster than or similar to that for temperate species, indicating cold compensation (Forster *et al.*, 1987; Davison *et al.*, 1988; Egginton and Davison, 1998). Repayment of an oxygen debt (half-time 20 minutes) (Forster *et al.*, 1987) faster than lactate depletion (Davison *et al.*, 1988) was observed after fatiguing exercise in the cryopelagic notothenioid *Pagothenia borchgrevinki*. In the sluggish Antarctic zoarcid *P. brachycephalum*, the recovery period after exercise (estimated from lactate removal) was much shorter (half-time 3 hours) than in the cold-acclimated North Sea species (half-time >24 hours) (van Dijk *et al.*, 1998; Hardewig *et al.*, 1998) (Figure 3.6). Enhanced aerobic capacity in Antarctic eelpout is indicated by slightly elevated activities of COX in white muscle and an aerobic scope enhanced over the one in the North Sea zoarcid (see above)

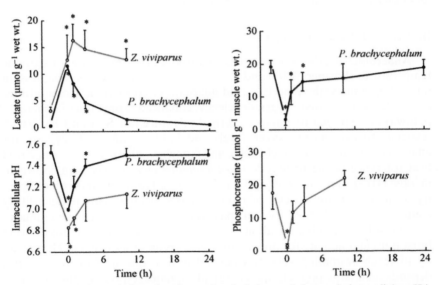

Fig. 3.6. Lactate accumulation, phosphocreatine depletion, and changes in intracellular pH in white muscle of cold-adapted and cold-acclimated eelpouts (*Zoarces viviparus*, North Sea, vs. *Pachycara brachycephalum*, Antarctic peninsula) during muscular exercise and subsequent recovery. Note the faster recovery in the Antarctic species (after Hardewig *et al.*, 1997; *indicates a significant difference from controls).

(Hardewig *et al.*, 1999a). Enhanced aerobic capacity allows a larger increase in excess postexercise oxygen consumption (EPOC) to pay off the oxygen debt. Starting from much lower resting metabolic rates (0.38 μmol O_2 g^{-1} h^{-1} in the eelpout at 0 °C, compared with 6.3 μmol O_2 g^{-1} h^{-1} in rainbow trout at 15 °C), EPOC and the rates of lactate clearance (3.3 μmol g^{-1} h^{-1}) and of creatine rephosphorylation (completed within 3 hours) in the Antarctic eelpout (Hardewig *et al.*, 1998) were similar to rates observed in trout (Scarabello *et al.*, 1992; Milligan, 1996). With respect to lactate clearance, the rate was faster than that observed in flounder acclimated to 11 °C (0.62 μmol g^{-1} h^{-1}) (Milligan and Wood, 1987). It appears that with down-regulated standard metabolism, the enhanced aerobic machinery is still effective and allows for a larger metabolic scope and faster postexercise recovery in the Antarctic eelpout. Faster repayment of a smaller oxygen debt was also visible in Greenland cod (*Gadus ogac*) compared with Atlantic cod (*G. morhua*) from the same area and acclimatized to the same temperature (Bushnell *et al.*, 1994). This indicates similar evolutionary trends in Arctic as in Antarctic fish. Fast recovery, thus, emerges as a consequence of cold-compensated aerobic capacity and appears beneficial in areas in which many predators like penguins and seals are warm blooded (endothermic) and, thus, not slowed down by cold temperatures.

B. Reduced Anaerobic Capacity?

Anaerobic metabolism becomes involved during exposure to extreme temperatures (Pörtner, 2001) and during muscular exercise. As such, it is an interesting issue with respect to cold adaptation. At extreme temperatures, and with the associated oxygen deficiency, enhanced expression of anaerobic pathways is expected to enhance thermal tolerance (see Section I.B). HIF-1 responds to decreasing oxygen levels and induces enhanced capacity for anaerobic metabolism (Hochachka and Somero, 2002). In fact, improved heat tolerance in the nematode *Caenorhabditis elegans* has been shown to involve HIF-1 (Treinin *et al.*, 2003) (Figure 3.3). However, anaerobic capacity is small in many species of polar fish (see below). This pattern likely contributes to the limited level of (passive) heat tolerance in these animals.

Cytosolic anaerobic metabolism in fish muscle is primarily used during burst muscular activity for prey capture and predator avoidance. The capacity to reach swimming velocities beyond critical during short-term bursts depends on a large pool of high-energy phosphates, especially phosphocreatine, followed by the activation of anaerobic glycolysis and lactate formation, and finally adenylate degradation (Kieffer, 2000). It also depends on the capacity to first accumulate and then remove end-products like lactate, as well as the ability to minimize and reverse associated pH disturbances.

The capacity for burst activity more than steady-state performance seems to be constrained in the permanent cold. Data predominantly collected in notothenioid fish suggest that in ways similar to those found in some eurythermal temperate fish (Sidell and Moerland, 1989), the capacity of anaerobic metabolism is not cold compensated (Egginton and Davison, 1998). The contribution of anaerobic glycolysis to energy production in Antarctic notothenioids is not as high as that found in cold-acclimated temperate species (Johnston et al., 1991a; van Dijk et al., 1998). Exhaustive exercise caused only a minor increase of lactate levels by about 1–3 μmol g^{-1} wet mass in the white musculature of Notothenia coriiceps and Pagothenia borchgrevinki (Dunn and Johnston, 1986; Davison et al., 1988) and was linked to reduced activities of glycolytic enzymes found in N. coriiceps. One concern with these data is that tissue lactate levels found in P. borchgrevinki were also high under control conditions (12–16 μmol g^{-1}), indicating incomplete recovery from previous stress, and that with product inhibition of lactate dehydrogenase (LDH), further lactate accumulation may have been prevented. Enzyme data in G. gibberifrons suggest higher capacities for lactate formation than in N. coriiceps in accordance with a more active mode of life (Dunn et al., 1989).

A trend toward reduced anaerobic glycolytic capacity and, accordingly, avoidance of exercise-induced intracellular acidosis in polar fish, is also mirrored in the functional properties of creatine kinase. Muscle-specific creatine kinase isoforms were isolated and sequenced for the icefish C. aceratus (Winnard et al., 2003). At 0.5 °C, this enzyme displayed K_m values for adenosine diphosphate (ADP) and phosphocreatine similar to other vertebrate creatine kinases; however, its activity was optimal at a pH of 7.6–7.7, slightly above intracellular pH values found in Antarctic fish in vivo (Moerland and Egginton, 1998; Bock et al., 2001). Much lower pH optima are found in vertebrates that rely on anaerobic glycolysis and experience associated acidosis during exercise. Sensitivity of muscular fibers to dihydrogen phosphate, which accumulates in larger quantities at high pH levels, is thought to be reduced in notothenioid muscle (Altringham and Johnston, 1985; Winnard et al., 2003). However, the detailed sequence and interaction of biochemical events during muscular exercise and their relevance for the energy status (Gibb's free energy of ATP hydrolysis) of notothenioid muscle awaits further study (for comparison, see Pörtner, 2002c; Pörtner et al., 1996).

Despite their reduced overall level of burst exercise capacity, short-term burst performance in Antarctic fish remained surprisingly independent of temperature between -1 and $+10$ °C (Wilson et al., 2001). This finding resembles earlier observations of temperature-independent burst swimming performance in salmon and herring discussed by Brett (1964). Another study by Johnson *et al.* (1996) identified a low Q_{10} of 1.2–1.3 for fast-start loco-motion in rainbow trout during acute temperature exposure. Larger Q_{10} values were observed for the twitch contraction kinetics of abdominal myo-tomes (1.9) and even more so in myofibrillar ATPase (2.9), indicating that thermal sensitivity was reduced when these molecular functions are integrated into the organism. Whereas temperature-dependent changes in myofibrillar protein isoforms were not observed in rainbow trout (Johnson *et al.*, 1996), such changes are part of the thermal acclimation response in common carp or goldfish (Johnson and Bennett, 1995; Wakeling *et al.*, 2000; Watabe, 2002). The respective situation in polar species deserves further investigation.

The finding of excessive aerobic muscle design at low-activity lifestyles would explain why anaerobic capacity is, as a trade-off, reduced in polar fish. Some variability in enzyme levels remains and relates to variable modes of life (Dunn *et al.*, 1989; Kawall *et al.*, 2002). In general, oxidative enzymes and creatine kinase, adenylate kinase, and AMP deaminase show relatively high degrees of cold compensation in Antarctic notothenioids. In contrast, glycolytic enzymes do not, at least in muscle (Dunn and Johnston, 1986; Johnston, 1987; Crockett and Sidell, 1990). Space constraints at high mito-chondrial densities, and thereby reduced levels of contractile proteins com-pared with temperate species, would imply space limitations for glycogen stores and glycolytic complexes required for extensive lactate formation. In accordance with these considerations, Guderley (1998) suggested that glyco-lytic enzyme activities increase in temperate species that do not show large adjustments in aerobic metabolism during cold acclimation (as in some salmonids), although they would decrease in species with a compensatory increase in oxidative capacity. A loss in the relative importance of anaerobic glycolysis in eurythermal fish in the cold is reflected in changing functional properties of white muscle LDH. These include a rise in Arrhenius activation energy seen during cold acclimation in eurythermal cod *G. morhua*, a finding that indicates an increase in kinetic barrier to flux through LDH (Zhakartsev *et al.*, 2003b; see below). Such shifts between metabolic pathways and functional enzyme properties may bear general importance not only for the preferred use of aerobic over anaerobic metabolism but also for the preferred use of lipids or protein over carbohydrates during aerobic exercise (Weber and Haman, 1996; see Section II.D).

A comparison of cold-acclimated temperate (North Sea) with Antarctic zoarcids (*Z. viviparus* vs. *P. brachycephalum*) supports a more differentiated picture with respect to the use of anaerobic metabolism. Zoarcids are benthic sluggish fish with a low degree of spontaneous activity. Surprisingly, both species displayed a similar anaerobic capacity in the cold (Figure 3.6). Both formed lactate at 0 °C ($11.5 \pm 0.7\,\mu\text{mol g}^{-1}$ muscle tissue) in similar amounts as seen in flounder acclimated to 11 °C (Milligan and Wood, 1987). Based on these data, a minimal glycolytic capacity does not appear as a general phenomenon in Antarctic fish. In accordance with the finding of cold-compensated anaerobic capacity in some, but not all, temperate freshwater fish (van Dijk *et al.*, 1998), the conclusion arises that cold compensation of anaerobic pathways is, in principle, possible in the polar cold. However, this possibility is not exploited in many notothenioid fish (Egginton and Davison, 1998). In contrast to strictly benthic zoarcids, notothenioids lead on average a more active life and, therefore, express aerobic metabolic design more strongly than moderately active temperate fish. In the zoarcids, the sit-and-wait mode of lifestyle is mirrored in a more moderate aerobic design (despite cold adaptation) that still leaves space for cold-compensated anaerobic metabolism.

For a comprehensive picture, the reduced mode of activity in Antarctic compared with temperate or tropical fishes needs to be considered. According to data compiled by Somero and Childress (1990), LDH activity scales positively with body size in temperate pelagic fishes but negatively in benthic sluggish fishes. Reduced levels of LDH capacity were also found with decreasing ambient light according to increasing depth of occurrence in the ocean. These findings indicate reduction of glycolytic capacity, with a reduction in overall locomotor activity. In Antarctic fishes, reduced motor activity combined with a cold-induced shift to enhanced levels of aerobic machinery most likely contributes to explain their reduced glycolytic capacity compared with temperate fishes.

The findings of moderate activity levels despite strongly enhanced aerobic tissue design features in notothenioid fish are reminiscent of a comparison among pelagic muscular (i.e., nonneutrally buoyant) squid like among loliginids and ommastrephids. Similar to cold-water fish, but for different reasons, squid also display greatly enhanced aerobic capacities. Compared with fish of similar lifestyle, these patterns have been explained by their costly mode of locomotion through jet propulsion (O'Dor and Webber, 1991). However, among squid, energy savings arise from the use of fins at low swimming speeds. In contrast to the more energetic jet swimmers of the open oceans, like the ommastrephid *Illex illecebrosus*, the more coastal fin swimmers, like the loliginid *Loligo pealei*, rely on aerobic metabolism at more moderate rates (O'Dor and Webber, 1991). Interestingly, they use the phosphagen system in burst swimming, but they do not use anaerobic glycolysis (for review, see Pörtner, 2002c; Pörtner and Zielinski, 1998). This comparison also emphasizes that energy savings in moderately active cold-stenothermal Antarctic fish impose constraints on muscular performance despite largely enhanced aerobic machinery. These constraints are mirrored in low SMRs and either uncompensated or even reduced levels of anaerobic metabolism and are reflected in both moderate aerobic and low anaerobic exercise capacities.

The following general pattern emerges: Muscle glycolytic capacity is minimized in cold-adapted, slow-cruising pelagic species at maximized aerobic design. In contrast, high-performance fish in tropical and temperate waters, which operate with lower mitochondrial densities than polar fish, use both aerobic and anaerobic metabolisms. Such high-performance life forms are scarce in polar areas, especially the Antarctic, if they even exist. One reason may be space constraints that cold temperatures impose on the muscle cell by large mitochondrial densities developing at the expense of myofibrillar density and, thus, force development. These constraints are less in cold-adapted, largely inactive benthic fish like the zoarcids in which glycolytic capacity is still expressed and fuels burst activity in attack or escape. When mitochondrial density is reduced in these species in parallel with decreased baseline energy demands, power generation and anaerobic bursts can take priority (Figure 3.7). This trend is still present in the Antarctic despite the emphasis on cold-compensated aerobic design. Overall, the pressure to enhance glycolytic capacity is likely alleviated in cold environments like the Antarctic and the deep sea where both ectothermic predators and prey are tied to low-velocity largely aerobic lifestyles. However, all ectotherms would then become easier prey for warm-blooded endothermic predators unless their hypometabolism goes hand in hand with strategies to hide in their complex environment.

Cold-compensated LDH activity was reported for brain of Antarctic fish (Kawall *et al.*, 2002), a finding that might be interpreted in a way that cold compensation of anaerobic metabolism occurs in nonmuscular tissues. However, lactate shuttling between astrocytes and neurons plays a major role in normal brain metabolism and signaling. In such a situation, cold-compensated glycolytic capacity may parallel the cold compensation of aerobic pathways. Such shuttling of lactate between aerobic and anaerobic pathways under aerobic conditions needs to be considered more widely, for example, in the heart and in other aerobic tissues (Brooks, 2002).

C. Larval Fish

In moderately active Antarctic fish, aerobic design is enhanced and anaerobic capacity reduced. This trade-off in cold adaptation should be most strongly expressed in larval fish, because small body size in itself requires an allometric increase in aerobic capacity and maximum speed depends even more on high levels of aerobic metabolism than in adults. In 2-mg larvae of temperate roach, *R. rutilus*, high standard and active metabolic rates demand three- to four-fold higher muscle mitochondrial densities

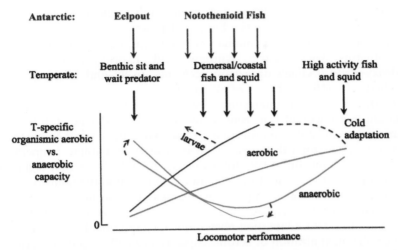

Fig. 3.7. Schematic depiction of whole-organism aerobic and anaerobic capacities (temperature specific, i.e., compared at the same temperature) contrasting the patterns of cold versus warm adaptation in groups displaying different levels of locomotor performance according to lifestyle. The figure illustrates how on a continuum between low and high levels of performance, the expression and use of anaerobic metabolic capacity is parabolic. Use of anaerobic metabolism is maximized in sluggish benthic species, minimized in moderately active aerobic cruisers (like the pelagic Antarctic notothenioids) and enhanced again in high-performance fish (and squid) at warmer temperatures (see text). Cold adaptation (broken arrows) elicits reduced performance levels at maximized aerobic design (blue vs. green lines), especially for cold-adapted larvae (see text). Because of cellular space constraints, cold adaptation of anaerobic capacity (magenta vs. red lines) in Antarctic fish occurs only in benthic sit-and-wait predators, whereas in moderately active Antarctic fish, anaerobic capacity appears reduced, as a trade-off in the maximization of aerobic design. Note that the figure does not differentiate between cold-adapted stenotherms and eurytherms (see Pörtner, 2002b).

than in larger (10-g) stages (Wieser, 1995). In muscle of smaller fish, in general, the increase in CS capacity parallels the progressive increase of mitochondrial densities (Somero and Childress, 1990). The resulting cellular space constraints should cause a minimization of anaerobic capacity. In fact, the capacity of muscle LDH was found to be reduced at small body size in pelagic fish. As a consequence, maximum lactate levels were about three to four times lower in white muscle of fatigued juvenile salmonids than in adult fish (Kieffer, 2000).

Conversely, baseline energy demand and SMR and, thus, the need to maximize aerobic machinery are reduced with increasing body size, thereby opening cellular space for enhanced myofibril densities and, thus, force generation. The reduction in mitochondrial density in larger fish also opens cellular space for glycolytic components (glycogen granules, enzyme complexes, buffer components). Enhanced power generation during burst activity becomes possible. Similar relationships prevail in endothermic homeotherms where citric synthase and hydroxy-acylcoenzyme A (hydroxy-acyl-CoA)-dehydrogenase scale negatively and LDH positively with body size (Hochachka *et al.*, 1988a,b).

In this context, an interesting observation is that maximum speed in Antarctic notothenioid fish seems constrained by cold temperatures more in larvae than in adults and was found to be two to three times below that of temperate and tropical larvae at their respective habitat temperatures (Archer and Johnston, 1989; Johnston *et al.*, 1991a). At first sight, this finding might either relate to the benthic origin of notothenioids or simply be due to the fact that high-performance predators do not exist for larval stages. However, the previous discussion strongly suggests that both small body size and cold adaptation elicit similar adaptational traits. In consequence, the maximization of aerobic capacity in small, including larval, fish likely reaches an earlier limit in the cold than in warmer waters. In adult (large) notothenioid fish, cold compensation is already supported by mitochondrial densities close to cellular space limitations. Compared with the example given for larval roach, three- to four-fold higher mitochondrial densities in larval than in adult notothenioids then appear unrealistically high, and this design limitation easily explains noncompensated aerobic scope and performance in larval notothenioid fish (Figure 3.7). These conclusions also indicate that energetic constraints on larval life are maximized in the cold, with the consequence that energy-saving strategies are maximized in larval compared to adult fish (see Section V).

D. Substrates in the Cold: Lipid Metabolism

In the cold, development of enhanced mitochondrial density goes hand in hand with the preferred use of lipids by mitochondria and, as a precondition, enhanced whole-body and intracellular storage of lipids. In temperate fish, preference for lipid catabolization by cold mitochondria is seen in striped bass with twofold higher rates observed at cold (5 °C) compared with warm (25 °C) acclimation temperatures (Sidell and Moerland, 1989). In rainbow trout slow muscle fibers, the trend to accumulate lipids seen during seasonal cold (Egginton *et al.*, 2000) correlates with increased capacity for β-oxidation, especially in red muscle (Guderley and Gawlicka, 1992).

As a consequence of high mitochondrial densities, Antarctic fish also display elevated capacities of mitochondrial β-oxidation as indicated by activities of 3-hydroxy-acyl-CoA-dehydrogenase and carnitine palmitoyl-transferase (CPT) (Sidell and Moerland, 1989). Transfer of acyl derivatives across the mitochondrial membrane, supported by CPT, rather than the capacity of β-oxidation, may exert rate-limiting control of lipid catabolism (Driedzic and Hochachka, 1978; Weber and Haman, 1996). Monoenoic fatty acids like 18:1 predominate in the lipid stored by notothenioids (whole animal) (Hagen et al., 2000), a finding in line with the preferred use of this fatty acid by notothenioid fish mitochondria (Sidell et al., 1995). Overall, acclimation and adaptation to cold leads to a shift to lipid catabolism.

It has been hypothesized that a higher adenylate energy charge with low levels of free ADP, AMP, and inorganic phosphate is maintained in tissues with elevated mitochondrial contents. This would alleviate the stimulation of glycolysis and favor the use of noncarbohydrate substrates (Holloszy and Coyle, 1984). Such a trend would be supported by the low-energy lifestyles in Antarctic fishes with limited use of anaerobic pathways. A shift from the glycogenolytic pathway, which is structurally associated with muscular fibrils, to the β-oxidation pathway located in the mitochondrial matrix is a logical consequence of increased mitochondrial and decreased myofibrillar volume fractions in a cell in the cold.

The capacity for β-oxidation is complemented by the mobilization of lipid stores in gut, liver, and adipose tissue. For transport in blood plasma, triglycerates and fatty acids are bound to albumin-like protein. The hydrolysis of triglycerates occurs by muscular lipase activity. The fatty acids are transferred through the membrane by nonesterified fatty acid transporters, and they are transported in cell fluids, again bound to protein. Studies by Metcalf et al. (1999a,b) demonstrate that the Antarctic toothfish (*Dissostichus mawsoni*), long-finned eels (*Anguilla dieffenbachii*), and short-finned eels (*Anguilla australis schmidtii*), from New Zealand, lack plasma albumin and use a high-density lipoprotein for palmitate transport in the plasma.

Complex transport mechanisms, similar to those identified in mammals (for review, see Hoppeler and Billeter, 1991), may be used to replenish depleted tissue stores from bloodborne chylomicrons and low-density lipoproteins. These are hydrolyzed by lipoprotein lipase (delivered from muscle) at the surface of the vascular bed. For transfer through the interstitial fluid, fatty acids are bound to an albumin-like protein. After diffusive transfer through the membrane, which is likely driven by a concentration gradient and not energy dependent, the transfer of fatty acids through the cytosol to mitochondria is facilitated by a small fatty acid–binding protein. Two isoforms of a cellular fatty acid–binding protein have been found in heart ventricle of Antarctic fish (Vayda et al., 1998). The role of all of these processes in fish is largely unexplored, particularly with respect to their thermal sensitivity.

In an ecological context, studies have emphasized the role of lipid storage in whole organisms in the cold and its use for energy storage, neutral buoyancy, and reproduction. As a general pattern at high latitudes, lipid stores in benthic herbivores are less than in either omnivores or carnivores, whereas in the pelagic both herbivores and omnivores/carnivores display enhanced lipid stores (Clarke and Peck, 1991). This is not the case in gelatinous (i.e., passive) plankton, indicating that the more elevated level of energy turnover/activity in the other pelagic forms may be crucial to cause the shift to lipid-based metabolism as seen in fish. In pelagic fish like *P. antarcticum*, extremely high lipid levels [up to 58% dry weight for the whole animal (Friedrich and Hagen, 1984; Hubold and Hagen, 1997)] are found in intermuscular and subcutaneous sacs. *P. antarcticum* lacks a swim bladder and, thus, maintains high lipid stores throughout all seasons to maintain neutral buoyancy (Hubold and Hagen, 1997). Its northern equivalent, the capelin (*Mallotus villosus*), possesses a swim bladder. It oscillates between lipid levels of 10 and 50% dry weight (whole body), mainly because of the seasonal provision of lipids to reproductive organs and processes. Intracellular lipid contents in oxidative muscle of notothenioids range between about 9% of dry weight in the demersal species *Gobionotothen nudifrons* and 37% in cryopelagic *Pagothenia borchgrevinki* (Sidell *et al.*, 1995) or 45.6% in *P. antarcticum* (Hubold, 1985).

Oscillation in lipid stores also reflects their use as an energy store during periods of restricted food supply in winter. Such patterns are also visible in zooplankton including larger pelagic invertebrates in polar seas (i.e., among euphausiids), which possess lipid stores between 46% dry weight in *Euphausia superba* and 56% in *Thysanoessa macrura* (Clarke and Peck, 1991; Hagen and Kattner 1998; Kattner and Hagen 1998). As a corollary, elevated lipid levels appear typical for polar fish and zooplankton, more so in pelagic than in benthic life forms, linked to higher levels of motor activity and SMRs. Although lipid stores in predators at the high end of the food chain will largely be supported by high lipid contents of, for example, zooplankton prey, the conclusion arises that both cellular lipid synthesis and catabolism are favored in the cold in all food-chain components.

Although the largest lipid contents have been found in pelagic fish, selective pressure toward lipid storage and aerobic lipid metabolism in the cold appears to prevail, even in fish that display low activity levels. This is emphasized by a comparison of benthic North Sea zoarcids (at 10 °C) and Antarctic zoarcids (at 0 °C) fed on the same shrimp (*Crangon crangon*) during laboratory maintenance. North Sea eelpout (*Z. viviparus*) maintained only 10% lipids per body dry weight, whereas Antarctic eelpout (*P. brachycephalum*) contained 33% (Brodte, 2001). This emphasizes the trend to synthesize and accumulate fat in cold-adapted fish fed on the same diet as temperate fish. A cold-induced shift to lipogenesis is also evident in temperate fish. Lipogenesis in male Gulf killifish, *Fundulus grandis*, was stimulated by cold temperature and led to lipid accumulation during cold exposure in autumn (Weld and Meier, 1985). Hepatocytes from cold-acclimated rainbow trout displayed significantly higher rates of fatty acid and sterol synthesis [measured as tritium incorporation at assay temperatures of 15 and 20 °C (Hazel and Sellner, 1979)] than hepatocytes from warm-acclimated rainbow trout.

The difference between Antarctic pelagic and benthic life forms suggests that maximized mitochondrial densities at elevated levels of motor activity as in pelagic notothenioid fish support enhanced lipid accumulations and their use as metabolic substrates. As reviewed by Pörtner (2002b), these patterns resemble to some extent metabolic features seen in high-performance scombrid fish like mackerel or tuna, as well as mammals, again reflecting similar patterns of metabolic organization during adaptation to cold and to elevated activity levels. This conclusion is not as clear with respect to the use of carbohydrates because differences in performance levels play a large role (see above, Figure 3.7). The lower performance levels in notothenioids and their excessive aerobic design imply less use of carbohydrates and low anaerobic capacity, in contrast to the situation in high-performance teleosts and mammals (Weber and Haman, 1996; Hoppeler and Weibel, 1998). With respect to the use of lipids, the picture may also have to be modified for elasmobranchs (and among invertebrates, cephalopods), which, in contrast to teleosts, use more protein than fat to fuel exercise metabolism. However, this does not influence the relationship between performance capacity and the use of carbohydrates and anaerobic pathways, at least in squid (Figure 3.7). In general, fuel preferences in endurance-adapted species, which operate aerobically at intermediate performance capacities, are suggested to shift from carbohydrates to lipids and proteins. These species will also obtain a much larger fraction of their energy from intramuscular substrate stores and a smaller fraction from bloodborne substrates than more sedentary fish (Johnston and Moon, 1980a,b; Weber and Haman, 1996). Only at the highest performance capacities will they display large anaerobic capacity associated with the use of significant glycogen stores (Figure 3.7). Once again, these life forms do not exist among Antarctic fish.

The following speculative scenario may support the shift toward enhanced fatty acid synthesis in (teleost) liver and muscle in the cold (Pörtner, 2002b): Elevated SMRs of pelagic zooplankton and fish (in contrast to benthic organisms) acclimated or adapted to cold are linked to excessive mitochondrial densities. This includes excess availability of mitochondrial intermediates like citrate. Especially at resting rates when other cellular costs are reduced and phosphorylation potential is high, excess citrate is exported from the mitochondria into the cytosol where net fatty acid synthesis occurs

(Goodridge, 1985). Excess citrate availability would be supported by excess mitochondrial CS capacities in the cold (Figure 3.5). Export into the cytosol occurs via the tricarboxylate anion carrier in exchange for malate that arises either from malic enzyme or from cytosolic malate dehydrogenase activities. Citrate is cleaved via ATP:citrate lyase to form acetyl-CoA and oxaloacetate (as a precursor for malate). Acetyl-CoA is carboxylated via acetyl-CoA carboxylase (ACC) to form malonyl-CoA and thereby fuels multifunctional fatty acid synthase and chain elongation.

NADPH is reoxidized during chain elongation and provided by the equilibrium enzymes glucose-6-P-dehydrogenase and malic enzyme. Accordingly, the fraction of glucose oxidized by the hexose monophosphate shunt (HMPS) pathway increases with cold acclimation in some species (Johnston and Dunn, 1987). However, control exerted via fatty acid synthase or citrate lyase is considered minimal, leaving ACC, which operates far from equilibrium, as the more important step in controlling fatty acid synthesis (Goodridge, 1985). ATP and bicarbonate are consumed during acetyl-CoA carboxylation. Elevated cellular bicarbonate levels would be provided by a cold-induced increase in pH, as expected from an alphastat pattern of intracellular pH regulation (see Section III). Excess citrate allosterically stimulates ACC.

A key role in the balance between lipogenesis and β-oxidation, for example, during transition to muscular exercise, is adopted by AMP-activated protein kinase (Winder and Hardie, 1999), which phosphorylates ACC and, thereby, causes a decrease in maximum velocity, decreases activation by citrate, and enhances inhibition by long-chain fatty acyl-CoA. Whether the degree of enzyme phosphorylation changes with temperature is unclear. However, lower levels of muscular ATP turnover associated with reduced activity periods in the cold, as in Antarctic stenotherms, increase the fraction of time available for lipogenesis.

Microsomal or mitochondrial chain elongation beyond C16, and especially the synthesis of triglycerides from glycerol 3-P, will also display thermal sensitivity; however, little is known about the degree of cold adaptation. If similar responses occur again during cold adaptation as during exercise training (investigated by Askeq *et al.*, 1975, in rats), cold adaptation may not only support lipogenesis via ACC but also cause an increase in glyceride synthase in muscle and adipose tissues. Cold-compensated ability to synthesize and store triglycerates would complement the capacity for enhanced lipid formation in muscle.

Finally, the use of fatty acids as substrates transported to and synthesized within muscle cells may be enforced by largely reduced rates of energy-dependent transport across cellular membranes, reflected by overall low rates of transepithelial and likely transmembrane Na^+/K^+-ATPase in the cold (see

Section III). Because cellular protein stores will be reduced as a result of enhanced mitochondrial densities, cellular lipid stores with their high-energy density would appear most suitable to replace protein as a substrate.

Together with increased lipid uptake from food, this complex picture would explain why tissue and whole-animal lipid stores are higher in the cold. Additional time for anabolism due to behavioral energy savings at high mitochondrial densities appears to support this shift. This scenario of a cold-induced shift to lipogenesis requires further experimental verification.

E. Biochemistry and Physiology of Cellular Oxygen Supply

Sidell and colleagues developed the view that lipid and membranes enhance intracellular oxygen solubility and diffusion velocity and, thereby, may play an important role in cellular oxygen supply in the cold (Tyler and Sidell, 1984; Egginton and Sidell, 1989; Londraville and Sidell, 1990; Sidell, 1998). This insight gains additional relevance in light of the hypothesis that thermal tolerance windows are set by oxygen limitations and that a cold-induced change in mitochondrial densities and functional capacities causes a shift of oxygen-limited thermal windows (see Sections I.A and B). In the cold, diffusional limitations for small molecules other than oxygen may not exist to the extent previously thought. According to Hubley *et al.* (1997) intracellular diffusion of high-energy phosphates does not become limiting during cold acclimation. However, gradients for creatine phosphate and Gibb's free energy of ATP hydrolysis were attenuated by rising mitochondrial volume density in goldfish acclimated from 25 to $5\,°C$.

A reduction in oxygen diffusion is reflected in a cold-induced fall of the Krogh diffusion constant KO_2 (by $1.6\%\,°C^{-1}$), which encompasses a drop in diffusion coefficient DO_2 (by $3\%\,°C^{-1}$), as well as an increase in oxygen solubility (by about $-1.4\%\,°C^{-1}$). A 42% lower Krogh constant would result for the cytosol of Antarctic fish (at $-1.8\,°C$), compared with that of a temperate fish at $25\,°C$ (Sidell, 1998). Work carried out on striped bass unequivocally demonstrates that lipid accumulation more than compensates for this drop in KO_2. KO_2 doubled between $25\,°C$ and $5\,°C$ because of a more than 10 times increase in the fractional cell volume filled with lipid and the associated doubling of cellular oxygen solubility. The diffusion coefficient DO_2 was more or less maintained because of the structural changes of the cell. With membranes being the preferred pathways of oxygen diffusion, Sidell (1998) argued that the enhanced mitochondrial density, together with the increasing level of lipid unsaturation, would support oxygen flux into the center of the cell.

It might appear that enhanced lipid accumulation in the cold might be primarily adaptive and driven by oxygen limitations. Alternatively, however, these patterns might be beneficial by-products of an increased mitochondrial proliferation and associated shift to lipid metabolism (see Section II.D). Some further insight arises from the scenario of a secondary drop in SMR during the evolution of polar fauna, which very likely occurred with elevated mitochondrial densities and some enhanced lipid contents already present as observed in extant eurytherms in the cold. With the drop in SMR, excess rather than too little oxygen became available in the cold, allowing for a reduction in ventilatory and circulatory effort and in the PO_2 gradient needed for oxygen flux (Pörtner, 2002b). In line with these considerations, low blood PO_2 levels of 30 mmHg were reported for *P. antarcticum* (Wöhrmann *et al.*, 1997), although this parameter may have been lowered as a result of sampling from stressed specimens (see Chapters 6 and 7, of this book). Excessive oxygen availability would strongly argue against the accumulation of lipid being driven by oxygen limitations. Energy savings would, however, result from enhanced diffusive oxygen supply and distribution in the cold and may be crucial for an understanding of the patterns observed. Such energy savings would be supported not only by lipid accumulation, but also by the 1.5-fold increase in oxygen solubility in the water between 20 and 0 °C.

In line with excessive oxygen supply in the cold, a loss of cardiac Mb, not known in cold-adapted eurytherms, occurred independently several times in Antarctic stenothermal fish, despite the presence of the respective DNA message. This would suggest secondary loss of function, most likely due to the fact that oxygen diffusion is enhanced to an extent by the accumulation of membrane lipids and lipid stores that Mb is no longer needed to enhance intracellular diffusibility of oxygen at high mitochondrial densities and low oxygen flux. At the same time, it may be the stability of the highly oxygenated natural environment that has contributed to the loss of Mb, which is seen as a safety measure during periods of extreme oxygen demand (Sidell, 1998). Loss of Mb would in fact support the energy-saving hypothesis.

A study by O'Brien and Sidell (2000) compared heart ventricles of three Antarctic fish species that do or do not express Hb and heart Mb. The red-blooded species *G. gibberifrons* expresses both Hb and Mb. Among the two icefishes, *Chionodraco rastrospinosus* expresses only Mb and *C. aceratus* neither. Mitochondrial density appeared to be negatively correlated with the presence of Mb. Both icefishes possess a large heart, 3.2 g kg^{-1} body mass in *C. aceratus*, 4 g heart kg^{-1} body mass in *C. rastrospinosus*, when compared with *G. gibberifrons* (0.7 g heart kg^{-1} body mass). *C. aceratus*, without Hb and Mb, possesses a larger volume density of 36% mitochondria compared with 20% in *C. rastrospinosus* (without Hb) and 16% in red-blooded *G. gibberifrons*. The surface density of inner mitochondrial membranes (cristae density), usually accepted as an indicator of aerobic capacity, was 1.5- to 1.7-fold lower per mitochondrial volume in the Mb-deficient than in the other two species. However, cristae density per gram of tissue was still larger by a factor of 1.55 or 1.68 in *C. aceratus* than in *G. gibberifrons* and *C. rastrospinosus*, respectively. Nonetheless, all three ventricles displayed very similar oxidative capacities per gram of tissue wet weight, indicated by almost identical levels of COX (18 U) and CS (12 U). Somewhat higher capacities of fatty acid transport into mitochondria were found in the two icefish species (indicated by carnitine palmitoyl transferase activities between 90 and 115 U in the two icefishes vs. about 60 in the red-blooded notothenioid), whereas hydroxy-acyl-CoA dehydrogenase was found at about 2.2 U in the two species with Mb, versus about 3 U in the Hb- and Mb-deficient icefish. Despite higher mitochondrial density, the Mb-deficient icefish possessed about twice the LDH capacity than the two other species, possibly indicating improved protection from oxygen deficiency in the Mb-deficient icefish heart. At the same time, this may indicate enhanced capacity of lactate uptake and use as a substrate of cardiac metabolism (see Section II.B).

The elevated heart weights in both icefishes may relate to increased cardiac output due to Hb-deficient blood (Tota *et al.*, 1991). The large difference in cristae surface densities between hearts in spite of similar aerobic capacity is noteworthy and contrasts the paradigm of a tight coupling between aerobic capacity and surface density (see above). The large cristae surface densities at low aerobic capacities in the Mb-deficient heart are in line with the Sidell (1998) hypothesis: that an enhanced mitochondrial network in the cold supports oxygen diffusion. In those cases in which Mb is lost, this occurs even at the expense of lower volume densities of myofibrils. The same conclusion has been developed by Dunn *et al.* (1989), who found that skeletal muscle fibers of demersal and pelagic icefishes (*C. aceratus* and *Pseudochaenichthys georgianus*, respectively), which also lack Mb, have higher mitochondrial densities (>50% in *P. georgianus*) but with lower oxygen consumption rates per unit volume of mitochondria (Johnston, 1987) than red-blooded notothenioids. This observation is in line with findings of low cristae densities at high mitochondrial volume fractions in icefish skeletal muscle (53% in the channichthyid *C. aceratus*), which contrast higher cristae densities at lower mitochondrial volume fractions in red-blooded notothenioid skeletal muscle (29% in *N. coriiceps*) (O'Brien *et al.*, 2003). In all investigated species, cristae density was conserved at about 9 m^2 g^{-1} muscle fresh mass. These patterns suggest that maximized mitochondrial densities and networks do support oxygen flux. If combined with minimized mitochondrial capacities and low costs of proton leakage, these patterns again reflect energy savings associated with enhanced diffusive oxygen supply.

The fact that Mb enhances cardiac performance in *C. rastrospinosus* compared with *C. aceratus* (Acierno *et al.*, 1997) suggests that the loss of

Mb is not only due to cold-enhanced oxygen supply. This loss may also be favored by the reduction in performance levels in Antarctic species and reflects a loss of functional scope with energy savings that allowed for the loss of pigments (see Sections II.A and B). This picture reflects enhanced stenothermy according to oxygen-limited thermal tolerance and is again supported by the potential sequence of evolutionary cold adaptation in Antarctic fishes: transition to stenothermal cold associated with a secondary reduction of metabolic rate at elevated mitochondrial densities, then loss of Hb, and, in some cases, Mb, balanced by an enhanced mitochondrial network at unchanged overall capacity. Excess oxygen supply firstly associated with high solubility in enhanced lipid stores and mitochondrial densities is then enhanced even further by reduced performance and oxygen demand. Both processes allowed for shallower PO_2 gradients and a reduction of ventilatory and circulatory effort. Excess metabolic capacity, combined with low energy turnover behavior, in turn supported further lipid accumulation (see above). Overall, the similarity of lipid metabolism patterns in high-performance and cold-adapted fish suggests that the principle pattern of lipid accumulation in the cold is not driven by oxygen limitations, but that enhanced diffusive oxygen supply in the cold is a secondary benefit of the enhanced accumulation of lipid stores and membranes, which supports energy savings in ventilatory and circulatory oxygen supply.

Further observations corroborate the prime role of energy savings in the permanent Antarctic cold. Both slow oxidative and fast glycolytic muscle fibers are fewer and larger in Antarctic fish and at reduced capillary density, compared with temperate and tropical species of similar mode of life (Johnston, 1987, 1989; Johnston *et al.*, 1988; Dunn *et al.*, 1989), with some variability between species (Egginton *et al.*, 2002). These features, found at a higher organizational level, likely became possible with enhanced oxygen supply due to cold-induced cellular lipid and membrane accumulation. The reduction in skeletal muscle cell number at increasing cell size is a feature that has become a plesiomorphic characteristic of Antarctic and sub-Antarctic notothenioids (Fernandez *et al.*, 2000; Johnston *et al.*, 2003) and likely involves further energy savings due to reduced total surface area of cellular membranes and associated ion exchange requirements. A trend for the development of red muscle fiber hypertrophy was also seen in cold-acclimated striped bass (Egginton and Sidell, 1989). These patterns again indicate excess oxygen availability and enhanced diffusive oxygen flux in the cold, even more so in Antarctic fish (overcompensation) than in cold-acclimated temperate fish. Egginton *et al.* (2002) emphasized the integrated role of temperature, capillary number, fiber size, mitochondrial volume fraction, diffusion constant, fiber composition, and capillary radius in setting the value of intracellular oxygen tension in perciform fish species (including notothenioids) across latitudes. Their study nicely illustrates how the absence of blood Hb in icefish leads to early hypoxemia at elevated temperatures and, thus, contributes to the level of stenothermy in this species. However, their analysis did not consider the likely variable levels of blood oxygen tensions across latitudes and depending on metabolic rate. This would appear as a crucial next step to be investigated in light of the concept of oxygen-limited thermal tolerance and the key role of circulatory capacity in fish in setting the widths of thermal tolerance windows (Figures 3.1 and 3.2).

As a general conclusion, the repeated secondary loss of Mb in Antarctic icefish cardiac muscle and its loss from oxidative skeletal muscle of the whole notothenioid family can be explained, once it is accepted that mitochondrial proliferation in the cold is primarily driven by capacity limitations of tissue function and secondarily modified to serve improved oxygen supply and, thereby, energy savings. Loss of Mb indicates that the cold-induced drop in oxygen diffusion was in fact overcompensated for by enhanced oxygen supply in the permanent cold, supporting large cell sizes, reduced ventilatory and circulatory effort, and thereby, led to energy savings at the expense of enhanced thermal sensitivity. This complex picture excludes diffusion limitations as a primary driving force of excessive mitochondrial proliferation; in contrast, enhanced energy savings due to enhanced oxygen diffusion may be crucial as a driving force. Accordingly, loss of Mb results from excess oxygen supply associated with the drop in oxygen demand during reduction of SMR and functional capacity. Furthermore, the loss of Hb and Mb likely comes with the benefit of reduced protein synthesis in the permanent cold and again associated energy savings. Overall, enhanced oxygen availability and enhanced diffusive oxygen supply combined with an emphasis on the use of energy-saving strategies at various levels of organization seem to be overarching principles of adaptation to life in the permanent cold of the marine Antarctic. The principle reasons for the emphasis of such strategies in the permanent cold are addressed in Section V.

III. MEMBRANE FUNCTIONS AND CAPACITIES: CONSTRAINTS IN ION AND ACID–BASE REGULATION

In general, cold acclimation and even more so adaptation to polar cold elicit increased lipid unsaturation and fluidity of cytoplasmic membranes in temperate and polar fish, indicating homeoviscous adaptation (Cossins, 1994). The same is true for mitochondrial membranes of goldfish and carp (van den Thillart and de Bruin, 1981; Wodtke, 1981b) and presumably, though not yet studied, for Antarctic fish mitochondria. Antarctic notothenioids and likely zoarcids (Bock et al., 2001) display elevated levels of polyunsaturated membrane fatty acid phospholipids with an increased use of phosphatidyl-ethanolamine over phosphatidyl-choline and to a lesser degree, higher levels of monounsaturates than temperate and warm-water fish (Hazel, 1995; Gracey et al., 1996; Storelli et al., 1998). High levels of unsaturated fatty acids are interpreted to support proton leakage rates through the inner mitochondrial membrane (Brand et al., 1992). At elevated temperatures, such enhanced proton leakage may cause a drop in the coupling of oxygen consumption to ATP synthesis, determined as ADP/O ratios, as seen in Antarctic ectotherm mitochondria (Hardewig et al., 1999b; Pörtner et al., 1999). Moreover, the level of polyunsaturates has been suggested to be correlated to metabolic rate, based on a comprehensive analysis of ectothermic reptiles, amphibians, and endothermic mammals. Here, the levels of polyunsaturated phospholipids (C20:3, 20:4, 22:6), especially docosahexanoic acid (22:6n–3), in cellular and mitochondrial membranes positively reflect the level of metabolic activity (Else and Wu, 1999; Hulbert and Else, 1999: Hulbert et al., 2002), with larger levels in endotherms than in ectotherms. Enhanced levels of 22:4n–6 rather than of 22:6n–3 fatty acids were found in brains of permanently endothermic vertebrates (mammals and birds) compared with cold-water and tropical fish (Farkas et al., 2000).

However, despite elevated levels of unsaturates and homeoviscous adaptation, the findings in Antarctic fish do not imply enhanced leakiness for ions or associated metabolic activity in Antarctic stenotherms at their ambient temperature. In fact, metabolic activity is reduced compared with temperate and tropical fish with an active lifestyle. This may become possible through a downregulation of ion channels, paralleled by low ion exchange capacities and reduced cell membrane surface/cell volume ratios of the larger myocytes (see Section II.F). Associated cost reductions would support the more sluggish mode of life of Antarctic stenotherms (Hochachka, 1988a,b; Thiel et al., 1996). As a global trend, the capacities of Na^+/K^+-ATPase are reduced in cold-water compared to tropical species (Pörtner et al., 1998) and, thereby, reflect the different levels of ion exchange and muscular activity. Preliminary evidence also suggests a difference in the energy cost of ion regulation between cold-adapted eurytherms and stenotherms. Part of the cost of ion regulation is in the flexibility of acid–base regulation, which is lower in stenotherms than in eurytherms (Pörtner and Sartoris, 1999). A more detailed picture of differences between cold-adapted stenotherms versus eurytherms is emerging (see below).

Despite an overall reduction in ion exchange capacity in the cold, some adjustment in ion exchange may be required to compensate for differential effects of temperature on active versus passive ion movements. Because of their different nature, passive ion fluxes along the electrochemical gradient (e.g., through ion channels) display less thermal sensitivity than carrier-mediated transport. Cold exposure may, thus, lead to an imbalance between these processes unless the organism is able to compensate (Hochachka, 1988a,b). Generally, two strategies have been described: Either ion pumps are upregulated during acclimation or adaptation to cold to match dissipative fluxes, or mechanisms are employed to reduce those fluxes (e.g., channel arrest). Both strategies may be used in fish. Cold-compensated activity of sarcoplasmic reticulum Ca^{2+}-ATPase was visible in cold-(including Antarctic) versus warm-water fish (McArdle and Johnston, 1980). In several temperate eurythermal teleosts like roach, *R. rutilus*, Na^+/K^+-ATPase activity was increased in hepatocytes and kidney during cold acclimation (Schwarzbaum *et al.*, 1992a,b) and was linked to an increased number of transporter molecules. Transporter activity was also increased in gills, however, even with a reduced number of pumps (Schwarzbaum *et al.*, 1991). In rainbow trout erythrocytes, cold acclimation led to an increase in total Na^+/K^+-ATPase activity without changing the number of ouabain binding sites, which reflect the number of transporter molecules (Raynard and Cossins, 1991).

In contrast, a reduction in ion leakage rates during cold acclimation, through reduced activity or density of ion channels, has been demonstrated in the more cold stenothermal freshwater fish *Salvelinus alpinus* (Schwarzbaum *et al.*, 1991). Such a strategy may also apply to cold-stenothermal Antarctic fish, in which reduced ion exchange capacities across gill epithelia may explain their higher serum osmolarity (Gonzalez-Cabrera *et al.*, 1995; Guynn *et al.*, 2002). Accordingly, the strategy used may depend on the width of the thermal tolerance window and associated levels of energy turnover, where higher costs and functional capacities are associated with wider windows of thermal tolerance (Pörtner, 2004).

The picture may be even more complex, first because of the existence of various Na^+/K^+-ATPase isoforms with likely different functional properties, and second because of the interaction between membrane lipids and membrane bound enzymes (Wodtke, 1981a,b) such as Na^+/K^+-ATPase. In temperate eelpout *Z. viviparus* (Lucassen *et al.*, unpublished), the expression of the α-subunit of Na^+/K^+-ATPase was increased in gills and liver of cold-acclimated specimens. However, the protein number as determined by two different antibodies remained constant in both tissues during cold acclimation (Figure 3.8). Kinetic properties of Na^+/K^+-ATPase were variable between tissues of the eelpout. Maximum activities were increased in gills regardless of assay temperature (Figure 3.9), whereas in liver, an increase in activity could be demonstrated only at low assay temperatures. When compared at the same assay temperature, increased activities of Na^+/K^+-ATPase at constant or falling protein numbers become at least partly

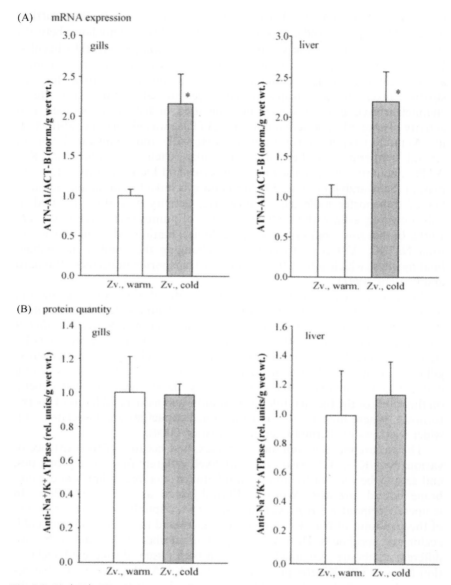

Fig. 3.8. Na$^+$/K$^+$-ATPase messenger RNA (mRNA) and protein levels in gills and liver of *Zoarces viviparus* acclimated to 3.0 °C (cold acclimated) and 13.5 °C (warm acclimated) for 2 months. (A) mRNA expression was analyzed using an RNase protection assay with species-specific RNA probes for ATN-A1 (Na$^+$/K$^+$-ATPase) relative to ACT-B (β-actin). (B) Quantification of Na$^+$/K$^+$-ATPase protein by immunodetection with anti–Na$^+$/K$^+$-ATPase antibodies

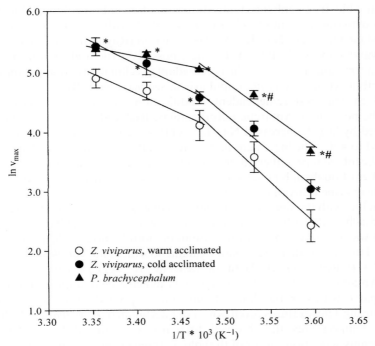

Fig. 3.9. Arrhenius plot of maximum Na^+/K^+-ATPase activity (v_{max}, μmol $h^{-1}g^{-1}$ wet weight) in gills of *Zoarces viviparus* and *Pachycara brachycephalum*. *Z. viviparus* was acclimated to $3.0\,°C$ (cold acclimated, filled circle) and $13.5\,°C$ (warm acclimated, open circle) for 2 months. *P. brachycephalum* (filled triangle) was adapted to $0\,°C$. The ouabain-sensitive ATPase activity was determined in membrane preparations of the respective tissues at 5, 10, 15, 20, and $25\,°C$. Values are means $\pm$SEM ($n = 5 - 7$). *Significantly different from the warm-acclimated control. #Values in *P. brachycephalum* significantly different from those in cold-acclimated *Z. viviparus*. Gills of cold-acclimated *Z. viviparus* possessed significantly higher Na^+/K^+-ATPase activities than warm-acclimated fish at all assay temperatures. *P. brachycephalum* displayed cold-compensated activities beyond those found in cold-acclimated *Z. viviparus* at low assay temperature (Lucassen *et al.*, unpublished).

(α6F, Takeyasu *et al.*, 1988). The amount was determined using a chemiluminescence reaction and a cooled charged couple device (CCD) camera system. Values are means $\pm$SEM ($n = 5$) and expressed relative to values found in the tissues of the warm-acclimated control group, which was set to 1. *Significant difference from the warm-acclimated control group (data from Lucassen *et al.*, unpublished). Whereas the mRNA of Na^+/K^+-ATPase was increased significantly in both tissues with cold acclimation, the protein number remained more or less unchanged in both tissues.

explainable from enhanced unsaturation of membrane lipids (linked to homeoviscous adaptation; Cossins *et al.*, 1981; Gibbs, 1995) and the associated facilitation of Na^+/K^+-ATPase function (Raynard and Cossins, 1991; Schwarzbaum *et al.*, 1992b). Further, possibly structural modifications of the enzyme during synthesis in a cold-adapted cellular environment may also contribute (Lucassen *et al.*, unpublished, see Section IV.B).

Such temperature-dependent modifications in protein structure may still be possible in Antarctic notothenioids. Reduced serum osmolality was elicited by largely increased Na^+/K^+-ATPase activity in *T. bernacchii* acclimated to 4 °C. The increment in Na^+/K^+-ATPase activity coincided with no significant change either in membrane fatty acid composition or in the number of ouabain binding sites, thereby indicating structural modifications of the enzyme (Guynn *et al.*, 2002).

In line with the suggested strategy of reduced ion leakage in Antarctic fish (Hochachka 1988a,b), maximal activity of gill Na^+/K^+-ATPase (assayed at 4 °C) was five times lower in the Antarctic notothenioid *T. bernacchii* acclimated to -1.5 °C than in the New Zealand notothenioid *Notothenia angustata* at 14 °C (Guynn *et al.*, 2002). In contrast, Na^+/K^+-ATPase activities in gills of the Antarctic eelpout *P. brachycephalum* indicate cold compensation when compared with warm-acclimated temperate eelpout *Z. viviparus* and at low measuring temperatures, even with cold-acclimated *Z. viviparus*. At higher assay temperatures, the observed difference was reduced due to the lower Arrhenius activation energy of the cold-adapted (*P. brachycephalum*) enzyme (Lucassen *et al.*, unpublished; Figure 3.9). It remains unclear whether the contrasting trends observed in notothenioids and zoarcids are due to factors other than temperature or a less stringent level of stenothermy in the Antarctic eelpout, which may implicate cold-compensated ion regulation capacities.

In this context, the downregulation of ion exchange in the cold may also affect lactate transport. Data available for Antarctic and temperate fish indicate that the release of lactate from fatiguing muscle is extremely limited. Although exceptionally high lactate levels were found in white muscle of Antarctic eelpout *P. brachycephalum* after exhaustive exercise (see above), very little lactate was found in the plasma (Hardewig *et al.*, 1998). In contrast, temperate species release significant lactate quantities into the blood, although in fish the bulk of lactate is retained in white muscle for gluconeogenesis *in situ* (Milligan, 1996). Less lactate was also found in the blood of exhausted cold-acclimated compared to warm-acclimated rainbow trout, although they had similar intracellular concentrations (Kieffer *et al.*, 1994). This indicates reduced transmembrane transfer of lactate and, most likely, non-release of associated protons in the cold (Hardewig *et al.*, 1998). In cold-adapted fish, rapid aerobic recovery likely occurs and compensates for these constraints. This may alleviate the need to release lactate and protons for rapid mitigation of the cellular acidosis (Figure 3.6). Once again, these patterns resemble those observed in squid in which non-release of the glycolytic end-product and associated proton quantities is also compensated for by rapid aerobic metabolic recovery *in situ* (Pörtner, 1994). An important question is whether the capacity of ion and acid–base regulation is large enough to compensate for exercise-induced acid–base disturbances or whether limitations at this level due to energy savings may contribute to constraints on anaerobic exercise in the permanent cold.

As mentioned earlier, the cost of acid–base regulation is involved in defining the cost of ion regulation. The cost and flexibility of acid–base regulation is reflected in its response to temperature. The alphastat hypothesis of acid–base regulation implies that pH regulation in poikilotherms should maintain a constant degree of protonation (α) of imidazole groups in proteins despite changes in body temperature. The constant dissociation state of histidine residues within proteins, especially in active sites of enzymes, is seen as a key factor in this process (Reeves, 1985). A shift in intracellular pH (pH_i) with changing body temperature may compensate for the temperature-dependent change of pK values of histidine imidazole groups ($\Delta pK/\Delta T \sim -0.018 /°C$). With some exceptions, an alphastat pattern of pH_i regulation has been found in several marine ectotherms (invertebrates and fish) from temperate latitudes (for a review, see Pörtner *et al.*, 1998), in which animals are exposed to wide diurnal and seasonal temperature fluctuations. Results of invasive studies performed in eurythermal North Sea eelpout (*Z. viviparus*) and more stenothermal Antarctic eelpout (*P. brachycephalum*) were confirmed by use of noninvasive ^{31}P-nuclear magnetic resonance (NMR) imaging. These data support the conclusion that a decrease in pH_i values upon warming, as predicted by the alphastat hypothesis, also occurs in cold-water fish (van Dijk *et al.*, 1997, 1999; Bock *et al.*, 2001; Sartoris *et al.*, 2003). In light of the hypothesis that an alphastat pattern of pH regulation may contribute to maintain metabolic flexibility, it is not surprising that this pattern is found only within the range of thermal tolerance of a species (Sommer *et al.*, 1997; Mark *et al.*, 2002).

In contrast to the eurythermal eelpout *Z. viviparus*, in which the contribution of passive mechanisms (response of cellular buffers) is small and the intracellular pH change is largely caused by active mechanisms (i.e., energy-dependent ion transport, van Dijk *et al.*, 1997), a large fraction of the temperature-induced pH change in stenothermal *P. brachycephalum* is caused by passive physicochemical buffering (-0.011 pH/°C, Pörtner and Sartoris, 1999). Such a large contribution of passive buffering is typically found in polar stenotherms (Pörtner and Sartoris, 1999), possibly due to maximized energy savings in this group and their limited thermal tolerance range. Otherwise, the mechanistic background of these patterns and the likely modified characteristics of ion and acid–base regulation of polar ectotherms is largely unknown.

IV. MOLECULAR PHYSIOLOGY OF METABOLIC FUNCTIONS

Changes in the functional properties and concentrations of enzymes and ion transporters observed in many studies of cold acclimation or adaptation warrant an evaluation of the underlying regulatory mechanisms. This includes studies of how the levels of functional proteins are adjusted, as well as analyses of how modifications of the cellular environment and of protein structure determine the kinetic properties of the enzyme.

A. Regulation of Gene Expression and Functional Protein Levels

As a model system of gene expression in the cold, the temperature-dependent expression of the LDH-B isoform, which is specific to aerobic red muscle, heart, and liver, has been examined intensively in the mummichog *Fundulus heteroclitus* (Powers and Schulte, 1998). In northern (cold-adapted) versus southern (more warm-adapted) populations, LDH-B showed a compensatory increase in enzyme concentration accompanied by increased mRNA levels, which were maintained even after long-term cold adaptation (Crawford and Powers, 1989). This difference was shown to be related to enhanced transcription rates in the north (Crawford and Powers, 1992). Schulte *et al.* (2000) found a stress-responsive regulatory element in southern populations in the upstream part of the LDH-B gene that was absent in the northern population and seems to be involved in setting different transcription rates. Therefore, the equilibrium enzyme LDH may reflect an example of evolutionary adaptation to different thermal environments via transcriptional control.

However, changing transcript levels are not always indicative of changing enzyme activities or functional protein levels. In the case of Na^+/K^+-ATPase of eelpout (*Z. viviparus*), cold acclimation led to enhanced mRNA levels in gills and liver without enhancing protein levels, as determined by means of antibodies (Lucassen *et al.*, unpublished; Figure 3.8). These increases in transcript levels are likely essential to just maintain functional protein levels.

Hardewig *et al.* (1999a) found that increased levels of COX (COX-2 and COX-4) mRNA in muscle of cold-acclimated *Z. viviparus* was accompanied by elevated enzyme activities (Figure 3.10). When comparing the ratios of mRNA levels and activities, it became clear that transcript levels of cold-acclimated eelpout were overcompensated relative to enzyme activities. At first glance, these results might argue for transcriptional control of enzyme

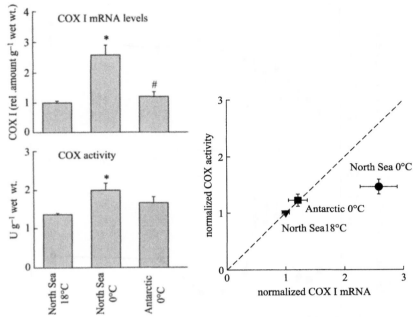

Fig. 3.10. Cytochrome *c* oxidase (COX) activity and expression of subunit I in the white musculature of eelpout from the North Sea (*Zoarces viviparus*, acclimated to cold or warm) and Antarctica (*Pachycara brachycephalum*). (Left) COX activity levels parallel the level of mitochondrial transcripts; however, values normalized to tissue wet mass indicate an over-proportional accumulation of messenger RNA (mRNA) in cold-acclimated *Z. viviparus*. (Adapted from Hardewig *et al.*, 1999a.)

activities; however, they also reflect loose relationships between mRNA and protein levels and, thus, indicate an additional role for translational and posttranslational processes.

In liver of *Z. viviparus*, CS activity started to increase immediately upon cold exposure, whereas the CS mRNA level displayed a delayed transient increase. Evidently, transcript levels did not parallel the change in enzyme activity (Figure 3.11; Lucassen *et al.*, 2003). During long-term acclimation to low temperature, mRNA even returned to control levels, whereas elevated enzyme quantities were maintained. This discrepancy, together with the delayed onset of transcript accumulation, indicates that enhanced transcript levels are no prerequisite for compensatory translation to take place. In both studies, the confamilial Antarctic *P. brachycephalum* demonstrated activity/ mRNA ratios for CS as well as COX similar to those of warm-acclimated *Z. viviparus*, indicating a balanced situation in the cold-adapted species

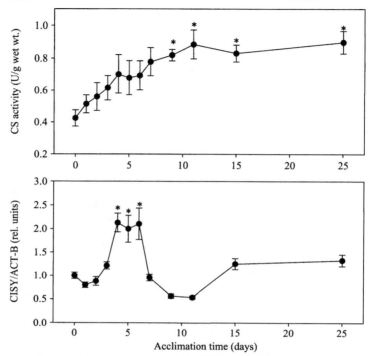

Fig. 3.11. Effects of temperature acclimation over time on activities (A) and messenger RNA (mRNA) levels for citrate synthase (B) in the liver of *Zoarces viviparus. Z. viviparus* from North Sea population kept at 10 °C were brought to 3.0 °C within 12 hours. Enzyme activities were assayed in liver of each fish at 3.5 °C. The relative amounts of citrate synthase mRNA (CISY) were determined by RNase protection assays. Data were corrected for loading differences using the β-actin signal (ACT-B). Values are means ±SEM and are expressed in units relative to the warm-acclimated control group (day 0), which was set to 1. The data were tested for outliers at the 95% significance level using Nalimov's test. *Significantly different from the warm-acclimated control group. Whereas enzyme activities rose progressively and reached a twofold increased level within about 10 days, the respective mRNA levels increased only transiently between days 4 and 6 (modified from Lucassen *et al.*, 2003).

(Hardewig *et al.*, 1999a; Lucassen *et al.*, 2003). Possibly, such ratios are kept at a minimum within the thermal optimum of a species.

Such phenomena are not restricted to eelpouts. Whereas Itoi *et al.* (2003) found a correlated increase of F_0F_1-ATPase activity and mRNA levels of nuclear-encoded subunits during cold acclimation of common carp, mRNA levels of mitochondrial-encoded subunits displayed threefold overcompensation. The nuclear encoded Δ^9-desaturase of carp could be induced by cold exposure, paralleled by increasing mRNA levels. However, the activity of this enzyme remained high after return of the respective mRNA to nearly control levels (Tiku *et al.*, 1996; Cossins *et al.*, 2002). Similar to *Z. viviparus*, an increase in CS activity preceded the increase in mRNA transcript during chronic stimulation of rabbit muscle (Seedorf *et al.*, 1986), which also elicits mitochondrial proliferation (Leary and Moyes, 2000; Hood, 2001). Also, COX activity was increased in rainbow trout muscle at unchanged mRNA levels (Battersby and Moyes, 1998). All of these examples suggest that mRNA levels and enzymatic activities are only loosely correlated; however, increased mRNA levels nonetheless accelerate the acclimation process and facilitate enzyme accumulation. These observations indicate an important role not only for transcriptional control, but also for either translational or posttranslational regulation during acclimation to cold. The regulatory patterns may differ between enzyme systems. Changes in the stability of mRNA and/or protein by changing ratios of synthesis over degradation may contribute to changing enzyme levels and activities.

B. Temperature Effects on Protein Structure and Function

Changing protein numbers can only partly compensate for a loss in activity during both seasonal acclimatization and permanent adaptation to cold, because of constraints in cellular space and energy availability. Structural modifications may also occur at the molecular level to improve enzyme operation in the cold. These modifications involve regions of a molecule that do not interact directly with ligands but indirectly influence the energetics of ligand binding and catalysis. Only minor changes in sequence will be required to adapt kinetic properties to different temperatures (Somero, 1997). In light of this, adaptive changes may be difficult to distinguish unless proteins of closely related species from different thermal habitats are compared.

However, some studies investigated orthologous proteins from less closely related species, using primary sequence data and their deduced amino acid sequence for comparative modeling of cold-adapted enzymes. Although in such cases, factors other than temperature may have to be considered, general trends have been postulated (for review, see Feller *et al.*, 1997; Marshall, 1997; and references therein). Cold-adapted enzymes tend to have fewer salt links, less interactions within the hydrophobic core, a reduction in the number of proline and arginine residues, a reduction in the hydrophobicity of the enzyme, and improved solvent interactions with a hydrophilic surface via additional charged side chains. In most cases, the catalytic and binding centers are not changed. Changes in the amino acid sequence of the enzyme may increase structural flexibility, sometimes at the expense of reduced thermal stability (see below). Therefore, it was hypothesized that a given degree of structural flexibility is required for enzyme function and that this is one of the conserved biological properties (Feller *et al.*, 1997; Marshall, 1997). Accordingly, the structural changes observed in cold-adapted enzymes contrast with those thought to increase the thermal stability of thermophilic proteins (Jaenicke, 1990). Each enzyme seems to display a specific pattern by using one or a combination of these altered interactions (Feller *et al.*, 1997).

Results of comparative studies of mammalian and fish cardiac troponin C are in line with these generalized conclusions. The higher affinity of Ca^{2+} binding in teleost relative to mammalian troponin could be ascribed to residues close but not within one of the low-affinity binding sites, increasing charge and hydrophilicity of this domain (Gillis *et al.*, 2003). As a result, mammalian and trout troponin C display similar conformations at their respective body temperatures (Blumenschein *et al.*, 2004).

When ectotherm protein structures are compared between temperate eurytherms and cold-adapted stenotherms, some structural differences are minute, some beyond the level of detection and some without functional consequences. For example, with nine exchanges in the primary sequence of troponin C from temperate (*O. mykiss*) compared to cold stenothermal icefish (*C. aceratus*), affinities for Ca^{2+} were indistinguishable (Gillis *et al.*, 2003). Fields and Somero (1998) confirmed small structural differences between orthologous LDHs, associated with altered kinetic properties. In contrast, different kinetic and structural properties according to functional requirements may exist for exactly the same enzyme protein, depending on different conformations according to temperature. In an early study by Ozernyuk *et al.* (1994), M4-LDH isoforms from cold-compared with warm-acclimated fish displayed enhanced thermal stability and a higher resistance to inactivation by urea. LDH with the same primary sequence but different kinetic behaviors and stabilities could be found among congeneric species of temperate water gobiid, as well as in Antarctic notothenioid species (Fields and Somero, 1997; Marshall *et al.*, 2000; Fields *et al.*, 2002). These differences in kinetic properties could be overcome through partial denaturation and renaturation. It was, therefore, concluded that different conformations do exist for the same enzyme protein (Ozernyuk *et al.*, 1994; Fields and Somero, 1997). Data obtained in Atlantic cod acclimated to either warm or cold temperatures indicate that this mechanism is effective not only on evolutionary time scales but also during seasonal temperature acclimatization (Zakhartsev *et al.*, 2004). The pathways determining these conformations have to be elucidated. Because cold-adapted and temperate species differ with respect to ion regulation and intracellular pH (Pörtner *et al.*, 1998; Pörtner and Sartoris, 1999; Guynn *et al.*, 2002), differences in the cellular milieu possibly contribute to the establishment of different conformations, which remain stable even after isolation of the protein from its native environment.

As a general rule, cold-adapted enzymes should display high catalytic efficiency associated with low thermal stability. Therefore, they are thought to optimize their catalytic efficiency (k_{cat}/K_m) by increasing turnover number, (k_{cat}), decreasing K_m (= increasing substrate affinity), or changing both (Feller *et al.*, 1997; D'Amico *et al.*, 2002). The interdependence of catalytic properties and thermal stability has already been questioned. The comparison of the properties of LDH orthologues from gobiid fishes also suggested that the kinetics and the thermal stability of an enzyme can evolve independently, and that different regions of a protein may be involved in affecting these two types of change (Fields and Somero, 1997). Using site-directed mutagenesis, the effects of single amino acid substitutions were followed (Holland *et al.*, 1997). Some differences between orthologues appeared to account for the difference in K_m values, whereas other substitutions seemed to influence the thermal stability of the enzymes. Therefore, it was concluded that evolutionary adaptation of proteins to temperature may independently affect kinetic properties and thermal stability. In support of this notion, directed evolution experiments resulted in new proteins, which again displayed patterns contrasting with the postulated trends (i.e., higher enzymatic activity at low temperature combined with enhanced thermal stability) (D'Amico *et al.*, 2002). As a corollary, no relationship between habitat temperature and the temperature threshold of heat denaturation could be observed in studies of closely related species from different thermal environments. Because the denaturation temperature is in most cases far beyond ambient temperature and the thermal limit of the whole organism, it remains questionable whether changes in this parameter reflect any clear correlation to the level of adaptation to cold in fishes.

For closely related organisms, the turnover number, k_{cat}, of LDH was found increased, whereas substrate affinity decreased (K_m increased) in parallel (Graves and Somero, 1982; Fields and Somero, 1997, 1998; Holland *et al.*, 1997), so K_m of the orthologous enzymes remained more or less unchanged when compared at habitat temperature. Therefore, K_m appears as one of the conserved parameters in enzyme function (Hochachka and Somero, 1984, 2002). This observation may extend to the organelle level: Maintenance of apparent K_m for total ADP was observed in short-horned sculpin (*Myoxocephalus scorpius*) red muscle mitochondria at falling acclimation temperatures (Guderley and Johnston, 1996).

As delineated so far, structural modifications and their kinetic consequences have mostly been investigated in soluble proteins. The question remains open whether the primary structures of membrane proteins, which may be influenced by membrane fluidity and composition, are affected in similar ways by cold adaptation. Because information on the three-dimensional structures of membrane proteins is very limited, conclusions are restricted to the analysis of primary structures. In zoarcid fish, the sequences of several membrane proteins, such as mitochondrial uncoupling protein (UCP2), or cellular membrane Na^+/K^+-ATPase and Na^+/H^+-antiporter have been analyzed (M. Lucassen, F. Mark, and H. O. Pörtner, unpublished observations). Protein and DNA sequences were found to be highly conserved (by ~98–99%), and there are no substitutions in the predicted active sites. Patterns as observed in soluble proteins are not evident from these comparisons between confamilial species. Because membrane proteins are also affected by the conditions of the surrounding milieu like pH, ionic composition, and membrane fluidity/composition (see above), it remains to be established whether any differences in primary structure between temperate and cold stenothermal membrane proteins are related to different body temperatures. The validity of these statements is emphasized by findings that membrane composition in fish is not only shaped by temperature but also by the level of ambient salinity (Cordier *et al.*, 2002).

The effect of temperature on biochemical reaction velocities under conditions of substrate saturation is adequately described by the Arrhenius equation ($k = A_e^{Ea/RT}$). Accordingly, any decrease in temperature will induce an exponential decrease of reaction velocity to an extent, which depends on the value of the Arrhenius activation energy Ea, which largely mirrors the enthalpy term $\Delta H^{\ddagger}$ of the Gibb's free energy of activation $\Delta G^{\ddagger}$:

$$\Delta H^{\ddagger} = E_A - RT$$

R = universal gas constant ($8.31434 \ J \ mol^{-1}K^{-1}$); T = temperature (K)

$$\Delta G^{\ddagger} = \Delta H^{\ddagger} - T\Delta S^{\ddagger}$$

It was postulated as a general rule that activation enthalpy in cold-adapted enzymes like LDH is lowered by structural modifications to counterbalance the expected decrease in reaction velocity (k) during cooling (Hochachka and Somero, 1984; Marshall, 1997; D'Amico *et al.*, 2002).

However, a unifying trend that Ea is reduced in the cold could not be confirmed in every example investigated. Some enzymes displayed higher Ea values after cold adaptation (Pörtner *et al.*, 2000). Others, including some that contribute to the control of metabolic flux like COX (seen in subpolar populations of *Arenicola marina*) or energy-dependent functions that improve overall tissue function in the cold like Na^+/K^+-ATPase (seen in the Antarctic *P. brachycephalum*), demonstrated the expected reduction in Ea values compared with those found in their temperate conspecifics or confamilials (Sommer and Pörtner, 2002; Lucassen *et al.*, unpublished). The latter category also includes a drop in Ea of myofibrillar ATPase (Johnston *et al.*, 1975).

To explain these contrasting patterns, the hypothesis was developed that the level of Ea reflects a trade-off between the enzyme concentration required to minimize diffusional limitations and the required facilitation or restriction of flux through specific reactions or pathways (Pörtner et al., 2000). A drop in Ea should be found mainly when flux is to be cold compensated. This may predominantly be the case in equilibrium enzymes with a lower contribution to metabolic control, like in LDH (Hochachka and Somero, 1984). However, even for LDH, the previous assumption of a uniform picture does not hold. In notothenioid fishes (Marshall et al., 2000), the temperate species demonstrated the lowest Ea below the Arrhenius break temperature (found at about 25 °C and far beyond habitat temperature), whereas the three Antarctic species demonstrated intermediate; and the sub-Antarctic species, the highest values. Cold-adapted specimens of the Norwegian coastal cod (G. morhua) population also displayed elevated levels of Arrhenius activation energies for LDH rather than the expected decrease (Zakartsev et al., 2004).

Other enzymes with a low activity, like glyceraldehyde-phosphate-dehydrogenase (GAPDH), phosphofructokinase (PFK), and isocitrate dehydrogenase (IDH), also display an increase in Ea in the cold (Pörtner et al., 2000). Such an increment may occur for two reasons. First, a cold-induced shift occurs from anaerobic to aerobic metabolic pathways. The rise in activation enthalpy for GAPDH, PFK, and LDH would reflect an enhanced kinetic barrier for anaerobic glycolysis in the cold and support such a shift to aerobic metabolism. Second, cold compensation of flux associated with cold-induced changes in expression patterns and increasing enzyme levels may be counteracted by rising Ea. Rising enzyme levels with maintained or even reduced flux may be essential to overcome cold-induced limitations in the diffusive flux of intermediates.

In fact, according to Pörtner et al. (2000), the reduction in standard metabolism among cold-adapted stenotherms, in contrast to cold-adapted eurytherms, may be linked to reduced mitochondrial capacities and increased Arrhenius activation energies of mitochondrial oxygen demand, especially proton leakage, and of flux-limiting enzymes in metabolism like IDH. Such a high kinetic barrier may support low metabolic flux in cold-adapted stenotherms, despite mitochondrial proliferation. Although Ea of overall metabolism appears to be reduced in active winter-acclimated animals with cold-compensated SMRs (Pörtner, 2002a; Zakartsev et al., 2003), high Ea values in Antarctic species reflect a high temperature dependence of metabolism and, in consequence, reduced heat tolerance (i.e., cold stenothermy of the whole organism). The rising oxygen demand during warming is mirrored in enhanced rates and cost of ventilation and circulation, which can be alleviated by exposure to hyperoxia (Mark et al., 2002). Mechanistic links between the various levels of function and organization, however, are only just emerging and should be elaborated in future research.

As a corollary, the changing kinetic properties and thermodynamics of enzyme function with temperature adaptation will be relevant for the response of metabolic complexes, cells, tissues, and finally whole organisms to temperature. The particular environment (the cellular milieu and other interacting proteins and macromolecules) contributes to a large extent to the pattern of temperature adaptation of a single molecule. The specific and contrasting strategies observed emphasize that molecular modifications must be interpreted in light of their functional consequences from the enzyme to the whole-organism level. In light of whole-organism adaptations to cold, enzymes that share control in metabolic flux are the most suitable candidates for studying structural and kinetic modifications, including changes in activation enthalpy, and their role in cellular and whole-organism metabolic control.

V. TRADE-OFFS IN ENERGY BUDGETS AND FUNCTIONAL CAPACITIES: ECOLOGICAL IMPLICATIONS

The previous sections have elaborated the molecular, organellar, cellular, and organismic aspects of metabolic performance with relevance to temperature adaptation and limitation. As a general pattern, the compensatory upregulation of enzyme capacities, of transmembrane ion exchange, and of mitochondrial functions suggests a high primary cost of cold adaptation. Low rates of standard metabolism in stenothermal polar, especially Antarctic, fishes indicate that this cost has secondarily been overcome despite maintenance of many structural characteristics of high energy turnover, aerobic systems. Building on this insight, this final section adds an analysis of the effects of cold adaptation on higher functions, other than exercise capacity, which are also relevant in determining survival, performance, and success of a species at the ecosystem level. Studies of higher functions may provide further access to overarching driving forces shaping adaptation to polar environments. The present analysis includes growth performance, reproduction strategies, and larval development. For growth performance, the underlying characteristics of protein synthesis and its capacity are addressed.

A. Protein Synthesis, Growth, and Standard Metabolism

The question arises how the level of energy turnover relates to growth and reproduction, as well as developmental traits seen in the permanent cold. At first glance, extreme seasonality of food availability in polar environments may enforce energy savings that free energy for the fueling of growth. However, a comparison of Antarctic stenotherms and sub-Arctic eurytherms suggests that these relationships may not fully explain the patterns observed. Temperature-induced trade-offs in energy budget may occur at the cellular and organismic levels and thereby support energy allocation to growth. In fact, available data indicate that a unifying principle may exist in which among related species, those with a lower level of SMR and, thus, lower baseline costs profit from higher growth rates. This principle may be enforced by permanently low ambient temperatures, as in the marine Antarctic. Higher growth rates may be advantageous because a larger body size may improve winter survival in juvenile fish, as shown in fish from temperate latitudes (Post and Parkinson, 2001).

Previous work on eurythermal cod (*G. morhua*) and eelpout (*Z. viviparus*) suggests that the rising cost of maintenance in eurytherms that are cold acclimated or cold adapted in a latitudinal cline occurs at the expense of a reduction in temperature-specific growth performance or in reproduction (Pörtner *et al.*, 2001). At the same time, the data available for cold-adapted eurytherms indicate that cold adaptation supports enhanced levels of tissue energy turnover suitable to support enhanced exercise capacity (Pörtner, 2002b; Figure 3.1). A trade-off between energy allocation to exercise capacity and growth performance may, thus, cause a shift to lower growth performance in sub-Arctic compared with temperate populations.

Such trade-offs also become apparent in global comparisons of Arctic and Antarctic fish, as well as within Antarctic stenotherms. The common ecological index of growth performance P (Pauly, 1979) describes the growth rate relative to how fast fish attain final size and their maximum achievable weight (Figure 3.12). When using this index, Dorrien (1993) found very low growth performance values of 0–1.5 in selected Arctic benthic fish species, whereas most stenothermal Antarctic notothenioids display *P* values of 1–3 (Kock and Everson, 1998), with the exception of *T. scotti* from the Weddell Sea (*P* = 0.71) and *D. mawsoni* from the Ross Sea (*P* = 4.01) (La Mesa and Vacchi, 2001). This overall pattern is in line with the suggested trade-off between SMR and growth performance, as well as the dependence of SMR on Arctic versus Antarctic temperature stability.

The data compiled by La Mesa and Vacchi (2001) suggest that such a trade-off may also be operative among Antarctic notothenioid fish species adapted to different ecological niches (Figure 3.12). The notothenioids are predominantly characterized by a total body length less than 45 cm, and most of them display benthic lifestyles like their common ancestor. Only a few notothenioids such as *P. antarcticum* (Boulenger) and *Pagothenia borchgrevinki* (Boulenger) have successfully invaded the midwater. An overview of age and growth in high Antarctic notothenioid fish revealed an overall increase of growth performance from pelagic to benthic lifestyles, again with the two

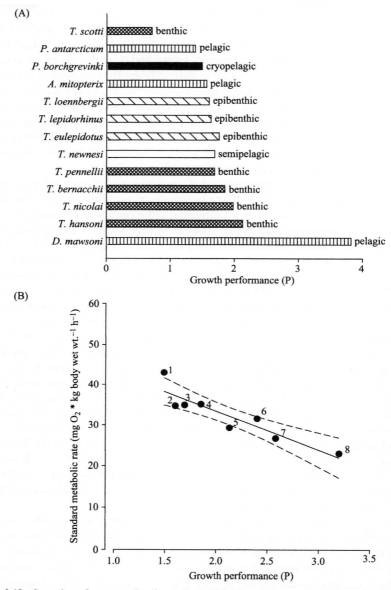

Fig. 3.12. Growth performance (P) of notothenioid fish in relation to (A) their mode of life in the high Antarctic zone and (B) their standard metabolic rate (SMR) in the high Antarctic and lesser Antarctic/sub-Antarctic zone. According to Pauly (1979), P ($= \log K + \log W_\infty$) describes the growth rate at the point of inflection of the size (mass) growth curve (K, year^{-1}, is a measure of how fast fish attain their final size, and W_∞ is the infinite weight of the fish, in grams).

exceptions, *T. scotti* and *D. mawsoni*. This may relate to the maximum body mass (W_∞) of the two species, which is very low for *T. scotti* (37 g) and exceptionally high for *D. mawsoni* (>100 kg). Along with the other pelagic species, *D. mawsoni* exhibits the lowest K values (La Mesa and Vacchi, 2001, K, year^{-1}, is a measure of how fast fish attain their final size). Finally, a comparison of *P* values with the levels of SMR in Figure 3.12 confirms that low growth and high SMR go hand in hand in Antarctic notothenioid fish.

These findings contrast the more intuitive explanation that the slower growth of Antarctic pelagic fish is a result of their energy-conserving mode of life and, thus, is directly affected by energy savings (Kunzmann and Zimmermann, 1992). Extensive energy savings in pelagic Antarctic fish include the use of lipids rather than an energetically more expensive swim bladder to reach near neutral buoyancy (Eastman and De Vries, 1982). Energy savings are also reflected in reduced ion exchange activities and lower costs of oxygen distribution. Here, energy savings mainly result from the shift to enhanced diffusive oxygen supply at shallower gradients of intracorporal PO_2 (see Section II.E). Furthermore, morphological features such as reduction of bones (Andriashev, 1985; Eastman 1985), reduced calcification of the skeleton (De Vries and Eastman, 1981), extensive lipid deposits (Eastman and De Vries, 1982), and morphometric parameters such as gill raker, fin, and body sizes (Ekau, 1988) are energy-saving adaptations to a midwater lifestyle. Nevertheless, despite these extensive energy savings, pelagic lifestyles seem more costly than benthic lifestyles as indicated by a higher SMR (Figure 3.12) (Priede, 1985; Dorrien, 1993; Zimmermann and Hubold, 1998). It may be concluded that pelagic fish species, because they are strictly tied to primary production and the pelagic food chain, experience more extreme seasonality of light and food availability. This may reflect a challenge to maximize energy intake during the short Antarctic summer.

At the same time, energy savings are likely an overarching principle of marine ectothermic life in the Antarctic. The extensive efforts of pelagic fish

---

(A) Graph redrawn using *P* values compiled by La Mesa and Vacchi (2001). (B) *P* values by La Mesa and Vacchi (2001) and Hubold (1991). SMR: Holeton (1970), Forster *et al.* (1987), Johnston *et al.* (1991b), Johnston and Battram (1993), Macdonald *et al.* (1988), Wells (1987) in Montgomery and Wells (1993), Morris and North (1984) in Thurston and Gehrke (1993), Zimmermann and Hubold (1998), Steffensen (2002). SMR values correspond to resting or SMRs, all expressed as mg O_2 kg body weight mass^{-1} h^{-1} calculated for a 100-g fish at temperatures close to freezing. SMR values of one species from various citations were summarized to one mean. Linear regression: $y = -9.6x + 52.6$ ($r = 0.91$); dashed lines indicate the 95% confidence interval. High-Antarctic species are (1) *Pagothenia borchgrevinki* (cryopelagic, active under sea ice), (2) *Trematomus loennbergii* (epibenthic), (3) *Trematomus pennellii* (benthic), (4) *Trematomus bernacchii* (benthic), (5) *Trematomus hansoni* (benthic). Lesser to sub-Antarctic species are (6) *Notothenia neglecta* (benthopelagic), (7) *Notothenia gibberifrons* (benthopelagic), (8) *Notothenia rossii* (benthopelagic).

to compensate for their more expensive mode of life then appear as an extreme example of this general constraint. In a counterintuitive way, the energy-saving strategies in the pelagic may not only result from seasonal food availability, but also from the obligatory trade-off between growth performance and SMR identified here. Among Antarctic fish, pelagic forms, thus, strive to minimize energy expenditures to achieve lower SMRs and, thus, higher growth rates, but are evidently not as successful in this strategy as benthic fish. Nonetheless, because of their higher degree of stenothermy in more stable climates, Antarctic fish display lower baseline cost and, thus, SMR than Arctic fish, as well as a larger fraction of energy remains for growth. The fact that this trade-off is operative at low to moderate exercise capacities and low levels of spontaneous activity, and that it is paralleled by further substantial energy savings (see above), emphasizes the confinement of polar marine ectotherms to an inexpensive life in the slow lane.

In a comparison of low versus high-energy turnover life forms, the patterns developed here for cold-adapted low-activity fish contrast with observations at the high end of the performance spectrum in fish and in coleoid cephalopods and in endotherms, in which the level of SMR correlates positively with growth performance (Lee, 1994; Brill, 1996). Excess resource availability may be a precondition for this contrasting strategy of maximizing growth by maximizing SMR.

As a note of caution, the trade-offs between growth rates and metabolic costs depicted here for cold-adapted low-activity fish may not easily explain all growth patterns and energy allocation strategies observed. These trade-offs were elaborated in eastern Atlantic cod populations and for Antarctic fish. They also match observations of reduced growth capacities at increasing latitudes in anchoveta along the coast of Chile (Castro *et al.*, 2001). For limnic large-mouth bass, modeling of energy allocation strategies also revealed that smaller size for age occurs in the north than in the south, consistent with field patterns for this species (Garvey and Marschall, 2003). Further examples exist among freshwater fish. However, such physiological differentiation according to latitude may not be universal for all species. For example, growth capacity may not clearly differ between northwestern cod populations in a latitudinal cline (Purchase and Brown, 2001), despite largely different temperatures. The growth model proposed here also contrasts higher growth capacities in northern than in southern populations of silverside (*Menidia menidia*) (Yamahira and Conover, 2002), which were suggested to compensate for the short growing season at high latitudes but would require lower metabolic rates than expected in a cold-adapted eurytherm. However, the metabolic and energetic background of such contrasting strategies has not been clearly elaborated. Explanations of potential differences between species or population-specific strategies to overcome thermal and seasonal constraints require consideration of temperature means and climate variability, as well as of the response to seasonal resource limitations (Hurst and Conover, 2003). The latter may define seasonal changes in energy allocation (e.g., fattening strategies in the fall) and, for Atlantic cod, may play a larger role on the western than the eastern side of the Atlantic.

Growth is a complex process that is subject to controls and limitations at various levels (Mommsen, 2001). In principle, however, growth and cellular functioning in tissues are closely related to protein synthesis and breakdown. Low rates of oxygen demand at low temperature, therefore, involve reduced protein turnover, supported by reduced synthesis of highly complex protein systems like Hb, Mb, ion pumps, and possibly other proteins of basal metabolism (Clarke, 1991). When excess energy and substrate are available, the protein synthesis machinery will be used for growth.

In light of energy savings, protein synthesis should be energy efficient to safeguard cellular functioning at the very low "operating temperatures" of Antarctica. Evidence exists that the cost of protein synthesis is unchanged regardless of temperature (Storch and Pörtner, 2003). At the same time, protein synthesis capacity should not be limiting to allow for maximized growth rates. With respect to capacity, increased RNA levels and elevated tissue RNA/protein ratios were interpreted to compensate for a cold-induced reduction in RNA translational efficiency *in vivo* (i.e., the capacity to synthesize protein per quantum RNA) (Whiteley *et al.*, 1996; Marsh *et al.*, 2001; Robertson *et al.*, 2001; Fraser *et al.*, 2002). Increased total RNA content in cold-adapted ectotherms (see also Figures 3.8 and 3.10) may simply be the result of low RNA turnover rates in the cold, leading to elevated steady-state RNA levels at no extra cost. This view is supported by the reduction of RNA levels seen in white muscle of 5 °C versus 0 °C acclimated Antarctic eelpout (Storch *et al.*, 2004). On top of passively elevated RNA levels, the RNA translation apparatus is cold compensated with enhanced catalytic efficiencies (Storch *et al.*, 2003, 2004), thereby supporting a high cost efficiency and high capacity. Largely cold-compensated protein synthesis capacities are complemented by cold-compensated cellular proliferation capacities in Antarctic fish (Brodeur *et al.*, 2003). Both processes are crucial to growth, and their cold compensation is in line with the suggested trend in polar fish to maximize growth performance at low baseline metabolic costs.

These patterns match current understanding that although annual growth of many polar species is low, maximum growth rates in Antarctic stenotherms can reach levels comparable to those found in the lower range for temperate species (Brey and Clarke, 1993; Arntz *et al.*, 1994; Peck, 2002). Kock and Everson (1998) compiled evidence that growth performance estimated by P is comparable in a number of Antarctic notothenioids and channichthyids to ecologically similar species from boreal and temperate waters. Stenothermal Antarctic eelpout (*P. brachycephalum*) also reach the same growth performance as a North Sea species (*Z. viviparus*) (Brodte, 2001). Not only among fish, but also among high-latitude Arctic and Antarctic invertebrates, low annual growth rates coincide with levels of growth performance as high as seen in species from lower latitudes (Dahm, 1999; Poltermann, 2000; Bluhm, 2001).

Overall, it appears that low standard metabolism at stable polar temperatures is a precondition for similar levels of growth performance at low and high latitudes, supported by cold-compensated capacities of protein synthesis as found in white muscle and gills of Antarctic scallops and eelpouts (Storch *et al.*, 2003, 2005) (Figure 3.13). Although growth capacity is cold-compensated in Antarctic stenotherms, this capacity may only be exploited during the short Antarctic summer. A capacity for the thermal adjustment of this process is evident in Antarctic eelpout (*P. brachycephalum*), in which protein synthesis capacity decreased during long-term acclimation from 0 to 5 °C (Storch *et al.*, 2005).

Finally, and as a corollary, this discussion allows us to reexamine three largely different explanations of slow annual growth rates at high latitudes: (1) the rate-limiting effect of low temperature, (2) seasonal resource limitation, and (3) rising costs of maintenance at the expense of a reduction in growth and in reproduction.

1. Current evidence presented above indicates that the ribosomal machinery has achieved a large degree of compensation for the rate-limiting effects of low temperature. Future studies are needed to quantify the extent of this compensation. However, it already appears that the rate-limiting effect of low temperature alone does not explain slow annual growth.

2. Seasonal resource limitation invokes no effect of temperature and has been a focal point of numerous studies regarding growth and seasonality at high latitudes. Both Clarke and North (1991) and Clarke and Peck (1991) suggested that although low temperature is likely to impose some general constraints on growth due to its rate-depressing effect on physiological systems, it is seasonal food availability and not temperature that normally limits growth rate. Meanwhile, hibernation patterns have been identified, in both invertebrates and fish, associated with low energy turnover, low protein synthesis capacities, and suspended growth during winter (Ashford and White, 1995; Lehtonen, 1996; North, 1998; North *et al.*, 1998; Brockington, 2001; Brockington and Peck, 2001). The environmental triggers of hibernation are unclear, but some examples indicate that hibernation occurs even with food being available. Hibernation may be due instead to a

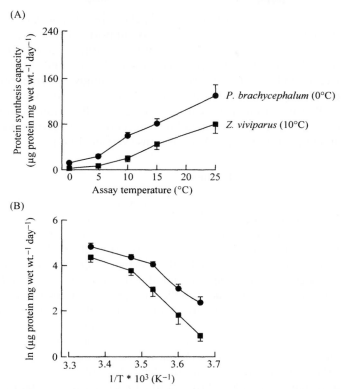

Fig. 3.13. *In vitro* protein synthesis capacities in white muscle of Antarctic eelpout *Pachycara brachycephalum* acclimated to 0 °C and of temperate eelpout *Zoarces viviparus* acclimated to 10 °C, measured in tissue lysates at various assay temperatures. Arrhenius plots of protein synthesis capacities display a significant break at the intersection of linear ($r = 0.99$ in the temperature range 10–25 °C) and second-order polynomial regressions ($r = 1.0$ in the temperature range 0–10 °C) for *P. brachycephalum* and as a linear regression ($r = 0.99$ in the temperature range 0–15 °C) for protein synthesis capacities in both tissues of *Z. viviparus* (values are means ±SE, *P. brachycephalum*, n = 5; *Z. viviparus*, n = 3). (based on data by Storch *et al.*, 2005.)

reduction of light and, to a minor extent, temperature. The use of hibernation strategies would, thus, contribute to explain the low average growth of Antarctic ectotherms.

3. Finally, the energy budget hypothesis discussed here and the suggested temperature-dependent trade-offs between lifestyle, exercise, and growth performance most consistently explain the patterns observed in cold-adapted stenotherms versus eurytherms. As a consequence of metabolic cold adaptation (see Section I) and of enhanced activity

levels, cold-adapted eurytherms and pelagic fish display elevated levels of tissue energy turnover suitable to support exercise capacity and spontaneous activity at the expense of reduced growth (Pörtner, 2002b) (Figures 3.1 and 3.12). The available data and the discussion above are in line with this argument. Accordingly, the fractions of the energy budget allocated to growth and to other components of SMR, as well as the mechanisms causing changes in allocation, need identification and quantification in polar stenotherms and eurytherms, as well as in temperate to tropical eurytherms (Wieser, 1989).

B. Reproduction and Development

Similar trade-offs in energy budgets may also operate at the level of fecundity and organismic development. Reduced fecundity was observed in sub-Arctic populations of cod (*G. morhua*) and eelpout (*Z. viviparus*) and was interpreted to reflect low temperatures, as well as elevated baseline metabolic costs in the eurythermal cold (Pörtner *et al.*, 2001). As a general hypothesis, Brey (1995) suggested that the fraction of assimilated energy used in reproduction (reproductive efficiency [RE]) remains independent of temperature. It seems coherent with this hypothesis that most Antarctic fish exhibit delayed maturity and do not start to reproduce before they reach 55–80% of their maximum observed body length, a size large enough to exploit abundant food resources (Kock and Everson, 1998; La Mesa and Vacci, 2001). Reproductive output and fecundity, thus, fall in the cold, as evidenced in many examples (Arntz *et al.*, 1994). To what extent the energetics of cold eurythermy versus stenothermy influences these patterns is unclear.

Drastically slowed larval development in Antarctic marine ectotherms compared with temperate species is a general rule with no known exception (Peck, 2002). This holds for fish (Clarke and North, 1991; White and Burren, 1992; Outram and Loeb, 1995) and invertebrates, such as nemerteans (Peck, 1993), brachiopods (Peck and Robinson, 1994), echinoderms, or crustaceans (Klages, 1993; Arntz *et al.*, 1994).

Data on developmental rates are again more abundant for invertebrates. The development from brood to hatch can last up to 25 months, as reported for the gastropods *Torellia mirabilis* and *Trophon cf. scotianus* (Hain and Arnaud, 1992). In Antarctic sea urchin embryos, protein turnover rates were found to be equivalent to those in temperate zone sea urchin embryos (Marsh *et al.*, 2001), but developmental rates were low (Bosch *et al.*, 1987; Stanwell-Smith and Peck; 1998; Marsh *et al.*, 1999). Hoegh-Guldberg and Pearse (1995) argued that rates of echinoderm development are near their maximum for a given ambient temperature at any site worldwide, and that temperature is the main regulator of development rate. It, thus, seems that cold-compensated protein synthesis capacities are exploited during growth but not during development.

Notothenioid larvae covering a wide range of body lengths and developmental stages have been observed with yolk reserves maintained, indicating that the level of yolk resorption may relate to early feeding success and food uptake or availability. Exogenous feeding before complete yolk resorption has been observed in some species (*Chionodraco rastrospinosus* and *Lepidonotothen larseni*) and is possibly characteristic of all notothenioids (Laman and Loeb, 1994). The relative utilization of available energy resources may be strongly balanced by environmental conditions and their variability.

Extended energy stores likely enhance larval survival at low temperatures (Marsh *et al.*, 1999; Marsh and Manahan, 2000). Whether limited food availability, reduced mobility, or both are responsible for the use of this strategy is unclear. Relative to their metabolic rates, Antarctic echinoderm larvae possess larger energy stores and planktotrophic larvae have higher nutrient transport capacities for amino acids dissolved in seawater when compared with larvae from temperate regions.

The ability of marine invertebrate larvae to delay metamorphosis even after becoming physiologically competent to metamorphose has been well documented for larvae in various phyla (Pechenik, 1990). Delaying metamorphosis has been found to decrease juvenile growth rates significantly and often dramatically in various species, particularly those that produce lecithotrophic larvae, which is very common in polar regions (Pechenik *et al.*, 1998; Pechenik, 1999). Marshall *et al.* (2003) suggested that larval maintenance, swimming, and metamorphosis all use energy from a common pool, and an increased allocation of resources to maintenance or swimming occurs at the expense of post-larval performance.

In light of the "trade-offs in energy budget" hypothesis developed above, uncompensated patterns of delay in larval development and the use of predominantly lecithotrophic larvae by cold stenotherms would become explainable. The excessive aerobic design of metabolism in fish larvae and the reduced performance levels in cold-adapted larvae indicate that the trade-offs, such as those between growth rate, SMR, and muscular activity levels seen in the adults, are even more profound in the larvae (see Section II.C). Because of these obligatory trade-offs, saving energy for growth is even more important for the larvae, with the following consequences: First, lecithotrophic passive larvae save energy by minimizing feeding and, thus, muscular activity. In the adult, however, enhanced storage supplies for lecithotrophic larvae occur at the expense of reduced and delayed fecundity. Second, larval development involves additional costs on top of those of maintenance, swimming, and growth. Energy savings may contribute to delay development even further in the cold. The general pattern emerges that although protein synthesis and growth capacities appear correlated (see above), the rates of higher functions like maturity, reproduction, fecundity, larval development, and spontaneous activity relevant for lifestyle are all reduced in stenotherms for the sake of freeing energy for maximized seasonal growth at minimal rates of overall energy turnover. As a corollary, hierarchies of energy allocation likely exist and clearly need elaboration.

VI. SUMMARY

Marine life in the polar, especially Antarctic, cold is characterized by a low to moderate pace at permanently low temperatures. This chapter sets out to identify the temperature-dependent trade-offs in the metabolic physiology, biochemistry, and functional genomics of cold adaptation in marine fish and to analyze the implications of these trade-offs for thermal tolerance and performance characteristics at organismal and ecological levels. During cold adaptation, shortage of aerobic capacity at an organismic level is compensated for by enhancing the functional capacity of cells and tissues through mitochondrial proliferation, an upregulation of aerobic enzyme activities and of cellular energy stores (lipids), and associated molecular and membrane adjustments. These compensatory adjustments parallel a downward shift of thermal tolerance windows and, in Antarctic fish, a reduction in baseline energy demands in the cold. These mechanisms also aid recovery rates from exhaustive exercise but are implemented at the expense of reduced anaerobic capacity, seen in moderately active life forms. Significant anaerobic capacity is found only at the low end of the activity spectrum (e.g., among benthic eelpout [Zoarcidae]).

Especially in demersal to pelagic Antarctic fishes, among notothenioids, low SMR and enhanced lipid stores, as well as a preference for lipid catabolism, indicate a high energy efficiency of aerobic metabolism at high ambient oxygen availability. In the lower range of performance levels, the same sustainable swimming velocities are achieved in temperate and Antarctic fish, supported by largely elevated numbers of mitochondria in red muscles of Antarctic fish. The accumulation of mitochondrial networks and lipid substrates supports enhanced diffusional oxygen supply and, thereby, further energy savings. ATP synthesis capacities of mitochondria are reduced and the associated drop in mitochondrial proton leakage contributes to low resting oxygen demand.

It appears that in a trade-off between space adopted by myofilaments, mitochondria, and sarcoplasmic reticulum, especially the need for more mitochondria for maintained functional capacity in the cold is a major constraint on maximum scope for activity and is linked to lower levels of muscular force per muscle cross-sectional area. Similarly, the capacity of the heart to maintain aerobic scope within the thermal tolerance window is limited, because of the space constraints within cardiomyocytes and the limits on the relative size of the heart. These trends in skeletal and cardiac muscle are synergistic and reflect the limitation of aerobic scope of the whole organism. These constraints are operative, especially in larvae and limit their level of activity and energy expenditure even more than in adults. Furthermore, limited functional capacity at reduced SMR contributes to reduced costs of cold tolerance but, as a trade-off, contributes to the narrow windows of thermal tolerance and low heat tolerance of Antarctic stenotherms.

Molecular studies indicate a selective upregulation of aerobic enzyme capacities (e.g., COX or CS) or transmembrane ion exchange (e.g., Na^+/K^+-ATPase) by transcriptional, translational, and likely, posttranslational control. Kinetic properties of enzyme proteins (especially LDH) are discussed in light of structural modifications at the protein level, with or without slight changes in the primary sequence of the molecule. Either downward or upward shifts of activation enthalpies may reflect their specific role in either facilitation (for functional capacity) or restriction of flux (for control of metabolic costs) through specific metabolic reactions and pathways.

Despite overall energy savings, protein synthesis capacities remain cold-compensated, supporting peak summer growth rates in Antarctic species comparable to those seen in temperate species. As a trade-off, baseline metabolic costs and, thus, SMR are reduced for the sake of a maximization of growth rates, especially in benthic fish. Energy-saving strategies are excessive in more active pelagic fish, in which the higher cost of living still causes lower growth at elevated SMRs. Growth maximization may also occur at the expense of reduced energy allocation to other higher functions. Overall, higher functions like exercise capacity and larval development display lower rates in the permanent cold, likely because energy efficiency needs to be enhanced and a maximum of energy allocated to individual growth and to storage supplies for lecithotrophic larvae. These trade-offs and the overproportional buildup of aerobic energy production machinery at low performance capacity may be a general constraint for all faunal elements of polar, especially Antarctic ecosystems, thereby explaining why high-performance (ectotherm) life is rare, if it exists at all, in the permanent cold.

REFERENCES

Acierno, R., Agnisola, C., Tota, B., and Sidell, B. D. (1997). Myoglobin enhances cardiac performance in antarctic icefish species that express the protein. *Am. J. Physiol.* **273,** R100–R106.

Altringham, J. D., and Johnston, I. A. (1985). Effects of phosphate on the contractile properties of fast and slow muscle fibres from an Antarctic fish. *J. Physiol.* **368**, 491–500.

Andriashev, A. P. (1985). A general review of the Antarctic bottom fish fauna. *In* "Proceedings of the 5th Cong. Eur. Ichthyol, Stockholm 1985" (Kullander, S. O., and Fernholm, B., Eds.), pp. 357–372. Department of Vertebrate Zoology, Museum of Natural History, Stockholm.

Angilletta, M. J., Niewiarowski, P. H., and Navas, C. A. (2002). The evolution of thermal physiology in ectotherms. *J. Therm. Biol.* **27**, 249–268.

Archer, S. D., and Johnston, I. A. (1989). Kinematics of labriform and subcrangiform swimming in the Antarctic fish. *Notothenia neglecta. J. Exp. Biol.* **143**, 195–210.

Archer, S. D., and Johnston, I. A. (1991). Density of cristae and distribution of mitochondria in the slow muscle fibers of Antarctic fish. *Physiol. Zool.* **64**, 242–258.

Arntz, W. E., Brey, T., and Gallardo, V. A. (1994). Antarctic zoobenthos. *Oceanogr. Mar. Biol. Ann. Rev.* **32**, 241–304.

Ashford, J. R., and White, M. G. (1995). An annual profile of growth in the otoliths of immature *Notothenia coriiceps* Richardson in relation to the environment at Signy Island, Antarctica. *Antarct. Sci.* **7**, 345–349.

Askeq, E. W., Dohm, G. L., Doub, W. H., Jr., Huston, R. L., and Van Natta, P. A. (1975). Lipogenesis and glyceride synthesis in the rat: Response to diet and exercise. *J. Nutr.* **105**, 190–199.

Axelsson, M., Agnisola, C., Nilsson, S., and Tota, B. (1998). Fish cardio-circulatory function in the cold. *In* "Cold Ocean Physiology" (Pörtner, H. O., and Playle, R., Eds.), pp. 327–364. Cambridge University Press, Cambridge.

Battersby, B. J., and Moyes, C. D. (1998). Influence of acclimation temperature on mitochondrial DNA, RNA, and enzymes in skeletal muscle. *Am. J. Physiol.* **275**, R905–R912.

Blier, P., and Guderley, H. (1988). Metabolic responses to cold acclimation in the swimming musculature of lake whitefish, *Coregonus clupeaformis. J. Exp. Zool.* **246**, 244–252.

Bluhm, B. (2001). Age determination in polar crustacea using the autofluorescent pigment lipofuscin. *Ber. Polarforsch./Rep. Polar Res.* **382**, 1–127.

Blumenschein, T. M. A., Gillis, T. E., Tibbits, G. F., and Sykes, B. D. (2004). Effect of temperature on the structure of trout troponin C. *Biochem.* **43**, 4955–4963.

Bock, C., Sartoris, F. J., Wittig, R. M., and Pörtner, H. O. (2001). Temperature-dependent pH regulation in stenothermal Antarctic and eurythermal temperate eelpout (Zoarcidae): An *in*-vivo NMR study. *Polar Biol.* **24**, 869–874.

Bosch, I., Beauchamp, K. A., Steele, M. E., and Pearse, J. S. (1987). Development, metamorphosis, and seasonal abundance of embryos and larvae of the antarctic sea urchin *Sterechinus neumayeri. Biol. Bull.* **173**, 126–135.

Boutilier, R. G., and St-Pierre, J. (2002). Adaptive plasticity of skeletal muscle energetics in hibernating frogs: Mitochondrial proton leak during metabolic depression. *J. Exp. Biol.* **205**, 2287–2296.

Brand, M. D., Steverding, D., Kadenbach, B., Stevenson, P. M., and Hafner, R. P. (1992). The mechanism of the increase in mitochondrial proton permeability induced by thyroid hormones. *J. Biochem.* **206**, 775–781.

Brett, J. R. (1964). The respiratory metabolism and swimming performance of young sockeye salmon. *J. Fish. Res. Bd. Can.* **21**, 1183–1226.

Brey, T. (1995). Temperature and reproductive metabolism in macrobenthic populations. *Mar. Ecol. Prog. Ser.* **125**, 87–93.

Brey, T., and Clarke, A. (1993). Population dynamics of marine benthic invertebrates in Antarctic and Subantarctic environments: Are there unique adaptations? *Antarct. Sci.* **5**, 253–266.

Brill, R. W. (1996). Selective advantages conferred by the high performance physiology of tunas, billfishes, and dolphin fish. *Comp. Biochem. Physiol.* **113A,** 3–15.

Brockington, S. (2001). The seasonal energetics of the Antarctic bivalve *Laternula elliptica* (King and Broderip) at Rothera Point, Adelaide Island. *Polar Biol.* **24,** 523–530.

Brockington, S., and Peck, L. S. (2001). Seasonality of respiration and ammonium excretion in the Antarctic echinoid *Sterechinus neumayeri*. *Mar. Ecol. Prog. Ser.* **219,** 151–168.

Brodeur, J. C., Calvo, J., Clarke, A., and Johnston, I. A. (2003). Myogenic cell cycle duration in *Harpagifer* species with sub-Antarctic and Antarctic distributions: Evidence for cold compensation. *J. Exp. Biol.* **206,** 1011–1016.

Brodte, E. (2001). Wachstum und Fruchtbarkeit der Aalmutterarten *Zoarces viviparus* (L) und *Pachycara brachycephalum* (Pappenheim) aus unterschiedlichen klimatischen Regionen Diploma Thesis, University of Bremen.

Brooks, G. A. (2002). Lactate shuttles in nature. *Biochem. Soc. Transact.* **30,** 258–264.

Burdon, R. H. (1987). Thermotolerance and the heat shock proteins. *In* "Temperature and Animal Cells" (Bowler, K., and Fuller, B. J., Eds.), Symposia of the Society of Experimental Biology, **21,** pp. 269–283.

Bushnell, P. G., Steffensen, J. F., Schurmann, H., and Jones, D. R. (1994). Exercise metabolism in two species of cod in Arctic waters. *Polar Biol.* **14,** 43–48.

Castro, L. R., Llanos, A., Blanco, J., Tarifeño, E., Escribano, R., and Landaeta, M. (2001). Influence of latitude variations in spawning habitat characteristics on the early life history traits of the anchoveta, *Engraulis ringens*, off northern and central Chile. *GLOBEC Report* **16,** 42–45.

Claireaux, G., Webber, D. M., Legardere, J.-P., and Kerr, S. R. (2000). Influence of water temperature and oxygenation on the aerobic metabolic scope of Atlantic cod (*Gadus morhua*). *J. Sea Res.* **44,** 257–265.

Clarke, A. (1991). What is cold adaptation and how should we measure it? *Am. Zool.* **31,** 81–92.

Clarke, A., and Peck, L. S. (1991). The physiology of polar marine zooplankton. *Polar Res.* **10,** 355–369.

Clarke, A., and North, A. W. (1991). Is the growth of polar fish limited by temperature? *In* "Biology of Antarctic Fish" (di Priso, G., Maresca, B., and Tota, B., Eds.), pp. 54–69. Springer Verlag, Berlin, Heidelberg.

Cordier, M., Brichon, G., Weber, J. M., and Zwingelstein, G. (2002). Changes in the fatty acid composition of phospholipids in tissues of farmed sea bass (*Dicentrarchus labrax*) during an annual cycle. Roles of environmental temperature and salinity. *Comp. Biochem. Physiol.* **133B,** 281–288.

Cossins, A. R. (1994). Homeoviscous adaptation of biological membranes and its functional significance. *In* "Temperature Adaptation of Biological Membranes" (Cossins, A. R., Ed.), pp. 63–76. Portland Press, London.

Cossins, A. R., Bowler, K., and Prosser, C. L. (1981). Homeoviscous adaptation and its effects upon membrane-bound proteins. *J. Therm. Biol.* **6,** 183–187.

Cossins, A. R., Murray, P. A., Gracey, A. Y., Logue, J., Polley, S., Caddick, M., Brooks, S., Postle, T., and Maclean, N. (2002). The role of desaturases in cold-induced lipid restructuring. *Biochem. Soc. Trans.* **30,** 1082–1086.

Crawford, D. L., and Powers, D. A. (1989). Molecular basis of evolutionary adaptation at the lactate dehydrogenase-B locus in the fish *Fundulus heteroclitus*. *Proc. Natl. Acad. Sci. USA* **86,** 9365–9369.

Crawford, D. L., and Powers, D. A. (1992). Evolutionary adaptation to different thermal environments via transcriptional regulation. *Mol. Biol. Evol.* **9,** 806–813.

Crockett, E. L., and Sidell, B. D. (1990). Some pathways of energy metabolism are cold adapted in Antarctic fishes. *Physiol. Zool.* **63,** 472–488.

Dahm, C. (1999). Ophiuroids (Echinodermata) of southern Chile and the Antarctic: Taxonomy, biomass, diet and growth of dominant species. *Sci. Mar. (Barc.)* **63**, 427–432.

Dalla Via, J., Huber, M., Wieser, W., and Lackner, R. (1989). Temperature related responses of intermediary metabolism to forced exercise and recovery in juvenile *Rutilus rutilus* (L.) (Cyprinidae: Teleostei). *Physiol. Zool.* **62**, 964–976.

D'Amico, S., Claverie, P., Collins, T., Georlette, D., Gratia, E., Hoyoux, A., Meuwis, M. A., Feller, G., and Gerday, C. (2002). Molecular basis of cold adaptation. *Philos. Trans. R. Soc. Lond. B Biol. Sci.* **357**, 917–925.

Davison, W., Forster, M. E., Franklin, C. E., and Taylor, H. H. (1988). Recovery from exhausting exercise in an Antarctic Fish, *Pagothenia borchgrevinki. Polar Biol.* **8**, 167–171.

De Vries, A. L., and Eastman, J. T. (1981). Physiology and ecology of notothenioid fishes of the Ross Sea. *J. R. Soc. NZ* **11**, 329–340.

Dorrien, C. V. (1993). Ecology and respiration of selected Arctic benthic fish species. *Ber. Polarforsch./Rep. Polar Res.* **125**, 1–104.

Driedzic, W. R., and Hochachka, P. W. (1978). Metabolism in fish during exercise. "Fish Physiology," Vol. VII. Academic Press, New York.

Dunn, J. F., and Johnston, I. A. (1986). Metabolic constraints on burst-swimming in the Antarctic teleost *Notothenia neglecta. Mar. Biol.* **91**, 433–440.

Dunn, J. F., Archer, S. D., and Johnston, I. A. (1989). Muscle fiber types and metabolism in post-larval and adult stages of notothenioid fish. *Polar Biol.* **9**, 213–223.

Eastman, J. T. (1985). The evolution of neutrally buoyant notothenioid fishes: Their specializations and potential interactions in the Antarctic marine food web. *In* "Antarctic Nutrient Cycles and Food Webs" (Siegfried, W. R., Condy, P. R., and Laws, R. M., Eds.), pp. 430–436 (IV. SCAR Symposium on Antarctic Biology, Wilderness, South Africa, 1983). Springer-Verlag, New York.

Eastman, J. T., and De Vries, A. L. (1982). Buoyancy studies of notothenioid fishes in McMurdo Sound, Antarctica. *Copeia* **2**, 385–393.

Egginton, S., and Davison, W. (1998). Effects of environmental and experimental stress on Antarctic fishes. *In* "Cold Ocean Physiology" (Pörtner, H. O., and Playle, R., Eds.), pp. 299–326. Cambridge University Press, Cambridge.

Egginton, S., and Sidell, B. D. (1989). Thermal acclimation of subcellular structures of fish muscle. *Am. J. Physiol.* **256**, R1–R10.

Egginton, S., Cordiner, S., and Skilbeck, C. (2000). Thermal compensation of peripheral oxygen transport in skeletal muscle of seasonally acclimatized trout. *Am. J. Physiol.* **279**, R375–R388.

Egginton, S., Skilbeck, C., Hoofd, L., Calvo, J., and Johnston, I. A. (2002). Peripheral oxygen transport in skeletal muscle of Antarctic and sub-Antarctic notothenioid fish. *J. Exp. Biol.* **205**, 769–779.

Ekau, W. (1988). Ecomorphology of nototheniid fish from the Weddell Sea, Antarctica. *Ber. Polarforsch./Rep. Polar Res.* **51**, 1–140.

Else, P. L., and Wu, B. J. (1999). What role for membranes in determining the high sodium pump molecular activity of mammals compared to ectotherms? *J. Comp. Physiol.* **169B**, 296–302.

Farkas, T., Kitajka, K., Fodor, E., Csengeri, I., Lahdes, E., Yeo, Y. K., Krasznai, Z., and Halver, J. E. (2000). Docosahexanoic acid–containing phospholipid molecular species in brains of vertebrates. *Proc. Natl. Acad. Sci. USA* **97**, 6362–6366.

Farrell, A. P. (1997). Effects of temperature on cardiovascular performance. *In* "Global Warming: Implications for Freshwater and Marine Fish" (Wood, C. M., and McDonald, D. G., Eds.), pp. 135–158. Cambridge University Press, Cambridge.

Feder, M. E., and Hofmann, G. E. (1999). Heat shock proteins, molecular chaperones and the stress response: Evolutionary and ecological physiology. *Annu. Rev. Physiol.* **61**, 243–282.

Feller, G., Arpigny, J. L., Narinx, E., and Gerday, C. (1997). Molecular adaptations of enzymes from psychrophilic organisms. *Comp. Biochem. Physiol.* **118A**, 495–499.

Fernandez, D. A., Calvo, J., Franklin, C. E., and Johnston, I. A. (2000). Muscle fibre types and size distribution in sub-Antarctic notothenioid fishes. *J. Fish Biol.* **56**, 1295–1311.

Fields, P., and Somero, G. (1997). Amino acid sequence differences cannot fully explain interspecific variation in thermal sensitivities of gobiid fish A4-lactate dehydrogenases (A4-LDHs). *J. Exp. Biol.* **200**, 1839–1850.

Fields, P. A., and Somero, G. N. (1998). Hot spots in cold adaptation: localized increases in conformational flexibility in lactate dehydrogenase A4 orthologs of Antarctic notothenioid fishes. *Proc. Natl. Acad. Sci. USA* **95**, 11476–11481.

Fields, P. A., Kim, Y. S., Carpenter, J. F., and Somero, G. N. (2002). Temperature adaptation in *Gillichthys* (Teleostii: Gobiidae) A(4)-lactate dehydrogenases: Identical primary structures produce subtly different conformations. *J. Exp. Biol.* **205**, 1293–1303.

Forster, M. E., Franklin, C. E., Taylor, H. H., and Davison, W. (1987). The aerobic scope of an Antarctic fish *Pagothenia borchgrevinki* and its significance for metabolic cold adaptation. *Polar Biol.* **8**, 155–159.

Franklin, C. E., Johnston, I. A., Batty, R. S., and Yin, M. C. (1996). Metabolic recovery in herring larvae following strenuous activity. *J. Fish Biol.* **48**, 207–216.

Fraser, K. P.P, Clarke, A., and Peck, L. S. (2002). Low-temperature protein metabolism: Seasonal changes in protein synthesis and RNA dynamics in the Antarctic limpet *Nacella concinna* Strebel 1908. *J. Exp. Biol.* **205**, 3077–3086.

Frederich, M., and Pörtner, H. O. (2000). Oxygen limitation of thermal tolerance defined by cardiac and ventilatory performance in the spider crab, *Maja squinado. Am. J. Physiol.* **279**, R1531–R1538.

Friedrich, C., and Hagen, W. (1984). Lipid contents of five species of notothenioid fish from high-Antarctic waters and ecological implications. *Polar Biol.* **14**, 359–369.

Garvey, J. E., and Marschall, E. A. (2003). Understanding latitudinal trends in fish body size through models of optimal seasonal energy allocation. *Can. J. Fish. Aquat. Sci.* **60**, 938–948.

Gibbs, A. (1995). Temperature, pressure and the sodium pump: The role of homeoviscous adaptation. *In* "Biochemistry and Molecular Biology of Fishes" (Hochachka, P. W., and Mommsen, T. P., Eds.), pp. 197–212. Elsevier Science, Amsterdam, Netherlands.

Giebelhausen, B., and Lampert, W. (2001). Temperature reaction norms of *Daphnia magna*: The effect of food concentration. *Freshw. Biol.* **46**, 281–289.

Gillis, T. E., Moyes, C. D., and Tibbits, G. F. (2003). Sequence mutations in teleost cardiac troponin C that are permissive of high Ca^{2+} affinity of site II. *Am. J. Physiol.* **284**, C1176–C1184.

Gonzalez-Cabrera, P. J., Dowd, F., Pedibhotla, V. K., Rosario, R., Stanley-Samuelson, D., and Petzel, D. (1995). Enhanced hypo-osmoregulation induced by warm-acclimation in antarctic fish is mediated by increased gill and kidney Na^+/K^+-ATPase activities. *J. Exp. Biol.* **198**, 2279–2291.

Goodridge, A. G. (1985). Fatty acid synthesis in eukaryotes. *In* "Biochemistry of Lipids and Membranes" (Vance, D. E., and Vance, J. E., Eds.), pp. 143–180. The Benjamin/Cummings Publishing Company, Menlo Park, California.

Gracey, A. Y., Logue, J., Tiku, P. E., and Cossins, A. R. (1996). Adaptation of biological membranes to temperature: Biophysical perspectives and molecular mechanisms. *In* "Animals and Temperature. Phenotypic and Evolutionary Adaptation" (Johnston,

I. A., and Bennett, A. F., Eds.), pp. 1–21. Cambridge University Press, CambridgeSociety for experimental Biology Seminar, Series 59.

Graves, J. E., and Somero, G. N. (1982). Electrophoretic and functional enzymic evolution in four species of eastern Pacific barracudas from different thermal environments. *Evolution* **36,** 97–106.

Guderley, H. (1998). Temperature and growth rates as modulators of the metabolic capacities of fish muscle. *In* "Cold Ocean Physiology" (Pörtner, H. O., and Playle, R., Eds.), pp. 58–87. Cambridge University Press, Cambridge.

Guderley, H., and Gawlicka, A. (1992). Qualitative modification of muscle metabolic organization with thermal acclimation of rainbow trout. *Onchorynchus mykiss. Fish Physiol. Biochem.* **10,** 123–132.

Guderley, H., and Johnston, I. A. (1996). Plasticity of fish muscle mitochondria with thermal acclimation. *J. Exp. Biol.* **199,** 1311–1317.

Guderley, H., and St-Pierre, J. (2002). Going with the flow or life in the fast lane: Contrasting mitochondrial responses to thermal change. *J. Exp. Biol.* **205,** 2237–2249.

Guynn, S., Dowd, F., and Petzel, D. (2002). Characterization of gill Na/K-ATPase activity and ouabain binding in Antarctic and New Zealand nototheniid fishes. *Comp. Biochem. Physiol.* **131**(A), 363–374.

Hagen, W., and Kattner, G. (1998). Lipid metabolism of the Antarctic euphausiid *Thysanoessa macrura* and its ecological implications. *Limnol. Oceanogr.* **43,** 1894–1901.

Hagen, W., Kattner, G., and Friedrich, C. (2000). The lipid compositions of high-Antarctic notothenioid fish species with different life strategies. *Polar Biol.* **23,** 785–791.

Hain, S., and Arnaud, P. M. (1992). Notes on the reproduction of high-Antarctic molluscs from the Weddell Sea. *Polar Biol.* **12,** 303–312.

Hardewig, I., van Dijk, P. L. M., and Pörtner, H. O. (1998). High energy turnover at low temperatures: Recovery from exercise in antarctic and common eelpout (Zoarcidae). *Am. J. Physiol.* **274,** R1789–R1796.

Hardewig, I., van Dijk, P. L., Moyes, C. D., and Pörtner, H. O. (1999a). Temperature-dependent expression of cytochrome-*c* oxidase in Antarctic and temperate fish. *Am. J. Physiol.* **277,** R508–R516.

Hardewig, I., Peck, L. S., and Pörtner, H. O. (1999b). Thermal sensitivity of mitochondrial function in the Antarctic notothenioid, *Lepidonotothen nudifrons. J. Comp. Physiol.* **169B,** 597–604.

Hazel, J. R. (1995). Thermal adaptation in biological membranes: Is homeoviscous adaptation the explanation? *Ann. Rev. Physiol.* **57,** 19–42.

Hazel, J. R., and Sellner, P. A. (1979). Fatty acid and sterol synthesis by hepatocytes from thermally acclimated rainbow trout (*Salmo gairdneri*). *J. Exp. Zool.* **209**(1), 105–114.

Heise, K., Puntarulo, S., Pörtner, H. O., and Abele, D. (2003). Production of reactive oxygen species by isolated mitochondria of the Antarctic bivalve *Laternula elliptica* (King and Broderip) under heat stress. *Comp. Biochem. Physiol.* **134C,** 79–90.

Hochachka, P. W. (1988a). Metabolic-, channel-, and pump-coupled functions: Constraints and compromises of coadaptation. *Can. J. Zool.* **66,** 1015–1027.

Hochachka, P. W. (1988b). Channels and pumps—Determinant of metabolic cold adaptation strategies. *Comp. Biochem. Physiol.* **90B,** 515–519.

Hochachka, P. W., and Somero, G. N. (1984). "Biochemical Adaptation." Princeton University Press, Princeton, New Jersey.

Hochachka, P. W., and Somero, G. N. (2002). "Biochemical Adaptation. Mechanism and Process in Physiological Evolution." Oxford University Press, Oxford.

Hochachka, P. W., Emmett, B., and Suarez, R. K. (1988). Limits and constraints in the scaling of oxidative and glycolytic enzymes in homeotherms. *Can. J. Zool.* **66,** 1128–1138.

Hoegh-Guldberg, O., and Pearse, J. S. (1995). Temperature, food availability, and the development of marine invertebrate larvae. *Am. Zool.* **35**, 415–425.

Holeton, G. F. (1970). Oxygen uptake and circulation by a hemoglobinless Antarctic fish (*Chaenocephalus aceratus* Lonnberg) compared with three red-blooded Antarctic fish. *Comp. Biochem. Physiol.* **34**, 457–471.

Holland, L. Z., McFall-Ngai, M., and Somero, G. N. (1997). Evolution of lactate dehydrogenase-A homologs of barracuda fishes (genus *Sphyraena*) from different thermal environments: Differences in kinetic properties and thermal stability are due to amino acid substitutions outside the active site. *Biochemistry* **36**, 3207–3215.

Holloszy, J. F., and Coyle, E. F. (1984). Adaptations of skeletal muscle to endurance training. *J. Appl. Physiol.* **56**, 831–838.

Hood, D. A. (2001). Invited Review: Contractile activity-induced mitochondrial biogenesis in skeletal muscle. *J. Appl. Physiol.* **90**, 1137–1157.

Hop, H., and Graham, M. (1995). Respiration of juvenile Arctic cod (*Boreogadus saida*): Effects of acclimation, temperature and food intake. *Polar Biol.* **15**, 359–367.

Hoppeler, H., and Billeter, R. (1991). Conditions for oxygen and substrate transport in muscles in exercising mammals. *J. Exp. Biol.* **160**, 263–283.

Hoppeler, H., and Weibel, E. R. (1998). Limits for oxygen and substrate transport in mammals. *J. Exp. Biol.* **201**, 1051–1064.

Hubley, M. J., Locke, B. R., and Moerland, T. S. (1997). Reaction-diffusion analysis of the effects of temperature on high-energy phosphate dynamics in goldfish skeletal muscle. *J. Exp. Biol.* **200**, 975–988.

Hubold, G. (1985). Stomach contents of the Antarctic silverfish *Pleuragramma antarcticum* from the southern and eastern Weddell Sea (Antarctica). *Polar Biol.* **5**, 43–48.

Hubold, G. (1991). Ecology of notothenioid fish in the Weddell Sea. *In* "Fishes of Antarctica. A Biological Overview" (Di Prisco, G., Pisano, E., and Clarke, A., Eds.), pp. 3–22. Springer-Verlag, Milano, Italia.

Hubold, G., and Hagen, W. (1997). Seasonality of feeding and lipid content in juvenile *Pleuragramma antarcticum* (Pisces: Nototheniidae) in the southern Weddell Sea. *In* "Antarctic Communities: Species, Structure and Survival" (Battaglia, B., Valenca, J., and Walton, D. W. H., Eds.), pp. 277–283. Cambridge University Press, Cambridge.

Hulbert, A. J., and Else, P. L. (1999). Membranes as possible pacemakers of metabolism. *J. Theor. Biol.* **199**, 257–274.

Hulbert, A. J., Rana, T., and Couture, P. (2002). The acyl composition of mammalian phospholipids: An allometric analysis. *Comp. Biochem. Physiol.* **132B**, 515–527.

Hurst, T. P., and Conover, D. O. (2003). Seasonal and interannual variation in the allometry of energy allocation in juvenile striped bass. *Ecology* **84**, 3360–3369.

Itoi, S., Kinoshita, S., Kikuchi, K., and Watabe, S. (2003). Changes of carp F_0F_1-ATPase in association with temperature acclimation. *Am. J. Physiol.* **284**, R153–R163.

Jaenicke, R. (1990). Protein structure and function at low temperatures. *Philos. Trans. R. Soc. Lond. B Biol. Sci.* **326**, 535–553.

Jobling, M. (1997). Temperature and growth: modulation of growth rate via temperature change. *In* "Global Warming: Implications for Freshwater and Marine Fish" (Wood, C. M., and McDonald, D. G., Eds.), pp. 225–253. Cambridge University Press, Cambridge.

Johnson, T. P., and Bennett, A. F. (1995). The thermal acclimation of burst escape performance in fish: An integrated study of molecular and cellular physiology and organismal performance. *J. Exp. Biol.* **198**, 2165–2175.

Johnson, T. P., Bennett, A. F., and McLister, J. D. (1996). Thermal dependence and acclimation of fast start locomotion and its physiological basis in rainbow trout (*Oncorhynchus mykiss*). *Physiol. Zool.* **69**, 276–292.

Johnston, I. A. (1982). Quantitative analyses of ultrastructure and vascularisation of the slow muscle of the anchovy. *Tissue Cell* **14**, 319–328.

Johnston, I. A. (1987). Respiratory characteristics of muscle fibres in a fish (*Chaenocephalus aceratus*) that lacks haem pigment. *J. Exp. Biol.* **133**, 415–428.

Johnston, I. A. (1989). Antarctic fish muscles—Structure, function and physiology. *Ant. Sci.* **1**, 97–108.

Johnston, I. A. (1990). Cold adaptation in marine organisms. *Philos. Trans. Roy. Soc. Lond.* **326B**, 655–667.

Johnston, I. A., Walesby, N. J., Davison, W., and Goldspink, G. (1975). Temperature adaptation in myosin of Antarctic fish. *Nature* **254**, 74–75.

Johnston, I. A., and Maitland, B. (1980). Temperature acclimation in crucian carp, *Carassius carassius* L., morphometric analyses of muscle fibre ultrastructure. *J. Fish Biol.* **17**, 113–125.

Johnston, I. A., and Moon, T. W. (1980a). Exercise training in skeletal muscle of brook trout (*Salvelinus fontinalis*). *J. Exp. Biol.* **87**, 177–194.

Johnston, I. A., and Moon, T. W. (1980b). Endurance exercise training in the fast and slow muscles of a teleost fish (*Pallachius virens*). *J. Comp. Physiol.* **135**, 147–156.

Johnston, I. A., and Dunn, J. (1987). Temperature acclimation and metabolism in ectotherms with particular reference to teleost fish. *Symp. Soc. Exp. Biol.* **41**, 67–93.

Johnston, I. A., Camm, J. P., and White, M. G. (1988). Specialisation of swimming muscles in the pelagic Antarctic fish,. *Pleuragramma antarcticum. Mar. Biol.* **100**, 3–12.

Johnston, I. A., Johnson, T. P., and Battram, J. C. (1991a). Low temperature limits burst swimming performance in Antarctic fish. *In* "Biology of Antarctic Fish" (di Prisco, G., Maresca, B., and Tota, B., Eds.), pp. 179–190. Springer-Verlag, Berlin.

Johnston, I. A., Clarke, A., and Ward, P. (1991b). Temperature and metabolic rate in sedentary fish from the Antarctic, North Sea and Indo-West Pacific Ocean. *Mar. Biol.* **109**, 191–195.

Johnston, I. A., and Battram, J. (1993). Feeding energetics and metabolism in demersal fish species from Antarctic, temperate and tropical environments. *Mar. Biol.* **115**, 7–14.

Johnston, I. A., Calvo, J., Guderley, H., Fernandez, D., and Palmer, L. (1998). Latitudinal variation in the abundance and oxidative capacities of muscle mitochondria in perciform fishes. *J. Exp. Biol.* **201**, 1–12.

Johnston, I. A., Fernandez, D. A., Calvo, J., Vieira, V., North, A. W., Abercromby, M., and Garland, T. (2003). Reduction in muscle fibre number during the adaptive radiation of notothenioid fishes: A phylogenetic perspective. *J. Exp. Biol.* **206**, 2595–2609.

Kattner, G., and Hagen, W. (1998). Lipid metabolism of the Antarctic euphausiid *Euphausia crystallorophias* and its ecological implications. *Mar. Ecol. Prog. Ser.* **170**, 203–213.

Kawall, H. G., Torres, J. J., Sidell, B. D., and Somero, G. N. (2002). Metabolic cold adaptation in Antarctic fishes: Evidence from enzymatic activities of brain. *Mar. Biol.* **140**, 279–286.

Kieffer, J. D. (2000). Limits to exhaustive exercise in fish. *Comp. Biochem. Physiol.* **126A**, 161–179.

Kieffer, J. D., Currie, S., and Tufts, B. L. (1994). Effects of environmental temperature on the metabolic and acid–base responses of rainbow trout to exhaustive exercise. *J. Exp. Biol.* **194**, 299–317.

Klages, M. (1993). Biology of the Antarctic gammaridean amphipod *Eusirus perdentatus* Chevreux, 1912 (Crustacea: Amphiphoda): Distribution, reproduction and population dynamics. *Antarc. Sci.* **5**, 349–359.

Kock, K. H., and Everson, I. (1998). Age, growth and maximum size of Antarctic notothenioid fish—revisited. *In* "Fishes of Antarctica. A biological Overview" (di Prisco, G., Pisano, E., and Clarke, A., Eds.), pp. 29–40. Springer-Verlag, Milano, Italia.

Korsmeyer, K. E., Dewar, H., Lai, N. C., and Graham, J. B. (1996). The aerobic capacity of tunas: Adaptation for multiple metabolic demands. *Comp. Biochem. Physiol.* **113A**, 12–24.

Kunzmann, A., and Zimmermann, C. (1992). *Aethotaxis mitopteryx*, a high-Antarctic fish with benthopelagic mode of life. *Mar. Ecol. Prog. Ser.* **88**, 33–40.

La Mesa, M., and Vacchi, M. (2001). Review. Age and growth of high Antarctic notothenioid fish. *Antarct. Sci.* **13**, 227–235.

Laman, E. A., and Loeb, V. J. (1994). RACER: Feeding incidence and yolk resorption in three species of Antarctic larval fishes. *Antarct. J. US* **29**, 162–164.

Lannig, G., Eckerle, L., Serendero, I., Sartoris, F. J., Fischer, T., Knust, R., Johansen, T., and Pörtner, H. O. (2003). Temperature adaptation in eurythermal cod (*Gadus morhua*): comparison of mitochondrial enzyme capacities in boreal & arctic populations. *Mar. Biol.* **142**, 589–599.

Lannig, G., Bock, C., Sartoris, F. J., and Pörtner, H. O. (2004). Oxygen limitation of thermal tolerance in cod, *Gadus morhua* L. studied by non-invasive NMR techniques and on-line venous oxygen monitoring. *Am. J. Physiol.* **287**, R902–R910.

Leary, S. C., and Moyes, C. D. (2000). The effect of bioenergetic stress and redox balance on the expression of genes critical to mitochondrial function. *In* "Environmental Stressors and Gene Responses" (Storey, K. B., and Storey, J. M., Eds.), pp. 209–230. Elsevier, Amsterdam, Netherlands.

Lee, P. G. (1994). Nutrition of cephalopods: fueling the system. *In* "Physiology of Cephalopod Molluscs. Lifestyle and Performance Adaptations" (Pörtner, H. O., O'Dor, R., and MacMillan, D. L., Eds.), pp. 35–51. Gordon and Breach Publishing, Basel.

Lehtonen, K. K. (1996). Ecophysiology of the benthic amphipod *Monoporeia affinis* in an open-sea area of the northern Baltic Sea: Seasonal variations in body composition, with bioenergetic considerations. *Mar. Ecol. Prog. Ser.* **143**, 87–98.

Londraville, R. L., and Sidell, B. (1990). Ultrastructure of aerobic muscle in Antarctic fishes may contribute to maintenance of diffusive fluxes. *J. Exp. Biol.* **150**, 205–220.

Lucassen, M., Schmidt, A., Eckerle, L. G., and Pörtner, H. O. (2003a). Cold induced mitochondrial proliferation in *Zoarces viviparus*: Changes in enzyme activities and mRNA levels. *Am. J. Physiol.* **258**, R1410–R1420.

MacDonald, J. A., Montgomery, J. C., and Wells, R. M. G. (1988). The physiology of McMurdo Sound fishes: Current New Zealand research. *Comp. Biochem. Physiol.* **90B**, 567–578.

Mark, F. C., Bock, C., and Pörtner, H. O. (2002). Oxygen limited thermal tolerance in Antarctic fish investigated by MRI and ^{31}P-MRS. *Am. J. Physiol.* **283**(5), R1254–R1262.

Marsh, A. G., and Manahan, D. T. (2000). Metabolic differences between "demersal" and "pelagic" development of the Antarctic sea urchin *Sterechinus neumayeri*. *Mar. Biol.* **137**, 215–221.

Marsh, A. G., Leong, P. K., and Manahan, D. T. (1999). Energy metabolism during embryonic development and larval growth of an Antarctic sea urchin. *J. Exp. Biol.* **202**, 2041–2050.

Marsh, A. G., Maxson, R. E., Jr., and Manahan, D. T. (2001). High macromolecular synthesis with low metabolic cost in Antarctic sea urchin embryos. *Science* **291**, 1950–1952.

Marshall, C. J. (1997). Cold-adapted enzymes. *Trends Biotechnol.* **15**, 359–364.

Marshall, C., Crossman, D., Love, C., McInnes, S., and Fleming, R. (2000). Structural Studies of Cold-Adaptation: Lactate Dehydrogenases from Notothenioid Fish. *In* "Antarctic Ecosystems: Models for a Wider Understanding" (Davison, W., and Williams, C. W., Eds.), pp. 123–133. Caxton Press, Christchurch, New Zealand.

Marshall, D. J., Pechenik, J. A., and Keough, M. J. (2003). Larval activity levels and delayed metamorphosis affect post-larval performance in the colonial ascidian. *Diplosoma listerianum*. *Mar. Ecol. Prog. Ser.* **246**, 153–162.

McArdle, H. J., and Johnston, I. A. (1980). Evolutionary temperature adaptation of fish sarcoplasmic reticulum. *J. Comp. Physiol.* **135**, 157–164.

Mente, E., Legeay, A., Houlihan, D. F., and Massabuau, J. C. (2003). Influence of oxygen partial pressure in protein synthesis in feeding crabs. *Am. J. Physiol.* **284,** R500–R510.

Metcalf, V. J., Brennan, S. O., and George, P. M. (1999a). The Antarctic toothfish (*Dissostichus mawsoni*) lacks plasma albumin and utilises high density lipoprotein as its major palmitate binding protein. *Comp. Biochem. Physiol.* **124B,** 147–155.

Metcalf, V. J., Brennan, S. O., Chambers, G., and George, P. M. (1999b). High density lipoprotein (HDL), and not albumin, is the major palmitate binding protein in New Zealand long-finned (*Anguilla dieffenbachii*) and short-finned eel (*Anguilla australis schmidtii*) plasma. *Biochim. Biophys. Acta* **1429,** 467–475.

Milligan, C. L. (1996). Metabolic recovery from exhaustive exercise in rainbow trout. *Comp. Biochem. Physiol.* **113A,** 51–60.

Milligan, C. L., and Wood, C. M. (1987). Muscle and liver intracellular acid–base and metabolite status after strenuous activity in the inactive, benthic starry flounder (*Platichthys stellatus*). *Physiol. Zool.* **60,** 54–68.

Mitchell, S. E., and Lampert, W. (2000). Temperature adaptation in a geographically widespread zooplankter, *Daphnia magna. J. Evol. Biol.* **13,** 371–382.

Moerland, T. S., and Egginton, S. (1998). Intracellular pH of muscle and temperature insight from *in vivo* ^{31}P-NMR measurements in a stenothermal Antarctic teleost (*Harpagifer antarcticus*). *J. Therm. Biol.* **23,** 275–282.

Mommsen, T. P. (2001). Paradigms of growth in fish. *Comp. Biochem. Physiol.* **129B,** 207–219.

Montgomery, J. C., and Wells, R. M. G. (1993). Recent advances in the ecophysiology of Antarctic notothenioid fishes: metabolic capacity and sensory performance. *In* "Fish Ecophysiology" (Rankin, J. C., and Jensen, F. B., Eds.), pp. 342–374. Chapman & Hall, London.

Morris, D. J., and North, A. W. (1984). Oxygen consumption of five species of fish from South Georgia. *J. Exp. Mar. Biol. Ecol.* **78,** 75–86.

Moyes, C. D., Mathieu-Costello, O. A., Brill, R. W., and Hochachka, P. W. (1992). Mitochondrial metabolism of cardiac and skeletal muscles from a fast (*Katsuwonas pelamis*) and a slow (*Cyprinus carpio*) fish. *Can. J. Zool.* **70,** 1246–1253.

Moyes, C. D., and Hood, D. A. (2003). Origins and consequences of mitochondrial variation in vertebrate muscle. *Ann. Rev. Physiol.* **65,** 177–201.

North, A. W. (1998). Growth of young fish during winter and summer at South Georgia, Antarctica. *Polar Biol.* **19,** 198–205.

North, A. W., White, M. G., and Trathan, P. N. (1998). Interannual variability in the early growth rate and size of the Antarctic fish *Gobionotothen gibberifrons* (Lonnberg). *Antarct. Sci.* **10,** 416–422.

O'Brien, K. M., and Sidell, B. D. (2000). The interplay among cardiac ultrastructure, metabolism and the expression of oxygen-binding proteins in Antarctic fishes. *J. Exp. Biol.* **203,** 1287–1297.

O'Brien, K. M., Skilbeck, C., Sidell, B. D., and Egginton, S. (2003). Muscle fine structure may maintain the function of oxidative fibres in haemoglobinless Antarctic fishes. *J. Exp. Biol.* **206,** 411–421.

O'Dor, R. K., and Webber, D. M. (1991). Invertebrate athletes: trade-offs between transport efficiency and power density in cephalopod evolution. *J. Exp. Biol.* **160,** 93–112.

Outram, D. M., and Loeb, V. J. (1995). RACER: Spatial and temporal variability in early growth rates of the antarctic silverfish (*Pleuragramma antarcticum*) around the antarctic continent. *Antarct. J. US* **30,** 172–174.

Ozernyuk, N. D., Klyachko, O. S., and Polosukhina, E. S. (1994). Acclimation temperature affects the functional and structural properties of lactate dehydrogenase from fish (*Misgurnus fossilis*) skeletal muscles. *Comp. Biochem. Physiol.* **107B,** 141–145.

Pauly, D. (1979). Gill size and temperature as governing factors in fish growth: A generalization of von Bertalanffy's growth formula. *Ber. Inst. Meereskd. Christian-Albrecht-Univ. Kiel* **63**, 1–156.

Pechenik, J. A. (1990). Delayed metamorphosis by larvae of benthic marine invertebrates: Does it occur? Is there a price to pay? *Ophelia* **32**, 63–94.

Pechenik, J. A. (1999). On the advantages and disadvantages of larval stages in benthic marine invertebrate life cycles. *Mar. Ecol. Prog. Ser.* **177**, 269–297.

Pechenik, J. A., Wendt, D. E., and Jarrett, J. N. (1998). Metamorphosis is not a new beginning. *Bioscience* **48**, 901–910.

Peck, L. S. (1993). Larval development in the Antarctic nemertean *Parborlasia corrugatus* (Heteronemertea: Lineidae). *Mar. Biol.* **116**, 301–310.

Peck, L. S. (2002). Ecophysiology of Antarctic marine ectotherms: Limits to life. *Polar Biol.* **25**, 31–40.

Peck, L. S., and Robinson, K. (1994). Pelagic larval development in the brooding Antarctic brachiopod. *Liothyrella uva. Mar. Biol.* **120**, 279–286.

Pennycuick, C. J. (1992). "Newton Rules Biology." Oxford University press, Oxford.

Poltermann, M. (2000). Growth, production and productivity of the Arctic sympagic amphipod *Gammarus wilkitzkii. Mar. Ecol. Prog. Ser.* **193**, 109–116.

Pörtner, H. O. (1994). Coordination of metabolism, acid–base regulation and haemocyanin function in cephalopods. *Mar. Fresh. Behav. Physiol.* **25**, 131–148.

Pörtner, H. O. (2001). Climate change and temperature dependent biogeography: Oxygen limitation of thermal tolerance in animals. *Naturwissenschaften* **88**, 137–146.

Pörtner, H. O. (2002a). Climate change and temperature dependent biogeography: Systemic to molecular hierarchies of thermal tolerance in animals. *Comp. Biochem. Physiol.* **132A**, 739–761.

Pörtner, H. O. (2002b). Physiological basis of temperature dependent biogeography: Trade-offs in muscle design and performance in polar ectotherms. *J. Exp. Biol.* **205**, 2217–2230.

Pörtner, H. O. (2002c). Environmental and functional limits to muscular exercise and body size in marine invertebrate athletes. *Comp. Biochem. Physiol.* **133A**, 303–321.

Pörtner, H. O. (2004). Climate variability and the energetic pathways of evolution: The origin of endothermy in mammals and birds. *Physiol. Biochem. Zool.* **77**, 959–981.

Pörtner, H. O., Webber, D. M., Boutilier, R. G., and O'Dor, R. K. (1991). Acid–base regulation in exercising squid (*Illex illecebrosus, Loligo pealei*). *Am. J. Physiol.* **261**, R239–R246.

Pörtner, H. O., Webber, D. M., O'Dor, R. K., and Boutilier, R. G. (1993). Metabolism and energetics in squid (*Illex illecebrosus, Loligo pealei*) during muscular fatigue and recovery. *Am. J. Physiol.* **265**, R157–R165.

Pörtner, H. O., Finke, E., and Lee, P. G. (1996). Metabolic and energy correlates of intracellular pH in progressive fatigue of squid (*Lolliguncula brevis*) mantle muscle. *Am. J. Physiol.* **271**, R1403–R1414.

Pörtner, H. O., and Zielinski, S. (1998). Environmental constraints and the physiology of performance in squids. *In* "Cephalopod Biodiversity, Ecology and Evolution" (Payne, A. I. L., Lipinski, M. R., Clarke, M. R., and Roeleveld, M. A. C., Eds.), **20**, pp. 207–221.

Pörtner, H. O., Hardewig, I., Sartoris, F. J., and van Dijk, P. L. M. (1998). Energetic aspects of cold adaptation; critical temperatures in metabolic, ionic and acid base regulation? *In* "Cold Ocean Physiology" (Pörtner, H. O., and Playle, R., Eds.), pp. 88–120. Cambridge University Press, Cambridge.

Pörtner, H. O., and Sartoris, F. J. (1999). Invasive studies of intracellular acid–base parameters: Quantitative analyses during environmental and functional stress. *In* "Regulation of Acid–Base Status in Animals and Plants" (Eggington, S., Taylor, E. W., and Raven, J. A., Eds.), pp. 69–98. Cambridge University Press, Cambridge.

Pörtner, H. O., Hardewig, I., and Peck, L. S. (1999). Mitochondrial function and critical temperature in the Antarctic bivalve *Laternula elliptica*. *Comp. Biochem. Physiol.* **124A,** 179–189.

Pörtner, H. O., van Dijk, P., Hardewig, I., and Sommer, A. (2000). Levels of metabolic cold adaptation: trade-offs in eurythermal and stenothermal ectotherms. *In* "Antarctic Ecosystems: Models for Wider Ecological Understanding" (Davison, W, and Howard Williams, C., Eds.), pp. 109–122. Caxton Press, Christchurch, New Zealand.

Pörtner, H. O., Berdal, B., Blust, R., Brix, O., Colosimo, A., De Wachter, B., Giuliani, A., Johansen, T., Fischer, T., Knust, R., Naevdal, G., Nedenes, A., Nyhammer, G., Sartoris, F. J., Serendero, I., Sirabella, P., Thorkildsen, S., and Zakhartsev, M. (2001). Climate effects on growth performance, fecundity and recruitment in marine fish: Developing a hypothesis for cause and effect relationships in Atlantic cod (*Gadus morhua*) and common eelpout (*Zoarces viviparus*). *Cont. Shelf Res.* **21,** 1975–1997.

Pörtner, H. O., Mark, F. C., and Bock, C. (2004). Oxygen limited thermal tolerance in fish? Answers obtained by Nuclear Magnetic Resonance techniques. *Respir. Physiol. Neurobiol.* **141,** 243–260.

Post, J. R., and Parkinson, E. A. (2001). Energy allocation strategy in young fish: Allometry and survival. *Ecology* **82,** 1040–1051.

Powers, D. A., and Schulte, P. M. (1998). Evolutionary adaptations of gene structure and expression in natural populations in relation to a changing environment: A multidisciplinary approach to address the million-year saga of a small fish. *J. Exp. Zool.* **282,** 71–94.

Prabhakar, N. R., and Kumar, G. K. (2004). Oxidative stress in the systemic and cellular responses to intermittent hypoxia. *Biol. Chem.* **385,** 217–221.

Priede, I. G. (1985). Metabolic scope in fishes. *In* "Fish Energetics: New Perspectives" (Tyler, P., and Calow, P., Eds.), pp. 33–64. Croom Helm, London, Sydney.

Purchase, C. F., and Brown, J. A. (2001). Stock-specific changes in growth rates, food conversion efficiencies, and energy allocation in response to temperature change in juvenile Atlantic cod. *J. Fish Biol.* **58,** 36–52.

Raynard, R. S., and Cossins, A. R. (1991). Homeoviscous adaptation and thermal compensation of sodium pump of trout erythrocytes. *Am. J. Physiol.* **260,** R916–R924.

Reeves, R. B. (1985). Alphastat regulation of intracellular acid–base state? *In* "Circulation, Respiration, and Metabolism" (Gilles, R., Ed.), pp. 414–423. Springer-Verlag, Berlin, Heidelberg.

Robertson, R. F., Whiteley, N. M., and Egginton, S. (1998). Cardiac and locomotory muscle mass in Antarctic fishes. *In* "Fishes of Antarctica. A Biological Overview." (di Prisco, G., Pisano, E., and Clarke, A., Eds.), pp. 197–204. Springer-Verlag, Milan, Italy.

Robertson, R. F., El-Haj, A. J., Clarke, A., Peck, L. S., and Taylor, E. W. (2001). The effects of temperature on metabolic rate and protein synthesis following a meal in the isopod *Glyptonotus antarcticus* Eights (1852). *Polar Biol.* **24,** 677–686.

Sartoris, F. J., Bock, C., and Pörtner, H. O. (2003). Temperature dependent pH regulation in eurythermal and stenothermal marine fish: an interspecies comparison using ^{31}P-NMR. *J. Therm. Biol.* **28,** 363–371.

Scarabello, M., Heigenhauser, G. J. F., and Wood, C. M. (1992). Gas-exchange, metabolite status and excess post-exercise oxygen consumption after repetitive bouts of exhaustive exercise in juvenile rainbow trout. *J. Exp. Biol.* **167,** 155–169.

Schulte, P. M., Glemet, H. C., Fiebig, A. A., and Powers, D. A. (2000). Adaptive variation in lactate dehydrogenase-B gene expression: role of a stress-responsive regulatory element. *Proc. Natl. Acad. Sci. USA* **97,** 6597–6602.

Schurmann, H., and Steffensen, J. F. (1997). Effects of temperature, hypoxia and activity on the metabolism of juvenile Atlantic cod. *J. Fish Biol.* **50,** 1166–1180.

Schwarzbaum, P. J., Wieser, W., and Niederstaetter, H. (1991). Contrasting effects of temperature acclimation on mechanisms of ionic regulation in a eurythermic and a stenothermic species of freshwater fish (*Rutilus rutilus* and *Salvelinus alpinus*). *Comp. Biochem. Physiol.* **98A**, 483–489.

Schwarzbaum, P. J., Niederstaetter, H., and Wieser, W. (1992a). Effects of temperature on the Na^+/K^+-ATPase and oxygen consumption in hepatocytes of two species of freshwater fish, roach (*Rutilus rutilus*) and brook trout (*Salvelinus fontinalis*). *Physiol. Zool.* **65**, 699–711.

Schwarzbaum, P. J., Wieser, W., and Cossins, A. R. (1992b). Species-specific responses of membranes and the Na^+/K^+ pump to temperature change in the kidney of two species of freshwater fish, roach (*Rutilus rutilus*) and arctic char (*Salvelinus alpinus*). *Physiol. Zool.* **65**, 17–34.

Seedorf, U., Leberer, E., Kirschbaum, B. J., and Pette, D. (1986). Neural control of gene expression in skeletal muscle. Effects of chronic stimulation on lactate dehydrogenase isoenzymes and citrate synthase. *Biochem. J.* **239**, 115–120.

Shaklee, J. B., Christiansen, J. A., Sidell, B. D., Prosser, C. L., and Whitt, G. S. (1977). Molecular aspects of temperature acclimation in fish: Contributions of changes in enzyme activities and isozyme patterns to metabolic reorganization in green sunfish. *J. Exp. Zool.* **201**, 1–20.

Sidell, B. D. (1980). Responses of goldfish (*Carassius auratus*, L.) muscle to acclimation temperature: Alterations in biochemistry and proportions of different fiber types. *Physiol. Zool.* **53**, 98–107.

Sidell, B. D. (1998). Intracellular oxygen diffusion: The roles of myoglobin and lipid at cold body temperature. *J. Exp. Biol.* **201**, 1118–1127.

Sidell, B. D., and Moerland, T. S. (1989). Effects of temperature on muscular function and locomotory performance in teleost fish. *Adv. Comp. Env. Physiol.* **5**, 115–155.

Sidell, B. D., Crockett, E. L., and Driedzic, W. R. (1995). Antarctic fish tissues preferentially catabolize monoenoic fatty acids. *J. Exp. Zool.* **271**, 73–81.

Somero, G. (1997). Temperature relationships; from molecules to biogeography. *In* "Handbook of Physiology Section 13. Comparative Physiology, II" (Dantzler, W., Ed.), pp. 1391–1444. Oxford University Press, Oxford.

Somero, G. N., and Childress, J. J. (1990). Scaling of ATP-supplying enzymes, myofibrillar proteins and buffering capacity in fish muscle: Relationship to locomotory habit. *J. Exp. Biol.* **149**, 319–333.

Sommer, A., Klein, B., and Pörtner, H. O. (1997). Temperature induced anaerobiosis in two populations of the polychaete worm Arenicola marina (L.). *J. Comp. Physiol.* **167B**, 25–35.

Sommer, A., and Pörtner, H. O. (2002). Metabolic cold adaptation in the lugworm *Arenicola marina*: Comparison of a North Sea and a White Sea population. *Mar. Ecol. Progr. Ser.* **240**, 171–182.

Sommer, A. M., and Pörtner, H. O. (2004). Mitochondrial function in seasonal acclimatisation versus latitudinal adaptation to cold, in the lugworm *Arenicola marina* (L.). *Physiol. Biochem. Zool.* **77**, 174–186.

Srere, P. A. (1985). Organization of proteins within the mitochondrion. *In* "Organized Multienzyme Systems. Catalytic Properties" (Welch, G. R., Ed.), pp. 1–61. Academic Press, New York, London.

Stankovic, A., Spalik, K., Kamler, E., Borsuk, P., and Weglenski, P. (2001). Recent origin of sub-Antarctic notothenioids. *Polar Biol.* **25**, 203–205.

Stanwell-Smith, D., and Peck, L. S. (1998). Temperature and embryonic development in relation to spawning and field occurrence of larvae of three Antarctic echinoderms. *Biol. Bull.* **194**, 44–52.

Steffensen, J. F., Bushnell, P. G., and Schurmann, H. (1994). Oxygen consumption in four species of teleosts from Greenland: No evidence of metabolic cold adaptation. *Polar Biol.* **14**, 49–54.

Steffensen, J. F. (2002). Metabolic cold compensation of polar fish based on measurements of aerobic oxygen consumption: Fact or artefact? Artefact! *Comp. Biochem. Physiol.* **132A**, 789–795.

Storch, D., and Pörtner, H. O. (2003). The protein synthesis machinery operates at the same expense in eurythermal and cold stenothermal pectinids. *Physiol. Biochem. Zool.* **76**, 28–40.

Storch, D., Heilmayer, O., Hardewig, I., and Pörtner, H. O. (2003). *In vitro* protein synthesis capacities in a cold stenothermal and a temperate eurythermal pectinid. *J. Comp. Physiol. B* **173**, 611–620.

Storch, D., Lannig, G., and Pörtner, H. O. (2005).Temperature-dependent protein synthesis capacities in Antarctic and temperate (North Sea) fish (Zoarcidae). *J. Exp. Biol.* **208**, 2409–2420

Storelli, C., Acierno, R., and Maffia, M. (1998). Membrane lipid and protein adaptations in Antarctic fish. *In* "Cold Ocean Physiology" (Pörtner, H. O., and Playle, R., Eds.), pp. 166–189. Cambridge University Press, Cambridge.

St-Pierre, J., Charest, P.-M., and Guderley, H. (1998). Relative contribution of quantitative and qualitative changes in mitochondria to metabolic compensation during seasonal acclimatisation of rainbow trout *Oncorhynchus mykiss. J. Exp. Biol.* **201**, 2961–2970.

Suarez, R. K., Lighton, J. R. B., Brown, G. S., and Mathieu-Costello, O. (1991). Mitochondrial respiration in hummingbird flight muscles. *Proc. Natl. Acad. Sci. USA* **88**, 4870–4873.

Takeyasu, K., Tamkun, M. M., Renaud, K. J., and Fambrough, D. M. (1988). Ouabain-sensitive Na^+ K^+-ATPase activity expressed in mouse L cells by transfection with DNA encoding the alpha-subunit of an avian sodium pump. *J. Biol. Chem.* **263**, 4347–4354.

Thiel, H., Pörtner, H. O., and Arntz, W. E. (1996). Marine life at low temperatures—A comparison of polar and deep-sea characteristics. *In* "Deep-Sea and Extreme Shallow-Water Habitats: Affinities and Adaptations" (Uiblein, F., Ott, J., and Stachowitsch, M., Eds.), pp. 183–219. Austrian Academy of Sciences, ViennaBiosystematics and Ecology Series 11.

Thurston, R. V., and Gehrke, P. C. (1993). Respiratory oxygen requirements of fishes: Description of OXYREF, a data file based on test results reported in the published literature. *In* "Fish Toxicology, and Water Quality Management. Proceedings of an International Symposium" (Russo, R. C., and Thurston, R. V., Eds.), pp. 98–108.

Tiku, P. E., Gracey, A. Y., Macartney, A. I., Beynon, R. J., and Cossins, A. R. (1996). Cold-induced expression of Δ^9-desaturase in carp by transcriptional and posttranslational mechanisms. *Science.* **271**, 815–818.

Tota, B., Agnisola, C., Schioppa, M., Acierno, R., Harrison, P., and Zummo, G. (1991). Structural and mechanical characteristics of the heart of the icefish *Chionodraco hamatus* (Lönnberg). *In* "Biology of Antarctic Fish" (DiPrisco, G., Maresca, B., and Tota, B., Eds.), pp. 204–219. Springer-Verlag, Berlin.

Tota, B., Cerra, M. C., Mazza, R., Pellegrino, D., and Icardo, J. (1997). The heart of the icefish as a paradigm of cold adaptation. *J. Therm. Biol.* **22**, 409–417.

Treinin, M., Shliar, J., Jiang, H., Powell-Coffman, J. A., Bromberg, Z., and Horowitz, M. (2003). HIF-1 is required for heat acclimation in the nematode. *C. elegans. Physiol. Genomics.* **14**, 17–24.

Trigari, G., Pirini, M., Ventrella, V., Pagliarani, A., Trombetti, F., and Borgatti, A. R. (1992). Lipid composition and mitochondrial respiration in warm- and cold-adapted sea bass. *Lipids* **27**, 371–377.

Tschischka, K., Abele, D., and Pörtner, H. O. (2000). Mitochondrial oxyconformity and cold adaptation in the polychaete *Nereis pelagica* and the bivalve *Arctica islandica* from the Baltic and White Seas. *J. Exp. Biol.* **203**, 3355–3368.

Tyler, S., and Sidell, B. D. (1984). Changes in mitochondrial disruption and diffusion distances in muscle of goldfish on acclimation to warm and cold temperatures. *J. Exp. Zool.* **232**, 1–9.

van den Thillart, G., and de Bruin, G. (1981). Influence of environmental temperature on mitochondrial membranes. *Biochim. Biophys. Acta* **640**, 439–447.

van Dijk, P. L. M., Hardewig, I., and Pörtner, H. O. (1997). Temperature dependent shift of pH_i in fish white muscle: Contribution of passive and active processes. *Am. J. Physiol.* **272**, R84–R89.

van Dijk, P. L. M., Hardewig, I., and Pörtner, H. O. (1998). Exercise in the cold: High energy turnover in Antarctic fish. *In* "Fishes of Antarctica. A Biological Overview" (di Prisco, G., Pisano, E., and Clarke, A., Eds.), pp. 225–236. Springer-Verlag, Berlin.

van Dijk, P., Tesch, C., Hardewig, I., and Pörtner, H. O. (1999). Physiological disturbances at critically high temperatures: A comparison between stenothermal antarctic and eurythermal temperate eelpouts (Zoarcidae). *J. Exp. Biol.* **202**, 3611–3621.

Vayda, M. E., Londraville, R. L., Cashon, R. E., Costello, L., and Sidell, B. D. (1998). Two distinct types of fatty acid-binding protein are expressed in heart ventricles of Antarctic teleost fishes. *Biochem. J.* **330**, 375–382.

Wakeling, J. M., Cole, N. J., Kemp, K. M., and Johnston, I. A. (2000). The biomechanics and evolutionary significance of thermal adaptation in the common carp, *Cyprinus carpio. Am. J. Physiol.* **279**, R657–R665.

Watabe, S. (2002). Temperature plasticity of contractile proteins in fish muscle. *J. Exp. Biol.* **205**, 2231–2236.

Weber, J. M., and Haman, F. (1996). Pathways for metabolic fuels and oxygen in high performance fish. *Comp. Biochem. Physiol.* **113A**, 33–38.

Weld, M. M., and Meier, A. H. (1985). Circadian and seasonal variations of lipogenesis in the Gulf killifish, *Fundulus grandis. Chronobiol. Int.* **2**, 151–159.

Wells, R. M. G. (1987). Respiration of Antarctic fish from McMurdo Sound. *Comp. Biochem. Physiol.* **88A**, 417–424.

White, M. G., and Burren, P. J. (1992). Reproduction and larval growth of *Harpagifer antarcticus* Nybelin (Pisces, Notothenioidei). *Antarct. Sci.* **4**, 421–430.

Whiteley, N. M., Taylor, E. W., and el, HajA. J. (1996). A comparison of the metabolic cost of protein synthesis in stenothermal and eurythermal isopod crustaceans. *Am. J. Physiol.* **271**, R1295–R1303.

Wieser, W. (1989). Energy allocation by addition and by compensation: an old principle revisited. *In* "Energy Transformations in Cells and Organisms" (Wieser, W., and Gnaiger, E., Eds.), pp. 98–105. Georg Thieme Verlag, Stuttgart.

Wieser, W. (1995). Energetics of fish larvae, the smallest vertebrates. *Acta Physiol. Scand.* **154**, 279–290.

Wilson, R. S., Franklin, C. E., Davison, W., and Kraft, P. (2001). Stenotherms at subzero temperatures: Thermal dependence of swimming performance in Antarctic fish. *J. Comp. Physiol.* **171B**, 263–269.

Winder, W. W., and Hardie, D. G. (1999). AMP-activated protein kinase, a metabolic master switch: Possible roles in Type 2 diabetes. *Am. J. Physiol.* **277**, E1–E10.

Winnard, P., Cashon, R. E., Sidell, B. D., and Vayda, M. E. (2003). Isolation, characterization and nucleotide sequence of the muscle isoforms of creatine kinase from the Antarctic teleost. *Chaenocephalus aceratus. Comp. Biochem. Physiol.* **134B**, 651–667.

Wodtke, E. (1981a). Temperature adaptation of biological membranes. The effects of acclimation temperature on the unsaturation of the main neutral and charged phospholipids in mitochondrial membranes of the carp (*Cyprinus carpio* L.). *Biochim. Biophys. Acta* **640**, 698–709.

Wodtke, E. (1981b). Temperature adaptation of biological membranes. Compensation of the molar activity of cytochrome *c* oxidase in the mitochondrial energy-transducing membrane during thermal acclimation of the carp (*Cyprinus carpio* L.). *Biochim. Biophys. Acta* **640**, 710–720.

Wöhrmann, A. P. A., Hagen, W., and Kunzmann, A. (1997). Adaptations of the Antarctic silverfish *Pleuragramma antarcticum* (Pisces: Nototheniidae) to pelagic life in high-Antarctic waters. *Mar. Ecol. Prog. Ser.* **151**, 205–218.

Wood, C. M., and McDonald, D. G. (Eds.) (1997). "Global Warming: Implications for Freshwater and Marine Fish." Cambridge University Press, Cambridge.

Yamahira, K., and Conover, D. O. (2002). Intra-vs. interspecific latitudinal variation in growth: Adaptation to temperature or length of the growing season? *Ecology* **83**, 1252–1262.

Zakhartsev, M. V., De Wachter, B., Sartoris, F. J., Pörtner, H. O., and Blust, R. (2003). Thermal physiology of the common eelpout (*Zoarces viviparus*). *J. Comp. Physiol. B* **173**, 365–378.

Zakhartsev, M., Johansen, T., Pörtner, H. O., and Blust, R. (2004). Effects of temperature acclimation on lactate dehydrogenase of cod (*Gadus morhua*): Fenetic, kinetic and thermodynamic aspects. *J. Exp. Biol.* **207**, 95–112.

Zimmermann, C., and Hubold, G. (1998). Respiration and activity of Arctic and Antarctic fish with different modes of life: A multivariate analysis of experimental data. *In* "Fishes of Antarctica. A Biological Overview" (di Prisco, G., Pisano, E., and Clarke, A., Eds.), pp. 163–174. Springer-Verlag, Italia.

Zummo, G., Acierno, R., Agnisola, C., and Tota, B. (1995). The heart of the icefish: bioconstruction and adaptation. *Braz. J. Med. Biol. Res.* **28**, 1265–1276.

4

ANTIFREEZE PROTEINS AND ORGANISMAL
FREEZING AVOIDANCE IN POLAR FISHES

ARTHUR L. DEVRIES
C.-H. CHRISTINA CHENG

   I. Introduction
  II. Freezing Challenge to Hyposmotic Teleost Fish in Ice-Laden Freezing Marine
      Environments
 III. Freezing Avoidance Strategies
      A. Migration to Ice-Free Habitats
      B. Undercooling in the Absence of Ice
  IV. Organismal Freezing Points
   V. Types of Antifreeze Proteins
      A. Antifreeze Glycoproteins
      B. Antifreeze Peptides
  VI. Noncolligative Lowering of the Freezing Point by Antifreeze Protein
      A. Freezing Behavior of Antifreeze Protein Solutions
      B. Adsorption-Inhibition Mechanism of Noncolligative Antifreeze Activity
 VII. Environmental Ice and Exogenous/Endogenous Ice in Polar Fish
      A. Endogenous Ice in Antarctic Fishes
      B. Fate of Endogenous Ice
VIII. The Integument as a Physical Barrier to Ice Propagation
  IX. Synthesis and Distribution of Antifreeze Proteins in Body Fluids
   X. Stability of Undercooled Fish Fluids Lacking Antifreeze Proteins
  XI. Serum Hysteresis Levels and Environmental Severity
      A. Constant Cold Extreme: Antarctic Marine Environments
      B. Thermal Variability: Arctic and North Subpolar Marine Environments
 XII. Mechanism of Organismal Freeze Avoidance

I. INTRODUCTION

Temperature is one of the key physical environmental factors governing
the distributions of living organisms. Although prokaryotes (eubacteria and
archaea) are able to populate the far extremes of thermal environments,

The Physiology of Polar Fishes: Volume 22
FISH PHYSIOLOGY

active metazoan eukaryote life is limited to a much narrower thermal range. For ectothermic animals unable to thermoregulate through metabolic means, the low end of the thermal range is represented by the polar regions. The year-round severe cold and absence of liquid water at the south polar region of Antarctica (land covered with an ~1-km ice sheet) has essentially excluded metazoan life. High-latitude polar oceans are also perennially frigid, but unlike terrestrial air temperatures that can plunge precipitously, polar seawater temperature seldom drops below its freezing point (f.p.) of −1.9°C because of the high heat capacity of water. The polar aquatic environments are found to sustain life of most of the major animal taxa, and some of them in large numbers. In the cold polar and northern cool temperate oceans, one of the major physiological/biochemical challenges for ectothermic animals is to generate sufficient energy for activity, growth, and reproduction at near-zero or subzero temperatures. It is well known that low temperatures depress the rate of reactions between molecules involved in energy-producing metabolic pathways and other biochemical and physiological processes. The presence of abundant invertebrate and vertebrate faunas in many parts of the polar seas indicates that these animals have adapted and met this low temperature challenge over evolutionary time. The largest marine vertebrate taxon is teleost fish, and polar species have provided many examples of biochemical and physiological adaptations to subzero temperatures (Hochachka and Somero, 2002). Although metabolic and biochemical adaptations are essential for life in the cold, the more direct and exigent environmental threat to survival for polar teleost fishes is freezing death. Marine teleost fishes have body fluids that are hyposmotic to seawater, which means they have a higher f.p. than seawater and would freeze in the −1.9°C polar marine environments. Unlike some other ectothermic animals that are freeze tolerant, freezing is always lethal in teleost fish (Scholander et al., 1957). Thus, the freezing polar seas, despite their much less extreme temperatures relative to land, present no less danger to the survival of teleost fish, the largest group of marine vertebrates. The ability to avoid freezing is important for teleosts in these environments before normal life can continue. To avoid freezing, many polar fishes have evolved antifreeze proteins (APs) that could lower the f.p. of their body fluids below ambient (DeVries, 1971). Since the discovery of the first fish antifreeze in the Antarctic fish in 1968 (DeVries and Wohlschlag, 1969; DeVries, 1970), significant advances have been made in the identification and characterization of new types of AP in other fish taxa and in the mechanism of action of these APs. Technological advances in protein chemistry have enabled very detailed descriptions of AP structures and formulation of hypotheses of how they bind to ice crystals and inhibit ice growth, leading to a large body of literature in these aspects (see reviews by Harding *et al.*, 1999, 2003; Davies *et al.*, 2002). In contrast, understanding the *in vivo* role of APs in freeze avoidance of the fish in their natural environments has lagged considerably behind. Ultimately, the biological importance of fish APs is their contribution to preserving organismal life and teleost diversity in otherwise uninhabitable freezing seas, and thus, how freezing avoidance is achieved in the natural environment decidedly warrants our thorough understanding. This chapter reviews freezing avoidance of polar fishes by means of APs, with particular attention to how antifreezes interface with the anatomy and physiology of the fish and in relation to the freezing marine environments they inhabit.

II. FREEZING CHALLENGE TO HYPOSMOTIC TELEOST FISH IN ICE-LADEN FREEZING MARINE ENVIRONMENTS

The equilibrium f.p. of a solution is determined by the concentration of dissolved solutes (molecules or ions) measured in osmolality (moles of osmotically active solutes per kilogram of water [Osm]). Oceanic seawater has a salt concentration generally equivalent to about 1030 mOsm, which depresses the colligative or equilibrium f.p. of water to −1.9°C. The equilibrium f.p. of a solution can be taken to be the melting temperature of the last ice crystal in a solution, which is also defined as the equilibrium melting point (m.p.). The blood and body fluids of marine teleost fishes are strongly hyposmotic to seawater. The blood salt content (mainly NaCl) of most marine teleost fishes is equivalent to about 300 mOsm and depresses the equilibrium f.p. of the blood to about −0.6 to −0.7°C (Black, 1951). Fishes living in the cold polar waters generally have a higher blood salt content than temperate-water fishes, with the highest found in some of the Antarctic notothenioid fishes and high latitude Arctic gadids (∼500–600 mOsm), and it accounts for most of the depression of the equilibrium f.p. of −1.0 to −1.1°C (DeVries, 1982; Denstad *et al.*, 1987; Enevoldsen *et al.*, 2003). However, these equilibrium f.p.s are still significantly higher than that of seawater (−1.9°C), and in the high latitude ice-covered polar seas, the fishes are undercooled (preservation of the liquid state at or below the equilibrium f.p. of blood) by as much as 0.9°C (DeVries and Wohlschlag, 1969). Although a difference of 0.9°C between the equilibrium f.p. of fish blood and seawater may appear numerically small, undercooling cannot exist and organismal freezing is certain if the freezing seawater temperature is accompanied by the presence of ice crystals in the water column. In ice-laden polar waters, fishes come in contact with ice crystals in the water column at their skin and gill surfaces and they ingest ice through food intake and drinking seawater. These ice crystals would provide the template that rapidly nucleates water molecules into the solid phase, leading to organismal freezing. Separation of environmental ice from undercooled body fluids by impermeable epithelial tissues can preserve the undercooled state, and intact skin of fish has been shown to provide this physical barrier (Turner *et al.*, 1985; Valerio *et al.*, 1992b). However, the skin can occasionally be breached by physical abrasion or by the attachment and burrowing of parasites, and the thin single-celled epithelium of the gill lamellae remains vulnerable to ice entry. Also, ice in the intestinal tract can lead to freezing of the hyposmotic intestinal fluid and consequently the fish. It takes only one nucleation site or entry of a single ice crystal to initiate freezing in an undercooled fish even if the amount of undercooling is small.

Freeze avoidance through the addition of colligative solutes in blood is rare for teleosts. There is only one known example of teleost fish in ice-laden environments, the rainbow smelt *Osmerus mordax*, which raises its serum osmolyte concentration until isosmotic with seawater by the synthesis of large amounts of glycerol in the winter (Raymond, 1992). The navaga cod *Eleginus navaga* from the inshore brackish habitats of the Russian White Sea also has a slightly elevated blood salt concentration and unidentified solutes, making it isosmotic with its brackish water environment (Christiansen *et al.*, 1995). A number of other polar fishes have evolved APs and rely on the APs' ability to inhibit ice growth in their blood to temperatures below that of ambient seawater ($-1.9°C$) to avoid freezing. APs circulate at high physiological concentrations when measured in mass per volume (mg/ml) but contribute very little to osmolal concentration because they are macromolecules. Thus, although APs substantially lower the temperature at which ice will grow (nonequilibrium f.p.), they affect the equilibrium f.p. or m.p. minimally as expected from colligative relationships (DeVries, 1971). The separation between the temperature of ice crystal growth or nonequilibrium f.p. and the equilibrium m.p., generally referred to as *thermal hysteresis*, is the hallmark of the fish APs. Interestingly, the rainbow smelt also synthesizes an AP, and thermal hysteresis was detected in navaga cod's serum, indicating the presence of an AP, along with elevated osmolyte concentrations.

III. FREEZING AVOIDANCE STRATEGIES

A. Migration to Ice-Free Habitats

Fishes not restricted by life histories or geographic confinement to icy freezing marine habitats can avoid freezing by seeking ice-free environments elsewhere. They either migrate to warmer offshore waters, a behavioral means used by some northern fishes, or occupy deep waters that are ice free due to pressure-dependent depression of the *in situ* f.p. [*in situ* f.p. is the f.p. at any given depth with the effect of pressure and other physical properties of seawater on f.p. taken into account (Millero, 1978)]. Examples of offshore migration include the longhorn sculpin *Myoxocephalus octodecemspinosus*, which moves off the coast into warmer deep water (4°C) during the winter (Leim and Scott, 1966), while the adult cunner *Tautogolabrus adspersus* moves offshore into the warmer coastal surface waters adjacent the Gulf Stream (Olla *et al.*, 1975). Interestingly, juvenile cunners remain in the shallow water but seek shelter under ledges and in burrows where they may avoid exposure to ice, or over winter in shallow burrows in the sediment in a state of torpor and may undercool at times as indicated by freezing of the fish when touched with a piece of ice (Olla *et al.*, 1975). The capelin *Mallotus villosus* is found in the sub-Arctic waters in the spring and summer where they spawn but move offshore during the winter (Narayanan *et al.*, 1995). In the high Arctic, Arctic char *Salvelinus alpinus* leave their resident streams and lakes in the summer and feed in the nearby marine waters but return to freshwater before the marine waters freeze (Dempson and Kristofferson, 1987). Some species of white fish (another salmonid) in the Bering Sea exhibit the same freeze-avoidance strategy.

B. Undercooling in the Absence of Ice

The ability of water samples to remain in the liquid state at temperatures below their equilibrium f.p. is referred to as *supercooling* or more correctly *undercooling*. Clean water in small volumes and free of particulate nucleators can be undercooled to −38.5°C before it spontaneously freezes, and this theoretical limit is termed the *temperature of homogeneous nucleation* (Angell, 1982). This illustrates that undercooling, though metastable, can exist as a persistent state. For non–AP-bearing polar marine fishes whose equilibrium f.p. based on blood salts and other osmolytes is −0.7°C, the amount of undercooling observed is at most only 1.2°C, because the equilibrium f.p. of even the most saline ocean water is about −1.9°C. In the absence of ice, fishes can be undercooled even more, as long as the bathing seawater does not spontaneously nucleate. In the laboratory, we could extensively undercool ice-free high-latitude Antarctic fishes by 7°C in a glycerol–seawater mixture for at least an hour, and the fish survived normally when returned to ambient seawater (DeVries and Cheng, 1992). Such extensive undercooling of ice-free specimens indicates the absence of exogenous and endogenous non-ice nucleators, and the stability of the undercooled state in fish. Some temperate fish such as killifish and cunner can be undercooled to −3°C (Scholander *et al.*, 1957), and thus, Arctic fishes very likely are also capable of extensive undercooling, although there have been no reported studies in the literature. Because of the stability of the under-cooled state in the absence of nucleators, some fish are able to employ modest undercooling as a means to avoid freezing. In the deep fjords of the Arctic region, undercooled fishes have been caught that quickly froze when brought to the surface and exposed to ice at −1.9°C (Scholander *et al.*, 1957). These undercooled deep-water fishes are in no danger of freezing as long as they remain at a depth where no ice formation occurs. In the Arctic, there are no thick ice shelves associated with coastal habitats that can generate extremely cold icy seawater at depth. F.p.–m.p. analyses of deep-water fishes in the Greenland Arctic indicate that they lack APs (Enevoldsen *et al.*, 2003), indicative of a freeze-avoidance strategy by undercooling in deep ice-free environment.

In contrast to the Arctic, ice can form even in the deep-water habitats (several hundred meters) of the high-latitude Southern Ocean because of the prevalence of ice shelves. Much of the Antarctic continental margin is associated with ice shelves, which are floating extensions of glaciers that flow from the immense inland ice sheet. The largest ice shelves, the Ross Ice Shelf of the Ross Sea and the Filchner-Ronne Ice Shelf of the Weddell Sea, have a thickness of 1 km and 600 m at their incipient edges, respectively. The ice shelf base provides a heat sink at these depths, and the water in contact with the shelf base may become as cold as −2.4°C (Nicholls and Makinson, 1998). Tidal currents could entrain this cold ice shelf water seaward and upward (Hunt *et al.*, 2003), and as it rises, it is undercooled with respect to the *in situ* equilibrium f.p. and could result in ice production deep in the water column (Foldvik and Kvinge, 1974). In McMurdo Sound, which is adjacent to the Ross Ice Shelf, this ice production is very likely involved in the formation of the abundant subice platelet layer on the underside of hard surface sea ice, as well as anchor ice mats on shallow bottoms (40 m) (Hunt *et al.*, 2003). Our McMurdo Sound hydrographic measurements indicate that during austral winter, freezing seawater can be found as deep as 175 m at times (DeVries, unpublished observations), and in the Weddell Sea, ice crystals have been collected in a self-closing net at depths of 250 m, indicating ice production at that depth (Dieckmann *et al.*, 1986). Conductivity, salinity, and depth profiles show that in the Weddell Sea, freezing isotherms (i.e., depth above which ice formation occurs) could be as deep as 500 m (Foldvik and Kvinge, 1974).

Despite ice formation in mid to deep Antarctic water, some undercooled fishes are found at depths that may not be encroached upon by ice derived from the cold ice shelf water. The snail fish *Paraliparis devriesi* lives at 600–700 m on the bottom of McMurdo Sound, Antarctica, where water temperature is at about −1.93°C. Its blood contains small amounts of AP, which results in a nonequilibrium blood f.p. of −1.3°C (Jung *et al.*, 1995), and thus, it is undercooled by 0.6 °C. It freezes when brought up through ice-laden surface waters, indicating inadequate antifreeze protection to survive in the icy surface condition. At its habitat depth of 700 m, it would have an *in situ* f.p. of -1.73 °C and is thus still undercooled by two-tenths of a degree. The presence of ice at the liparid's habitat is unlikely because the base of the distal (seaward) edge of the Ross Ice Shelf at the transition to annual sea ice over McMurdo Sound diminishes to less than 500 m deep and presumably will not cause ice formation below that depth. The two species of Antarctic zoarcid fishes in the sound, *Lycodichthys dearborni* and *Pachycara brachycephalum* that live at about 500–600 m, have significant amounts of AP, and nonequilibrium f.p.s near that of the seawater (-1.9 °C) or slightly above. When brought to the surface, many will begin to freeze. In contrast, the predominant Antarctic fish group, the fully AP-fortified notothenioid fishes caught throughout the water column of the McMurdo Sound, have nonequilibrium f.p.s slightly below that of seawater (DeVries, 1988) and do not freeze when brought through ice-laden surface water with the exception of the deep-dwelling *Trematomus loennbergii* that occasionally freezes at surface. Thus, the Antarctic zoarcid and liparid fishes are the two major fish groups that lack sufficient AP to avoid freezing in high Antarctic shallow water and exist in an undercooled state at their deep-water habitats. The lack of complete protection in these two taxa may be related to the possibility that they are recent immigrants to the Antarctic continental shelf, and in the deeper offshore waters, they would not require APs. Liparids and zoarcids have a worldwide distribution, and many members are associated with deep continental shelf waters and some ocean basins.

The small amount of undercooling observed in various deep-water Arctic and Antarctic fishes appears to be metastable for the life of the fish as long as they remain at ice-free depths. In addition, the apparent stability of the undercooled state indicates a lack of exogenous ice in their deep-water environments or any efficient endogenous ice nucleators.

IV. ORGANISMAL FREEZING POINTS

Of utmost importance for teleost survival in ice-laden freezing environments is that the organismal f.p. (i.e., the temperature at which ice will propagate through the fish fluids) must be lower than ambient water temperature. The "organismal f.p." is defined here as the temperature at which a fish will freeze in the presence of exogenous or endogenous ice. Systematic whole fish f.p. determinations have not been done for most polar fishes because of logistical difficulties. We were able to determine the organismal f.p. for a number of Antarctic notothenioid fishes at the McMurdo Sound field research station. The measurement involves lowering the temperature of ice-free seawater in which the specimen is kept to below $-1.9\,^\circ$C, which can be achieved by adding appropriate amounts of glycerol to the seawater to increase solute concentration and lower seawater equilibrium f.p. We found that a glycerol/seawater mixture does not alter the blood osmolality or the intrinsic equilibrium f.p. of the fish for short periods (1–2 hours). Fish specimens caught from the wild were first held at $1\,^\circ$C seawater for a day to melt out any ice that may be present on or within the fish. The specimen is then transferred to the glycerol/seawater bath and periodically touched with a piece of ice on the skin as the temperature of the bath is gradually lowered. Organismal f.p. is the temperature at which inoculative freezing of the specimen occurs, which is extremely rapid with the specimen turning opaque and rigid. This undercooling/nucleation approach can also be used to assay whether wild-caught specimens carry endogenous or exogenous ice derived from the environment at various times of the year. If ice is associated with the fish somewhere, it will spontaneously freeze when organismal f.p. is reached without manually touching it with ice. With this technique, shallow water Antarctic notothenioid fishes of McMurdo Sound are found to have organismal f.p.s that are a few tenths of a degree below the ambient temperature of $-1.9\,^\circ$C (Table 4.1). Very similar organismal.f.p.s are obtained when using environmental fishes that harbor endogenous or exogenous ice crystal.

Comparisons of blood serum nonequilibrium f.p.s with organismal f.p.s indicate that the former are a relatively good proxy for the latter. However, the blood nonequilibrium f.p.s in the case of the Antarctic fishes strongly depend on the technique used for their determination. Earlier studies used the f.p. osmometer, which undercools samples at a preset cooling rate before freezing is initiated. The activity of some APs turned out to be

Table 4.1

Environmental Temperature, Presence of Ice, Organismal Freezing Points, and Blood Melting/Freezing Points for Some Polar Fishes

Species	Environmental temperature (°C)	Ice present	Organismal freezing point	Blood freezing point	Blood melting point
Pagothenia borchgrevinki	-1.9	Yes	-2.3	-2.75	-1.1
Trematomus bernacchii	-1.9	Yes	-2.2	-2.50	-1.14
Lycodichthys dearborni	-1.9	No	-1.9	-1.9	-0.9
Eleginus gracilis	-1.8	Yes	-1.9	-2.1	-1.1
Myoxocephalus verrucosus	-1.8	Yes	-2.0	-2.3	-0.9

cooling rate dependent, so their nonequilibrium f.p.s were substantially underestimated (Pearcy, 1961; DeVries and Wohlschlag, 1969), and some APs failed to show any antifreeze activity at all when rapidly frozen (Fletcher et al., 1981; Schrag and DeVries, 1982). Nonequilibrium f.p.s obtained by observing the growth of a polycrystalline seed ice that spans the inner diameter of a $10 \mu l$ capillary tube half full of serum (DeVries, 1986) corresponded more closely with organismal f.p.s than those obtained by other f.p. determination techniques (DeVries, 1988). Nonequilibrium f.p.s determined with the Clifton nanoliter osmometer/cryoscope using very small ($\leq 5 \mu m$ diameter) single ice crystal gave values a few tenths of a degree lower than those obtained with the capillary tube method for Antarctic notothenioids. The antifreeze activity of notothenioid blood is strongly dependent on the cooling rate and size of the test seed ice crystal, due to the presence of a second antifreeze component besides the well-studied major antifreeze glycoprotein (AFGP) in these fish. This second antifreeze is a 16-kDa peptide that occurs in low concentration in the blood and in itself has low hysteresis activity but synergistically enhances the activity of the major circulating AFGPs in notothenioid fish (Jin, 2003). With most of the Arctic fishes, the nonequilibrium f.p. (determined with a Clifton nanoliter osmometer) is a good proxy for their organismal f.p. because they lack other APs that may impart a seed ice crystal size–dependent activity.

V. TYPES OF ANTIFREEZE PROTEINS

The first fish AP was discovered in the Antarctic nototheniid fishes (Notothenioidei, family Nototheniidae) from McMurdo Sound and was identified as a series of glycoproteins (DeVries and Wohlschlag, 1969; DeVries, 1970). The first peptide antifreeze was identified in the winter flounder *Pseudopleuronectes americanus* (Duman and DeVries, 1974, 1976). Three additional types of peptide antifreeze have subsequently been discovered in northern hemisphere fishes and are well characterized. These different types of APs differ in protein sequence, secondary and tertiary structures (Figure 4.1). In most cases, near-identical APs have evolved in unrelated taxa, whereas different types are present in closely related taxa, epitomizing the processes of convergent and independent evolution (Cheng, 1998).

A. Antifreeze Glycoproteins

The Antarctic notothenioid fishes and several northern and Arctic gadid cods synthesize alanine-rich glycopeptide antifreeze (AFGPs) molecules (DeVries and Wohlschlag, 1969; DeVries *et al.*, 1970; Raymond *et al.*,

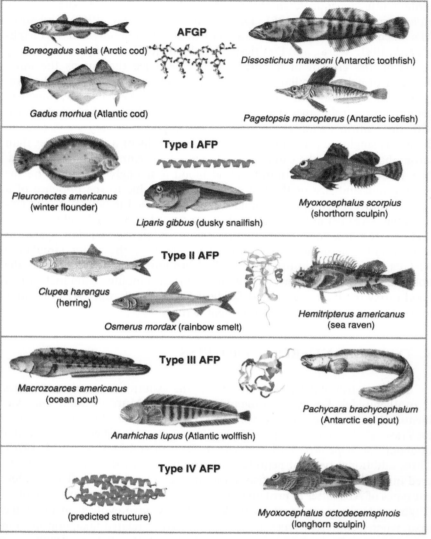

Fig. 4.1. Antifreeze protein structures shown along with the corresponding fish species. The structure of the gadid and notothenioid antifreeze glycoproteins (AFGPs), as well as the longhorn sculpin antifreeze peptide (AFP) are from structural modeling. The structures of type I, II, and III AFPs are experimentally solved by X-ray crystallography and/or nuclear magnetic resonance spectroscopy.

1975; Osuga and Feeney, 1978; Van Voorhies *et al.*, 1978; Fletcher *et al.*, 1981). The AFGP peptide backbone consists of Ala/Pro-Ala-Thr repeats with each Thr residue O-glycosylated by a disaccharide (*N*-acetyl-D-galactosamine-(α-1-3)-galactose (Shier *et al.*, 1972; Shier and DeVries, 1975). The different sizes are made up of different numbers of this basic glycotripeptide repeat and vary from 4 to 56 repeats (Cheng, 1996). Early characterizations identified eight distinct sizes on polyacrylamide gel electrophoresis and were named AFGP 1–8, with AFGP 1 being the largest size and the AFGP 8 the smallest (Table 4.2) (DeVries *et al.*, 1970). The two smallest sizes, AFGP 7 and AFGP 8, are the most abundant isoforms, comprising about two-thirds of circulating AFGPs in most notothenioid species. These eight sizes have been subcategorized into two groups, with the large molecular mass ones (AFGPs 1–5) having a peptide backbone of only Ala-Ala-Thr repeats, and the smaller ones (AFGPs 6–8) having a Pro occasionally replacing the first Ala of the tripeptides. Better gel resolution and protein visualization techniques subsequently revealed many more bands within each of these two groups (Figure 4.2), but the historical numbering system persists. Besides size heterogeneity, the small AFGPs, though appearing as single bands on gel electrophoresis, were found to be heterogeneous in composition, as the Pro-for-Ala replacement at the first alanine, can occur at one or more of the tripeptide repeats (DeVries *et al.*, 1971; O'Grady *et al.*, 1982b). Recent AFGP gene sequences showed that the large size AFGPs also contain some Pro-for-Ala substitutions, and the coding sequence for a large size isoform with as many as 88 tripeptide repeats (equivalent to ∼54 kDa) was found (Cheng and Chen, 1999). Together, the variation in the number of tripeptide repeats leading to length isoforms and the variation in the placement of the Pro residue leading to compositional isoforms result in a high degree of

Table 4.2
Molecular Weights of Glycoprotein Antifreezes Isolated from the
Blood of the Antarctic *Pagothenia borchgrevinki*[a]

Glycoprotein (Nr)	MW (daltons)
1	33,700
2	28,000
3	21,500
4	17,000
5	10,500
6	4200–7900
7	3500
8	2600

[a]Molecular weights determined by ultracentrifugation and polyacrylamide electrophoresis.

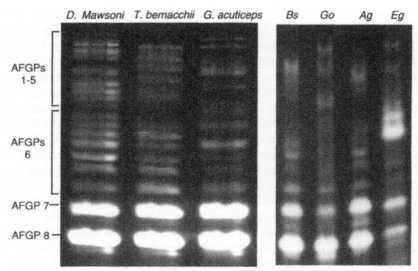

Fig. 4.2. Antifreeze glycoproteins (AFGPs) from Antarctic notothenioid fish (left) and northern codfishes (right) on gradient polyacrylamide gel electrophoresis showing the size heterogeneity. The AFGPs were labeled with a fluorescent tag (fluorescamine), and the fluorescent intensity of each isoform is representative of physiological concentration. AFGPs 6 and above resolved into multiple bands. The notothenioid AFGPs were run in triplicates. The cod species are as follows: *Bs, Boreogadus saida; Ag, Arctogadus glacialis; Go, Gadus ogac; Eg, Eleginus gracilis.*

protein heterogeneity within the AFGP family despite the simplicity of the tripeptide monomer.

Although not as well characterized in the Arctic gadids, the cod AFGPs also occur as a heterogeneous family of length isoforms (Figure 4.2), but generally the largest AFGP sizes are smaller than their Antarctic counterparts, and the saffron cod *Eleginus gracilis* has two very prominent isoforms in the AFGP 6 size range (Figure 4.2). The cod AFGPs are nearly identical in primary structure including Pro for Ala substitutions to the Antarctic version, the major difference being some of the Thr residues in the tripeptide repeats are replaced by an Arg or Lys residue (O'Grady *et al.*, 1982c; Chen *et al.*, 1997b).

Both the Antarctic notothenioid and the northern cod AFGPs are encoded by large polyprotein genes, each of which encodes a series of AFGP molecules linked in tandem by small cleavable spacer amino acid residues (Chen *et al.*, 1997a,b; Cheng and Chen, 1999). Detailed analyses of the AFGP genes from these two unrelated groups of fish show that they have evolved their respective AFGP from different genomic origin. The Antarctic notothenioid AFGP gene was derived from a trypsinogen-like serine protease gene, and its evolution was estimated to have occurred at about 7–15 million years ago (MYA), which correlated with the time estimate of the onset of sea-level glaciation of the Antarctic waters (Chen et al., 1997a). Although the evolutionary ancestry of the cod AFGP gene remains to be determined, it was not derived from a trypsinogen-like protease gene (Chen et al., 1997b). Thus, the AFGPs of Antarctic notothenioid fish and the northern cod arose by convergent evolution and represent a rare case of protein sequence convergence.

B. Antifreeze Peptides

A number of Arctic and north temperate fishes and two species of Antarctic fish synthesize one of four distinct types of antifreeze peptide (AFP). Northern winter flounder, plaice, sculpin, and snailfish have Ala-rich helical peptides referred to as type I AFPs (Duman and DeVries, 1976; DeVries and Lin, 1977a; Hew et al., 1985; Evan and Fletcher, 2001). There are usually three to four isoforms, with molecular weights of 3000–4000 Da in the flat fishes and sculpin, whereas the snailfish AFP is larger, about 9500 Da. A second Ala-rich antifreeze in winter flounder exists in very low concentration in the blood but has much greater activity and is larger in size, about 16,700 Da (Marshall and Davies, 2004). Type II AFPs are cysteine rich β-structured peptides of about 14,000–17,000 Da found in sea raven, smelt, and herring (Slaughter et al., 1981; Ewart and Fletcher, 1990; Ng and Hew, 1992; Ewart and Fletcher, 1993), and are related to the carbohydrate-binding domain of Ca^{2+}-dependent and Ca^{2+}-independent lectins with which they share about 30% protein sequence identity (Ewart et al., 1992; Ewart and Fletcher, 1993). Type III AFPs are small globular AFPs found in the Atlantic Ocean pout (Hew et al., 1984) and wolffish (Scott et al., 1988), and Antarctic eelpouts (Schrag et al., 1987; Cheng and DeVries, 1989). Their sizes are about 6500–7000 Da except for a 14,000-Da two-AFP domain variant that is also present in the Antarctic eelpout Lycodichthys dearborni (Wang et al., 1995b). They are related to the C-terminus of sialic acid synthase based on sequence similarity (Baardsnes and Davies, 2001; Consortium, 2001). Type IV AFP from the longhorn sculpin Myoxocephalus octodecemspinosus has a molecular mass of 12,000 Da, shares about 20% sequence identity with members of the exchangeable apolipoprotein super-family (Deng et al., 1997; Zhao et al., 1998), and is predicted to have a helix bundle structure (Deng and Laursen, 1998). The physiological significance or the role of type IV AFP in longhorn sculpin with respect to preventing organismal freezing is unclear because its very low in vivo concentrations produce insignificant hysteresis for freezing avoidance. When concentrated in vitro, it shows appreciable hysteresis and, therefore, can be considered an

AFP on a mechanistic basis. The AFPs as a rule have far fewer isoforms than the AFGPs. The least heterogeneous is the type III AFP of the Antarctic eelpout *Pachycara brachycephalum*, which exists as one predominant isoform (Cheng and DeVries, 1989). The distinct evolutionary origins of the AFPs are reflected in their distinct protein sequences and higher order structures, and their common ice binding and ice growth inhibition abilities exemplify functional convergence (Cheng, 1998).

VI. NONCOLLIGATIVE LOWERING OF THE FREEZING POINT BY ANTIFREEZE PROTEIN

Addition of nonvolatile solutes to pure water lowers the equilibrium f.p. of the solution linearly in a concentration-dependent manner, by 1.858°C/1000 mOsm of dissolved solute particles. This is one of the colligative properties of solution and a consequence of Raoult's Law governing the vapor pressure of solution. APs occur in substantial physiological concentrations (mass/volume), but these translate into very minor osmolal concentrations because of their macromolecular nature. For example, the approximately 7000-Da type III AFP of the Antarctic eelpout *P. brachycephalum* circulates at about 20 mg/ml in the blood, but this is equivalent to an osmotic concentration of only 3 mOsm and will effectuate only 0.0056°C of equilibrium f.p. depression based on colligative relationships. A seed ice crystal in a 20-mg/ml type III AFP solution will start to melt at this colligative m.p. of −0.0056°C, but it will not grow until the temperature is lowered to about −1.0°C, substantially lower than the colligative f.p (or m.p.). This freezing temperature is a noncolligative or nonequilibrium f.p., signifying the noncolligative, concentration-independent basis of the f.p.-depressing mechanism by APs. Thermal hysteresis, the measure for antifreeze activity, refers to this temperature difference between the equilibrium m.p. and the noncolligative f.p. in the presence of AP.

A. Freezing Behavior of Antifreeze Protein Solutions

The unusual lowering of the nonequilibrium f.p., but not the colligative or equilibrium m.p., first observed in AFGP-bearing Antarctic notothenioid fish serum, lead to detailed studies of the freezing behavior of solutions of purified APs and evaluations of the techniques to achieve accurate determination of their nonequilibrium f.p.s (DeVries, 1971; Raymond and DeVries, 1972). F.p. osmometers were first used for measuring blood nonequilibrium f.p.s and provided close but underestimated values in most cases (Pearcy, 1961). They were unreliable for estimating f.p.s of blood fortified with-

AFGPs (DeVries and Wohlschlag, 1969; Fletcher *et al.*, 1981, 1982a). In fact AFGPs 7 and 8, the smallest sized and most abundant AFGPs in nototheniod fish, showed no nonequilibrium f.p. depression when examined with an f.p. osmometer (Raymond and DeVries, 1972; Schrag and DeVries, 1982). It was also found that for some of the APs, the nonequilibrium f.p. depends strongly on the rate at which the AP sample was frozen (Raymond and DeVries, 1972; Schrag and DeVries, 1982), and the inaccurate f.p. determinations by the Fiske or Advanced® f.p. osmometer were due to a fast preset cooling rate. Later studies used slow cooling of small volumes of AP in 10-μl capillary tubes seeded with a small ($\sim$25 μm diameter) polycrystalline ice seed, and visual observation of the freezing and melting process of the seed ice crystal in the AP sample (Schrag and DeVries, 1982; DeVries, 1986). AFGPs 7 and 8 were found to have substantial antifreeze activity by this method (Schrag *et al.*, 1982). Most newer studies use the Clifton nanoliter osmometer/cryoscope, which allows controlled melting of the polycrystalline seed ice into a small single crystal. The temperature-dependent growth of the single seed ice crystal can be accurately accessed, and the effect of APs on ice crystal growth morphology can be visualized and captured with digital images (Fletcher *et al.*, 2001).

The freezing behavior of purified AFGPs and type I winter flounder AFP has been most thoroughly studied. These two APs show both similarities and differences in their effect on crystal growth habit. Very large single crystals in the presence of AFGPs 1–5 show no growth on the prism planes but a small amount on the basal plane, resulting in pitting of the surface in the "hysteresis gap" (Raymond *et al.*, 1989); however, a very small seed ice crystal (10 μm) in a 20-mg/ml solution of AFGPs 1–5 showed no detectable microscopic growth until the temperature was lowered to the nonequilibrium f.p. of about -1.2°C (Figure 4.3). Right below this temperature, a rapid burst of fine ice spicules propagate from the seed ice, with their long axis parallel to the c-axis, the non-preferred axis of growth for ice (Figure 4.3) (Raymond and DeVries, 1977; DeVries, 1982, 1986; Knight *et al.*, 1984). With the same concentration of winter flounder type I AFP, microscopic growth occurs within the hysteresis gap, and the resulting discoid seed crystal eventually forms a hexagonal bipyramid at which point growth stops (Chakrabartty *et al.*, 1989). Upon cooling below the hysteresis gap, ice spicules rapidly grow, but they tend to be thicker than those seen with the AFGPs. Likewise, with the small AFGPs 7 and 8, hexagonal bipyramids form in the hysteresis gap and coarse spicules form at the nonequilibrium f.p. In general, most of the AFPs show growth in the hysteresis gap, giving rise to hexagonal bipyramids at least at lower concentrations (2–10 mg/ml) (Raymond *et al.*, 1989; Chao *et al.*, 1995; Evan and Fletcher, 2001). Though not thoroughly documented, similar ice growth morphologies are observed in the native

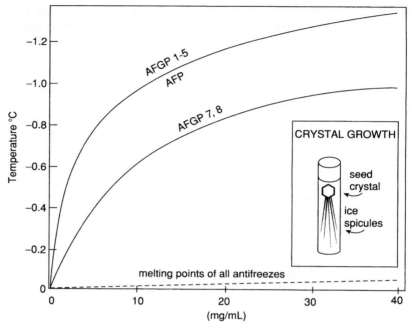

Fig. 4.3. Freezing temperatures and melting points of aqueous solutions of antifreeze glycoproteins (AFGPs) and selected antifreeze proteins (APs) (type III AFP) as a function of concentration. The freezing temperature of AFGPs 7 and 8 are only about one-third to two-third those of the large AFGPs, 1–5. All melting points are very similar. At temperatures below the hysteresis gap, rapid spicular ice growth occurs (as shown in the inset).

serum of some of the AFGP-bearing cods and AFP-bearing fishes where blood nonequilibrium f.p.s are higher than $-1.5°C$. In the case of the AFGP-bearing Antarctic notothenioid fishes and high Arctic gadids, although their small AFGPs (AFGPs 7 and 8) in pure solution yield hexagonal bipyramids over a wide range of concentrations, there is no indication of the pyramidal ice growth morphology in the native serum regardless of the magnitude of the hysteresis. There is only spicular growth from the small irregularly shaped seed crystals at temperatures below the hysteresis gap. Apparently the strong ice growth inhibition effects of the large AFGPs are sufficient to prevent any hexagonal bipyramid growth within the hysteresis gap. Electron micrographs of replicas of the spicular ice needles formed in the presence of both AFGPs and AFPs reveal that the needles are hexagonal in cross section and the blunt ends of some have pyramidal faces (Wilson *et al.*, 2002), suggesting that AP absorption occurs on the prism planes or orientations close to them.

A small amount of the most effective fish antifreeze, the large size group AFGPs 1–5, completely prevents microscopic ice growth in the blood, which one would expect if they were to prevent any ice expansion that causes tissue damage and interferes with physiological processes. This is particularly important for the high-latitude Antarctic notothenioid fishes because many of them carry endogenous ice for much of the year (Tien, 1995; Cziko *et al.*, 2005a). In those fishes that show hexagonal bipyramidal ice growth in the native serum, it may be that ice rarely enters into their body fluids, and thus, less potent APs are adequate to protect them from the occasional ice crystal that does enter. Investigations to determine the extent of endogenous "iciness" in the fish species with an AP that allows ice growth in the hysteresis gap would be worthwhile and informative.

B. Adsorption-Inhibition Mechanism of Noncolligative Antifreeze Activity

Despite their diversity in composition, structure, and size, all APs function by binding to ice and arresting its growth within the fluids of the fish. The mechanism most generally accepted is that of adsorption of AP molecules to ice, leading to inhibition of ice growth (Raymond and DeVries, 1977; Raymond *et al.*, 1989; Knight *et al.*, 1991). This initial model postulates that APs bind to ice through hydrogen bonding and involves a lattice match between regularly spaced hydrogen-bonding moieties in the protein and the water molecules in the ice crystal lattice (DeVries, 1984; Knight *et al.*, 1991). Extensive structural studies in the past 2 decades on the adsorption process to decipher the amino acid residues that comprise the ice-binding sites in different AFPs and the biophysical and energetic bases of the adsorption process have led to the revision of the hydrogen-bonding hypothesis for ice binding in some AFPs. These studies have been reviewed in great detail (Haymet *et al.*, 1999; Brown and Sonnichsen, 2002; Harding *et al.*, 2003), which the reader can consult for an in-depth discussion. A brief description of the adsorption-inhibition mechanism is provided here.

In vitro ice-binding studies revealed that different APs recognize and preferentially adsorb to specific crystallographic planes in the ice crystal (Knight *et al.*, 1991; Knight and DeVries, 1994), consistent with a different docking requirement by the structural binding sites specific to each AP, and the crystallographic orientation in the ice crystal lattice that best fits this requirement. The ice crystallography of antifreeze adsorption was examined experimentally by growing a single ice crystal of known orientation in low AP concentrations (0.1% w/v) into an ice hemisphere, followed by ice surface sublimation to determine the location and orientation of adsorbed antifreeze molecules (Knight *et al.*, 1991; Knight and DeVries, 1994; Wierzbicki *et al.*, 1996). Ice growth is not arrested at these low concentrations, and AP molecules bound to specific crystallographic planes of the growing ice surface become incorporated into the ice as the ice grows over them. Sublimation of the final ice hemisphere removes the surface layers of water molecules, and the regions where the bound antifreeze molecules appear as an etched surface like ground glass, while the rest of the hemisphere's surface is clear and smooth. In general, the diverse APs bind to low index planes of ice (i.e., primary and secondary prism planes) and pyramidal planes originating on prism planes, and the alignment direction for adsorbed molecules could be determined with reasonable confidence for APs that have an elongated structure—the notothenioid AFGPs (Knight *et al.*, 1993; Knight and DeVries, 1994) and type I AFP of the flat fishes and sculpin (Knight *et al.*, 1991; Wierzbicki *et al.*, 1996).

The AFGPs are thought to exist in solution either as an expanded polyproline type of helix (Bush and Feeney, 1986) or as an extended random coil structure (Raymond *et al.*, 1977; Brown and Sonnichsen, 2002). In the polyproline helix conformation, the disaccharides would be positioned on one side of the molecule, and this orientation would allow their hydroxyl groups to hydrogen bond to the surface of ice, resulting in the binding of the AFGP to ice. The small AFGPs 7 and 8 were found to adsorb to the prism planes $\{10\bar{1}0\}$ of ice and were deduced to align along the a-axes. The spacing of the tripeptide repeats (9.31Å) is about twice the periodicity of the a-axes (4.519Å). Thus, hydrogen bonding presumably occurs at the hydroxyls of the Thr-linked disaccharides of each repeat with alternative water molecules along the a-axes (Knight *et al.*, 1993). Type I AFP of winter flounder and plaice are near-perfect helices. They are found to adsorb on the pyramidal planes $\{20\bar{2}1\}$ of ice along the $<01\bar{1}2>$ direction, which has a periodicity that closely matches the lengths of the 11-residue repeats (16.5Å) of these AFPs, and binding presumably occurs at the four repeated polar Thr residues (Knight *et al.*, 1991). The helical type I AFP and AFGPs are both amphipathic, which would position the putative icebinding moieties linearly on one side of the molecule for alignment to the planar surface of ice. Chemical modifications of the hydroxyls of the AFGP disaccharides lead to loss of thermal hysteresis, which supports the hydrogen-bonding premise of ice binding (Shier *et al.*, 1972). However, for the winter flounder AFP, conservative replacement of Thr with Ser (preserving the putative icebinding hydroxyl) leads to substantial loss of antifreeze activity, whereas replacement of Thr by the nonpolar Val leads to less decrease in activity, which suggests that hydrogen bonding may not be the predominant force for binding (Chao *et al.*, 1997; Haymet *et al.*, 1999). It was also observed that the nonpolar Ala-rich face plus the adjacent Thr is a perfectly conserved region in type I AFP isoforms, and that modification of the immediate Ala neighbor of Thr affects ice binding (Baardsnes *et al.*, 1999). In addition, replacing Ala residues on the hydrophobic side of the helical type I AFP of sculpin antifreeze with charged Lys residues lead to loss of activity (Baardsnes *et al.*, 2001). Collectively these results indicate an important role for a hydrophobic interaction in the binding of type I AFP to ice.

Type II and type III AFP are globular proteins that do not contain repetitive sequence, so the ice-lattice match model does not appear to apply. The ice-binding residues of type II AFP have not been definitively identified but are inferred by site mutagenesis to be the same as the homologue of a carbohydrate-binding site in herring AFP (Ewart *et al.*, 1998) or by computational analysis to be a surface patch of 19 residues (Cheng *et al.*, 2002). The three-dimensional structure of type III AFP has been well resolved and the putative ice-binding residues have been identified and mapped in detail through extensive site-directed mutagenesis of selected residues (Jia *et al.*, 1996; Yang *et al.*, 1998; Ko *et al.*, 2003). Some of these studies indicate that the arrangement of the ice-binding residues constitutes a flat surface that matches with the planar surface of ice for binding to occur through hydrogen bonding (Yang *et al.*, 1998). Another contends that the putative binding surface of type III is relatively hydrophobic and the tight packing of the side chains on this surface precludes effective hydrogen bonding to ice (Sonnichsen *et al.*, 1996). The ice crystallography of type III AFP adsorption is not entirely resolved, showing binding primarily to the prism faces but also at orientations between prism and basal, which could not be correlated to any particular binding motif in the protein. Thus, despite extensive literature on the structure–function relationship of type III AFP, the precise nature of how it binds to ice is not definitive. The structure of type IV AFP has not been experimentally determined, but its primary sequence has a tandem 11-amino acid repeat pattern (Cheng, 1998), and the protein is predicted to be a helix bundle (Davies and Sykes, 1997; Deng and Laursen, 1998). The mode of type IV AFP binding remains to be determined.

The essential element of the inhibition mechanism is that APs adsorb to specific crystal planes and incompletely cover them (Wilson *et al.*, 1993), and thus, ice growth can occur only between the adsorbed AP molecules, resulting in growth fronts that are highly curved (Raymond and DeVries, 1977) (Figure 4.4). Highly curved growth fronts are less stable relative to planar fronts, and the growth of these fronts will cease once their local f.p. is equivalent to the undercooling of the solution. Further cooling will lead to an increase in curvature, but no visible bulk ice growth is seen until the nonequilibrium f.p. is reached. Conceptually, this process can be visualized as water molecules in highly curved fronts having fewer neighbors to hydrogen bond to than those on a planar front (DeVries, 1984). Therefore, at any given temperature within the hysteresis gap, the water molecules will spend more time in the liquid phase than in the solid phase. In order for them to remain in the highly curved front, energy must be taken out of the system, which can be accomplished by lowering the bulk temperature, equivalent to

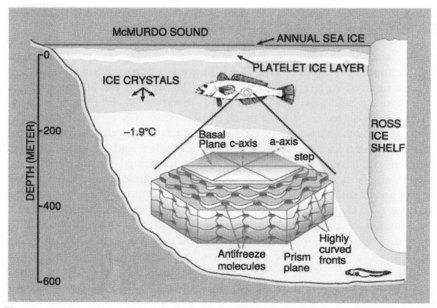

Fig. 4.4. A schematic of a notothenioid fish in McMurdo Sound, Antarctica, in ice-laden freezing seawater near the Ross Ice Shelf. An ice crystal is shown within the fish with antifreeze glycoprotein (AFGP) molecules adsorbed onto its prism planes, altering the crystal planes surfaces and preventing growth of the crystal.

lowering the f.p. of the solution. The generally accepted inhibition mechanism is explained on the basis of the Kelvin effect, and Brown and Sonnichsen (2002) give a thorough discussion of this mechanism.

VII. ENVIRONMENTAL ICE AND EXOGENOUS/ENDOGENOUS ICE IN POLAR FISH

The adsorption-inhibition mechanism presupposes the presence of ice somewhere in the body fluids of fish, such that the APs would be functionally and physiologically pertinent (DeVries, 1988). In other words, APs can only exert their antifreeze effect if an ice crystal is present to which they can bind, resulting in ice growth inhibition, which in turn prevents organismal freezing. The source of ice is the perennially freezing marine environments of both polar oceans, and the shallow coastal waters of the north temperate oceans that freeze in the winter months. Exogenous ice expectedly is acquired via contact association on surface tissues (integuments and gills), and ingestion through diet and seawater drinking (the gastrointestinal [GI] tract lumen can be considered a continuation of the external surface). Ice that somehow becomes acquired in other compartments of the fish is considered endogenous ice, and the route of its entry is unclear.

A. Endogenous Ice in Antarctic Fishes

In the near-shore waters of McMurdo Sound, Antarctica, notothenioid fishes are exposed to minute ice crystals in the water column for much of the year. To test for presence of ice, various tissues and fluids from environmental specimens were sampled under strict temperature control to prevent inadvertent melting and introduced into physiological saline that was undercooled to $-5°C$. Samples that contained ice would quickly nucleate and freeze the undercooled saline. As expected, surface tissues—skin, gills, and intestinal fluid—were found to contain ice. Surprisingly the deep-seated spleen also contained ice, but no other internal tissues or fluids tested positive for ice (Tien, 1995) (Table 4.3). The presence of ice in the deep-seated spleen indicates that ice must somehow enter across the body surface into the blood circulation and become lodged in the spleen because teleost spleens filter blood and sequester blood cells (Fänge and Nilsson, 1985; Franklin et al., 1993). There is also the possibility that when ice enters and becomes covered with adsorbed AFGP molecules, it is recognized as foreign and phagocytosed by immune cells or macrophages that return in the circulation to the spleen, because teleost spleen is also known to be the main site of destruction of aged blood cells and other effete blood cells (Fänge and Nilsson, 1985). If ice is transported to the spleen by blood, paradoxically, ice has not been detected in blood samples from fish that tested positive for splenic ice. However, this may be a false negative because the volume of blood sample withdrawn for testing is a minute fraction of the vascular volume, and the chance of it containing ice is probabilistically minute, especially if the total number of ice crystals in the blood circulation is small.

The cryopelagic nototheniid *Pagothenia borchgrevinki* forages in the subice platelet layer habitats where the water is coldest and ice most abundant, and all individuals tested positive for both exogenous and splenic ice. Experimental estimates of the number of splenic ice crystals in this species by use of the nucleation of undercooled saline assay of serial dilutions of splenic cell suspensions showed a range of 5–88 ice crystals per fish depending on the time of year (between October and January) (Cziko et al., 2005b). Thus, the frequency of ice entry into the circulation is not high, consistent with the probabilistic lack of ice detection in blood samples. There also appears to be

Table 4.3
Presence or Absence of Ice in Selected Tissues in the
Shallow-Water Antarctic, *Pagothenia borchgrevinki*[1]

Tissue or fluid	Present or absent
Skin	+
Gills	+
Stomach	+
Spleen	+
Muscle	−
Liver	−
Kidney	−
Heart	−
Brain	−
Blood	−
Ocular fluid	−
Urine	−
Intestinal fluid	+
Bile	−

[a] Presence or absence of ice was determined by nucleation of undercooled (−4.5°C) physiological saline using approximately 200 mg of environmental fish tissue or fluid.

a species difference in the spleen "iciness" among notothenioid fishes. McMurdo shallow water benthic species *Trematomus bernacchii* and *Trematomus hansoni* inhabit mats of anchor ice at 20 m or above and would be expected to have similar ice and low temperature exposure as *P. borchgrevinki*, but only about 60% of individuals that tested positive for exogenous ice are positive for endogenous ice in the spleen (DeVries, unpublished observations). The reason for this difference from *P. borchgrevinki* is unknown but may be related to resistance to ice propagation across their surface tissues.

The rate of ice acquisition in the environment has been examined by returning ice-free (by warming in 1 °C seawater) notothenioid fishes to their respective ice-laden habitats. Within a matter of hours, they acquire ice on their skin and gills and, within 1–2 days, in their intestinal fluid. However, it takes several days for the appearance of ice in the spleen. Furthermore, the kinetics of splenic ice acquisition is not a linear accumulation with time; no fish might reacquire splenic ice for 3 days, but on the fifth day, 90% might test positive for splenic ice. The sporadic acquisition of ice in the spleen suggests that the environmental conditions that lead to the accumulation of endogenous ice are unique and probably are not present on a daily basis. Freezing conditions in aquaria with the addition of very small ice crystals failed to introduce endogenous ice even over extended periods (Tien, 1995), and thus, the exact environmental conditions that lead to ice accumulation in the spleen are not completely understood.

B. Fate of Endogenous Ice

Presumably, ice enters the notothenioid fish through a breach in the gills, skin, or intestinal epithelial lining, and its growth in the interstitial fluids would be arrested by adsorption of AFGPs. Regardless of how ice eventually reaches the spleen, a steady accumulation of growth-arrested ice crystals in the spleen throughout the lifetime of the fish, which may span 10–15 years, would eventually create physiological problems for the fish if there is no mechanism to remove endogenous ice. Until recently, thermal melting of endogenous ice was thought impossible because the water of McMurdo Sound was believed to be perennially at or near its f.p. of −1.9°C (Littlepage, 1965), which is below the equilibrium m.p. of ice in the fish (−1.0 to −1.1°C). High-resolution multiyear temperature records of shallow-water habitats (9–40 m), however, revealed temperature fluctuations and warming episodes during the austral summer months (January–March), reaching peak water temperatures of about −0.5°C, and the total time of temperature excursions above −1.1°C, the temperature at which endogenous ice could melt, in the order of days (Hunt et al., 2003) (Figure 4.5). Thus, thermal melting as a mechanism to eliminate endogenous ice in the local fish during those warming episodes is a distinct possibility. However, not all individuals sampled during the early austral summer are ice free, suggesting endogenous ice may be melted later in the austral summer, may persist until the next austral summer, or there are other biological means of ridding them. In addition, the temperatures of deeper waters of the sound (100 m) do not rise above −1.5°C (Littlepage, 1965), which means that ice acquired in deeper water fish will not melt. There is some preliminary evidence for uptake of AFGP-adsorbed ice crystals by macrophages in notothenioid fish, so a disposal mechanism involving macrophages may exist. Macrophages are known to generate free radicals and acidify their lysosomes by proton pumping accompanied by chloride entry (Sonawane et al., 2002), so endocytosed ice may be melted in endosomes through ion transport to elevate the local m.p. The fate of endogenous ice in these fish is an interesting problem that awaits investigation albeit a technically difficult one.

Whether north polar fishes acquire endogenous ice like the Antarctic fish is not known. It is expected they will at least acquire exogenous ice during winter months. Unlike the thermally isolated Antarctic marine environment, northern oceans are in open communication and influenced by southern oceanic currents and, therefore, undergo large seasonal variation in water temperatures. Summer surface water temperatures could reach 8°C even in

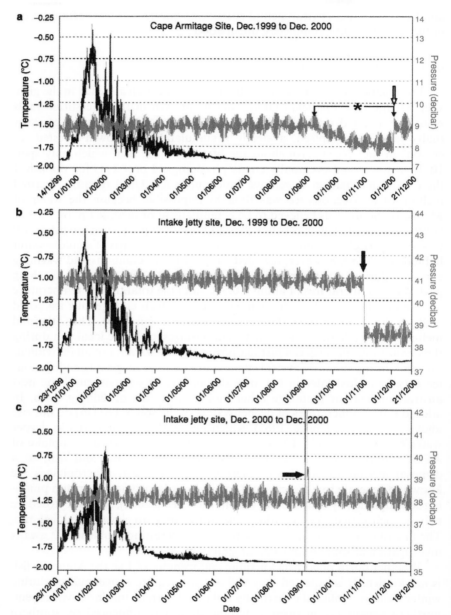

Fig. 4.5. Year-round seawater temperature (black trace) and benthic pressure (grey trace) recorded every 15 minutes from two McMurdo Sound inshore sites. (a) Cape Armitage site at 9m in 1999–2000. The depression in the pressure record (*) was caused by ice accumulation on

high latitudes (Enevoldsen *et al.*, 2003), so thermal melting could readily eliminate endogenous ice if fish acquires it in the winter months.

VIII. THE INTEGUMENT AS A PHYSICAL BARRIER TO ICE PROPAGATION

It is quite certain that the endogenous ice in the spleen of Antarctic notothenioid fishes is derived from entry of environmental ice through one or more of the surface tissues, because ice-free specimens returned to ice-laden water reacquire splenic ice. What is not certain is which particular tissue is the weak link. At the cellular level, it is well documented that the cell membrane constitutes a physical barrier to ice propagation. For some cells, an undercooling of several degrees is required for ice to pass through pores in the cell membrane (Mazur, 1966). Thus, for polar fishes where the undercooling is small, the driving force for propagation through cell membranes would be small. Propagation of ice through the corneal epithelial tissue (the clear head skin over the eye and corneal epithelium underneath) into the 0.5°C undercooled ocular fluid does not occur in Antarctic notothenioid fishes at its environmental temperature of -1.9°C (Turner *et al.*, 1985). The largest external surface—the skin, which consists of multiple cellular layers overlain with scales in most species and by mucous in all—would be expected to constitute a resistant physical barrier to ice entry. With winter flounder and ocean pout, 1.0°C of undercooling relative to the blood f.p. (-1.5°C) is required for *in vitro* trans-skin ice propagation for skin derived from AFP-fortified winter specimens (Valerio *et al.*, 1992b). In summer when winter flounder loses its AFPs, only 0.5°C undercooling was required, and addition of AFPs to the serosal side of skin resulted in greater undercooling needed for trans-skin ice propagation (Valerio *et al.*, 1992b). These results suggest that the presence of APs in the interstitial spaces of the skin is an important component of the resistance to ice propagation across the skin.

A number of AFP-bearing fishes have been found to produce a "skin-type" AFP that may have a role in preventing ice entry. Winter flounder, shorthorn sculpin, and longhorn sculpin skin-type AFP is type I in the Ala-rich feature but distinct from the circulatory AFP synthesized by the liver in these species (Gong *et al.*, 1996; Low *et al.*, 1998, 2001). The lack of secretory

the logger causing it to float, and dislodging the ice (⇩) returned the logger to its original depth. (b) The McMurdo Station saltwater intake jetty site at 40 m in 1999–2000, and (c) the same site in 2000–2001. In (b) (⬇) indicates when the logger was repositioned by divers, and in (c), (➡) indicates the logger was retrieved for downloading and then repositioned. (Reprinted from Hunt *et al.*, 2003.)

signal sequences in the skin-type AFP-coding sequences indicates that they are intracellular proteins and may function intracellularly, which is inconsistent with the known function of APs in protecting extracellular fluids from freezing (Fletcher *et al.*, 2001; Low *et al.*, 2001). In the case of the winter flounder skin AFP, though without a secretory signal, surprisingly it appeared to be found outside the cells also (Murray *et al.*, 2003). However, it is not clear how immunohistochemical detection will differentiate between interstitial fluid AFP derived from circulation (liver-type) versus the skin type, which are both Ala rich and have similar sequence stretches that would constitute similar epitopes, resulting in cross-reacting antibodies. Various suggestions have been raised as to what the biological significance of the skin-type AFP may be, but its actual role as a freezing prevention protein is far from certain and requires more definitive evidence. For example, there has been no estimation of what physiological concentrations of these skin-type AFPs may be, ostensibly because it is difficult to make such measurements. In all the reported studies, the activity of skin-type AFP was measured by concentrating isolated native protein or heterologously expressed recombinant protein to serum AFP levels, which may be far above the physiological concentration in the skin. If they are of insufficient *in vivo* concentrations to inhibit ice growth, then their role as antifreeze in fish is dubious. Adding to these uncertainties are two observations. First, skin-type AFP is not necessarily related to the blood type. Winter flounder and shorthorn sculpin skin-type AFPs resemble the liver type structurally, all small Ala-rich molecules, and thus conceivably could have been derived from a common antecedent with structural potentials to become an AFP. However, the longhorn sculpin skin-type AFP is unrelated to its own blood AFP in sequence (which has an inferred apolipoprotein ancestry) but is similar to the Ala-rich shorthorn sculpin and winter flounder skin-type AFP (Low *et al.*, 2001). Why a given species would evolve two distinct AFPs and express each in a different tissue type, though not impossible, defies a parsimonious explanation and suggests a possible separate function for one of the AFPs. The other observation is that skin-type AFP is also found in fish that has no blood AFP, casting doubt on a freeze-prevention role. Skin thermal hysteresis indicative of the presence of an AFP was detected in the cunner *Tautogolabrus adspersus* (Valerio *et al.*, 1990), which has no blood antifreeze, indicating no requirement for the freeze-avoidance function in this fish and consistent with the known behavioral means employed by the adults and juveniles to avoid freezing (Olla *et al.*, 1975). Similarly for longhorn sculpin, whose physiological serum AFP levels are too insignificant to confer freeze avoidance (Deng *et al.*, 1997), and which avoids freezing by moving offshore to warmer deep water (Leim and Scott, 1966), has a putative skin-type AFP. A survey of teleost fishes will be instructive, and if putative skin-type "AFP" is found to be prevalent in both AP and non-AP-bearing fishes, its presence may have a primary non-antifreeze function, and its ice-binding activity may be incidental to its structural characteristics. Skin antimicrobial agents are short peptides prevalent in fish (Cole *et al.*, 1997), and it is certainly worth investigating whether the putative skin AFP has antimicrobial properties.

The uncertain role of skin-type AFP notwithstanding, the skin in itself is anatomically an effective barrier to ice entry (Turner *et al.*, 1985; Valerio *et al.*, 1992b). The question remains as to how shallow-water Antarctic notothenioid fishes derive their endogenous ice. Given that *P. borchgrevinki* spleens contain only a few to less than about 100 ice crystals, even though it lives in the iciest habitat (the subice platelet layer), ice entry must not be highly frequent. Perhaps the site of entry is the more vulnerable gill lamellae epithelium, or the intestinal epithelium, where a single layer of cells on the apical side separates the ice-laden water from the blood, and the ice-carrying intestinal fluid from the interstitial fluid respectively. Lesions in these epithelia could come about either by damage from ingestion of hard chitinous prey, some of which may exit through the gills, or as it passes through the intestine. Another possibility is the presence of epi-parasites on the gills and intestinal parasites that create lesions during attachment to the epithelium. The gills and skin of notothenioid fishes are commonly infested by parasitic copepods, and their digestive tracts often heavily by cestode and nematode worms (Zdzitowiecki, 1998). Physically induced and parasite lesions are possible sites of ice entry, but these are only logical inferences, and definitive studies should be performed to verify the validity of the supposition.

IX. SYNTHESIS AND DISTRIBUTION OF ANTIFREEZE PROTEINS IN BODY FLUIDS

The largest fluid compartment in fish is blood. Because ice crystals could enter into the blood circulation, as demonstrated for the Antarctic notothenioid fish, fortification of the hypoosmotic blood serum by AP against freezing is critical. Liver is believed to be the site of synthesis and secretion of blood APs, and from the blood circulation, AP diffuses down its concentration gradient through leaky capillary junctions into other fluid compartments including the cerebral spinal fluid, extradural fluid, interstitial fluid, and peritoneal and pericardial fluids (Ahlgren *et al.*, 1988) (Table 4.4). A strong AFP mRNA signal is found in liver RNA of fishes that express their respective type of AFP (Hew and Yip, 1976; Lin, 1979; Lin and Long, 1980; Hew *et al.*, 1988; Ewart and Fletcher, 1993; Wang *et al.*, 1995a), consistent with the liver being the major synthesis and secretory source of circulatory

Table 4.4

Tissue and Fluid Distribution of Antifreeze Glycopeptides in the Antarctic Notothenioids
Pagothenia borchgrevinki and *Dissostichus mawsoni* based on Fluorescamine Polyacrylamide Gel
Electrophoresis and Distribution of Radiolabeled AFGPs and Polyethylene Glycol

Tissue or fluid	Presence of antifreeze glycopeptides
Blood plasma	+
Extradural	+
Peritoneal	+
Pericardial	+
Bile	+
Intestinal fluid	+
Aqueous humor	−
Vitreous humor	−[a]
Endolymph	−[a]
Urine	−

[a]Trace of AFGP-8 in these fluids.

AFP. Northern codfishes also have AFGP mRNA expression in their liver, similar to the AFP-bearing fish (Chen *et al.*, 1997b). Although it has been reported that the liver synthesizes the AFGPs in the Antarctic notothenioids (Hudson *et al.*, 1979; Haschemeyer and Mathews, 1980; O'Grady *et al.*, 1982a), our studies could not demonstrate definitive presence of AFGP mRNA in the liver of notothenioid fish (Figure 4.6), so it becomes unclear whether liver is the source of blood AFGPs in this Antarctic group.

The other fluid compartment that faces direct and imminent freezing threat is the fluid of the GI tract because it is exposed to frequent incoming dietary ice. The swallowed food and seawater is isosmotic with seawater in the esophagus where there is no digestion but becomes increasingly hyposmotic as it transits and becomes digested in the rest of the GI tract. In the stomach, the addition of fluids, acid secretion, and the small amount of protein digestion expectedly will begin to dilute the ingested seawater. We have found expression of AFGP mRNA in the anterior portion of the stomach wall, as well as presence of AFGP in the stomach fluid of notothenioid fishes, indicating AFGPs are secreted into the stomach to prevent freezing. In the small intestine, where the bulk of digestion occurs, large amounts of fluids including pancreatic bicarbonate and enzymes, bile, and intestinal mucosal secretions are added, further lowering solute concentration. In order to ion and osmoregulate, marine teleosts actively transport NaCl from the ingested seawater into the blood across the intestinal epithelium and ultimately out into the ambient water through chloride cells in the

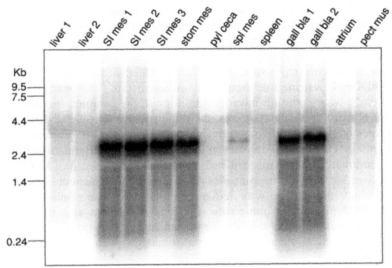

Fig. 4.6. Northern blot analysis for antifreeze glycoprotein (AFGP) messenger RNA (mRNA) in various tissues of *Dissostichus mawsoni*. About 8 μg of total RNA was loaded per lane and hybridized to a P^{32}-labeled AFGP coding sequence probe. Tissues: liver 1, surface layer of liver; liver 2, internal portion of liver; SI mes 1, 2, 3, pancreatic mesentery associated with the anterior, middle, and posterior section, respectively, of the small intestine; stom mes, pancreatic mesentery associated with stomach; pyl ceca, pyloric ceca stripped of pancreatic mesentery; spl mes, pancreatic mesentery associated with spleen; spleen, internal portion of spleen; gall bla 1, gallbladder with surface mesentery scraped; gall bla 2, gallbladder with surface mesentery intact; atrium; pect mus, pectoral muscle. AFGP mRNA signal (hybridized mRNA bands at about 2.5–3.0 kb) is found in all tissues that contain or have associating pancreatic tissue. Gallbladder (gall bla 1) has exocrine pancreas infiltration (Eastman and DeVries, 1997) and shows intense signal even when surface mesentery was scraped off.

gill (Evans, 1993). Absorption of NaCl from the intestinal fluid leaves behind a fluid isosmotic to the blood but hyposmotic (equilibrium f.p. about −1.0°C) to seawater and will likely freeze as it contains ice crystals. The presence of AFGP in intestinal fluids has been demonstrated (O'Grady *et al.*, 1983). AFGP concentrations in the fluid of the anterior portion of the small intestine are lower but become concentrated as water is reabsorbed along the tract toward the rectum where the fluid nonequilibrium f.p. is around −2.2°C, well below that of ambient seawater temperature (O'Grady *et al.*, 1983). Our studies indicate that a major source of intestinal AFGPs in notothenioid fishes is the pancreas, where there is very high AFGP mRNA expression (Figure 4.6), and a full complement of AFGPs are present in the pancreatic fluid that can be sampled in sufficient quantity from the large notothenioid *D. mawsoni* for analysis (Cheng *et al.*, 2005). The AFGPs secreted by the anterior stomach presumably exit the stomach undegraded (AFGPs are acid resistant) and add to their concentration in the small intestine. Pancreatic expression of AP has also been verified in the pancreatic tissue of AFGP-bearing northern cods and in various AFP-bearing species, and in cases where enough intestinal fluid could be sampled, the protein is present (Cheng *et al.*, 2005). Thus, the phylogenetically diverse AP-bearing fishes have converged on a common physiological solution, pancreatic synthesis and secretion of AP, to prevent intestinal freezing. In the notothenioid fishes examined, there is no evidence to suggest that intestinal AFGPs are reabsorbed, so they are presumed to be excreted with growth-arrested ice crystals they bind to or with undigested material. The persistently high AFGP mRNA signal in pancreatic tissues is consistent with the need to replenish intestinal loss of AFGPs. This is an energetic costly process but has to be maintained, as the alternative, freezing death, is the ultimate selection force.

If notothenioid fish liver expresses little or no AFGP mRNA, why hepatic synthesis of the protein could be measured in the early studies (Hudson *et al.*, 1979; O'Grady *et al.*, 1982a) is puzzling. A possible explanation is the presence of pancreatic tissues that might have infiltrated into the liver parenchyma along the hepatic portal vein and biliary duct system (Eastman and DeVries, 1997). Teleost pancreas is not a discrete organ, but a diffuse tissue scattered within mesenteries associated with all the components of the GI tract and non-GI tract organs in the abdominal cavity (Eastman and DeVries, 1997), and in some teleosts, it infiltrates into the liver, forming hepatopancreas islets (Hinton *et al.*, 1972; Kurokawa and Suzuki, 1995). Thus, the observed liver synthesis of AFGPs in the early studies may have its origin in the liver-associated pancreatic tissue and should be revisited with histological and immunohistological examinations. The question remains as to where the site is of synthesis for the circulatory AFGPs in the notothenioid fishes. The liver, which is the major source of secretory proteins in vertebrates, does not appear to express AFGP mRNA. Except for the anterior stomach and pancreatic tissue, thus far no other tissues or organs examined appear to express AFGP mRNA based on Northern blot analysis (Figure 4.6). If anterior stomach and pancreas are the sources of circulatory AFGPs in notothenioid fish, what may the anatomical or physiological route be that will transport the protein into blood circulation? Investigations into these fundamental questions may yet reveal unique and interesting physiology associated with the Antarctic notothenioid fishes.

X. STABILITY OF UNDERCOOLED FISH FLUIDS LACKING ANTIFREEZE PROTEINS

Some of the polar fishes' body fluids are nearly devoid of APs and are undercooled between 0.5 and 1°C relative to the seawater temperature (Dobbs and DeVries, 1975; Turner *et al.*, 1985) (Table 4.4). The urine and endolymph of Antarctic notothenioid fishes contain only trace amounts of the small AFGPs (Dobbs and DeVries, 1975; Ahlgren *et al.*, 1988). AFGPs essentially are absent in the urine of all high-latitude Antarctic notothenioid fishes, because of the aglomerular kidneys of these species where urine formation involves only secretion (Dobbs *et al.*, 1974; Eastman and DeVries, 1986). A few non-Antarctic notothenioid fishes that lack AFGPs appear to have a few glomeruli that may reflect the ancestral condition rather than a functional necessity of these structures (Eastman and DeVries, 1986). The bladder urine of the high-latitude Antarctic notothenioids is undercooled by about 1°C. It is enclosed within a thick muscular body wall covered by skin, and the interstitial spaces of both are well fortified with AFGPs (Ahlgren *et al.*, 1988). The only opening to the exterior is through the urinary sphincter and it is occluded by mucus. Only during micturition could ice enter, and propagation up the flowing stream of urine seems highly unlikely. There are no reported incidences of captured fishes with frozen urinary bladders, only in an occasional aquarium specimen whose bladder was catheterized for experimental purposes and exposed to small ice crystals in the flow through seawater. Thus, the body wall appears to constitute an excellent barrier and the modest urine undercooling can persist, even considering the periodic opening of the urethral sphincter. In contrast to the aglomerular notothenioids, the AFP-bearing Antarctic eelpouts *Lycodichthys dearborni* and *Pachycara brachycephalum* have kidney glomeruli, but no filtration occurs because the filtration barrier is thick, and in most nephrons, the Bowman's capsule does not appear to connect to the tubule. They are, thus, functionally aglomerular with urine formation resulting from tubular secretion, and do not lose their AFP into the formative urine (Eastman *et al.*, 1979).

In the case of northern and Arctic fishes, few comprehensive studies have been done, but the winter flounder does have a glomerular kidney and urine formation involves filtration and reabsorption from the formative urine (Petzel and DeVries, 1980). The flounder type I AFP is small and would be expected to filter through into the primary urine in the Bowman's space. However, flounder AFP molecules are anionic and are largely prevented from passing through the anionic basement membrane of the capillaries because of charge–charge repulsion, resulting in their conservation in the circulation (Petzel and DeVries, 1980; Boyd and DeVries, 1983). Other studies indicate that there are substantial AFPs in the urine of the winter flounder, sea raven, cod, and ocean pout (Fletcher *et al.*, 1989). It is possible that renal conservation mechanisms cannot achieve complete conservation, and a small amount of AP is lost to the urine. Teleost bladder wall is known to actively reabsorb salt and water (Demarest, 1984) and the high concentration of AFP observed in the bladder urine of these fishes may be the consequence of concentrating the small amounts of AFP from a large volume of ureteral urine. Even considering the relatively low glomerular filtration rates in most marine fishes, if the APs are not largely conserved at the filtration barrier, the fish would be unable to maintain the necessary protective blood levels of AFPs because the synthesis rates are not expected to be high, particularly at the lower temperatures. In the case of the winter flounder where filtration rates are 0.5 ml/kg/hr (Petzel and DeVries, 1979), if AFPs were filtered and not reabsorbed, all its circulating concentration of AFP would be lost in 48 hours. Although AP has been reported in the urine of cod and ocean pout (Fletcher *et al.*, 1989), histological examinations indicate that their kidneys are largely functionally aglomerular (Eastman *et al.*, 1987). The AP in the urine perhaps represents a small amount of leakage through the nephron and becomes concentrated because of the salt and water reabsorption processes in the bladder and not because of loss by filtration. More definitive research is needed to document kidney filtration rates of AFPs and to determine whether relatively high AP levels reported in bladder urine are an artifact of the concentration process of ureteral urine in the bladder.

The endolymph of the Antarctic notothenioid fishes lacks hysteresis even when the most sensitive techniques are employed, a finding consistent with fluid secretion by active transport of ions followed by osmotic influx of water into the semicircular canals. The undercooled endolymph in the semicircular canals is buried deep within the skull and surrounded by tissues fortified with AFGPs, so it is highly unlikely that it would ever be nucleated by ice. The ocular fluids on the other hand are separated from the environment by a relatively thin transparent head skin and cornea. The vitreous and aqueous fluids do contain small amounts of the small AFGPs, as determined by gel electrophoresis and hysteresis measurements (Turner *et al.*, 1985), but not enough to prevent freezing. However, the ocular fluid undercooling (0.5°C) is small, and the cornea and the overlying transparent head skin constitute an effective barrier to the inward propagation of ice at −1.9°C (Turner *et al.*, 1985). These two tissues lack visible vascularization, but their interstitium is fortified with AFGPs. Collectively, these factors prevent ice nucleation of the undercooled ocular fluids.

XI. SERUM HYSTERESIS LEVELS AND
ENVIRONMENTAL SEVERITY

The most commonly used indicator of the relative level of APs in fish blood is the difference between the equilibrium m.p. and nonequilibrium f.p., termed thermal hysteresis (Duman and DeVries, 1975; DeVries, 1982). In both Arctic and Antarctic fishes, there is a strong correlation between environmental temperature and ice abundance and the magnitude of the serum thermal hysteresis of the fish (Duman and DeVries, 1975; DeVries, 1982). However, the biogeographies of the Antarctic and Arctic Oceans are drastically different, directly affecting the synthesis and circulating levels of antifreeze in the AP-bearing Antarctic and northern fishes.

A. Constant Cold Extreme: Antarctic Marine Environments

Antarctica is a lone south polar continent surrounded by the vast Southern Ocean. The Southern Ocean, though abutting all other oceans in the Southern Hemisphere, is in effect isolated from them by the massive clockwise moving Antarctic Circumpolar Current (ACC), which spans 200–1000 km and reaches the sea floor (Foster, 1984). The ACC is both a thermal barrier, decoupling the warm subtropical gyres from the cold Antarctic waters, and a physical barrier against migration of fish in either direction. The unrestricted flow of the ACC presumably began around 25 MYA, leading to the thermal isolation and subsequent glaciation of Antarctica and its water, reaching present-day conditions around mid-Miocene (10–15 MYA) (Kennett, 1977, 1982). Thus, the Antarctic waters, though spanning a vast latitudinal range (~55°S to 78°S), are the coldest in the world, extremely cold stable at high latitudes, and experience only minor variations at the lower latitudes, with peak austral summer temperature at about −0.5°C and 2°C, respectively (Clarke, 1987; Hunt et al., 2003). Correlated with these constant frigid conditions is the constitutive presence of AFGPs in the endemic notothenioids, and at very high circulatory concentrations in the high-latitude species in McMurdo Sound (78°S) (DeVries et al., 1970; Ahlgren and DeVries, 1984). Variations in environmental exigency for fishes in the Southern Ocean do exist and are related to the presence or absence of nearby ice shelves and to the depth of the fish habitat, both of which affect the extent of iciness in the water column that's already at freezing temperatures.

Water temperature and ice formation reach their extremes in waters adjacent to the large ice shelves in the Southern Ocean. This is the case with the marine environment of McMurdo Sound, Antarctica (78°S 166°E),

Table 4.5

Comparison of Depth, Environmental Temperature, Presence or Absence of Ice,
Blood Freezing/Melting Points, and Hysteresis in Polar Fish

Species	Depth (m)	Water	Ice present	Blood (FP) (°C)	Blood (MP) (°C)	Hysteresis (°C)
High Antarctic McMurdo Sound						
Pagothenia borchgrevinki	3–30	−1.93	+	−2.7	−1.1	1.6
Trematomus bernacchii	10–250	−1.93	+	−2.5	−1.1	1.4
Trematomus loennbergii	200	−1.93	−	−2.2	−1.1	1.1
Lycodichthys dearborni	500	−1.93	−	−1.9	−1.0	1.0
Antarctic Peninsula						
Notothenia coriiceps	2–300	−1.8	+,−	−2.2	−1.0	1.2
Chaenocephalus aceratus	30–300	−1.0	−	−1.5	−0.9	0.6
High Arctic						
Boreogadus saida	2	−1.8	+	−2.2	−1.1	1.1
Myoxocephalus verrucosus	3–50	−1.8	+,−	−2.3	−0.9	1.4
Eleginus gracilis	5	−1.8	+	−2.1	−1.0	1.1

which is under the influence of the largest Antarctic ice shelf nearby, the Ross Ice Shelf, where tidal advections of extremely cold ice shelf water likely contribute to recurring ice formation in the water column (Hunt *et al.*, 2003). Much of McMurdo Sound is covered with hard surface sea ice throughout the year, as thick as 3–4 m at some locations. McMurdo notothenioid fish species are found in the platelet ice layer under the surface hard ice, in shallow waters, and in the deepest part of the sound (750 m), and the water temperatures spanning this huge depth vary by only a few hundredths of a degree. This provides an unparalleled opportunity to examine the effects of depth and environmental ice condition on the amount of serum hysteresis in closely related fish (Table 4.5). The cryopelagic notothenioid *P. borchgrevinki* that is closely associated with the subice platelet layer has the greatest serum hysteresis (1.6°C), as this water column habitat is the coldest, and the subice platelet layer is continually growing, making it also the iciest habitat. Benthic species such as *T. bernacchii*, which live on mats of anchor ice in shallow water (30 m), have slightly less hysteresis (1.4°C), whereas the deep-water (300 –500 m) species such as *T. loennbergii* have less serum hysteresis (1.1°C), consistent with a lower frequency of encountering ice at depth. In fact *T. loennbergii* will freeze when brought to the surface when icy freezing conditions exist in the surface waters, indicative of lower serum hysteresis levels. Similarly, the McMurdo Sound eelpout *Lycodichthys dearborni*, which live at 500–600 m bottoms, have similar depth-related lower serum hysteresis (1.0°C) and suffer freezing death when brought up to icy surface water like *T. loennbergii*. In contrast, the Antarctic Peninsula marine environment near Anvers Island (64°45′S 64°03′W) is relatively milder due to its lower latitudinal position and to the absence of a nearby ice shelf that may trigger ice nucleation in the water column. Consistent with these less extreme conditions, the common local shallow-water notothenioid species *Notothenia coriiceps* and *Chaenocephalus aceratus* have substantially lower serum hysteresis of 1–2°C and 0.6°C, respectively (DeVries, 1988), as compared to their McMurdo Sound shallow-water counterparts (Ahlgren and DeVries, 1984) (Table 4.5).

B. Thermal Variability: Arctic and North Subpolar Marine Environments

Unlike the isolated Antarctic, the Arctic has no north central polar land mass, and the Arctic Ocean is covered by dynamic multilayer pack ice. The Arctic Ocean and north cool-temperate oceans are open to temperate oceanic influences and experience much greater annual changes of temperatures. Northern fish populations are not restricted hydrographically (only by life histories), and some codfishes in particular are well known to make migrations across large geographic distances and, thus, temperature clines (Howe, 1991; Kurlansky, 1997). Correlated with these variables are latitudinal and seasonal variations in antifreeze levels in different AP-bearing northern fishes. There are two high Arctic gadids, the polar cod *Boreogadus saida* that occurs around Spitzbergen (78° 00N, 17°03′E), and the ice cod (also known as East Siberian cod) *Arctogadus glacialis* that is known to occur in the Arctic basin (Cohen *et al.*, 1990) and has been caught in substantial numbers in the fjords near Uummannaq, west Greenland (70°N, 51°W) (Præbel and Ramlov, 2004). These coastal waters are freezing (−1.9°C) and icy in the winter and warm up to 1–2°C in the summer, but *A. glacialis* has been documented to be associated with a cold layer of water (−1°C) at 20–50 m even in July in the Uummannaq Fjord (Jordan *et al.*, 2001; Præbel and Ramlov, 2004). Thus, polar cod and ice cod can be considered the north polar counterparts of the Antarctic notothenioids in being stenothermal and in the extent of exposure to environmental severity. Winter serum hysteresis of these two high Arctic gadids has not been reported, but *B. saida* caught in mid-January and kept in 0.6°C aquarium water has a mean serum hysteresis of 1.15°C (Cheng, Christiansen, and Fevolden, unpublished observations), indicating that the value is likely higher at −1.9°C water in the wild. The AFP-bearing high Alaskan Arctic sculpin *Myoxocephalus verrucosus*, a permanent shallow-water inhabitant, has very high serum hysteresis of 1.4°C (Table 4.5).

Much larger thermal changes occur at the lower northern latitude marine environments. From 41°N to 46.5°N along the north Atlantic coast between

Shinnecock Bay, New York, and Newfoundland, Canada, where populations of two cod species *Microgadus tomcod* and *Gadus morhua* have been
studied (Hew *et al.*, 1981; Fletcher *et al.*, 1982b; Reisman *et al.*, 1984),
shallow water falls to only about −1.2 °C, is ice-covered in the winter under
the influence of the cold Labrador current, and rises to as high as 20 °C to
12 °C, respectively in the summer (Petzel *et al.*, 1980; Fletcher *et al.*, 1985a).
The coastal habitats of saffron cod *E. gracilis* in the Alaskan Arctic, at Nome
(about 65 °N) and Kotzebue (67.5 °N) are more severe, freezing up in early
November and remaining so for almost 7 months. Ice breaks up toward the
end of May, and the surface water can warm up to as high as 12 °C in
midsummer. These latitudinal differences in environmental severity correlate
with the maximum winter hysteresis levels of these boreal cods. The peak
winter serum thermal hysteresis is about 0.66 °C for the Newfoundland
Atlantic cod, *G. morhua* (Fletcher *et al.*, 1982b), 0.5 °C and 0.68 °C for the
tomcod *M. tomcod* from Shinnecock Bay, NY, and Nova Scotia, respectively (Duman and DeVries, 1974; Fletcher *et al.*, 1982b; Reisman
et al., 1984), and about 1.1 °C for the Alaskan saffron cod *E. gracilis*
(Raymond *et al.*, 1975) and the Labrador Fjord cod *Gadus ogac* (Van
Voorhies *et al.*, 1978).

Aside from latitudinal differences in peak winter levels in lower northern
latitudes, blood antifreeze levels often cycle with seasonal variations in water
temperature, including the Newfoundland *G. morhua* (Fletcher *et al.*,
1982b), Shinnecock Bay *M. tomcod*, and grubby sculpin (Reisman *et al.*,
1984, 1987), type III AFP of ocean pout (Fletcher *et al.*, 1985b) and *Zoarces
viviparus* off the coast of Denmark (Sørensen and Ramløv, 2001), and type I
AFP of the north Atlantic winter flounder (Petzel *et al.*, 1980; Fletcher *et al.*,
1985a). For these fishes, generally bloodborne antifreeze appears or its level
begins to rise during autumn, usually between October and November
before any danger of freezing, reaching peak levels in the winter months
(January to March) when water temperatures are at their lowest, and then
declines to either lower levels, as in the ocean pout (Fletcher *et al.*, 1985b), or
completely disappears, as in the winter flounder sculpin, the two cods and
the viviparous zoarcid (Petzel *et al.*, 1980; Fletcher *et al.*, 1985a; Reisman
et al., 1987; Sørensen and Ramløv, 2001) when water temperature rises in
April or May. The exact timing of the rise, peak, and decline phases differs
between species and between different geographic populations within the
same species (Fletcher *et al.*, 1985a,b).

Annual antifreeze level cycles have not been reported for the high Arctic
species. However, Greenland cod *G. ogac* and polar cod *B. saida* caught
during August in Disko Bay of West Greenland (70 °N) have substantial
serum thermal hysteresis, 0.72 °C and 0.87 °C respectively, but Atlantic cod
G. morhua caught in the same water showed no hysteresis (Enevoldsen *et al.*,

2003). In addition, *B. saida* caught in mid-January and kept at 6°C aquarium water for 21/2 months has a mean serum hysteresis of 0.92°C (Cheng, Christiansen, and Fevolden, unpublished observations). The ice cod *A. glacialis* also maintains high levels of hysteresis (1°C) in the summer (Præbel and Ramlov, 2004). These data indicate that the seasonal cycles of AFGP levels of high Arctic cods differ from cods that principally populate lower latitudes such as *G. morhua* and *M. tomcod*. The presence of substantial serum hysteresis in the summer in the Greenland cod, polar cod, and ice cod correlates with the northerly range of their habitats and, thus, the degree and duration of environmental severity they encounter. *G. ogac* occurs mostly between 50°N and 72°N, in the coastal water of West Greenland and scattered northeast and northern coastal areas of North America (Cohen *et al.*, 1990). *B. saida* and *A. glacialis* are by far the most northerly cod species. Their distribution is circum-Arctic, widely found in the north polar basin and along the ice edge, and rarely below 60°N (Cohen *et al.*, 1990; Fevolden *et al.*, 1999). Their high summer (August) antifreeze levels indicate either little loss in antifreeze levels in the summer or that upregulation of AFGP production has already commenced in anticipation for winter.

Serum hysteresis values have been used as a proxy of circulating concentrations of APs in the blood of fishes, but the quantitative relationship between the two is not strictly linear, because different AP isoforms within a species could have different antifreeze activities (Schrag *et al.*, 1982; Kao *et al.*, 1985; Marshall and Davies, 2004). In addition, the proportion of different isoforms in the blood is often difficult to determine. This is especially true with the high degree of AFGP protein heterogeneity in the case of the Antarctic notothenioid fishes, where some have a different ratio of the more potent large AFGP isoforms to the less active small isoforms (Jin, 2003). Only in the case where a single isoform exists can the hysteresis value be used to accurately estimate the concentration of circulating AFP in the fish, and the only known example is that of the Antarctic eelpout *P. brachycephalum* (Cheng and DeVries, 1989). Thus, unless techniques are developed that can accurately determine the concentration of APs in the blood, the magnitude of serum hysteresis remains the best indicator of the antifreeze potential in the fish and has correlated well with the severity of the environment in terms of low temperature and ice abundance.

XII. MECHANISM OF ORGANISMAL FREEZE AVOIDANCE

Past hypotheses of organismal freezing avoidance suggested that the cornea and skin fortified with APs comprise a significant barrier to ice propagation into fish (Turner *et al.*, 1985; Valerio *et al.*, 1992b). Our documentation of endogenous ice in Antarctic notothenioids suggests that the skin and other epithelial surfaces—the gills and lining of the gut—are not absolute barriers, because ice must have transited one or more of these surfaces to eventually arrive at the spleen. Upon reintroduction of ice-free fish to their natural environment, endogenous ice in spleen takes several days to appear, so ice entry is not as fast as one would predict based on the ice abundance in the environment. Nonetheless, and importantly, this experiment demonstrates that ice can and does enter into fish under natural conditions. The length of time for splenic ice to appear and the relatively small numbers of ice crystals that the spleen carries indicate that the epithelial surfaces can retard ice entry but are not absolute barriers. Thus, freeze avoidance of fishes in the extreme environment of McMurdo Sound most likely involves both resistance to entry of ice at the epithelial surfaces and adsorption to the occasional ice crystal that enters by AFGPs followed by inhibition of its growth. The surface epithelial barriers, therefore, constitute an effective first line of defense, and the adsorption of APs on the few crystals that do enter and inhibition of their growth are an equally important second line of defense when the epithelial barriers are breached.

Atlantic cod larvae of the northern stock have an equilibrium f.p. of $-0.88\,^{\circ}$C (determined from larval homogenate) and no hysteresis but can resist inoculative freezing when touched by ice down to $-1.35\,^{\circ}$C, indicating integumental resistance to ice entry (Valerio *et al.*, 1992a). Gill lamellae in cod larvae only appear 5 weeks posthatch (Hunt von Herbing *et al.*, 1996a,b), so the more vulnerable single-cell gill epithelium is absent for that period. More evidence that the fish integument can be a very effective barrier comes from freezing studies of the McMurdo notothenioid fish larvae of various species in ice-laden seawater. Immediately upon hatching in austral Winter (late August to early September), the larvae of the naked dragonfish *Gymnodraco acuticeps* swim from their nests on the rocks directly to the subice platelet layer at the underside of the surface hard ice, where ice crystals are abundant and the water temperature is at its coldest for the year. Microsampling of the blood of these larvae showed it has very low levels of hysteresis, with a nonequilibrium f.p. of only $-1.2\,^{\circ}$C, and thus are undercooled by $0.7\,^{\circ}$C with respect to ambient seawater, and theoretically should freeze. Paradoxically the larvae sought out the iciest freezing habitat, the subice layer, presumably to avoid predation. Underwater observations showed that these larvae do not freeze, and laboratory-freezing experiments indicate that they are resistant to freezing to temperatures below $-2\,^{\circ}$C. The logical and perhaps only conclusion that can be drawn is that their integumental surfaces including gill epithelium are absolute barriers to ice entry at this life cycle stage (Cziko *et al.*, 2005a). Only at 22 weeks after hatch do the blood hysteresis levels reach those of the adults, suggesting that they cannot undercool when early juvenile stages are reached, become prone to ice entry, and require full complement of AFGPs for protection. Intestinal fluid hysteresis of dragonfish larvae is initially twice that of the blood, indicating perhaps that the gut lining is a more imminent site of ice entry, because they begin feeding as soon as they hatch, so intestinal fluid freezing by dietary ice must be prevented. In contrast to the low blood hysteresis in the Antarctic dragonfish larvae, the larvae of Z. *viviparus* from the Roskilde Fjord, Denmark, have almost adult levels of antifreeze activity upon parturition in February (Sørensen and Ramløv, 2002).

In the case of developing fish embryos, Atlantic cod eggs, with a mean f.p. of $-0.78°C$ (egg homogenate) and no hysteresis, can resist freezing to $-4.0°C$ if mechanically disturbed or to as low as -10 to $-17°C$ if undisturbed (Valerio *et al.*, 1992a). Thus, the chorion must be highly impermeable to inoculative freezing, so the ooplasm can undercool by such large magnitudes. Similar egg chorion resistance to ice propagation also exists in the developing embryos of the Antarctic notothenioid, the dragonfish *G. acuticeps.*

Avoidance of freezing for polar fish throughout their life cycle including fertilization, larval development, and adult stages is of great importance because if frozen they die and are eliminated from the gene pool. Much remains to be learned about how the integumental resistance to ice entry at the various epithelial surfaces and about the fate of ice that does occasionally enter because some year-round freezing marine environments such as the high Antarctic preclude thermal melting.

ACKNOWLEDGMENT

We gratefully acknowledge the U.S. National Science Foundation, Office of Polar Programs, for grant support to A. L. D and C.-H. C. C. for our research in both the Antarctic and the Arctic described in this chapter.

REFERENCES

Ahlgren, J. A., Cheng, C.-H. C., Schrag, J. D., and DeVries, A. L. (1988). Freezing avoidance and the distribution of antifreeze glycopeptides in body fluids and tissues of antarctic fish. *J. Exp. Biol.* **137,** 549–563.

Ahlgren, J. A., and DeVries, A. L. (1984). Comparison of antifreeze glycoproteins from several Antarctic fishes. *Polar Biol.* **3,** 93–97.

Angell, C. A. (1982). Supercooled water. *In* "Water: A Comprehensive Treatise" (Franks, F., Ed.), Vol. 7, pp. 1–81. Plenum Press, New York.

Baardsnes, J., and Davies, P. L. (2001). Sialic acid synthase: The origin of fish type III antifreeze protein? *TIBS* **24,** 468–469.

Baardsnes, J., Jelokhani-Niaraki, M., Kondejewski, L. H., Kuiper, M. J., Kay, C. M., Hodges, R. S., and Davies, P. L. (2001). Antifreeze protein from shorthorn sculpin: Identification of the ice-binding surface. *Protein Sci.* **10**, 2566–2576.

Baardsnes, J., Kondejewski, L. H., Hodges, R. S., Chao, H. M., Kay, C., and Davies, P. L. (1999). New ice-binding face for type I antifreeze protein. *FEBS Lett.* **463**, 87–91.

Black, V. S. (1951). Some aspects of the physiology of fish. *Univ. Toronto Stud. Biol. Ser.* **71**, 53–89.

Boyd, R. B., and DeVries, A. L. (1983). The seasonal distribution of anionic binding sites in the basement membrane of the kidney glomerulus of the winter flounder *Pseudopleuronectes americanus. Cell Tissue Res.* **234**, 271–277.

Brown, J. D., and Sonnichsen, F. D. (2002). The structure of fish antifreeze proteins. *In* "Fish Antifreeze Proteins" (Ewart, K. V., and Hew, C. L., Eds.), pp. 109–138. World Scientific, Singapore.

Bush, C. A., and Feeney, R. E. (1986). Conformation of the glycotripeptide repeating unit of antifreeze glycoprotein of polar fish as determined from the fully assigned proton N.M.R. spectrum. *Int. J. Pept. Protein Res.* **28**, 386–397.

Chakrabartty, A., Yang, D. S., and Hew, C. L. (1989). Structure–function relationship in a winter flounder antifreeze polypeptide: II. Alteration of the component growth rates of ice by synthetic antifreeze polypeptides. *J. Biol. Chem.* **264**, 11313–11316.

Chao, H. M., Deluca, C. I., and Davies, P. L. (1995). Mixing antifreeze protein types changes ice crystal morphology without affecting antifreeze activity. *FEBS Lett.* **357**, 183–186.

Chao, H. M., Houston, M. E., Hodges, R. S., Kay, C. M., Sykes, B. D., Loewen, M. C., Davies, P. L., and Sonnichsen, F. D. (1997). A diminished role for hydrogen bonds in antifreeze protein binding to ice. *Biochemistry* **36**, 14652–14660.

Chen, L., DeVries, A. L., and Cheng, C. C.-H. (1997a). Evolution of antifreeze glycoprotein gene from a trypsinogen gene in Antarctic notothenioid fish. *Proc. Natl. Acad. Sci. USA* **94**, 3811–3816.

Chen, L., DeVries, A. L., and Cheng, C. C.-H. (1997b). Convergent evolution of antifreeze glycoproteins in Antarctic notothenioid fish and Arctic cod. *Proc. Natl. Acad. Sci. USA* **94**, 3817–3822.

Cheng, C.-H., and Chen, L. (1999). Evolution of an antifreeze glycoprotein. *Nature* **40**, 443–444.

Cheng, C.-H. C. (1996). Genomic basis for antifreeze glycopeptide heterogeneity and abundance in Antarctic notothenioid fishes. *In* "Gene Expression and Manipulation in Aquatic Organisms" (Ennion, S. J., and Goldspink, G, Eds.), **58**, pp. 1–20. Cambridge University Press, Cambridge.

Cheng, C.-H. C. (1998). Evolution of the diverse antifreeze proteins. *Curr. Opin. Genet. Dev.* **8**, 715–720.

Cheng, C.-H. C., and DeVries, A. L. (1989). Structures of antifreeze peptides from the antarctic eel pout, *Austrolycichthys brachycephalus. Biochim. Biophys. Acta* **997**, 55–64.

Cheng, C.-H. C., Logue, J., Cziko, P., and Chen, L. (2005). Prevention of intestinal freezing by pancreatic expression of antifreeze proteins: A common mechanism in all antifreeze-bearing fish. *J. Biol. Chem.* (manuscript submitted).

Cheng, Y., Yang, Z., Tan, H., Liu, R., Chen, G., and Jia, Z. (2002). Analysis of ice-binding sites in fish type II antifreeze protein by quantum mechanics. *Biophys. J.* **83**, 2202–2210.

Christiansen, J. S., Chernitsky, A. G., and Karamushko, O. V. (1995). An Arctic teleost fish with a noticeably high body fluid osmolality: A note on the navaga, *Eleginus navaga* (Pallas 1811), from the White Sea. *Polar Biol.* **15**, 303–306.

Clarke, A. (1987). Seasonality in the Antarctic marine environment. *Comp. Biochem. Physiol.* **90**, 461–473.

Cohen, D. M., Inada, T., Iwamoto, T., and Scialabba, N. (1990)."Gadiform Fishes of the World." FAO Fisheries Synopsis, No. 125 Vol. 10. United Nations, Rome.

Cole, A. M., Peddrick, W., and Diamond, G. (1997). Isolation and characterization of pleur-ocidin, an antimicrobial peptide in the skin secretions of winter flounder. *J. Biol. Chem.* **272**, 12008–12013.

Consortium, I. H. G. S. (2001). Initial sequencing and analysis of the human genome. *Nature* **409**, 860–921.

Cziko, P., Cheng, C.-H. C., and DeVries, A. L. (2005a). Freezing avoidance in Antarctic larval dragon fish. Manuscript in preparation.

Cziko, P., Hunt, B. M., Præbel, K., and DeVries, A. L. (2005b). Endogenous ice in McMurdo Sound fishes. Manuscript in preparation.

Davies, P. L., Baardsnes, J., Kuiper, M. J., and Walker, V. K. (2002). Structure and function of antifreeze proteins. *Phil. Trans. R. Soc. Lond.* B **357**, 927–935.

Davies, P. L., and Sykes, B. D. (1997). Antifreeze proteins. *Curr. Opin. Struct. Biol.* **7**, 828–834.

Demarest, J. R. (1984). Ion and water transport by the flounder urinary bladder: Salinity dependence. *J. Physiol.* **246**, F395.

Dempson, J. B., and Krístofferson, A. H. (1987). Spatial and temporal aspects of the ocean migration of anadromous Arctic char. *Am. Fish. Soc. Symp.* **1**, 340–357.

Deng, G., Andrews, D. W., and Laursen, R. A. (1997). Amino acid sequence of a new type of antifreeze protein from the long horn sculpin *Myoxocephalus octodecimspinosis. FEBS Lett.* **402**, 17–20.

Deng, G., and Laursen, R. A. (1998). Isolation and characterization of an antifreeze protein from the longhorn sculpin, *Myoxocephalus octodecimspinosis. Biochim. Biophys. Acta* **1388**, 305–314.

Denstad, J.-P., Aunaas, T., Borseth, J. F., Aaset, A. V., and Zachariassen, K. E. (1987). Thermal hysteresis antifreeze agents in fishes from Spitsbergen waters. *Polar Res.* **5**, 171–174.

DeVries, A. L. (1970). Freezing resistance in antarctic fishes. *In* "Antarctic Ecology." (Holdgate, M. W., Ed.), pp. 320–328. Academic Press, New York.

DeVries, A. L. (1971). Glycoproteins as biological antifreeze agents in Antarctic fishes. *Science* **172**, 1152–1155.

DeVries, A. L. (1982). Biological antifreeze agents in cold water fishes. *Comp. Biochem. Physiol.* **73A**, 627–640.

DeVries, A. L. (1984). Role of glycopeptides and peptides in inhibition of crystallization of water in polar fishes. *Trans. Roy. Soc. Lond.* B **304**, 575–588.

DeVries, A. L. (1986). Glycopeptide and peptide antifreeze—Interaction with ice. *In* "Methods in Enzymology" (Colowick, S. P., and Kaplan, N. O., Eds.), Vol. 127, pp. 293–303. Academic Press, New York.

DeVries, A. L. (1988). The role of antifreeze glycopeptides and peptides in the freezing avoid-ance of Antarctic fishes. *Comp. Biochem. Physiol.* **90B**, 611–621.

DeVries, A. L., and Cheng, C.-H. C. (1992). The role of antifreeze glycopeptides and peptides in the survival of cold-water fishes. *In* "Water and Life: Comparative Analysis of Water Relationships at the Organismic, Cellular and Molecular Levels" (Somero, G. N., Osmond, C. B., and Bolis, C. L., Eds.), pp. 301–315. Springer-Verlag, Berlin Heidelberg.

DeVries, A. L., Komatsu, S. K., and Feeney, R. E. (1970). Chemical and physical properties of freezing point depressing glycoproteins from Antarctic fishes. *J. Biol. Chem.* **245**, 2901–2908.

DeVries, A. L., and Lin, Y. (1977a). Structure of a peptide antifreeze and mechanism of adsorption to ice. *Biochim. Biophys. Acta* **495**, 388–392.

DeVries, A. L., Vandenheede, J., and Feeney, R. E. (1971). Primary structure of freezing point-depressing glycoproteins. *J. Biol. Chem.* **246**, 305–308.

DeVries, A. L., and Wohlschlag, D. E. (1969). Freezing resistance in some Antarctic fishes. *Science* **163**, 1073–1075.

Dieckmann, G., Rohardt, G., Hellmer, H., and Kipfstuhl, J. (1986). The occurrence of ice platelets at 250m depth near the Filchner Ice Shelf and its significance for sea ice biology. *Deep*-Sea Res. **33,** 141–148.

Dobbs, G. H., and DeVries, A. L. (1975). Renal function in Antarctic teleost fishes: Serum and urine composition. *Mar. Biol.* **29,** 59–70.

Dobbs, G. H., Lin, Y. L., and DeVries, A. L. (1974). Aglomerularism in Antarctic fish. *Science* **185,** 793–794.

Duman, J. G., and DeVries, A. L. (1974). Freezing resistance in winter flounder, *Pseudopleuronectes americanus. Nature* **247,** 237–238.

Duman, J. G., and DeVries, A. L. (1975). The role of macromolecular antifreezes in cold water fishes. *Comp. Biochem. Physiol.* **52A,** 193–199.

Duman, J. G., and DeVries, A. L. (1976). Isolation, characterization, and physical properties of protein antifreezes from the winter flounder, *Pseudopleuronectes americanus. Comp. Biochem. Physiol.* **53B,** 375–380.

Eastman, J. T., Boyd, R. B., and DeVries, A. L. (1987). Renal corpuscle development in boreal fishes with and without antifreezes. *Fish Physiol. Biochem.* **4,** 89–100.

Eastman, J. T., and DeVries, A. L. (1986). Renal glomerular evolution in Antarctic notothenioid fishes. *J. Fish Biol.* **29,** 649–662.

Eastman, J. T., and DeVries, A. L. (1997). Morphology of the digestive system of Antarctic nototheniid fishes. *Polar Biol.* **17,** 1–13.

Eastman, J. T., DeVries, A. L., Coalson, R. E., Nordquist, R. E., and Boyd, R. B. (1979). Renal conservation of antifreeze peptide in Antarctic eelpout, *Rhigophila dearborni. Nature* **282,** 217–218.

Enevoldsen, L. T., Heiner, I., DeVries, A. L., and Steffensen, J. F. (2003). Does fish from the Disko Bay area of Greenland possess antifreeze proteins during the summer? *Polar Biol.* **26,** 365–370.

Evan, R. P., and Fletcher, G. L. (2001). Isolation and characterization of type I antifreeze proteins from Atlantic snailfish (*Liparis atlanticus*) and dusky snailfish (*Liparis gibbus*). *Biochim. Biophys. Acta* **1547,** 235–244.

Evans, D. H. (1993). Osmotic and ionic regulation. *In* "The Physiology of Fishes" (Evans, D. H., Ed.). CRC Press, Boca Raton.

Ewart, K. V., and Fletcher, G. L. (1990). Isolation and characterization of antifreeze proteins from smelt (*Osmerus mordax*) and Atlantic herring (*Culpea harengus harengus*). *Can. J. Zool.* **68,** 1652–1658.

Ewart, K. V., and Fletcher, G. L. (1993). Herring antifreeze protein: Primary structure and evidence for a C-type lectin evolutionary origin. *Mol. Mar. Biol. Biotech.* **2,** 20–27.

Ewart, K. V., Li, Z., Yang, D. S. C., Fletcher, G. L., and Hew, C. L. (1998). The ice-binding site of Atlantic herring antifreeze protein corresponds to the carbohydrate-binding site of C-type lectins. *Biochem.* **37,** 4080–4085.

Ewart, K. V., Rubinsky, B., and Fletcher, G. L. (1992). Structural and functional similarity between fish antifreeze proteins and calcium-dependent lectins. *Biochim. Biophys. Res. Commun.* **185,** 335–340.

Fänge, R., and Nilsson, S. (1985). The fish spleen: Structure and function. *Experientia* **41,** 152–158.

Fevolden, S. E., Martinez, I., and Christiansen, J. S. (1999). RAPD and scnDNA analyses of polar cod, *Boreogadus saida* (Pisces, Gadidae) in the north Atlantic. *SARSIA* **84,** 99–103.

Fletcher, G. L., Addison, R. F., Hew, C. L., and Slaughter, D. (1982a). Antifreeze proteins in the Arctic shorthorn sculpin (*Myoxocephalus scorpius*). *Arctic* **35,** 302–306.

Fletcher, G. L., Haya, K., King, M. J., and Reisman, H. M. (1985a). Annual antifreeze cycles in Newfoundland, New Brunswick and Long Island winter flounder *Pseudopleuronectes americanus*. *Mar. Ecol Prog. Ser.* **21**, 205–212.

Fletcher, G. L., Hew, C. H., and Davies, P. L. (2001). Antifreeze proteins of teleost fishes. *Annu. Rev. Physiol.* **63**, 359–390.

Fletcher, G. L., Hew, C. H., and Joshi, S. B. (1981). Isolation and characterization of antifreeze glycoproteins from the frost fish, *Microgadus tomcod*. *Can. J. Zool.* **60**, 348–355.

Fletcher, G. L., Hew, C. L., Li, X., Haya, K., and Kao, M. H. (1985b). Year-round presence of high levels of plasma antifreeze peptides in a temperate fish, ocean pout (*Macrozoarces americanus*). *Can. J. Zool.* **63**, 488–493.

Fletcher, G. L., King, M. J., Kao, M. H., and Shears, M. A. (1989). Antifreeze proteins in the urine of marine fish. *Fish Physiol. Biochem.* **6**, 121–127.

Fletcher, G. L., Slaughter, D., and Hew, C. L. (1982b). Seasonal changes in the plasma levels of glycoprotein antifreeze, Na$^+$, Cl$^-$, and glucose in Newfoundland Atlantic cod, *Gadus morhua*. *Can. J. Zool.* **60**, 1851–1854.

Foldvik, A., and Kvinge, T. (1974). Conditional instability of sea water at the freezing point. *Deep-Sea Res.* **21**, 169–174.

Foster, T. D. (1984). The marine environment. *In* "Antarctic Ecology" (Laws, R. M., Ed.), Vol. 2, pp. 345–371. Academic Press, London.

Franklin, C. E., Davison, W., and McKenzie, J. C. (1993). The role of the spleen during exercise in Antarctic teleost, *Pagothenia borchgrevinki*. *J. Exp. Biol.* **174**, 381–386.

Gong, Z., Ewart, K. V., Hu, Z., Fletcher, G. L., and Hew, C. L. (1996). Skin antifreeze protein genes of the winter flounder, *Pleuronectes americanus*, encode distinct and active polypeptides without the secretory signal and pro-sequences. *J. Biol. Chem.* **271**, 4106–4112.

Harding, M. M., Anderberg, P. I., and Haymet, A. D. J. (2003). Antifreeze glycoproteins from polar fish. *Eur. J. Biochem.* **270**, 1381–1392.

Harding, M. M., Ward, L. G., and Haymet, A. D. J. (1999). Type I antifreeze proteins: Structure-activity studies and mechanism of ice growth inhibition. *Eur. J. Biochem.* **264**, 653–665.

Haschemeyer, A. E. V., and Mathews, R. W. (1980). Antifreeze glycoprotein synthesis in the Antarctic fish *Trematomus hansoni* by constant infusion *in vivo*. *Physiol. Zool.* **53**, 383–393.

Haymet, A. D. J., Ward, L. G., and Harding, M. M. (1999). Winter flounder antifreeze proteins: Synthesis and ice growth inhibition of analogues that probe the relative importance of hydrophobic and hydrogen-bonding interactions. *J. Am. Chem. Soc.* **121**, 941–948.

Hew, C. L., Joshi, S., Wang, N. C., Cao, M.-H., and Ananthanarayanan, V. S. (1985). Structures of shorthorn sculpin antifreeze polypeptides. *Eur. J. Biochem.* **151**, 167–172.

Hew, C. L., Slaughter, D., Fletcher, G. L., and Joshi, S. (1981). Antifreeze glycoproteins in the plasma of Newfoundland Atlantic cod (*Gadus morhua*). *Can. J. Zool.* **59**, 2186–2192.

Hew, C. L., Slaughter, D., Joshi, S., Fletcher, G. L., and Ananthanarayanan, V. S. (1984). Antifreeze polypeptides from the Newfoundland ocean pout, *Macrozoarces americanus*: Presence of multiple and compositionally diverse components. *J. Comp. Physiol.* **B155**, 81–88.

Hew, C. L., Wang, N.-C., Joshi, S., Fletcher, G. L., Scott, G. K., Hayes, P. H., Buettner, B., and Davies, P. L. (1988). Multiple genes provide the basis for antifreeze protein diversity and dosage in the ocean pout, *Macrozoarces americanus*. *J. Biol. Chem.* **263**, 12049–12055.

Hew, C. L., and Yip, C. (1976). The synthesis of freezing-point depression protein of the winter flounder *Pseudopleuronectes americanus* in *Xenopus laevis* oocytes. *Biochim. Biophys. Res. Commun.* **71**, 845–850.

Hinton, D. E., Snipes, R. L., and Kendall, M. W. (1972). Morphology and enzyme histochemistry in liver of largemouth bass (*Micropterus salmoides*). *J. Fish Res. Board Can.* **29**, 531–534.

Hochachka, P. W., and Somero, G. N. (2002). "Biochemical Adaptation: Mechanism and Process in Physiological Evolution." Oxford University Press, New York.

Howe, G. J. (1991). Biogeography of gadoid fishes. *J. Biogeogr.* **18**, 595–622.

Hudson, A. P., DeVries, A. L., and Haschemeyer, A. E. V. (1979). Antifreeze glycoprotein biosynthesis in Antarctic fishes. *Comp. Biochem. Physiol.* **62B**, 179–183.

Hunt, B. M., Hoefling, K., and Cheng, C.-H. C. (2003). Annual warming episodes in seawater temperatures in McMurdo Sound in relationship to endogenous ice in notothenioid fish. *Antarct. Sci.* **15**, 333–338.

Hunt von Herbing, I., Miyake, T., Hall, B. K., and Boutilier, R. G. (1996a). Ontogeny of feeding and respiration in larval Atlantic cod *Gadus morhua* (Teleostei, Gadiformes). I. Morphology. *J. Morphol.* **227**, 15–35.

Hunt von Herbing, I., Miyake, T., Hall, B. K., and Boutilier, R. G. (1996b). Ontogeny of feeding and respiration in larval Atlantic cod *Gadus morhua* (Teleostei, Gadiformes). II. Function. *J. Morphol.* **227**, 37–50.

Jia, Z. C., Deluca, C. I., Chao, H. M., and Davies, P. L. (1996). Structural basis for the binding of a globular antifreeze protein to ice. *Nature* **384**, 285–288.

Jin, Y. (2003). Freezing avoidance of Antarctic fishes: The role of a novel antifreeze potentiating protein and the antifreeze glycoproteins. PhD Thesis, University of Illinois, Urbana-Champaign.

Jordan, A. D., Jurngersen, M., and Steffensen, J. F. (2001). Oxygen consumption of East Siberian cod: No support for the metabolic cold adaptation theory. *J. Fish Biol.* **59**, 818–823.

Jung, A., Johnson, P., Eastman, J. T., and DeVries, A. L. (1995). Protein content and freezing avoidance properties of the subdermal extracellular matrix and serum of the Antarctic snailfish, *Paraliparis devriesi. Fish Physiol. Biochem.* **14**, 71–80.

Kao, M. H., Fletcher, G. L., Wang, N. C., and Hew, C. H. (1985). The relationship between molecular weight and antifreeze polypeptide activity in marine fish. *Can. J. Zool.* **64**, 578–582.

Kennett, J. P. (1977). Cenozoic evolution of Antarctic glaciation, the circum-Antarctic Ocean, and their impact on global paleoceanography. *J. Geophys. Res.* **82**, 3843–3860.

Kennett, J. P. (1982). "Marine Geology." Prentice-Hall, New Jersey.

Knight, C. A., Cheng, C.-H. C., and DeVries, A. L. (1991). Adsorption of helical antifreeze peptides on specific ice crystal surface planes. *Biophys. J.* **59**, 409–418.

Knight, C. A., and DeVries, A. L. (1994). Effects of a polymeric, non-equilibrium "antifreeze" upon ice growth from water. *J. Crystal Growth* **143**, 301–310.

Knight, C. A., DeVries, A. L., and Oolman, L. D. (1984). Fish antifreeze protein and the freezing and recrystallization of ice. *Nature* **308**, 295–296.

Knight, C. A., Driggers, E., and DeVries, A. L. (1993). Adsorption of fish antifreeze glycopeptides to ice, and effects on ice crystal growth. *Biophys. J.* **64**, 252–259.

Ko, T. P., Robinson, H., Gao, Y.-G., Cheng, C.-H. C., DeVries, A. L., and Wang, A. H.-J. (2003). The refined crystal structure of an eel pout Type III antifreeze protein RD1 at 0.62A resolution reveals structural microheterogeneity of protein and solvation. *Biophys. J.* **84**, 1228–1237.

Kurlansky, M. (1997). "Cod: A Biography of the Fish That Changed the World." Penguin Books, New York.

Kurokawa, T., and Suzuki, T. (1995). Structure of the exocrine pancreas of flounder (*Paralichthys olivaceus*): Immunological localization of zymogen granules in the digestive tract using anti-trypsinogen antibody. *J. Fish Biol.* **46**, 292–301.

Leim, A. H., and Scott, W. B. (1966). "Fishes of the Atlantic Coast of Canada." Fish Research Board Canada, Ottawa.

Lin, Y. (1979). Environmental regulation of gene expression. *J. Biol. Chem.* **254**, 1422–1426.

Lin, Y., and Long, D. J. (1980). Purification and characterization of winter flounder antifreeze peptide messenger ribonucleic acid. *Biochemistry* **19**, 1111–1116.

Littlepage, J. L. (1965). Oceanographic investigations in McMurdo Sound, Antarctica. *In* "Biology of the Antarctic Seas II" (Llano, G. A., Ed.), Vol. 5, pp. 1–37. American Geophysical Union, Washington, DC.

Low, W. K., Lin, Q., Stathakis, C., Miao, M., Fletcher, G. L., and Hew, C. L. (2001). Isolation and characterization of skin-type, type I antifreeze polypeptides from the longhorn sculpin, *Myoxocephalus octodecemspinosus. J. Biol. Chem.* **276**, 11582–11589.

Low, W. K., Miao, M., Ewart, K. V., Yang, D. S., Fletcher, G. L., and Hew, C. L. (1998). Skin-type antifreeze protein from the shorthorn sculpin, *Myoxocephalus scorpius*: Expression and characterization of Mr 9700 recombinant protein. *J. Biol. Chem.* **273**, 23098–23103.

Marshall, C. B., and Davies, P. L. (2004). Hyperactive antifreeze protein in a fish. *Nature* **429**, 153.

Mazur, P. (1966). Physical and chemical basis of injury in single-celled miro-organisms subjected to freezing and thawing. *In* "Cryobiology" (Meryman, H. T., Ed.), pp. 214–310. Academic Press, London.

Millero, F. J. (1978). Freezing point of seawater—Eighth report of the joint panel of oceano-graphic tables and standards. *In* UNESCO Technical Papers in Marine Science 28 pp. 29–31. UNESCO, Paris.

Murray, H. M., Hew, C. H., and Fletcher, G. L. (2003). Spatial expression patterns of skin-type antifreeze protein in winter flounder (*Pseudopleuronectes americanus*) epidermis following metamorphosis. *J. Morphol.* **257**, 78–86.

Narayanan, S., Carscadden, J., Dempson, J. B., O'Connell, M. F., Prinsenberg, S., Reddin, D. G., and Shackell, N. (1995). Marine climate off Newfoundland and its influence on salmon (*Salmo salar*) and capelin (*Mallotus villosus*). *In* "Climate Change and Northern Fish Populations" (Beamish, R. J., Ed.), Vol. 121, pp. 461–474. Can. Spec. Publ. Fish. Aquat. Sci., Ottawa.

Ng, N. F. L., and Hew, C. L. (1992). Structure of an antifreeze polypeptide from the sea raven: Disulfide bonds and similarity to lectin binding proteins. *J. Biol. Chem.* **267**, 16069–16075.

Nicholls, K. W., and Makinson, K. (1998). Ocean circulation beneath the western Ronne Ice Shelf, as derived from *in situ* measurements of water currents and properties. *In* "Ocean, Ice and Atmosphere: Interactions at the Antarctic continental margin." (Jacobs, S. S., and Weiss, R. F., Eds.), Vol. 75, pp. 301–318. American Geophysical Union, Washington DC.

O'Grady, S. M., Clarke, A., and DeVries, A. L. (1982a). Characterization of glycoprotein antifreeze biosynthesis in isolated hepatocytes from *Pagothenia borchgrevinki. J. Exp. Zool.* **220**, 179–189.

O'Grady, S. M., Ellory, J. C., and DeVries, A. L. (1982b). Protein and glycoprotein antifreezes in intestinal fluid of polar fishes. *J. Exp. Biol.* **98**, 429–438.

O'Grady, S. M., Ellory, J. C., and DeVries, A. L. (1983). The role of low molecular weight antifreeze glycopeptides in the bile and intestinal fluid of Antarctic fishes. *J. Exp. Biol.* **104**, 149–162.

O'Grady, S. M., Schrag, J. D., Raymond, J. A., and DeVries, A. L. (1982). Comparison of antifreeze glycopeptides from Arctic and Antarctic fishes. *J. Exp. Zool.* **224**, 177–185.

Olla, B. L., Bejda, A. J., and Martin, D. (1975). Activity, movements, and feeding behavior of the cunner, *Tautogolabrus adspersus*, and comparison of food habits with young tautog, *Tautoga onitis*, off Long Island, New York. *Fishery Bull. Ill.* **733**, 895–900.

Osuga, D. T., and Feeney, R. E. (1978). Antifreeze glycoproteins from Arctic fish. *J. Biol. Chem.* **253**, 5338–5343.

Pearcy, W. G. (1961). Seasonal changes in osmotic pressure in flounder sera. *Science* **134**, 193–194.

Petzel, D. H., and DeVries, A. L. (1979). Renal handling of peptide antifreeze in northern fishes. *The Bulletin, Mount Desert Island Biological Laboratory* **19**, 17–19.

Petzel, D. H., and DeVries, A. L. (1980). Renal handling of anionic and cationic antifreeze peptides in the glomerular winter flounder. *The Bulletin, Mount Desert Island Biological Laboratory* **20**, 17.

Petzel, D. H., Reisman, H. M., and DeVries, A. L. (1980). Seasonal variation of antifreeze peptide in the winter flounder, *Pseudopleuronectes americanus*. *J. Exp. Zool.* **211**, 63–69.

Præbel, K., and Ramløv, H. (2005). Antifreeze activity in the gastrointestinal fluids of *Arctogadus glacialis* is dependent upon food type. *J. Exp. Biol.* (in press).

Raymond, J. A. (1992). Glycerol is a colligative antifreeze in some northern fishes. *J. Exp. Zool.* **262**, 347–352.

Raymond, J. A., and DeVries, A. L. (1972). Freezing behavior of fish blood glycoproteins with antifreeze properties. *Cryobiology* **9**, 541–547.

Raymond, J. A., and DeVries, A. L. (1977). Adsorption-inhibition as a mechanism of freezing resistance in polar fishes. *Proc. Natl. Acad. Sci. USA* **74**, 2589–2593.

Raymond, J. A., Lin, Y., and DeVries, A. L. (1975). Glycoprotein and protein antifreezes in two Alaskan fishes. *J. Exp. Zool.* **193**, 125–130.

Raymond, J. A., Radding, W., and DeVries, A. L. (1977). Circular dichroism of protein and glycoprotein fish antifreeze. *Biopolymers* **16**, 2575–2578.

Raymond, J. A., Wilson, P. W., and DeVries, A. L. (1989). Inhibition of growth of non-basal planes in ice by fish antifreeze. *Proc. Natl. Acad. Sci. USA* **86**, 881–885.

Reisman, H. M., Fletcher, G. L., Kao, M. H., and Shears, M. A. (1987). Antifreeze proteins in the grubby sculpin, *Myoxocephalus aenaeus* and the tomcod, *Microgadus tomcod*: Comparisons of seasonal cycles. *Env. Biol. Fishes* **18**, 295–301.

Reisman, H. M., Kao, M. H., and Fletcher, G. L. (1984). Antifreeze glycoprotein in a southern population of Atlantic tomcod, *Microgadus tomcod*. *Comp. Biochem. Physiol.* **78A**, 445–447.

Scholander, P. F., Van Dam, L., Kanwisher, J. W., Hammel, H. T., and Gordon, M. S. (1957). Supercooling and osmoregulation in Arctic fish. *J. Cell. Comp. Physiol.* **49**, 5–24.

Schrag, J. D., Cheng, C.-H. C., Panico, M., Morris, H. R., and DeVries, A. L. (1987). Primary and secondary structure of antifreeze peptides from arctic and antarctic zoarcid fishes. *Biochim. Biophys. Acta* **915**, 357–370.

Schrag, J. D., and DeVries, A. L. (1982). The effects of freezing rate on the cooperativity of antifreeze glycopeptides. *Comp. Biochem. Physiol.* **74A**, 381–385.

Schrag, J. D., O'Grady, S. M., and DeVries, A. L. (1982). Relationship of amino acid composition and molecular weight of antifreeze glycopeptides to non-colligative freezing point depression. *Biochim. Biophys. Acta* **717**, 322–326.

Scott, G. K., Hayes, P. H., Fletcher, G. L., and Davies, P. L. (1988). Wolffish antifreeze protein genes are primarily organized as tandem repeats that each contain two genes in inverted orientation. *Mol. Cell. Biol.* **8**, 3670–3675.

Shier, W. T., and DeVries, A. L. (1975). Carbohydrate of antifreeze glycoproteins from an Antarctic fish. *FEBS Lett.* **54**, 135.

Shier, W. T., Lin, Y., and DeVries, A. L. (1972). Structure and mode of action of glycoproteins from Antarctic fishes. *Biochim. Biophys. Acta* **263**, 406–413.

Slaughter, D., Fletcher, G. L., Ananthanarayanan, V. S., and Hew, C. H. (1981). Antifreeze proteins from the sea raven, *Hemitripterus americanus*. *J. Biol. Chem.* **256**, 2022–2026.

Sonawane, N. D., Thiagarajah, J. R., and Verkman, A. S. (2002). Chloride concentration in endosomes measured using a ratioable fluorescent Cl⁻ indicator. *J. Biol. Chem.* **277**.

Sonnichsen, F. D., Deluca, C. I., Davies, P. L., and Sykes, B. D. (1996). Refined solution structure of type III antifreeze protein: Hydrophobic groups may be involved in the energetics of the protein-ice interaction. *Structure* **4**, 1325–1337.

Sørensen, T. F., and Ramløv, H. (2001). Variations in antifreeze activity and serum inorganic ions in the eelpout *Zoarces viviparus*: Antifreeze activity in the embryonic state. *Comp. Biochem. Physiol.* **130A**, 123–132.

Sørensen, T. F., and Ramløv, H. (2002). Maternal-fetal relations in antifreeze production in the eelpout *Zoarces viviparus*. *Cryo*-Lett. **23**, 183–190.

Tien, R. (1995). Freezing avoidance and the presence of ice in shallow water Antarctic fishes. University of Illinois, Urbana-ChampaignPhD Thesis,.

Turner, J. D., Schrag, J. D., and DeVries, A. L. (1985). Ocular freezing avoidance in antarctic fishes. *J. Exp. Biol.* **118**, 121–132.

Valerio, P. F., Goddard, S. V., Kao, M. H., and Fletcher, G. L. (1992a). Survival of northern Atlantic cod (*Gadus morhua*) eggs and larvae when exposed to ice and low temperature. *Can. J. Fish. Aquat. Sci.* **49**, 2588–2595.

Valerio, P. F., Kao, M. H., and Fletcher, G. L. (1990). Thermal hysteresis activity in the skin of the cunner, *Tautogolabrus adspersus*. *Can. J. Zool.* **68**.

Valerio, P. F., Kao, M. H., and Fletcher, G. L. (1992b). Fish skin: An effective barrier to ice crystal propagation. *J. Exp. Biol.* **164**, 135–151.

Van Voorhies, W. V., Raymond, J. A., and DeVries, A. L. (1978). Glycoproteins as biological antifreeze agents in the cod *Gadus ogac* (Richardson). *Physiol. Zool.* **51**, 347–353.

Wang, X., DeVries, A. L., and Cheng, C.-H. C. (1995a). Genomic basis for antifreeze peptide heterogeneity and abundance in an Antarctic eel pout: Gene structures and organization. *Mol. Mar. Biol. Biotech.* **4**, 135–147.

Wang, X., DeVries, A. L., and Cheng, C.-H. C. (1995b). Antifreeze peptide heterogeneity in an Antarctic eel pout includes an unusually large major variant comprised of two 7 kDa type III AFPs linked in tandem. *Biochim. Biophys. Acta* **1247**, 163–172.

Wierzbicki, A., Taylor, M. S., Knight, C. A., Madura, J. D., Harrington, J. P., and Sikes, C. S. (1996). Analysis of shorthorn sculpin antifreeze protein stereospecific binding to (2–10) faces of ice. *Biophys. J.* **71**, 8–18.

Wilson, P. W., Beaglehole, D., and DeVries, A. L. (1993). Ellipsometry determination of glycopeptide antifreeze adsorption to ice. *Biophys. J.* **64**, 1878–1884.

Wilson, P. W., Gould, M., and DeVries, A. L. (2002). Hexagonal shaped spicules in frozen fish antifreeze solutions. *Cryobiology* **44**, 240–250.

Yang, D. S., Hon, W., Bubanko, S., Xue, Y., Seetharaman, J., Hew, C. L., and Sicheri, F. (1998). Identification of the ice-binding surface on a type III antifreeze protein with a "flatness function" algorithm. *Biophys. J.* **74**, 2142–2151.

Zdzitowiecki, K. (1998). Diversity of Digenea, parasites of fishes in various areas of the Antarctic. *In* "Fishes of Antarctica, a Biological Overview" (di Prisco, G., Pisano, E., and Clarke, A., Eds.), pp. 87–94. Springer-Verlag, Milano.

Zhao, Z., Deng, G., Lui, Q., and Laursen, R. A. (1998). Cloning and sequencing of cDNA encoding the LS-12 antifreeze protein in the longhorn sculpin, *Myoxocephalus octodecimspinosis*. *Biochim. Biophys. Acta* **1382**, 177–180.

RESPIRATORY SYSTEMS AND METABOLIC RATES

JOHN F. STEFFENSEN

I. The Respiratory System
   A. Gill Structure and Morphometrics
   B. Cutaneous Oxygen Uptake
II. Metabolic Rates
   A. Effect of Temperature on Metabolism
   B. Concept of Metabolic Cold Adaptation
   C. Oxygen Consumption of Exercising Fish
III. Conclusion
   A. Addendum

I. THE RESPIRATORY SYSTEM

Fish living in the cold waters found at the poles are exposed to extreme temperatures, which have important biochemical and physiological consequences. Although the solubility of oxygen in cold water is very high, the cold ambient temperatures can potentially depress metabolic and physiological processes that are needed to move oxygen from water to the tissues where it can be used to support aerobic metabolism.

Because of the effects of temperature on the rate of biochemical reactions, fish exposed to cold water will have reduced metabolic rates. But what consequences will this have? Do they have smaller gill surface areas because they do not need to transfer a large amount of oxygen? Do they rely more on cutaneous oxygen uptake than temperate or tropical fishes? In this chapter, first we explore components of the respiratory system in polar fish and several adaptations that allow them to maximize their aerobic performance in these harsh conditions.

As mentioned, it is well known that the metabolic rate of fish decreases with temperature (Clarke and Johnston, 1999). However, based on early experiments by August Krogh (1914) on goldfish (*Carassius auratus*) acutely

The Physiology of Polar Fishes: Volume 22
FISH PHYSIOLOGY

exposed to cold, the concept of metabolic cold adaptation (MCA) was born. This ecological/physiological concept puts forth the idea that polar fish have a higher metabolic rate than predicted when compared to fish from temperate regions measured at the same temperature. However, do polar fish really have a higher metabolic rate than should be expected when compared to fish from temperate or tropical areas? Despite a great deal of new information that now appears to contradict this idea, the principle still seems to persist. In the second part of this chapter, we synthesize the literature on MCA and try to put the whole controversy into perspective.

A. Gill Structure and Morphometrics

Most fish rely mainly on gills for the exchange of respiratory gases, although some obtain a considerable fraction (up to 35%) of oxygen across the skin (Wells, 1986; Rombough, 1998a). Gas exchange in the gills occurs at the secondary lamellae, and in the few nototheniidae that have been studied (Steen and Berg, 1966; Westermann et al., 1984), the general gill morphology is not that different from that found in temperate-water species.

In some Antarctic fishes such as the hemoglobin-less channichthyid *Chaenocephalus aceratus*, the surface area of the gill is reportedly relatively low (Hughes, 1972). Berg and Steen's (1966) morphological studies on preserved specimens of what was believed to be *C. aceratus* (but most likely was *Pseudochaenichthys georgianus*) and *Champsocephalus esox* led to the conclusion that there were no specific parameters in the gill structure that correlated with the lack of hemoglobin. Others, however, have reported large gill areas in Channichthyidae (Ruud, 1954; Jakubowski and Byczkowska-Smyk, 1970). Although variability likely exists among the species studied, it is important to remember that the appropriate comparison for Antarctic species should be a relatively inactive temperate species because among temperate species lamellar surface area varies significantly with activity level (Hughes, 1966).

Kunzmann (1990) summarized a number of morphometric studies of both red-blooded and hemoglobin-less Antarctic fish, compared them with temperate and tropical species, and concluded that icefish (Channichthyidae) gill surface area and diffusion distances are similar to those of any other relatively inactive fish. The unit gill surface area was reported to be 100–400 mm^2 g^{-1}, which is within the range of most inactive marine teleosts (Table 5.1). Kunzmann (1990) measured the unit gill area for *Pleuragramma antarcticum* and the more sluggish *Notothenia gibbifrons* at 105 and 67 mm^2 g^{-1}, respectively. For comparative purposes, it should be noted that the very sluggish goosefish (*Lophiodes*) and the highly active mackerel

(*Scomber*) have gill surface areas of 51 and 2551 mm^2 g^{-1}, respectively (Schmidt-Nielsen, 1975).

Although gill surface clearly has an impact on oxygen extraction, the efficient transfer of oxygen from water to blood also depends on diffusion distances in the gas exchanger. This parameter has been studied in only a few polar species. The thickness of the diffusion barrier in the secondary lamellae of notothenioid gills has been reported to vary from 1 to 6 μm (Steen and Berg, 1966; Westermann et al., 1984; Kunzmann, 1990), which is similar to the diffusion distances reported for five species of relatively inactive pleuronectids and soleids (2.8–5.6 μm) according to Eastman (1993). In comparison, highly active fish such as tunas can have a diffusion barrier as small as 0.6–1.2 μm (Hughes, 1984). In conclusion, therefore, neither the surface area nor the diffusion barrier in notothenioid secondary lamellae is significantly different among comparable teleosts.

Because the oxygen-carrying capacity of the blood is also low and has a very low affinity, oxygen extraction across the gills is low in channichthyids. Holeton (1970), thus, reported the oxygen extraction from the water pumped over the gills to be only 8% on average. In the respiratory system, oxygen consumption, ventilation, and oxygen extraction are related to each other via the Fick principle, which states

$$VO_2 = V_g \times [O_2] \times \% \text{ extraction from water stream,} \qquad (1)$$

where VO_2 is oxygen consumption (mg O_2 kg^{-1} hr^{-1}), V_g is gill water flow (ml O_2 kg^{-1} hr^{-1}), and $[O_2]$ is oxygen content in inspired water.

The Fick principle would predict that in the face of a low oxygen extraction, gill water flow must be relatively high to maintain a reasonable oxygen uptake. This has indeed been shown to be the case by Hemmingsen and Douglas (1977) and Holeton (1970), who found V_g to be 197 ml min^{-1}, which is as high as that found for fish with a higher metabolic rate living at a higher temperature. By comparison, plaice (*Pleuronectes platessa*) has an oxygen extraction of 69% and a ventilatory flow of 89 ml min^{-1} at 10 °C (Steffensen et al., 1981). In addition, because most of the oxygen is physically dissolved in the plasma in hemoglobin-less channichthyids, the oxygen-carrying capacity of the blood is reduced to approximately 10% of that of other notothenioids (Holeton, 1970). In the cardiovascular system, oxygen content of the blood and cardiac output are also related to each other by the Fick principle:

$$VO_2 = Q \times [(A - V)O_2], \qquad (2)$$

where VO_2 is oxygen consumption (mg O_2kg^{-1} hr^{-1}), Q is cardiac output (ml/kg/hr), and $[(A-V)O_2]$ is oxygen content in arterial and venous blood, respectively. Once again, the Fick principle would predict that if the

Table 5.1
Number of Gill Lamellae per Millimeter of Filament Length NL (mm, One Side), Unit Gill Areas (UGA), and Water-Blood Distances (WBD) from Published Sources

	Weight (g)	NL/mm	UGA (mm^2 g)	Reference	WBD (μm)	Reference
Active fish						
Katsuwonis pelamis	3258	31.8	1350	Muir and Hughes (1969)	0.6	Hughes (1970)
Scomber scombrus	182	31	1158	Gray (1954)		
Trachurus trachurus	26	38.5	783	Hughes (1966)	2.2	Hughes (1970)
Sluggish fish						
Lophius piscatorus	6392	11	196	Gray (1954)		
Anguilla anguilla	428	19	302	Gray (1954)		
Opsanus tau	251	10.9	132	Hughes and Gray (1972)	5	Muir and Hughes (1969)
Pleuronectes platessa	200		99	De Jager (1977)		
Limanda limanda	200		89	De Jager (1977)		
Antarctic fish						
Chaenocephalus aceratus	~1000	~9.5	170	Jakubowski (1982)		
Chaenocephalus aceratus	1040	8	420	Steen and Berg (1966)	6	
Chaenocephalus aceratus	~1000	~9.5	108	Hughes (1972)		
Chaenocephalus aceratus	1000		120	Holeton (1976)		

Species	Mb (g)	NL/mm	UGA	Reference	WBD	Reference
Pseudochaenichthys georgianus	1040		372	Holeton (1976)		
Channichthys rugosus	450	12	134	Jakubowski et al. (1969)		
Champsocephalus esox	66[a]	9	867	Steen and Berg (1966)	6	Steen and Berg (1966)
Notothenia neglecta	2200[a]		318	Holeton (1976)		
Notothenia rossii	2	12	230	Holeton (1976)		
Notothenia rossii	280	7.5		Westermann et al. (1984)	1.6	Westermann et al. (1984)
Notothenia gibberifrons	100	13	67	Westermann et al. (1984)		
Notothenia tessalate	50[a]	~9	518	Kunzmann (1990)	2	Steen and Berg (1966)
Trematomus newnesi	75		300	Steen and Berg (1966)		
Pagothenia borchgrevinki	150		200	Jakubowski et al. (1974)		
Pleuragramma antarcticum	20	21	105	Jakubowski et al. (1974)		
Gymnodraco acuticeps	~100	~14		Kunzmann (1990)	3.3	Kunzmann (1990)
Gymnodraco acuticeps			ca. 250	Jakubowski et al. (1974)	2.6	Eastman and Hikida (1991)

[a]Observation for a single fish.

NL/mm, number of lamellae per millimeter of filament; UGA (mm^2 g), unit gill areas; WBD (μm), water–blood-distance.

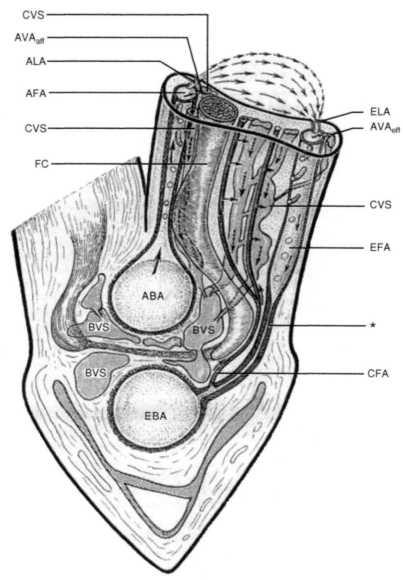

Fig. 5.1. Diagram illustrating the main nonrespiratory vessels in icefish gill arch cross-section. The basal part of a gill filament is drawn transparently. Respiratory lamellae omitted to demonstrate internal vasculature of filament. Blood flow direction indicated by arrows. ABA, afferent branchial artery; ALA, afferent lamellar arteriole; AVA_{aff}, arteriovenous anastomosis between the efferent filament artery (AFA) and the CVS; AVA_{eff}, arteriovenous anastomosis between the

blood has a low oxygen-carrying capacity, metabolism can be maintained only by a relatively high cardiac output (see Chapter 6, in this volume). This, too, has been documented by Hemmingsen et al., (1972), who reported the cardiac output in C. aceratus to be 99–153 ml kg^{-1}. This is significantly higher than the 61 ml min^{-1} reported by Holeton (1970) for the same species. Both Holeton (1970) and Hemmingsen et al. (1972) calculated the cardiac output from the Fick equation, but if this approach is used in animals with a high cutaneous oxygen uptake, overestimates of the calculated cardiac output can result. Even so, such errors could not account for the fact that cardiac output several-fold higher than common values reported for hemoglobin-containing fishes by Satchell (1971). For instance, cardiac output is three to five times larger in P. georgianus and Champsocephalus aceratus compared to Atlantic cod (Hemmingsen and Douglas, 1977; Hemmingsen, 1991).

Once oxygen transits the epithelium making up the secondary lamellae, it must be picked up by the blood for delivery to the tissues. In general, oxygen transfer across fish gills is believed to be perfusion limited, not diffusion limited (Randall and Daxboeck, 1984). Hence, an increase in oxygen uptake can only be achieved by an increase in gill blood flow, as predicted by the Fick equation (Equation 2). Vogel and Koch (1981) studied the morphology of gill vessels of formalin-preserved Champsocephalus gunnari, C. aceratus, and P. georgianus and concluded that the general vascular architecture of icefish gills conforms largely to the well-studied teleostean scheme, although branchial arteries had large diameters and thin walls and the marginal channels appeared exceptionally large (Figure 5.1). In a study of corrosion casts of icefish gills by Rankin and Tuurala (1998), it was likewise concluded that the only special feature of their gills is the large size of the blood vessels and particularly the prominent and continuous marginal channels (Figure 5.2). In addition, they studied the effect of changes in afferent and efferent blood pressure on gill resistance in isolated perfused gill arches and found that increasing perfusion rate did not change gill resistance (Rankin and Tuurala, 1998). Reducing efferent pressure, in contrast, increased the gill resistance. Further, they found that noradrenaline caused large increases in the thickness of the lamellar blood space and increased lamellar height, despite a greatly reduced afferent pressure (Rankin and Tuurala ,1998). This suggests that modulation of pillar cell active tension might be involved in control of lamellar perfusion. In red-blooded fish, there may not be any

---

efferent filament artery (EFA) and the CVS; BVS, branchial venous system; CFA, central filament artery; CVS, central venous sinus; EBA, efferent branchial artery; ELA-efferent lamellar arteriole; FC, filament cartilage; *, sphincter segment of the efferent artery. (Used, with permission, from Vogel and Koch, 1981.)

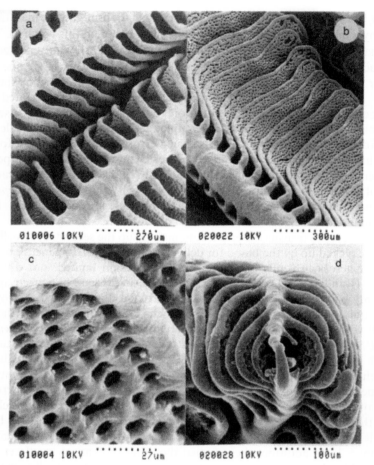

Fig. 5.2. Acrifix casts of papaverine-treated *C. hamatus* gills showing (a) the large filament arteries, (b) the large rectangular lamellae, and (c) the large marginal channels. (d) An end-on view of a filament cast. (From Rankin and Tuurala, 1998.)

advantage in increasing lamellar width to more than the diameter of a erythrocyte if diffusion distance should be optimized. In icefish, however, an increase in oxygen demand will require that much larger blood volumes be pumped through the gills because of the limited oxygen-carrying capacity of the blood, and hence greatly enhanced possibilities for pillar cell relaxation may have evolved (Rankin and Tuurala, 1998). This idea is consistent with icefish having large blood volumes, low blood pressures, and very large hearts and cardiac outputs in comparison with other fish (Johnston *et al.*, 1983; see Chapter 6, in this volume).

In addition to the "primary" circulatory system that delivers blood to the tissues, most (if not all) teleost fishes have a "secondary" circulatory system that circulates blood with an extremely reduced red blood cell content. Typically, the hematocrit of the primary circulatory system ranges from 10 to 30%, whereas the hematocrit of the secondary system is more typically 1–5% (Steffensen and Lomholt, 1992). The secondary circulatory system is considered to be analogous to the lymphatic system in mammals (Vogel and Clavietz, 1981; Vogel, 1985), and it is found primarily in the gills, gut, and fins but is most extensive in the outer surfaces as the skin, where it appears to be involved in osmoregulation and nutrient supply to the skin (Steffensen and Lomholt, 1992). Although difficult to conclusively establish, the volume of the secondary system is thought to range between 10 and 40% of the volume of the primary system of rainbow trout (Bushnell et al., 1998) and Atlantic cod (Gadus morhua) (Skov and Steffensen, 2003). Although the extent and size of the secondary circulatory system in polar species is unknown, the lack of red blood cells in the Channichthyidae certainly begs questions about its size and function in this very unusual group of fish. There are indications, however, of a secondary circulation in this group. In most teleosts, the secondary system vessels appear as small nutritive anastomoses in the gills. Vogel and Koch (1981) have shown interarterial anastomoses in the gills of C. gunnari Lonnberg, C. aceratus, and P. georgianus, so one might conclude that some sort of secondary system also occurs in these fish.

Oxygen transport at the gills can also be affected by disease. In the epidermis of some temperate-water fishes, particularly northern hemisphere gadoids and pleuronectids, a form of tumor is sometimes found, typically as skin lesions or as bilateral swellings of the pseudo-branches (Brooks et al., 1969; Alpers et al., 1977). These tumors, known as X-cell tumors, have also been shown in the gills of zoarcid eel pouts (Desser and Kahn, 1982). The tumors are characterized by the affected tissue being packed with large spherical cells surrounded by supporting cells. The origin of X-cells is unknown, hence the name, but it is suggested that they are either protozoan parasites, with neuroendocrine origin or virally transformed tissues (Brooks et al., 1969; Alpers et al., 1977; Desser and Kahn, 1982). X-cell disease has also been shown to exist in gills of approximately 15% of the Antarctic Pagothenia borchgrevinki caught in McMurdo (Franklin and Davison, 1988; Davison and Franklin, 2003), but it is not common in this species outside of Antarctic waters. The X-cells obstruct blood flow in gill lamellae (Figure 5.3), and hence limiting oxygen uptake (Davison et al., 1990). Not surprisingly, the perfusion limitation in the gill can have a severe impact on the physiology of the affected individuals. Of the 903 specimens examined by Davison (1998), 22% had the disease. The X-cell–infected fish had a relative gill mass approximately 72% higher than that of unaffected fish, an 8%

Fig. 5.3. Longitudinal sections through gill filaments of *Pagothenia borchgrevinki.* (A) Normal healthy fish. (B) X-cell–affected fish. (From Davison, 1998.)

reduction in condition factor, and a 15% reduction in average body mass. Their exercise performance seemed to be affected because elevated heart rate recovered much slower after exercise than in healthy fish (Davison and Franklin, 2003).

In conclusion, the morphology and morphometric measurements of notothenioid fish gills are similar to those of sluggish temperate marine species with low levels of activity.

B. Cutaneous Oxygen Uptake

A substantial portion of a fish's oxygen uptake (up to 30%) has been demonstrated to occur across surface area of the skin rather than the gills in several freshwater and marine teleosts (Berg and Steen, 1965; Kirsch and

Nonnotte, 1977; Nonnotte and Kirsch, 1978; Nonnotte, 1981; Steffensen *et al.*, 1981). Initially, cutaneous respiration was suggested to be limited to a few species of fish that were normally exposed to unusual conditions that hampered gas exchange at the gill or caused hypoxia (e.g., an eel moving in wet grass) (Krogh, 1904). Now, it seems clear that cutaneous gas exchange is probably important for many species of fish and especially for larval fishes (Rombough, 1998a). What role cutaneous oxygen plays in supporting metabolism in the entire fish has not been established. On the one hand, cutaneous oxygen uptake in the plaice *P. platessa* has been reported to be as high as 27% of the total oxygen consumption (Steffensen and Lomholt, 1981). However, it has also been suggested that cutaneous oxygen uptake is used only to support metabolism of the skin and is not made available to the general circulation (Nonnotte and Kirsch, 1978; Steffensen and Lomholt, 1985).

It does seem clear, however, that cutaneous oxygen uptake is particularly important during embryonic and larval development (Rombough and Moroz, 1990; Rombough, 1998a, b). Measurements of the surface areas of the yolk sac, the fins, the head and trunk, and gill lamellae of 3.7 days post-hatched Chinook salmon (*Oncorhynchus tshawytscha*) showed that the cutaneous surfaces accounted for 96% of the total external surface area available for respiratory gas exchange (Rombough, 1990). The relative importance of the skin as an oxygen exchanger begins to decline as the gills become functional. As the fish grows, the branchial surface area increased relatively more than the cutaneous surface area and was 1.7 times the latter at 180 days post-hatching.

The contribution of cutaneous oxygen uptake in polar fishes might be increased relative to temperate species because their overall metabolic rate is reduced with decreasing temperature, hematocrit is reduced in the blood, and oxygen solubility in plasma and water is increased at low temperature (DeJours, 1975). Direct measurements of cutaneous oxygen uptake in the scaleless zoarcid Antarctic icequab (*Rhigophila dearborni*) carried out by Wells (1986) showed that it accounted for approximately 35% of the total oxygen uptake. Wells (1987) likewise reported that the cutaneous oxygen uptake of the postopercular skin in the notothenioids *Trematomus bernacchii* and *Pagothenia borchgrevinki* accounted for 17 and 9% of the total oxygen uptake, respectively. In the channichthyid *C. aceratus*, the cutaneous oxygen uptake has been estimated to be as high as 40% of total oxygen consumption (Hemmingsen and Douglas, 1970). This value, however, may be an overestimate because it was extrapolated from a measurement made only on tail skin (Eastman, 1993).

Holeton (1975) has questioned whether cutaneous oxygen uptake can be really that high based on several points. First, the skin receives only a fraction of the cardiac output, and what it receives is well-oxygenated blood with little capacity to absorb more. Second, the gills are highly organized structures with efficient counter-current oxygen exchange equipped with a branchial pump ensuring convection over the gills compared to the skin's passive flow. Third, the blood–water diffusion barrier is at least an order of magnitude greater for the skin of *C. aceratus* compared with gill lamellae. Even in the Antarctic *P. borchgrevinki* with the X-cell disease that obstructs gill blood flow, the gill would be effective enough to supply the tissues with oxygen without relying on cutaneous respiration (Davison *et al.*, 1990).

In conclusion, it does not seem likely that cutaneous oxygen consumption plays a relatively more important role in Antarctic fishes than for other fish.

II. METABOLIC RATES

Fish, like other animals, need a supply of chemical energy to power various metabolic functions. The overall use of chemical energy is typically referred to as *energy metabolism*. Because most animals generate adenosine triphosphate (ATP) by oxidation of food, energy metabolism can be measured by monitoring the rate of oxygen consumption. Even so, because energy can also be obtained by anaerobic fermentation (i.e., without oxygen), oxygen consumption cannot always be used as a measure of total energy metabolism (see Chapter 3).

Metabolic rates can be divided into three categories. The basal or standard metabolic rate is the minimum energy expenditure required to keep the fish alive. To reduce all extraneous energy expenditures, standard metabolic rate is measured in a postabsorptive, nonreproductive resting fish. In addition, the fish should not be undergoing somatic growth or energy storage. Resting metabolism is the best practical and hence commonly used estimate that can be made of basal metabolism, but basal and resting metabolism are not identical (Clarke and Fraser, 2004). A second measure of metabolism, routine metabolism, is measured in fish showing normal or spontaneous activity and not necessarily in a postabsorptive state. Because measurement conditions are not so strictly defined, measurements of routine metabolism can vary considerably because of unquantified activity levels and feeding states. Although it is not as useful a measure as standard metabolic rate for measurement of particular physiological functions such as osmoregulation or cost of transport, routine metabolism is commonly used by ecologists as an index of energy expenditure in the field. Active metabolism is the highest rate of energy expenditure and normally occurs during high-speed sustained swimming, which is energetically very costly. Active metabolism is usually measured in a fish forced to swim at a maximum sustained (i.e., aerobic) speed in a swimming respirometer. The maximum oxygen consumption of a fish attained under these conditions is equivalent to the "aerobic capacity" for mammals. Swimming at even higher speeds may be possible but will most likely exceed the capacity of the respiratory–circulatory system and the fish will be forced to rely on anaerobic metabolism and eventually fatigue.

The ratio between standard and active metabolic rate is termed the *scope for activity* and can be as high as 14 (Korsmeyer *et al.*, 2002). Because of the relatively uncontrolled nature of the measurement, routine metabolic rate can in principle vary between standard and active metabolic rate. For some teleosts, the highest rates of oxygen consumption are not during swimming, but during digestion after a feeding event. This may be particularly prevalent in bottom-dwelling or slow-moving teleosts that are sit-and-wait ambush predators. The difference between standard metabolic rate and maximum metabolic rate during digestion is termed *specific dynamic action* (SDA) or *heat increment*. In other cases, the maximum metabolic rate might be measured after exercise, after a critical swimming speed trial, at a relatively low swimming speed, and when recovering from exhaustion (Schurmann and Steffensen, 1997).

Because metabolic rate is influenced by a wide range of factors (temperature, body mass, activity level, feeding state, etc.), there has been quite a lot of discussion on how best to determine the standard metabolic rate correctly (Krogh, 1914; Holeton, 1974; Steffensen *et al.*, 1994). Although some factors can (and should) be controlled in the lab (temperature, activity, light), others cannot. For instance, because of practical constraints in carrying out field work, most polar fish have been studied during the summer months when the fish are in a reproductive state. This may have resulted in elevated oxygen consumption values compared to metabolic rates at the same temperature during the winter.

There have been at least three attempts to review the literature and derive standard metabolism for fish, in general. In 1974, Altman and Dittmer reviewed 365 records of metabolism in 34 species and calculated the average standard metabolic rate to be 89 ± 34 (SD) mg O_2 kg^{-1} hr^{-1}. However, no attempt was made to correct for the diverse experimental temperatures, although the 34 fish species were mainly temperate and temperature-acclimated fish were studied. Thurston and Gehrke (1993) collated oxygen consumption, temperature, and activity state data from more than 2000 publications and published them in the OXYREF database. In the most complete study, Clarke and Johnston (1999) compiled resting (standard) oxygen consumption data from 138 studies and 69 species living over a temperature range of approximately 40 °C and calculated a scaling factor to correct for body mass effects. They found that resting metabolic rate (R_b:

mmol O_2 h^{-1}) was related to body mass (M: wet mass, g) by $R_b = aM^b$, where a is a constant and b the scaling exponent. The mean scaling exponent, b, for the 69 individual species was 0.79. The general equation for all 69 teleost was $\ln R_b = 0.80 (\ln M) - 5.43$ (Clarke and Johnston, 1999).

A. Effect of Temperature on Metabolism

In most cases, increasing temperature accelerates biochemical and physiological processes (at least within ecologically relevant temperature range). Therefore, standard metabolism of fish will increase with elevated temperature. This acute response of metabolism is believed to be the direct effect of temperature on the different metabolic processes. The change in metabolic rate that occurs for every 10 °C change in temperature is termed Q_{10} and is usually between 2 and 3 for most teleosts (Holeton, 1974). After an acclimation period at the higher temperature, metabolism usually stabilizes at a new but lower acclimated value, somewhere between the original and the acute values. Conversely, fish exposed to an acute temperature decrease would initially decrease metabolism, whereas during acclimation, metabolism would increase to a level between the original and the acute. Consequently, Q_{10} values for temperature-acclimated fish tend to be between 1 and 2. Thus, results of studying and comparing metabolic rates of fish exposed to different temperatures will obviously depend on whether the fish are acclimated or not. How long complete acclimation to a new temperature takes is not well established, but it is probably temperature dependent and may require weeks, depending on the extent of the change.

The effect of a temperature on the standard metabolic rate has been studied for many species (Beamish, 1964; Beamish and Mookherjii, 1964; Brett, 1964; Fry, 1971; Clarke and Johnston, 1999; Clarke and Fraser, 2004). Ege and Krogh (1914) appear to be the first to systematically study the effect of temperature on fish metabolism. Using goldfish, *Carassius carassius*, they measured oxygen consumption at temperatures ranging from 0 to 28 °C. It is important to note that the fish were anesthetized, not temperature acclimated, and exposed to acute temperature changes. Ege and Krogh (1916) calculated Q_{10} values of up to 9.8 at temperatures ranging from 0 and 5 °C, whereas it was only 2.2 at temperatures between 23 and 28 °C. Their results, in combination previous experiments by Krogh (1914) have become known as Krogh's "standard" goldfish (Krogh, 1916). The normal temperature range for goldfish is 10–30 °C. Nevertheless, Beamish and Mookherjii (1964) later showed that goldfish do not have such high Q_{10} values, but more "normal" values between 2 and 3 at temperatures ranging from 10 to 30 °C.

When Clarke and Johnston (1999) incorporated the effect of temperature with scaling of metabolic rate with body mass, they found that the relationship between resting oxygen consumption and environmental temperature for a 50-g fish was curvilinear and probably fit an Arrhenius model best. Therefore, the relationship between metabolic rate and temperature in fish is identical mathematically to the description of the thermal behavior of reaction rate and the equilibrium constant by Arrhenius (1915) and van't Hoff (1896). The Arrhenius model fit to all 69 species showed a lower thermal sensitivity of resting metabolism (mean $Q_{10} = 1.83$ over the temperature range from 0 to 30 °C) compared to a typical value of 2.40 (median) from within-species studies (Clarke and Johnston, 1999). Accordingly, resting metabolism, R_b, of the 69 teleost species, corrected to a standard body mass of 50 g, could be described as $lnR_b = 15.7 - 5.02\,K^{-1}$, where T is absolute temperature ($10^3\,K$).

B. Concept of Metabolic Cold Adaptation

As mentioned in the introduction, MCA describes the concept that polar fish adapted to living in very cold waters have an unusually high metabolic rate when compared with temperate-water species after correction to similar temperatures with Q_{10} values. Scholander et al. (1953) was among the first to investigate metabolic adaptation in Arctic and tropical Arctic and tropical fish, and extrapolated their metabolic rate results to a common temperature using Q_{10} values from Krogh's "standard" goldfish. They concluded that the Arctic forms showed very marked adaptation to cold water because their metabolism was 30–40 times higher than that of tropical fish extrapolated to polar temperatures. Based on this finding, the authors stated that it was obvious that cold-water species were adapted to maintain a high metabolic rate in spite of the low temperature. Further reports of elevated metabolic rates in polar fish came when Wohlschlag (1960) reported that Antarctic notothenioids from McMurdo Sound also had higher oxygen consumption rates, some 5–10 times higher, than temperate fishes extrapolated to subzero temperatures, and the concept of *MCA* was first coined in this publication. A series of studies by Wohlschlag (1960, 1962, 1964a,b,c) further explored MCA and established that it is a well-known physiological and ecological adaptive mechanism (Brett and Groves, 1979; Wells, 1987; Eastman, 1993).

However, Holeton (1973, 1974) questioned whether MCA in polar fish was a reality or the result of inaccurate measurements resulting from methodological problems or inappropriate extrapolations. He was particularly concerned about the length of time allowed for acclimating the fish to the experimental conditions, possible disturbance of the fish during the measurement cycle, and whether the fish had been allowed enough time to digest the last meal—all factors known to increase metabolic rate. As an example,

Holeton (1970) showed that the metabolic rate of Arctic charr, *Salvelinus alpinus*, could fluctuate sixfold within a few hours. With these potential problems in mind, Holeton carefully measured oxygen consumption of 11 Arctic fish species from Resolute Bay and compared the results to those previously published for Arctic and temperate species. The results of his studies showed that none of the investigated species, except possibly the Arctic cod, *Boreogadus saida*, had metabolic rates as high as had been previously described for metabolically cold-adapted Arctic and Antarctic species. In addition, Holeton (1974) noticed that for about 48 hours after capture and handling of the fish, there was a marked decrease in oxygen consumption before the fish settled their metabolism to relatively stable levels, an interesting result that was either not seen or ignored in previous studies. This initial elevated metabolic rate was ascribed to either handling stress, oxygen debt, possibly digestion of food, or a combination (Holeton, 1974). It is well known that metabolism may be elevated above standard levels for weeks after feeding in fish at low temperatures, so experiments should be planned with this in mind. In his landmark papers, Holeton also criticized the validity of Krogh's "standard" goldfish and, by extension, the derived Q_{10} factors, by comparing their data with results from Beamish and Mookherjii (1964). In doing so, it was clear that Krogh's oxygen consumption values were significantly elevated (up to three times), particularly at the higher temperatures, compared to those of Beamish and Mookherjii, thereby explaining why Ege and Krogh (1914) found the unusually high Q_{10} of 9.8 at the lower temperature range. Interestingly, even though Ege and Krogh themselves alluded to the high Q_{10} value as "obviously wrong," it continued to be used extensively by others to compare oxygen consumption of temperate fish with polar fish in support of MCA concept (Scholander *et al.*, 1953; Wohlschlag 1960, 1962, 1964a,b,c).

Holeton (1970) also pointed out the importance of comparing fish whose mass, ecology, and behavior were similar, none of which had been controlled for in previous studies. These ideas have been further supported by the work of DeVries and Eastman (1981), as well as others (Wells, 1987; McDonald *et al.*, 1987). However, despite the criticisms leveled by Holeton and summarized by the title of his 1974 paper "Metabolic Cold Adaptation of Polar Fish: Fact or Artifact?" the concept of MCA continues to have traction, particularly regarding notothenioids from McMurdo Sound (McDonald *et al.*, 1987; Wells, 1987; Eastman, 1993). The adaptive significance of MCA remains unclear because it is difficult to understand why an elevated standard metabolic rate should benefit polar fish, especially when their maximum metabolic rate is similar to that of temperate fish. MCA effectively compresses metabolic scope for activity in polar fish (Dunbar, 1968). On the other hand, if standard oxygen consumption is elevated in polar fish, it may be a consequence of increased energy expenditures necessitated by other adaptive changes such as maintaining high concentration of antifreeze glucoprotein by synthesis by the liver, tubular secretion in the kidneys rather than filtration, and high energetic costs to power membrane ion pumps, as suggested by McDonald *et al.* (1987).

MCA was reviewed and reexamined by Wells (1987) and McDonald *et al.* (1987, 1988). In an effort to eliminate problems with the measurement process itself, Wells (1987) repeated the experiments by Wohlschlag on notothenioids from McMurdo Sound with flow-through respirometers designed to reduce handling stress, hypoxia, and buildup of excretory products. Unfortunately, the methodology generated other problems, particularly regarding exponential washout and time lag (Steffensen, 1989), which made the respirometers relatively insensitive to short-term changes in activity-induced metabolism. Wells (1987) found that even though the oxygen consumption values were not as large as those of Wohlschlag (1964c), they were still higher than expected, leading him to conclude that MCA was still a valid concept in *Trematomus* and *Pagothenia*. This work has been referred to by others (McDonald *et al.*, 1987; Eastman, 1993) as proof that MCA is a fact.

The experiments mentioned in this chapter used either closed respirometry (Ege and Krogh, 1914; Wohlschlag, 1960, 1961, 1964) or open (flow-through) respirometry (Holeton, 1974; Wells, 1987). Closed respirometry involves placing a fish in a closed container in which oxygen is measured and oxygen concentration is continuously (not necessarily constantly) decreasing over time, allowing a calculation of metabolic rate. However, fish are never exposed to constant levels of oxygen. Open respirometry involves a fish placed in a container with a constant water flow-through, and oxygen concentration is measured at the inlet and outlet. Metabolic rate is measured from the difference in the oxygen levels for inlet and outlet water and the water flow. Typically, the fish are exposed to constant oxygen levels. Both techniques have benefits and drawbacks (see Steffensen [1989] for a complete discussion). Automated respirometry systems and so-called "stop-flow" or intermittent respirometry became available and cost effective with the advent of personal computers and data acquisition systems around 1980 (Steffensen *et al.*, 1984). Accurate control of flush pumps allowed accurate closed-system respirometry while avoiding the problem of accumulation of waste products associated with maintaining a fish for long periods in sealed chambers because oxygen consumption was measured over a short period (5 minutes), followed immediately by a 3–5 minute flush period when freshwater was introduced. Computer control and oxygen data acquisition also allowed for remote measurements of oxygen consumption over a period of days without the experimenter being in the lab.

Intermittent respirometry revealed that the observation by Holeton (1970) of elevated oxygen consumption for up to 48 hours after transfer to the respirometer was due to handling stress. This is an important issue because many previous metabolic rate studies on polar fish (Wohlschlag, 1960, 1964a,b,c) had failed to acclimate fish to the chamber and allow the stress effects on oxygen consumption to subside. In addition, measurements of oxygen consumption of a rainbow trout, *Oncorhynchus mykiss*, over several days revealed sporadic fluctuations in resting metabolic rate, as well as the effects of initial handling stress (Steffensen, 2002). Figure 5.4 shows that immediately after being transferred to the respirometer, the oxygen consumption was about 400 mg O_2 kg^{-1} hr^{-1} or nearly as high as maximum uptake during sustained swimming (Bushnell *et al.*, 1984). During the next 9–10 hours, oxygen consumption decreased to a baseline level of about 50 mg O_2 kg^{-1} hr^{-1} for the next 6–7 hours (standard metabolic rate)—that is eight times lower than the initial measurement! Oxygen consumption even started to fluctuate when the natural ambient light levels increased in the lab during the day. If this experiment had been performed with a closed respirometer without repeated flushes (Krogh, 1914; Wohlschlag, 1960), it would not have been possible to discriminate clearly between periods with handling stress and quiet resting. In addition, the fish would have been exposed to

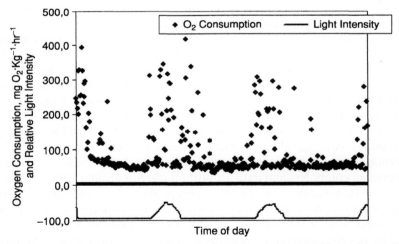

Fig. 5.4. Example of oxygen consumption of rainbow trout (*Oncorhynchus mykiss*) with a body weight of 392 g measured at 10 °C from December 23 to 26. Solid line indicates the relative light intensity. Each point represents the oxygen consumption measured during 5 minutes, followed by 5 minutes flushing (1 point every 10 minutes). See text for further details. (From Steffensen, 2002.)

progressive hypoxia. In contrast, nonautomated flow-through respirometry (Wells, 1987) would dampen oxygen content changes and average metabolic rate measurements over time, a problem that may be greatest in daylight hours given the fluctuations seen during daylight. In Figure 5.4, the average oxygen consumption from 8:30 to about 22:30 would have been approximately $124\,mg\ O_2\ kg^{-1}\ hr^{-1}$, or about 2.5 times higher than the standard metabolic rate. (See Steffensen, 1989, for more details on problems with respirometry.)

Even though Holeton (1974) was cognizant of many of the problems involved in accurately measuring standard metabolic rate, his studies suggest that quite possibly *B. saida* might have a higher than expected oxygen consumption. In an effort to evaluate this possibility, experiments were performed on three gadoid species with the same thermal history, ecology, and behavior using intermittent respirometry (Steffensen *et al.*, 1984). The study site in Greenland (69′11N, 53′30W) encompassed the northernmost limit of distribution of the mainly temperate-water Atlantic cod *Gadus morhua*, the southernmost distribution of the Arctic cod *B. saida*, and the center of distribution of the Greenland cod *Gadus uvak*. All of the species of Greenland fishes showed an initial increase in oxygen consumption at the start of the measurements (Figure 5.5). After several hours, however, oxygen consumption steadied and rarely fluctuated. These steady-state oxygen consumption measurements showed no significant differences between the Atlantic and exclusively polar *B. saida*. Hence, there was no evidence of MCA, as suggested by Holeton (1974). One species, the Arctic fish the *Arctogadus glacialis*, which is found at subzero temperatures and in association with ice only, still remained to be investigated. Jordan *et al.* (2000) succeeded in obtaining some specimens from Uummannaq Bay, West Greenland (70′42N, 52′00W) and has shown that they, too, do not have elevated metabolic rates compared to what could be expected when compared to temperate fish, when using a Q_{10} of 2.40, which is a typical within-species acclimation study value published by Clarke and Johnston (1999). Hence, to our knowledge, there is still no reliable proof of MCA in any of the Arctic fish. The data from the aforementioned studies, as well as data from others for Arctic and Antarctic fish, can be found in Table 5.2. Data are all mass corrected to a 100-g animal for easier comparison.

The remaining candidate fish species for MCA are the notothenioids from McMurdo Sound in Antarctica. One distinct difference with McMurdo Sound and most other locations in the Arctic and Antarctic is that the environmental temperature here is very stable and rarely increases above $-1.2\,°C$. Hence, the fish from this area are more stenothermal than those from any other area, and they may not be able to acclimate to higher temperatures. Standard metabolic rates were measured in several species of

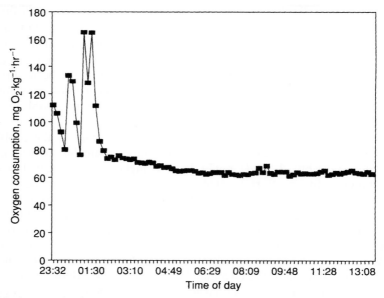

Fig. 5.5. An example of oxygen consumption of Greenland cod (*Gadus uvak*) starved for 3 days. During the initial 2–3 hours, oxygen consumption is elevated due to experimental handling stress and the researchers being present in the laboratory. When left alone, oxygen consumption decreased to a steady level after about 6–8 hours. (From Steffensen *et al.*, 1994.)

notothenioids from McMurdo Sound. Figure 5.6 is an example of oxygen consumption of a *T. bernacchii* with a body mass of 110 g measured at −1.1 °C. Like most fish, this individual showed an initial elevation in oxygen consumption at about 85 mg O_2 kg^{-1} hr^{-1}. In contrast to most other fishes, however, it never settled down, and oxygen consumption fluctuated between 17 and 70 mg O_2 kg^{-1} hr^{-1}. If one uses the method of fitting the raw data frequency distribution to a double normal distribution as described by Steffensen *et al.* (1984), this fish would have a standard metabolic rate of 17.3 mg O_2 kg^{-1} hr^{-1}. The use of slow-responding flow-through respirometers, or closed respirometry over long periods, would have averaged all the measurements and resulted in a standard metabolic rate over the initial 9 hours of 44.6 mg O_2 kg^{-1} hr^{-1}, or more than twice the calculated standard metabolic rate.

The standard metabolic rate of *T. bernacchii* with an average body mass of 115.7 g was 25.6, or 27.4 ±6.9 mg O_2 kg^{-1} hr^{-1} when weight corrected to a 100-g fish (Steffensen and DeVries, in preparation). In contrast, Wohlschlag (1960) reported a value of 84.8 mg O_2 kg^{-1} hr^{-1} (weight corrected to 100 g) for this species, or more than three times higher than Steffensen and DeVries

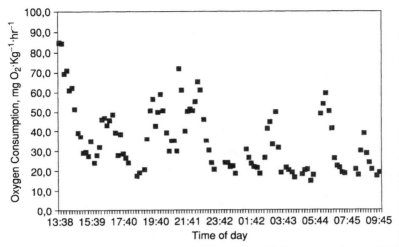

Fig. 5.6. Example of oxygen consumption of the notothenioid *Trematomus bernacchii* at an experimental temperature of $-1.0\,°C$. Body weight, 110 g. Data from Steffensen and DeVries (in preparation). (From Steffensen, 2002.)

(in preparation), while Wells (1987) reported a value of $55.7\,\text{mg}\,O_2\,\text{kg}^{-1}\,\text{hr}^{-1}$ (weight corrected to 100 g). The reason for these previously higher values most likely can be ascribed to methodology issues and the lack of fast-responding respirometers with high temporal resolution.

Similarly, *Trematomus hansoni* was reported earlier to have a standard metabolic rate of 65.9 (Wohlschlag, 1964) and $70.2\,\text{mg}\,O_2\,\text{kg}^{-1}\,\text{hr}^{-1}$ (Wells, 1987) (weight corrected to 100 g). However, Steffensen and DeVries (in preparation) found a value of $22.4 \pm 4.3\,\text{mg}\,O_2\,\text{kg}^{-1}\,\text{hr}^{-1}$. *P. borchgrevinki* was reported to have a standard metabolic rate of 105.1 and 49.2 by Wohlschlag (1964) and Wells (1987), respectively, but only $28.2 \pm 5.4\,\text{mg}$ $O_2\,\text{kg}^{-1}\,\text{hr}^{-1}$ by Steffensen and DeVries (in preparation) (all fish weights corrected to 100 g). The conclusion of this study on Antarctic notothenioids was that they do not have any significantly higher standard oxygen consumption than temperate fish extrapolated to similarly low subzero temperatures and, hence, do not support the theory of MCA based on aerobic oxygen consumption being unusually high in polar fish.

Even lower metabolic rates of notothenioids, as well as some arctic fish, were published by Zimmerman and Hubold (1998). They meticulously and elegantly measured metabolism with intermittent respirometry and activity with infrared camera simultaneously during long time periods under well-controlled conditions (Figure 5.7). In addition to a notably low standard metabolic rate for several species (Table 5.2), they found that fluctuations

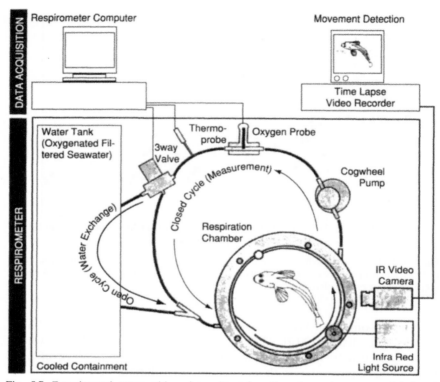

Fig. 5.7. Experimental setup with a intermittent-flow-through respirometer and infrared camera for simultaneous measurement of metabolism and activity. (From Zimmermann and Hubold, 1998.)

were mostly associated with the fish's activity. They also concluded that the knowledge of a fish's ecotype and activity is essential for a reasonable comparison of metabolic rates. In addition, they conclude that this was not always taken into account in previous investigations, resulting in contradictions about the adaptations of polar fish (Zimmerman and Hubold, 1998).

Even before these accurate measures of oxygen consumption, Clarke (1983, 1991) was critical of the concept of MCA and stated that metabolism is a particularly misleading indicator of temperature compensation because it represents the sum of many processes that may react differently to temperature. He suggested that "the use of respiration rate to assess temperature compensation should be abandoned forthwith" (Clarke, 1991) and argued for the use of other indicators to investigate temperature adaptation and

MCA. Clarke apparently changed his view, however, as Clarke and Johnston (1999) suggested that the only valid test of MCA was to control for phylogeny and confine the analysis to perciformes and compare the resting metabolic rates of polar notothenioids with those of nonpolar perciformes. When they did so, they found no evidence that resting metabolic rates of notothenioids were different than those predicted from the relationship between metabolic rate and temperature of nonpolar perciforme fishes (Clarke and Johnston, 1999); hence, neither provided support for MCA.

Several other investigations have dealt with different aspects of oxygen consumption of polar fishes. Among those are Ralph and Everson (1968) who measured the metabolism of several Antarctic fish; values can be found in Table 5.2. Everson and Ralph (1970) also determined the respiratory metabolism of the hemoglobin-free icefish *C. aceratus*. Morris and North (1984) measured oxygen consumption of five species of fish from South Georgia and pointed out that it could fluctuate considerably during the day. Saint-Paul *et al.* (1988) measured the acclimation effects on routine oxygen consumption of the Antarctic fish *Pogonophryne scotti*. Johnston *et al.* (1991) and Johnston (1993) compared feeding and metabolism in Antarctic, temperate, and tropical sedentary fish and found that routine metabolism of several tropical hawk fishes at 25 °C was approximately three times higher than that for the Antarctic *Notothenia coriiceps* at 0 °C. Hop and Graham (1995) measured respiration of juvenile *B. saida* and the effects of acclimation, temperature, and food intake and showed that chronic starvation significantly decreased metabolism (Table 5.2), which otherwise is ignored by most investigators. Further, Karamushko and Christiansen (2002) investigated the aerobic scope and resting metabolism in the oviferous and post-spawning Barents Sea capelin *Mallotus mallotus* and found that metabolism was about 30% higher in the oviferous, maybe caused by the production of gonads (vitellogenesis), metabolically active eggs, or effects of the hormonal state associated with reproduction. This is the only study showing how important reproductive status is to metabolic rate in polar fish.

Although current experiments basically reject MCA as a valid concept, there is no doubt that polar fish are physiologically adapted to the low temperatures in many ways (see other chapters in this volume). Therefore, it seems reasonable to conclude that either these adaptations are probably not so energetically costly that they result in an elevated standard metabolic rate or they are compensated for by reductions in other energy-requiring processes. Unfortunately, measurements of oxygen consumption during antifreeze synthesis remain to be carried out. Regardless, the traditional graphical expression of MCA, with polar fish having an increased metabolism compared to extrapolated values for temperate and tropical species (Figure 5.8), should cease.

Table 5.2

Metabolic Rates of Arctic and Antarctic Fishes Recalculated to 100 g

Species	Weight (g)	VO₂ (Mg O₂ kg hr)	VO₂ (100 g) (Mg O₂ kg hr)	Temperature (°C)	Location	Activity/ habitat	Reference
Antarctic							
Family							
Nototheniidae							
Pagothenia borchgrevinki			48.7	−1.5	M	Active cryopelagic	Wohlschlag, 1960, 1964c; MacDonald *et al.*, 1987
Pagothenia borchgrevinki	108.3	48.4	49.2*	−1.5	M	- '' -	Wells, 1987
Pagothenia borchgrevinki	48	25.9 (min)	22.4*		M	- '' -	Zimmermann and Hubold, 1998
Pagothenia borchgrevinki	10.5	26.7	28.2	−1.1	M	- '' -	Steffensen and DeVries, in preparation
Trematomus bernacchii			50.5	−1.5		benthic	Wohlschlag, 1964c; MacDonald *et al.*, 1987
Trematomus bernacchii	178.1	49.6	55.7*	−1.5	M	- '' -	Wells, 1987
Trematomus bernacchii	55.2	12.5	11.1	0.16	M	- '' -	Zimmermann and Hubold, 1998
Trematomus bernacchii	115.7	25.6	27.4	−1.1	M	- '' -	Steffensen and DeVries, in preparation
Trematomus hansoni			59.6	−1.5	M	benthic predator	Wohlschlag, 1964c; and MacDonald *et al.*, 1987
Trematomus hansoni			25.9	+3.0	SS	- '' -	Morris and North, 1984; MacDonald *et al.*, 1987
Trematomus hansoni	65.1	145.4	70.2*	−1.5	M	- '' -	Wells, 1987

Trematomus hansoni	111.3	22	22.4	−1.1	M	benthic predator	Steffensen and DeVries, in preparation
Trematomus centronotus	43.2	183.3	48.8*	−1.5	M	sedentary benthic	Wells, 1987
Trematomus loennbergii			36.5	−1.5	M	deepwater benthic	Wohlschlag, 1964c; MacDonald *et al.*, 1987
Trematomus loennbergii	34.8	158.1	38.1*	−1.5	M	-"-	Wells, 1987
Trematomus nicolai	114.2	34.4	35.3*	−1.5	M	deepwater benthic	Wells, 1987
Trematomus newnesi	127.1	41	42.8	−1.1	M		Steffensen and DeVries, in preparation
Notothenia coriiceps			94.6	0	AP	Active pelagic	Hemmingsen *et al.*, 1969; MacDonald *et al.*, 1987
Notothenia nudifrons			42.5	0	AP	Less active	Hemmingsen *et al.*, 1969; MacDonald *et al.*, 1987
Notothenia nudifrons			67.4	3.0	SS	-"-	Morris and North, 1984; MacDonald *et al.*, 1987
Notothenia gibberifrons			68.8	0	AP	Pelagic	Hemmingsen *et al.*, 1969; MacDonald *et al.*, 1987
Notothenia gibberifrons	438	23	30.9*	0.5	SS	-"-	Holeton, 1970
Notothenia neglecta	978	28.7	45.3*	0.5	SS	Pelagic	Holeton, 1970
Notothenia neglecta			87.3	−0.5 to +1.7	SS	-"-	Ralph and Everson, 1968; MacDonald *et al.*, 1987
Notothenia rossii			42	−0.5 to +1.7	SS	Active	Ralph and Everson, 1968; MacDonald *et al.*, 1987
Notothenia rossii			63.9	3.0	SS	-"-	Morris and North, 1984; MacDonald *et al.*, 1987

(continued)

Table 5.2 (*continued*)

Species	Weight (g)	VO_2 (Mg O_2 kg hr)	VO_2 (100 g) (Mg O_2 kg hr)	Temperature (°C)	Location	Activity/ habitat	Reference
Family **Channichthyidae**							
Channichthys aceratus			56.2	−0.5 to +1.7	SS	Sedentary benthic	Ralph and Everson, 1968; MacDonald *et al.*, 1987
Channichthys aceratus	566–2160		35.8	1.0	AP	- " -	Hemmingsen and Douglas, 1970; MacDonald *et al.*, 1987
Channichthys aceratus	1250	23.5	38.9*	0.5	SS	- " -	Holeton, 1970
Channichthys rhinoceratus			170	+3 to +6.7	K	Benthic post-spawning	Hureau *et al.*, 1977; MacDonald *et al.*, 1987
Pagetopsis macropterus			23.4	0	M	Sedentary	Hemmingsen *et al.*, 1969; MacDonald *et al.*, 1987
Pseudochaenichthys georgianus			31.8	+0.5 to +2.0	AP	Pelagic predator	Hemmingsen & Douglas 1977 & MacDonald *et al.*, 1987
Family **Bathydraconidae**							
Gymnodraco acuticeps	74.4	46.7	44.0*	−1.5	M	Benthic predator	Wells, 1987
Gymnodraco acuticeps	87.2	34.4	33.5	−0.62	M	- " -	Zimmermann and Hubold, 1998
Family Zoarcidae							
Rhigophila dearborni			19.4	−1.5	M	Sluggish benthic	Wohlschlag, 1964c; MacDonald *et al.*, 1987
Rhigophila dearborni	32.2	16.8	13.4*	−1.5		Sluggish benthic	Wells, 1987

Arctic

Family Gadidae							
Boreogadus saida	0.7–122		40.9	−1.5	R	Pelagic/cryopelagic	Holeton, 1974
Boreogadus saida	53.0	95.6	84.2	4.0	WG	-"-	Steffensen et al., 1994
Boreogadus saida	14.35	51.03	34.6*	0.4	R	-"-	Hop, 1995 (lt)
Boreogadus saida	20.2	67.79	39.2*	1.2	R	-"-	Hop, 1995 (st)
Boreogadus saida	110.6	17.2 (min)	17.6	2.02	S	-"-	Zimmermann and Hubold, 1998
Arctogadus glacialis	601.5	40.9	59.0	2.0	WG	Cryopelagic	Drud Jordan et al., 2001
Gadus morhua	155.1	61.0	66.3	4.0	WG	Benthic	Steffensen et al., 1994
Gadus ogac	180.9	64.8	72.8	4.0	WG	Benthic	Steffensen et al., 1994
Family Cottidae							
Average of five species	0.1–577		28.3	−1.5	R	Inactive benthic	Holeton, 1974
Myoxocephalus scorpius	128.1	45.4	47.7	4.0	R	Inactive benthic	Steffensen et al., 1994
Myoxocephalus scorpius	33.3	16.3 (min)	13.1	2.82	S	-"-	Zimmermann and Hubold, 1998
Family Zoarcidae							
Average of three species	0.1–147		15.4	−1.5	R	Inactive benthic	Holeton, 1974
Family Anarhichadidae							
Anarhichas minor	27.7	35.3	27.3	0.72	S	Inactive benthic	Zimmermann and Hubold, 1998
Family Salmonidae							
Salvelinus alpinus	0.1–286		23.4	2.0	R	Active freshwater	Holeton, 1973

Mean weight-specific metabolic rates raw and/or recalculated for a standard 100 g fish, using the weight relationships in the cited references, except where noted. When two references appear, the first indicates the original work, and the second a previous recalculation to 100 g. In other cases, *indicates that a weight exponent of 0.8 was used according to Clarke (1999).

Locations: AP, Antarctic Peninsula; K, Kerguelen; M, McMurdo Sound; R, Resolute; S, Spitzbergen; SS, Scotia Sea; WG, West Greenland. min., minimum metabolic rate measured in several experiments; five species of Cottidae = *Artediellus uncinatus*, *Gymnocanthus tricuspis*, *Icelus bicornis*, *Icelus spatula*, and *Myoxocephalus scorpius*; 3 species of Zoarcidae = *Gymnelis viridis*, *Lycodes turneri*, *Lycodes mucosus*; lt, long-term starved fish; st, short-term starved fish.

229

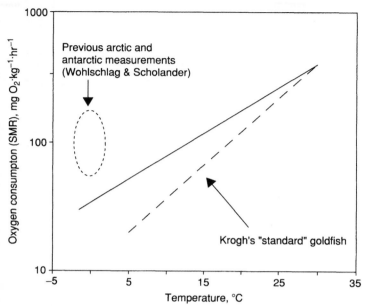

Fig. 5.8. Traditional interpretation of metabolic cold adaptation: that polar fishes have a higher than expected metabolic rate. Solid line represents tropical and temperate species at their normal temperatures and extrapolated to −2 °C. Dotted oval represents previous Arctic and Antarctic measurements by Scholander (1954), Wohlschlag (1960, 1964a, 1964b, 1964c), and Wells (1987). Dotted line represents Krogh's standard goldfish (Krogh, 1916; Holeton, 1974). See the text for further details. (From Steffensen, 2002.)

C. Oxygen Consumption of Exercising Fish

Based primarily on enzymatic studies, Crockett and Sidell (1990) suggested that polar fish might have an expanded aerobic scope when compared to temperate-water fish. In contrast, Pörtner (2002) reported that the activity level and maximum performance of polar fish may be reduced compared to those from warmer environments. Do polar fish have a larger scope than temperate fish? Forster *et al.* (1987) reported that both the aerobic scope and the net cost of transport of the Antarctic fish *P. borchgrevinki* were similar to those of other fish. Similarly, Bushnell *et al.* (1994) measured the metabolic scope of the temperate-water Atlantic cod and Greenland cod (both in Greenland at 4 °C) and found no difference in critical swimming speed and either standard or active metabolic rate, with metabolic scope being 2–3 in both species. In comparison, rainbow trout have a metabolic scope of 7.5 (Wieser, 1985). Zimmerman and Hubold (1998) further reported maximum scopes for activity for the Arctic sluggish *Myoxocephalus scorpius* and active *B. saida* to be 2.9 and 8.4, respectively, whereas the Antarctic *T. bernacchii* and *P. borchgrevinki* had values up to 2.2 and 5.0, respectively.

In addition to these studies, burst swimming was studied in the Antarctic fish *Notothenia neglecta* by Dunn and Johnston (1986), who found that the sedentary benthic fish *N. neglecta* did not produce high levels of lactic acid during exercise. In the closely related pelagic active *P. borchgrevinki*, Davison *et al.* (1988) concluded that it did not perform well at high speeds because of an inability of the white myotomal muscles to produce ATP by anaerobic glycolysis. Consequently, only low levels of lactic acid were produced, which were rapidly broken down. The authors of both studies, however, speculated that the inability to produce lactic acid could be an artifact due to long holding periods (+6 months).

Oxygen limitations and a depressed aerobic scope should be a associated with low temperature environments (Pörtner, 2002). The cold-induced reduction in aerobic capacity, however, appears to be compensated for to some extent at the cellular level by elevated mitochondrial densities, accompanied by molecular and membrane adjustments for the maintenance of muscle function (Pörtner, 2002). Polar fish, therefore, can potentially have an aerobic metabolism similar to those of high-performance swimmers in warmer waters. They only reach low performance levels, however, despite taking aerobic design to an extreme (Pörtner, 2002).

In a study on muscle metabolism and growth in Antarctic fishes, Johnston (2003) concluded that at -1 to $0\,°C$, the oxygen consumption of isolated mitochondria per milligram of mitochondrial protein showed no evidence of upregulation relative to mitochondria from temperate and tropical perciforme fishes. The mitochondria content of slow muscle fibers in Antarctic notothenioids is toward the upper end of the range reported for teleosts with similar lifestyles. High mitochondrial densities facilitate ATP production and oxygen diffusion through the membrane lipid compartment of the muscle fiber. Adequate oxygen flux in the large diameter muscle fibers of notothenioids is likely possible because of the reduced metabolic demand and enhanced oxygen solubility associated with the low temperature (Johnson, 2003).

In conclusion, it seems as if polar fish, Antarctic and the few studied Arctic, have metabolic scopes similar to that of temperate fish with a similar lifestyle.

III. CONCLUSION

The controversy over MCA started 90 years ago with crude experiments by Ege and Krogh of a single goldfish and was extended by Q_{10} calculations

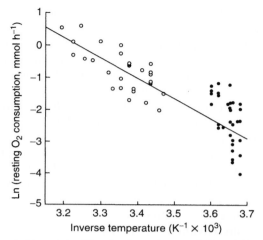

Fig. 5.9. Result from desktop study of data for the resting metabolism of 69 species of perciforme fish. The solid circles represent Antarctic notothenioids and the open circles different temperate and tropical species. The data are presented as an Arrhenius plot, with the line representing a fit to all the nonpolar fish. (From Clarke and Johnston, 1999.)

(Scholander *et al.*, 1953) and more measurements (Wohlschlag, 1960, 1961, 1964a,b,c) in the 1950s–1960s. In 1974, Holeton posed some critical questions about the concept and suggested that MCA may have been an artifact, a view unsupported by further measurements. Current measurements of oxygen consumption suggest that Holeton was correct, and MCA is an artifact. The current view, based on data from 69 species, is that all teleost fish metabolism simply scale with temperature, as shown in Figure 5.9 for perciformes (Clarke and Johnston, 1999), so polar fish do not have a resting metabolic rate any higher than predicted from the overall rate–temperature relationship established for temperate and tropical species (Clarke and Johnston, 1999; Clarke and Fraser, 2004). This conclusion has consequences for the metabolic maintenance costs of the many adaptations shown by polar fishes, possibly indicating that the energetic costs of antifreeze synthesis and maintenance of high muscle mitochondrial densities are compensated for by reduction in other energy requiring processes. Certainly, gill structure and morphometrics show no unusual features.

A. Addendum

There has been some confusion concerning the common name for *Boreogadus saida*. In North America, it is called the *Arctic cod*, and in Europe the *polar cod*. In North America, a polar cod is equivalent to

Arctogadus glacialis, which in the United Kingdom is called *Arctic cod* and in Norway and Denmark *ice cod*. Two species of *Arctogadus* have been described earlier, the *A. glacialis* and the *A. boresovi*, or East Siberian ice cod. Recent investigations have shown that they are the same species, *A. glacialis* (Møller *et al.*, 2002; Jordan *et al.*, 2003).

ACKNOWLEDGMENTS

The authors want to thank the Nordic Arctic Research Program (NARP), the Nordic Research Academy (NORFA), the Danish Natural Science Research Council, the Carlsberg Foundation, and the Elisabeth and Knud Petersen Foundation for financial support to carry out research in the Arctic and Antarctic. Two anonymous referees and Dr. Peter G. Bushnell are gratefully acknowledged for comments on the manuscript.

REFERENCES

Alpers, C. E., Cain, B. B., Myers, M., Welling, S. R., Poore, M., Bagshaw, J., and Dave, C. J. (1977). Pathologic anatomy of pseudobranch tumors in Pacific cod *Gadus macrocephalus*. *J. Natn. Cancer Inst.* **59**, 377–398.

Arrhenius, S. (1915). Quantitative Laws in Biological Chemistry. Bell, London.

Beamish, F. W. H. (1964). Respiration of fishes with special emphasis on the standard oxygen consumption. II. Influence of weight and temperature on respiration of several species. *Can. J. Zool.* **42**, 177–188.

Beamish, F. W. H., and Mookherjii, P. S. (1964). Respiration of fishes with special emphasis on standard oxygen consumption. I. Influence of weight and temperature on respiration of goldfish *Carassius auratus* L. *Can. J. Zool.* **42**, 161–175.

Berg, T., and Steen, J. B. (1965). Physiological mechanisms for aerial respiration in the eel. *Comp. Biochem. Physiol.* **15**, 469–484.

Brett, J. R. (1964). The respiratory metabolism and swimming performance of young sockeye salmon. *J. Fish. Res. Bd. Can.* **21**, 1183–1226.

Brett, J. R., and Groves, T. D. D. (1979). Physiological energetics. *In* "Fish Physiology" (Hoar, W. S., and Randall, D. J., Eds.), Vol. VIII, pp. 279–352. Academic Press, New York.

Brooks, R. E., McArn, G. E., and Wellings, S. R. (1969). Ultrastructural observations on an unidentified cell type found in epidermal tumors of flounders. *J. Natl. Cancer Inst.* **43**, 97–109.

Bushnell, P. G., Steffensen, J. F., and Johansen, K. (1969). Oxygen consumption and swimming performance in hypoxia-acclimated rainbow trout, *Salmo gairdneri*. *J. Exp. Biol.* **113**, 225–235.

Bushnell, P. G., Steffensen, J. F., Schurmann, H., and Jones, D. R. (1994). Exercise metabolism in two species of cod in Arctic waters. *Pol. Biol.* **14**, 43–48.

Bushnell, P. G., Conklin, D. J., Duff, D. W., and Olson, K. R. (1998). Tissue and whole-body extracellular, red blood cell and albumin spaces in the rainbow trout as a function of time: A reappraisal of the volume of the secondary circulation. *J. Exp. Biol.* **201**, 1381–1391.

Clarke, A. (1983). Life in cold water: The physiological ecology of polar marine ectotherms. *Oceanogr. Mar. Biol. Annu. Rev.* **21**, 341–453.

Clarke, A. (1991). What is cold adaptation and how should we measure it? *Am. Zool.* **31**, 81–92.

Clarke, A., and Johnston, N. M. (1999). Scaling of metabolic rate with body mass and temperature in teleost fish. *J. Applied Ecol.* **68**, 893–905.

Clarke, A., and Fraser, K. P. P. (2004). Why does metabolism scale with temperature? *Functional Ecol.* **18**, 243–251.

Crockett, E. L., and Sidell, B. D. (1990). Some pathways of energy metabolism are cold adapted in Antarctic fishes. *Phys. Zool.* **63**, 472–488.

Davison, W. (1998). X-cell gill disease in *Pagothenia borchgrevinki* from McMurdo Sound, Antarctica. *Pol. Biol.* **19**, 17–23.

Davison, W., Forster, M. E., Franklin, C. E., and Taylor, H. H. (1988). Recovery from exhausting exercise in an Antarctic fish, *Pagothenia borchgrevinki. Pol. Biol.* **8**, 167–171.

Davison, W., Franklin, C. E., and Carey, P. W. (1990). Oxygen uptake in the Antarctic teleost *Pagothenia borchgrevinki.* Limitations imposed by X-cell disease. *Fish Physiol. Biochem.* **8**, 69–78.

Davison, W., and Franklin, C. E. (2003). Hypertension in *Pagothenia borchgrevinki* caused by X-cell disease. *J. Fish Biol.* **63**, 129–136.

De Jager, S., Smit-Onel, M. E., Videler, J. J., Van Gils, B. J. M., and Uffink, E. M. (1977). The respiratory area of the gills of some teleost fishes in relation to their mode of life. *Bijdr. Dierkd.* **46**, 199–205.

DeJours, P. (1975). "Principles of Comparative Respiratory Physiology." North-Holland, Amsterdam.

Desser, S. S., and Kahn, R. A. (1982). Light and electron microscopy observations on pathological changes in the gills of the marine fish *Lycodes lavalaei* Vladyskov snf Tremblay associated with the proliferation of an unidentified cell. *J. Fish Dis.* **5**, 351–364.

DeVries, A. L., and Eastman, J. T. (1981). Physiology and ecology of notothenioid fishes of the Ross Sea. *J. R. Soc. NZ* **4**, 329–340.

Dunbar, M. J. (1968). "Ecological Development in Polar Regions: A Study in Evolution." Prentice-Hall, Englewood Cliffs, New Jersey.

Dunn, J. F., and Johnston, I. A. (1986). Metabolic constraints on burst-swimming in the Antarctic teleost *Notothenia neglecta. Mar. Biol.* **91**, 433–440.

Eastman, J. T. (1993). "Antarctic Fish Biology. Evolution in a Unique Environment." Academic Press, San Diego.

Eastman, J. T., and Hikida, R. S. (1991). Skin structure and vascularization in the Antarctic notothenioid fish *Gymnodraco acuticeps. J. Morphol.* **208**, 347–365.

Ege, R., and Krogh, A. (1914). On the relation between the temperature and the respiratory exchange in fishes. *Int. Rev. Hydrobiol. Hydrograph* **7**, 48–55.

Everson, I., and Ralph, R. (1970). Respiratory metabolism of *Xheanocephalus aceratus.* (Holdgate, M. W., Ed.). "Antarctic Ecology" Academic Press.

Franklin, C. E., and Davison, W. (1988). X-cells in the gills of an Antarctic teleost, *Pagothenia borchgrevinki. J. Fish. Biol.* **32**, 341–353.

Fry, F. E. J. (1971). The effect of environmental factors on the physiology of fish. *In* "Fish Physiology" (Hoar, W. S., and Randal, D. J., Eds.), Vol. 6, pp. 1–98. Academic Press, New York.

Forster, M. E., Franklin, C. E., Taylor, H. H., and Davidson, W. (1987). The aerobic scope of an Antarctic fish, *Pagothenia borchgrevinki* and its significance for metabolic cold adaptation. *Pol. Biol.* **8**, 155–159.

Gray, I. E. (1954). Comparative study of the gill area of marine fishes. *Biol. Bull.* **107**, 219–225.

Hemmingsen, E. A. (1991). Respiratory and cardiovascular adaptations in haemoglobin-free fish: Resolved and unresolved problems. *In* "Biology of Antarctic Fish" (di Prisco, G., Maresca, B., and Tota, B., Eds.), pp. 191–203. Springer-Verlag.

Hemmingsen, E. A., Douglas, E. L., and Grigg, G. C. (1969). Oxygen consumption in an Antarctic hemoglobin-free fish, *Pagestopsis macropterus,* and in three species of Notothenia. *Comp. Biochem. Physiol.* **29**, 467–470.

Hemmingsen, E. A., and Douglas, E. L. (1970). Respiratory characteristics of the hemoglobin-free fish *Chaenocephalus aceratus*. *Comp. Biochem. Physiol.* **33**, 733–744.

Hemmingsen, E. A., Douglas, E. L., Johansen, K., and Millard, R. W. (1972). Aortic blood flow and cardiac output in the haemoglobin-free fish *Chaenocephalus aceratus*. *Comp. Biochem. Physiol.* **43A**, 1045–1051.

Hemmingsen, E. A., and Douglas, E. L. (1977). Respiratory and circulatory adaptations to the absence of hemoglobin in channichthyid fishes. *In* "Adaptations within Antarctic Ecosystems: Proceedings of the Third SCAR Symposium on Antarctic Biology" (Llano, G. A., Ed.). Smithsonian Institution, Houston, Texas.

Holeton, G. F. (1970). Oxygen uptake and circulation by a hemoglobin-less Antarctic fish (*Chaenocephalus aceratus* Lonnberg) compared with three red-blooded Antarctic fish. *Comp. Biochem. Physiol.* **34**, 457–471.

Holeton, G. F. (1973). Respiration of Arctic char (*Salvelinus alpinus*) from a high Arctic lake. *J. Fish. Res. Board. Can.* **30**, 717–723.

Holeton, G. F. (1974). Metabolic cold adaptation of polar fish: Fact or artefact? *Physiol. Zool.* **47**, 137–152.

Holeton, G. F. (1975). Respiration and morphometrics of hemoglobinless Antarctic icefish. *In* "Respiration of Marine Organisms: Proceedings of the Marine Section, First Maine Biomedical Science Symposium. South Portland" (Cech, J. J. Jr., Bridges, D. W., and Horton, D. B., Eds.), pp. 198–211. The Research Institute of the Gulf of Maine.

Holeton, G. F. (1976). Respiratory morphometrics of white and red blooded Antarctic fish. *Comp. Biochem. Physiol.* **54A**, 215–220.

Hop, H., and Graham, M. (1995). Respiration of juvenile Arctic cod (*Boreogadus saida*): Effects of acclimation, temperature, and food intake. *Pol. Biol.* **15**, 359–367.

Hughes, G. M. (1966). The dimensions of fish gills in relation to their function. *J. Exp. Biol.* **45**, 177–195.

Hughes, G. M. (1970). Morphological measurements of the gills of fishes in relation to their respiratory function. *Folia Morphol (Prague)* **18**, 78–95.

Hughes, G. M. (1972). Distribution of oxygen tension in the blood and water along the secondary lamella of the icefish gill. *J. Exp. Biol.* **56**, 481–492.

Hughes, G. M. (1984). General anatomy of fish gills. *In* "Fish Physiology" (Hoar, W. S., and Randall, D. J., Eds.), Vol. XA, pp. 1–72. Academic Press, London.

Hughes, G. M., and Gray, I. E. (1972). Dimensions and ultrastructure of toadfish gills. *Biol. Bull. (Woods Hole)* **143**, 150–161.

Hureau, J. C., Petit, D., Fine, J. M., and Marneux, M. (1977). New cytological, biochemical, and physiological data on the colorless blood of the Antarctic Channichthyidae. *In* "Adaptations within Antarctic Ecosystems: Proceedings of the Third SCAR Symposium on Antarctic Biology" (Llano, G. A., Ed.), pp. 459–477. Smithsonian Institution, Washington, DC.

Jakubowski, M., and Byczkowska-Smyk, K. (1970). Respiratory surfaces of the white-blooded fish *Chaenichthys rugosus*, Regan. *Pol. Arch. Hydrobiol.* **17**, 273–281.

Jakubowski, M. (1982). Dimensions of respiratory surfaces of the gills and skin in the Antarctic white-blooded fish, *Chaenocephalus aceratus* Lonnberg (Channichthyidae). *Z. Fur Mikros. Anat. Forsch.* **96**(1), 145–156.

Jakubowski, M., Byczkowska-Smyk, W., and Mikhalev, Y. (1969). Vascularization and size of the respiratory surfaces in the white-blooded fish *Chaenichthys rugosus*. *Zool. Pol.* **19**, 503–515.

Johnston, I. A., Fitch, N., Zummo, G., Wood, R. E., Harrison, P., and Tota, B. (1983). Morphometric and ultrastructural features of the ventricular myocardium of the haemoglobin-less icefish *Chaenocephalus aceratus*. *Comp. Biochem. Physiol.* **76A**, 475–480.

Johnston, I. A., Clarke, A., and Ward, P. (1991). Temperature and metabolic rate in sedentary fish from the Antarctic, North Sea and Indo-Pacific Ocean. *Mar. Biol.* **109**, 191–195.

Johnston, I. A. (1993). Feeding energetic and metabolism in demersal fish species from Antarctic, temperate and tropical environments. *Mar. Biol.* **115**, 7–14.

Johnston, I. A. (2003). Muscle metabolism and growth in Antarctic fishes (suborder Notothenioidei): Evolution in a cold environment. *Comp. Biochem. Physiol. B* **136**, 701–713.

Jordan, A. D., Jungersen, M., and Steffensen, J. F. (2000). Oxygen consumption of the East Siberian cod: No support for the metabolic cold adaptation theory. *J. Fish Biol.* **59**, 818–823.

Jordan, A. D., Møller, P. R., and Nielsen, J. G. (2003). Revision of *Arctogadus* Dryagin, 1932 (Teleostei: Gadidae). *J. Fish Biol.* **62**, 1339–1352.

Karamushko, L. I., and Christiansen, J. S. (2002). Aerobic scaling and resting metabolism in oviferous and post-spawning Barents Sea capelin *Mallotus villosus villosus* (Muller, 1776). *J. Exp. Mar. Biol. Ecol.* **269**, 1–8.

Kirsch, R., and Nonnotte, G. (1977). Cutaneous respiration in three freshwater teleosts. *Respir. Physiol.* **29**, 339–354.

Korsmeyer, K., Steffensen, J. F., and Herskin, J. (2002). Energetics of rigid-body swimming, undulatory swimming, and gait transition in parrotfish (*Scarus schlegeli*) and triggerfish (*Rhinecanthus aculeatus*). *J. Exp. Biol.* **205**, 1253–1263.

Krogh, A. (1904). Some experiments on the cutaneous respiration of vertebrate animals. *Scan. Arch. Physiol.* **16**, 348–357.

Krogh, A. (1914). The quantitative relation between temperature and standard metabolism in animals. *Int. Z. Phys. Chem. Biol.* **1**, 491–508.

Krogh, A. (1916). "The Respiratory Exchange of Animals and Man." Longmans, Green and Co., London.

Kunzmann, A. (1990). Gill morphometrics of two Antarctic fish species *Pleuragramma antarcticum* and *Notothenia gibbifrons*. *Pol. Biol.* **11**, 9–18.

Macdonald, J. A., Montgomery, J. C., and Wells, R. M. G. (1987). Comparative physiology of Antarctic fishes. *Adv. Mar. Biol.* **24**, 321–388.

Møller, P. R., Jordan, Anders, D., Gravlund, P., and Steffensen, J. F. (2002). Phylogenetic position of the cryopelagic codfish *Arctogadus*, Djargin, 1932 based on partial mitochondrial cytochrome b sequences. *Pol. Biol.* **25**, 342–349.

Morris, D. J., and North, A. W. (1984). Oxygen consumption of five species of fish from South Georgia. *J. Exp. Mar. Biol. Ecol.* **78**, 75–86.

Muir, B. S., and Hughes, G. M. (1969). Gill dimensions for three species of tuna. *J. Exp. Biol.* **51**, 271–287.

Nonnotte, G., and Kirsch, R. (1978). Cutaneous respiration in seven sea-water teleosts. *Respir. Physiol.* **35**, 111–118.

Pörtner, H. (2002). Physiological basis of temperature-dependent biogeography: Trade-offs in muscle design and performance in polar ectotherms. *J. Exp. Biol.* **205**, 2217–2230.

Ralph, R., and Everson, I. (1968). The respiratory metabolism of some Antarctic fish. *Comp. Biochem. Physiol.* **17**, 299–307.

Randall, D. J., and Daxboeck, C. (1984). Oxygen and carbon dioxide transport across fish gills. *In* "Fish Physiology" (Hoar, W. S., and Randall, D. J., Eds.), pp. 263–314. Academic Press, New York.

Rankin, J. C., and Tuurala, H. (1998). Gills of Antarctic fish. *Comp. Biochem. Physiol.* **119A**, 149–163.

Rombough, P. J. (1998a). Respiratory gas exchange, aerobic metabolism and effects of hypoxia during early life. *In* "Fish Physiology" (Hoar, W. S., and Randall, D. J., Eds.), Vol. XIA, pp. 59–161. Academic press, San Diego.

Rombough, P. J. (1998b). Partitioning of oxygen uptake between the gills and skin in fish larvae. A novel method for estimating cutaneous oxygen uptake. *J. Exp. Biol.* **201**, 1763–1769.

Rombough, P. J., and Moroz, B. M. (1990). The scaling and potential importance of cutaneous and branchial surfaces in respiratory gas exchange in young salmon (*Oncorhynchus tshawytscha*). *J. Exp. Biol.* **154**, 1–12.

Ruud, J. T. (1954). Vertebrates without erythrocytes and blood pigments. *Nature* **1973**, 848–850.

Saint-Paul, U., Hubold, G., and Ekau, W. (1988). Acclimation effects on routine oxygen consumption of the Antarctic fish *Pogonophryne scotti*. *Pol. Biol.* **9**, 125–128.

Satchell, G. H. (1971). "Circulation in Fishes." Cambridge University Press, London.

Schmidt-Nielsen, K. (1975). "Animal Physiology; Adaptation and Environment," P. 699. Cambridge University Press.

Scholander, P. F., Flagg, W., Walters, V., and Irving, L. (1953). Climatic adaptation in Arctic and tropical poikilotherms. *Physiol. Zool.* **26**, 67–92.

Schurmann, H., and Steffensen, J. F. (1997). Effects of temperature, hypoxia and activity on the metabolism of Atlantic cod, *Gadus morhua. J. Fish Biol.* **50**, 1166–1180.

Skov, P., and Steffensen, J. F. (2003). The blood volumes of the primary and secondary circulatory system in the Atlantic cod *Gadus morhua* L., using plasma bound Evans blue and compartmental analysis. *J. Exp. Biol.* **206**, 591–599.

Steen, J. B., and Berg, T. (1966). The gills of two hemoglobin-free fishes compared to those of other teleosts—With a note on severe anemia in an eel. *Comp. Biochem. Physiol.* **18**, 517–526.

Steffensen, J. F., and DeVries, A. Oxygen consumption of 4 species of notothenioids from Antarctica: No unusually high metabolism. In preparation.

Steffensen, J. F., Lomholt, J. P., and Johansen, K. (1981). Gill water flow and O_2 extraction during graded hypoxia in two ecologically distinct species of flatfish, the flounder, *Platichthys flesus*, and the plaice, *Pleuronectes platessa. Env. Biol. Fish.* **7**, 157–163.

Steffensen, J. F. (1989). Some errors in respirometry of aquatic breathers: How to avoid and correct for them. *Fish Physiol. Biochem.* **6**, 49–59.

Steffensen, J. F. (2002). Metabolic cold adaptation of polar fish based on measurements of aerobic oxygen consumption: Fact or artefact? Artefact! *Comp. Biochem. Physiol. A* **132**, 789–795.

Steffensen, J. F., Bushnell, P. G., and Schurmann, H. (1994). Oxygen consumption in four species of teleosts from Greenland: No evidence of metabolic cold adaptation. *Pol. Biol.* **14**, 49–54.

Steffensen, J. F., Johansen, K., and Bushnell, P. G. (1994). An automated swimming respirometer. *Comp. Biochem. Physiol.* **79A**, 473–476.

Steffensen, J. F., Lomholt, J. P., and Johansen, K. (1981). The relative importance of skin oxygen uptake in the naturally buried plaice, *Pleuronectes platessa*, exposed to graded hypoxia. *Respir. Physiol.* **44**, 269–275.

Steffensen, J. F., and Lomholt, J. P. (1985). Cutaneous oxygen uptake and its relation to skin blood perfusion and ambient salinity in the plaice, *Pleuronectes platessa. Comp. Biochem. Physiol.* **81A**, 373–375.

Steffensen, J. F., and Lomholt, J. P. (1992). The secondary vascular system. *In* "Fish Physiology" (Hoar, W. S., Randall, D. J., and Farrell, A. P., Eds.), Vol. XIIA, pp. 185–217. Academic Press, San Diego.

Thurston, R. V., and Gehrke, P. C. (1993). Respiratory oxygen requirements of fishes: Description of OXYREF, a data file based on test results reported in the published literature. *In* "Fish Physiology, Toxicology, and Water Quality Management: Proceedings of an International Symposium, Sacramento, California, USA, September 18–19, 1990" (Russo, R. C., and Thurston, R. V., Eds.), pp. 95–108. US Environmental Protection Agency EPA/600/R93/157.

Van't Hoff, J. H. (1896). "Studies of Chemical Dynamics." Williams & Norgate, London.

Vogel, W. O. P. (1985). Systemic vascular anastomoses, primary and secondary vessels in fish, and the phylogeny of lymphatics. *Alfred Benzon Symp. Copenhagen* **21**, 143–159.

Vogel, W. O. P., and Clavietz, M. (1981). Vascular specialization in fish, but no evidence for lymphatics. *Z. Naturf.* **36C**, 490–492.

Vogel, W. O. P., and Koch, K.-H. (1981). Morphology of gill vessels in icefish. *Arch. Fisch. Wiss.* **31**, 139–150.

Wells, R. M. G. (1986). Cutaneous oxygen uptake in the Antarctic icequab, *Rhigophila dearborni* (Pisces: Zoarcidae). *Pol. Biol.* **5**, 175–179.

Wells, R. M. G. (1987). Respiration of Antarctic fish from McMurdo Sound. *Comp. Biochem. Physiol.* **88A**, 417–424.

Westermann, J. E. M., Barber, D. L., and Malo, L. G. (1984). Observations on gills of pelagic and demersal juvenile *Notothenia rossii*. *Br. Antarct. Surv. Bull.* **65**, 81–89.

Wieser, W. (1985). Developmental and metabolic constraints of the scope for activity in young rainbow trout (*Salmo gairdneri*). *J. Exp. Biol.* **118**, 133–142.

Wohlschlag, D. E. (1960). Metabolism of an Antarctic fish and the phenomenon of cold adaptation. *Ecology* **41**(2), 287–292.

Wohlschlag, D. E. (1962). Antarctic fish growth and metabolic differences related to sex. *Ecology* **43**, 589–597.

Wohlschlag, D. E. (1964a). An Antarctic fish with unusually low metabolism. *Ecology* **44**, 557–564.

Wohlschlag, D. E. (1964b). Respiratory metabolism and ecological characteristics of some fishes in McMurdo Sound, Antarctica. *In* "Biology of the Antarctic Seas" (Lee, M. O., Ed.), Vol. 1, pp. 33–62. American Geophysical Union, Washington, DC.

Wohlschlag, D. E. (1964c). Respiratory metabolism and growth of some Antarctic fishes. *In* "Bioloie Antarctique" (Scientific Committee on Antarctic Research, Proceedings of the 1st Symposium on Antarctic Biology, Paris) (Carrick, R., Holdgate, M. W., and Prevost, J., Eds.), pp. 489–502. Hermann, Paris.

Zimmerman, C., and Hubold, G. (1998). Respiration and activity of Arctic and Antarctic fish with different modes of life: A multivariate analysis of experimental data. *In* " Fishes of Antarctica: A Biological Review" (di Prisco, G., Pisano, E., and Clarke, A., Eds.), pp. 163–174. Springer-Verlag, Italia, Milano.

FURTHER READING

DeVries, A. L., and Cheng, C-H. C. (1989). The role of antifreeze glycopeptides and peptides in the survival of cold-water fishes. *In* "Water & Life" (Somero, G., *et al*, Eds.), pp. 301–315. Springer-Verlag.

6

THE CIRCULATORY SYSTEM AND ITS CONTROL

MICHAEL AXELSSON

I. The Cardiovascular System in the Cold
  A. Blood Viscosity and the Effect on Cardiac Work
  B. Physiological Regulation of Hematocrit by the Spleen
  C. Vascular Resistance and Vascular Geometry
II. Anatomy and Control of the Fish Heart
  A. Cardiac Myocytes
  B. Cardiac Performance
  C. Control of Heart Rate
  D. Control of Cardiac Contractility and Stroke Volume
III. The Branchial and Systemic Vasculature
  A. Control of the Branchial Vasculature
  B. Control of the Systemic Vasculature
IV. Integrated Cardiovascular Responses
  A. Stress
  B. Exercise
  C. Hypoxia
V. Summary

I. THE CARDIOVASCULAR SYSTEM IN THE COLD

Temperature affects the physiology of poikilotherm organisms such as fish by affecting the rate of chemical reactions in two general ways: It affects the rate of chemical reactions by a factor of 2–3 for every 10 °C change in temperature, and it affects the dynamics of noncovalent interaction, which is important for biological structures. These effects are reflected in many physiological processes, including the cardiovascular system at both the metabolic and the morphological level. The heart, like the other musculature, is a chemomechanical converter, and both its excitability and mechanical performance are affected by temperature and temperature changes. As in other tissues, adaptive mechanisms operate by reshaping it metabolically

The Physiology of Polar Fishes: Volume 22
FISH PHYSIOLOGY

and morphologically to optimize its performance and counteract the effects of temperature. Data are limited on cardiovascular effects of temperature for fish (mostly teleosts) living in polar environments, as is the case for many temperate fish species. Nevertheless, it still seems as if the cardiovascular biology of different species to some extent mirrors the thermal tolerance limits for each group. Eurythermal groups show a remarkable ability to compensate for the effects of seasonal and daily changes in environmental temperature. Stenothermal groups that live within a narrower temperature limit and do not face large seasonal or daily fluctuations in environmental temperature have developed cardiovascular adaptations for optimal function within a restricted thermal window. Regardless, if the thermal independence is due to evolutionary adaptation or acclimatory compensation, the overall goal is to maintain an effective chemomechanical transformation and cardiovascular control mechanism. The aspects of cardiocirculatory enantiostasis (conserved function) span from effects of temperature on physical characteristics to interactions between different cells up to the integrated response of the entire animal.

This chapter considers known aspects of the circulatory physiology of polar fishes, addressing a central question: Is the circulatory system of polar fishes fundamentally different from that of temperate fish species? Unfortunately, the cardiovascular data for polar living fishes are limited to very few species, and any generalities are advanced with caution. Moreover, in the absence of any cardiovascular information on cartilaginous polar fishes, the focus here is necessarily on teleost species.

A. Blood Viscosity and the Effect on Cardiac Work

With a decrease in water temperature, the viscosity of the blood increases. Blood viscosity at $0\,°C$ is about 40% higher in the red-blooded *Trematomus bernacchii* than it would be at $10\,°C$. The direct importance of blood viscosity in circulatory physiology is clearly illustrated by reference to Poiseuille's Law, $Q = \pi \Delta P r^4 / 8 \, \eta l$ (Q = flow, π = 3.14, ΔP = pressure gradient, r = vessel radius, η = viscosity, l = length of vessel). Furthermore, because the circulatory system can be simply represented by Ohm's Law (blood pressure = product of blood flow and vascular resistance), the equation can, thus, be rewritten for resistance as $R = 8 \, \eta l / \pi r^4$. Thus, vascular resistance changes in direct proportion to changes in η. In addition, η, through its direct effects on vascular resistance, affects the workload of the heart directly (Farrell, 1984; Macdonald and Wells, 1991). Thus, a 40% increase in η will increase cardiac workload by the same percentage, given no other change.

Many factors affect the viscosity of the blood: (a) the number of red blood cells (RBCs), which is typically described by the percentage of packed RBCs in blood, the hematocrit (Hct), (b) an increase in the viscosity of the plasma per se, (c) an increase in the stiffness of the RBCs, making them less deformable, and (d) alterations to blood velocity, which through the shear rate dependency of blood viscosity, can greatly affect blood viscosity (see below). Among these four factors, the most important factor is the number of RBCs. One of the adaptations seen in polar fishes is a reduced Hct and even more pronounced a reduction of mean cell hemoglobin (Hb) concentration (MCHC) (Egginton, 1996; Davison et al., 1997; Axelsson et al., 1998).

At cold temperatures, a halving of Hct approximately halves η, which is an important compensation to maintain or even lower cardiac workload compared to temperate species living at higher temperatures.

Many cold living fish species have antifreeze compounds dissolved in the plasma (see Chapter 4). These antifreeze compounds can be present at such a high concentration (antifreeze glycopeptides = 3% w/v in the plasma of both Antarctic and Arctic species) that they increase the plasma viscosity (Macdonald and Wells, 1991; Eastman, 1993). The data available for northern cold species show no uniform pattern. Blood viscosity levels of the shorthorn sculpin (*Myoxocephalus scorpius*), longhorn sculpin (*Myoxocephalus octodecemspinosus*), and Arctic char (*Salvelinus alpinus*) are lower compared with the winter flounder (*Pseudopleuronectes americanus*) (Graham and Fletcher, 1985; Graham et al., 1985) (Table 6.1). Even if the total plasma protein concentration was not different between these species, the difference in viscosity is attributed to an unspecified difference in plasma protein composition between the sculpins and the winter flounder (Graham and Fletcher, 1985).

Another factor that affects the whole blood viscosity is the RBC membranes. The RBC membrane contains phospholipids that affect the membrane fluidity, and together with intracellular microtubular structures that have effects on the membrane rigidity, they determine the shear elastic modulus (rigidity) of the RBCs (Lecklin et al., 1995). A decrease in temperature is likely to reduce RBC deformability, and this will increase blood viscosity. Polar fishes, like temperate species, could compensate for this by altering the lipid composition in their membranes and by changing the properties of the intracellular microtubular components. Nervous tissues in ectotherms in cold habitats have a higher proportion of unsaturated fatty acids, which are more fluid at low temperatures than saturated fatty acid of the same molecular size. This cellular process is termed *homeoviscous adaptation* and is found in Antarctic fish species (Macdonald et al., 1987;

Table 6.1

Blood Viscosity in Various Fish Species at Two Shear Rates (22.5 and 225 s⁻¹) Are Listed with Test Temperature, Hematocrit (Hct), and Hemoglobin Concentration ([Hb])

Species	Temperature °C	Hct (%)	[Hb] (g l⁻¹)	Viscosity (22.5 s⁻¹)	Viscosity (225 s⁻¹)	Reference
Cold-Water Living Species						
Chionodraco kathleenae (Antarctic)	−1.8	1.2	0	4	3.6	Wells et al., 1990
Cryodraco antarcticus (Antarctic)	−1.8	1.2	0	4.7	3.8	Wells et al., 1990
Chaenocephalus aceratus (Antarctic)	0		0		3	Hemmingsen and Douglas, 1972
Pseudochaenichthys georgianus (Antarctic)	0		0		2.5	Hemmingsen and Douglas, 1972
Pagothenia borchgrevinki (Antarctic)	−1.8	17.5			6.5	Macdonald and Wells, 1991
Trematomus bernacchii (Antarctic)	−1.8	7.6–15.4[a]	18.3–31.6	4.9–11.2	4.4–6.8	Wells et al., 1990
Salvelinus alpinus (Arctic)	0	27		11.5		Graham et al., 1985
Myoxocephalus octodecemspinosus (Arctic)	0	18[b]		6		Graham et al., 1985
Temperate Species						
Myoxocephalus scorpius	0	21[b]		7		Graham et al., 1985
Pseudopleuronectes americanus	0	23	49	22[b]		Graham and Fletcher, 1983
	15			6.5[b]		
Parapercis colias	0	20			6[b]	Macdonald and Wells, 1991
	20				3[b]	

Underlined species are white-blooded Antarctic fish species.

[a]Range of Hct for camulated and acute sampled fish.

[b]Data extrapolated from graphs.

Storelli *et al.*, 1998). Another factor that affects the RBC deformability is the MCHC. Antarctic species have lower MCHCs compared to temperate species, thereby reducing intracellular viscosity and increasing RBC deformability (Egginton, 1996) (Table 6.2).

Whole blood acts as a non-newtonian fluid. As a result, its viscosity is exponentially related to shear rate, which is a function of blood velocity. Thus, as shear rate (and blood velocity) increases, η decreases exponentially. The effect of shear rate on plasma and blood from Arctic and Antarctic

Table 6.2

Normal Values for Hemoglobin Concentration ([Hb]), Hematocrit (Hct), Mean Cell Hemoglobin Concentration (MCHC), and Mean Cell Hemoglobin (MCH) for Antarctic and Sub-Antarctic Fish Species

Species	[Hb] (g l⁻¹)	Hct (%)	MCHC (g l⁻¹)	MCH (pg)	Reference
Notothenia					
Trematomus hansoni	35	34	108	47	Wells *et al.*, 1980
Trematomus newnesi	26	39	61	27	Wells *et al.*, 1980
Trematomus bernacchii	24	21	143	33	Wells *et al.*, 1980
	18	8	242	67	Wells *et al.*, 1990
	28	14	201		Davison *et al.*, 1994
	19	11	150–180		Davison *et al.*, 1995
Trematomus centronotus	29	19	166	54	Wells *et al.*, 1980
Trematomus nicolai	23				Wells *et al.*, 1980
Trematomus loennbergii	31	24	130		Wells *et al.*, 1980
Dissostichus mawsoni	43	27	160	44	Wells *et al.*, 1980
		24			Macdonald and Wells, 1991
Pagothenia borchgrevinki	38	32	134	53	Wells *et al.*, 1980
Pagothenia borchgrevinki	26	15	174		Wells *et al.*, 1984
Pagothenia borchgrevinki	33	15	226		Franklin *et al.*, 1993
Notothenia coriiceps	63	24	252	28	Lecklin *et al.*, 1995
Chionodraco hamatus		1.23			Macdonald and Wells, 1991
Cryodraco antarcticus		1.25			Macdonald and Wells, 1991
Eleginops maclovinus		30			Agnisola *et al.*, 1997
Patagonotothen tessellata		28			Agnisola *et al.*, 1997
Paranotothenia magellanica		32			Agnisola *et al.*, 1997
Bathydraconidae					
Gymnodraco acuticeps	22	24	98	27	Wells *et al.*, 1980
Harpagiferidae					
Histiodraco velifer	9	20	45		Wells *et al.*, 1980
Zoarcidae					
Rhigophila dearborni	37	21	174		Wells *et al.*, 1980

species is appreciable, and at low shear rates, the difference between the red-blooded and white-blooded species is high (50–100%) (Graham and Fletcher, 1985; Wells, 1990; reviewed by Macdonald and Wells, 1991) (Figure 6.1 and Table 6.1). Therefore, despite the benefit of reducing Q to reduce cardiac workload, the selection pressure favoring a high flow rate is likely quite high if one also considers the benefit of maintaining a high shear rate.

As discussed earlier, high Hct and or high MCHC, in combination with low temperature, lead to a high total blood viscosity that will add to the workload of the heart (Macdonald and Wells, 1991). Reported Hct levels for Antarctic fish are often in the range of 8–15% (*Pagothenia borchgrevinki* and *Trematomus bernacchii*) (Wells *et al.*, 1990; Franklin *et al.*, 1993; Davison *et al.*, 1994), which is about half of the values commonly reported for temperate fish species. MCHC is also lower (Egginton, 1997a; Egginton and

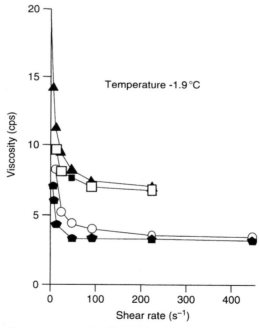

Fig. 6.1. The effect of temperature on the effect of shear rate on viscosity of the blood. The filled triangles and squares represent data for *Pagothenia borchgrevinki* (triangles) and *Trematomus bernacchii* (squares) at 20% hematocrit, and the open circles represent the mean value for plasma from *P. borchgrevinki and T. bernacchii*. The filled diamonds show mean values for blood from *Cryodraco antarcticus* and *Chionodraco hamatus*. (Redrawn from Macdonald and Wells, 1991.)

Davison, 1998; Gallaugher and Farrell 1998) (Table 6.2). The reduction in Hct is extreme in icefish (family Channichthyidae suborder Notothenioidei), a group endemic to the Antarctic and sub-Antarctic regions because they lack RBCs due to a deletion of β-globin locus and have only a small number of white blood cells for their immune function (Cocca et al., 1995). As a result, blood viscosity at -1.86 to $0\,^{\circ}C$ is lower in the icefish species, compared with red-blooded Antarctic species T. bernacchii and P. borchgrevinki. In addition, their blood is essentially independent of shear rate (Wells et al., 1990; Macdonald and Wells, 1991), thereby solving two problems related to cold temperature (Table 6.1). The trade-off with Hb-free blood is a reduction in the oxygen-carrying capacity of blood, and only physically dissolved oxygen is transported. For icefish, the major compensation for a low Hct is a high cardiac output, about 10 times higher compared to red-blooded cold living species (Table 6.3). This compensation provides no benefit to cardiac workload unless there is a concurrent decrease in overall vascular resistance, which is the case in the icefish family (see below).

The absence of RBCs also means that total oxygen-carrying capacity of the blood is much lower. This has several ramifications, some of which are addressed in later sections and other chapters (see Chapter 7). Of importance is that in resting temperate fish species, only about one-third of the arterial O_2 is removed from the blood during its passage through capillaries. This means that oxygen extraction can increase during periods of increased oxygen demand (aerobic exercise) (Kiceniuk and Jones, 1977) or during recovery from burst swim (Farrell and Clutterham, 2003). The oxygen reserve in white-blooded fish species is lower, but this is at least in part compensated for by a larger blood volume and a higher cardiac output (Acierno et al., 1995). Furthermore, certain red-blooded polar fishes are remarkably adept at increasing Hct rapidly under stressful situations by releasing erythrocytes from the spleen (Table 6.4). Although splenic contraction increases blood viscosity, the benefit is that arterial oxygen-carrying capacity and venous O_2 reserve are increased exactly at a time when they are most beneficial. In the next section, we focus on the physiological regulation of Hct by the spleen and the consequences for the animals.

B. Physiological Regulation of Hematocrit by the Spleen

Rapid changes in Hct are possible in many animals by splenic contraction. In fact, this RBC release can be so rapid that Hct can be elevated appreciably by poor blood sampling techniques (see Tables 6.2 and 6.4).

Although resting Hct in the red-blooded species of the family Nototheniidae is lower compared to that of temperate fish species (see Table 6.2), some species show substantial changes in Hct during stress and exercise (see

Table 6.3

A Comparison of Cardiac Output and Vascular Resistance Values for Selected Antarctic Notothenioids and Temperate Teleosts

Species	Stroke volume (ml kg^{-1})	Cardiac output (ml min^{-1} kg^{-1})	Systemic vascular resistance (Pa min kg ml^{-1})	Reference
Cold-Water Living Species				
Chaenocephalus aceratus[a]	6.5	118	21	Hemmingsen et al., 1972
Chaenocephalus aceratus[a]		66–104		Hemmingsen and Douglas, 1972
Chaenocephalus aceratus[a]	4.4	61	25	Holeton, 1970
Pseudochaenichthys georgianus	5.8–7.8	67–94		Hemmingsen and Douglas, 1977
Chionodraco hamatus[b]	8.5	210		Tota et al., 1991[c]
Pagothenia borchgrevinki[a]	1.40	29.6	159	Axelsson et al., 1992
Trematomus. bernacchii[a]	1.67	17.6	197	Axelsson et al., 1992
Notothenia coriiceps[c]	0.27	6.7	477	Egginton, 1997b
Eleginops maclovinus[b]	0.45	19.5		Agnisola et al., 1997
Patagonotothen tessellata[b]	0.3	21.3		Agnisola et al., 1997
Paranotothenia magellanica[b]	0.36	21.1		Agnisola et al., 1997
Temperate Species				
Ophiodon elongatus[a]	0.37	11.2	343	Farrell, 1981
Hemitripterus americanus[a]	0.34	15.4		Farrell et al., 1982
Gadus morhua[a]	0.45	17.3	188	Axelsson, 1988
Oncorhynchus mykiss[a]	0.41	17.6	234	Bernier et al., 1999
Anguilla australis	0.22	11.4		Hughes et al., 1981

[a]*In vivo* data.

[b]*In vitro* data basal conditions.

[c]*In vivo* data based on microspheres.

Table 6.4
Effects of Different Stressors on Hematocrit (% Hct Increase) for a Number of
Antarctic Species and Some Temperate Species

Species	Hct increase (%)	Stressor	Reference
Cold-Water Living Species			
Pagothenia borchgrevinki	110	Exercise	Davison *et al.*, 1988
Pagothenia borchgrevinki	93[a]	Exercise	Franklin and Davison, 1993
Pagothenia borchgrevinki	101	Enforced exercise	Wells *et al.*, 1984
Trematomus bernacchii	67[a]	Hypoxia	Davison and Franklin, 1994
	30[a]	Exercise	
Trematomus bernacchii	55	Hypoxia	Davison, 2001
Notothenia neglecta	−5[a]	Enforced exercise	Egginton *et al.*, 1991
Notothenia rossii	76[a]	Enforced exercise	Egginton *et al.*, 1991
Dissostichus mawsoni	71	Surgery	Wells *et al.*, 1984
Temperate Species			
Alosa sapidissima	9	Migration	Leonard and McCormick, 1999
Oncorhyncus mykiss	23[a]	Air exposure	Pearson and Stevens, 1991
Oncorhyncus mykiss	3–5	Exercise (sustained and U_{crit})	Gallaugher *et al.*, 1992
Seriola quinqueradiata	12[a]	Hypoxia	Yamamoto *et al.*, 1983

[a]Value compensated for hemoconcentration and cell swelling.

Table 6.4). For example, resting Hct in *P. borchgrevinki* increased by 2.4-fold from 15% to 35% during strenuous exercise (Franklin, 1993). Even when compensations were made for RBC swelling and hemoconcentration, an almost twofold increase in Hct could be attributed to the release of RBCs from the spleen. This increase in Hct was estimated to increase blood viscosity from about 6 to 11 cps. A similar increase in Hct (from 15% up to 31%) was found after severe stress *in P. borchgrevinki*, and in one specimen of *Dissostichus mawsoni*, Hct increased from 25% up to 43% after surgery (Wells *et al.*, 1984). Large increases in Hct have been observed in another Antarctic species *T. bernacchii*, with a 66% increase in Hct in response to hypoxia and acute temperature change and a 31% increase after stressful exercise (chasing) (Davison, 1994).

In temperate species, splenic increases in Hct are generally proportionately smaller, but then Hct is already higher and fish apparently have an optimal maximum Hct (Yamamoto *et al.*, 1983; Pearson and Stevens, 1991; Gallaugher *et al.*, 1992, 1995; Leonard and McCormick, 1999) (Table 6.4).

In most vertebrates, including fish, the spleen is controlled by adrenergic excitatory mechanisms via either adrenergic innervation or circulating catecholamines (Nilsson, 1994). The capsular and trabecular smooth muscles regulate the volume of the spleen, and thus the amount of erythrocytes stored/released. Contractions promote RBC release, thereby increasing Hct. In polar fish, but not necessarily temperate species, it seems likely that circulating catecholamines play a minor role in effecting the large increases in Hct. The reason for this conclusion is the relatively modest and slow increases in plasma catecholamine levels observed during hypoxia and intensive exercise in the Antarctic species tested so far (see below) relative to the more rapid and larger increase in Hct. Nervous control of the spleen is via autonomic postganglionic nerve fibers from the celiac ganglion that reach the spleen via the splanchnic (elasmobranchs and teleosts) or splenic (tetrapods) nerve. In the Atlantic cod *Gadus morhua*, the innervation of the spleen seems to be both adrenergic and cholinergic, and acetylcholine causes a contraction of the spleen (Nilsson and Grove, 1974; Winberg *et al.*, 1981). In contrast, the spleen in the Antarctic bald notothenioid *P. borchgrevinki* is controlled entirely by cholinergic muscarinic mechanisms. Injection of atropine decreased Hct from 18.6% to 6.6% and increased spleen somatic index from 0.60 to 0.89 (Nilsson *et al.*, 1996). Unfortunately, other polar fish species have not been studied regarding the nervous and humoral controls of the spleen, and it is, therefore, hard to say whether the cholinergic dominance in *P. borchgrevinki* is an effect of evolutionary lineage rather than the extreme environmental temperature (Egginton *et al.*, 2001).

Splenic contraction does allow for a rapid increase in oxygen-carrying capacity of the blood during exercise and stress, but this increase carries with it a proportionate increase in cardiac workload (Egginton, 1996). However, most polar fishes studied do not appear to have compensated for this by increasing cardiac ventricular mass.

The relative mass of the ventricle in polar fishes expressed as percentage of body mass has a similar range as most temperate fish, with the exception of the tuna and the icefish family (*Channichthyidae*) where the relative heart mass is similar to the relative heart mass found in smaller mammals (0.4%) (Tota *et al.*, 1991) (Figure 6.2). Even if the relative ventricle mass is similar between tunas, icefish, and smaller mammals, there is another distinction that needs to be taken into account and that is the pressure generating capacity of the ventricle. The ventricles of tunas are capable of generating blood pressures approaching those found in mammals, whereas the icefish ventricle is classified as a low-pressure high-volume pump (Figure 6.3). The icefish heart generates a far lower arterial blood pressure (1.5–3.0 kPa), but a higher stroke volume compared to most temperate fish species (see Table 6.3). La Place's Law (T = PR (2 × W), where T = wall tension, R = radius, P = pressure, and

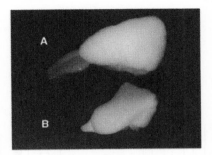

Fig. 6.2. A comparison of hearts from (A) the icefish *Pagetopsis macropterus* (body mass 89 g, total heart mass 0.43 g) and (B) the red-blooded Antarctic fish *Pagothenia borchgrevinki* (body mass 80 g, total heart mass 0.23 g), to illustrate the larger ventricle of the icefish. The yellow appearance of the icefish heart is due to lack of myoglobin in the myocardium, which is plentiful in *P. borchgrevinki*. Icefish courtesy of Professor A. L. De Vries.

W = wall thickness) dictates that for the heart to generate a larger stroke volume with the same afterload, the myocardial wall thickness (muscle wall tension) must be greater. Thus, by reducing arterial blood pressure, not only is cardiac workload reduced, but the need to increase ventricular wall thickness is also reduced. The icefish heart is a type I heart (Tota *et al.*, 1983), where the spongy type of myocardium acts as a "multichambered arrangement where a number of lacunae act as small auxiliary pumps" (for further discussion, Zummo *et al.*, 1995). The fact that stroke volume and not heart rate compensates for the reduction on blood oxygen-carrying capacity seems to suggest that reaction rates remain a limiting factor at polar temperatures.

Routine cardiac output in icefish is markedly higher ($\sim$6–10 times) compared with both polar and temperate teleosts (see Table 6.3). Compared with red-blooded Antarctic species, this difference represents a near-perfect compensation for the loss of RBCs in terms of arterial oxygen convection. The oxygen-carrying capacity of Hb-free blood is 0.25 mmol l^{-1} in *Chaenocephalus aceratus* and is about one-tenth of the 2.6 mmol l^{-1} found for dorsal aortic blood of *Notothenia coriiceps* and *Notothenia rossii* (Egginton, 1997b). Nevertheless, the oxygen-carrying capacity in *Notothenia* species is still about four times lower compared with temperate fish species and six times lower compared to tuna (for references, see Gallaugher and Farrell, 1998).

Total ventricular mass and protein content increase when temperate fish species are exposed to cold temperatures, and anemia can induce cardiac compensations (Kent *et al.*, 1988; Graham and Farrell, 1989, 1990; Tiitu and Vornanen, 2002). Clearly, the plasticity of ventricular mass, which is also expressed in sexual maturing male trout (Farrell *et al.*, 1988; Thorarensen and Davie, 1996), provides a genetic locus in fish in which natural selection could have acted. The trade-off to a larger ventricle may be a decrease in

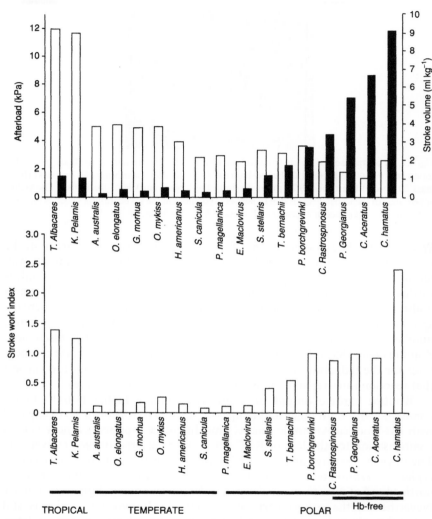

Fig. 6.3. A comparison of routine cardiac afterload (ventral aortic blood pressure, open bars), stroke volume (closed bars) (upper panel) and stroke work (lower panel) for selected tropical (TROP), temperate, and polar fishes. References: *Thunnus albacares* and *Katsuwonus pelamis*: Bushnell and Brill, 1992; *Anguilla australis*: Hipkins, 1985; *Ophiodon elongatus*: Farrell, 1981; *Gadus morhua*: Axelsson and Nilsson, 1986; *Oncorhynchus mykiss*: Kiceniuk and Jones, 1977; Farrell *et al.*, 1988; *Hemitripterus americanus*: Farrell *et al.*, 1985; *Scyliorhinus canicula*: Short *et al.*, 1977; *P. magellanica* and *Eleginops maclovinus*: Agnisola, 1997; *Scyliorhinus stellaris*: Farrell and Jones, 1992; *T. bernacchii* and *Pagothenia borchgrevinki*: Axelsson *et al.*, 1992; *Chionodraco rastrospinosus*: Acierno, 1997; *Pseudochaenichthys georgianus*: Hemmingsen and Douglas, 1977; *C. aceratus*: Hemmingsen *et al.*, 1972; *C. hamatus*: Tota *et al.*, 1991. (Modified from Axelsson *et al.*, 1998.)

cardiac scope (see below), but as noted earlier, this may not be critical in polar environments.

C. Vascular Resistance and Vascular Geometry

In the red-blooded Antarctic species tested so far, the systemic vascular resistance is similar to that measured in temperate species (see Table 6.3). However, in the icefish group, the systemic vascular resistance is one-fifth compared with the red-blooded cold living species, as well as other temperate fish species (see Table 6.3).

Although plasma (blood) η is low in icefish (see Table 6.1), it is at best only twofold lower than that for red-blooded notothenioids under comparable conditions (see Table 6.1). Therefore, the low systemic vascular resistance is likely a combination of a low blood viscosity and larger diameter capillaries (for discussion, Hemmingsen and Douglas, 1972; Wells, 1990; Feller and Gerday, 1997; Egginton and Rankin, 1998).

Re-inspection of Poiseuille's Equation shows that vascular resistance is directly influenced by vascular geometry. Several vascular modifications are possible that could decrease vascular resistance: (a) Resistance is directly proportional to vessel length, (b) resistance is also inversely proportional to vessel radius to the fourth power, and (c) the number of vessels connected in parallel (capillary density) affects the vascular resistance of the tissue (directly proportional to the total cross-sectional area). Given that a small change in vessel radius could have a much larger effect on vascular resistance than the blood viscosity, one might expect changes in vessel radius at frigid temperatures. However, it is important to recognize that an increase in capillary diameter may not be an adaptation simply to an increase in η. With a high cardiac output and low arterial oxygen content, oxygen delivery to the tissue could potentially be limited because of (a) too rapid capillary transit time (limited by reaction kinetics and diffusivity) and (b) a rapid decrease in the oxygen partial pressure gradient as oxygen is removed from capillary. Large-bore capillaries would compensate by increasing the pool of available blood and decreasing capillary transit time. Indeed, Egginton et al. (2002) calculated that to keep the minimum PO_2 above zero in the icefish muscle fibers, their capillary radius has to be at least 50% larger compared to red-blooded species. Increased capillary density would have the same effects.

II. ANATOMY AND CONTROL OF THE FISH HEART

In comparison with temperate fish species, the gross anatomy of the heart in polar fishes appears to be unremarkable. The heart is located posterior to the gills and is protected to some degree by the pectoral girdle.

The four chambers (sinus venosus, atrium, ventricle, and bulbus arteriosus) are serially arranged within a membranous pericardium much like other teleosts. Similarly, the pericardium is subdivided into an outer part called the *parietal pericardium* that adheres to the surrounding tissue and the inner part called the *visceral pericardium* or *epicardium*. However, the pericardial fluid, which is an ultrafiltrate from the plasma, does contain antifreeze glycoproteins in many polar species and, thus, gives the heart antifreeze protection (DeVries, 1971). The parietal pericardium adheres to the surrounding tissue, and the heart, thus, sits in a more or less ridged structure, which affects the mechanics of the heart. Because of this, some polar species [*Chionodraco hamatus* (Tota *et al.*, 1991), *Eleginops maclovinus*, *Paranotothe-nia magellanica*, and *Patagonotothen tessellata* (Agnisola *et al.*, 1977)] can achieve a *vis*-à-fronte (suction filling) cardiac filling, much the same way as it has been described for temperate species of teleosts and elasmobranchs (Satchell, 1991; Farrell and Jones, 1992; Altimiras and Axelsson, 2004).

Fish hearts have been characterized according to the arrangements of the muscle fibers and presence of coronary circulation (Tota *et al.*, 1983). The hearts of the notothenioids studied so far, like those of most other teleosts, are type I, which is a fully trabeculated ventricle, lacking a coronary vasculature and its associated compact myocardium. This "spongy" myocardium greatly increases the surface area to ventricular volume ratio (compared to the compact type of myocardium found in some high-performance fish species such as tunas, mammals, and birds, for example) and, thereby, increases the area of contact between the myocytes and their blood supply, which is the venous blood pumped through the ventricle lumen. This anatomical arrangement may be particularly important for polar fishes with reduced or no RBCs. As noted earlier, the low Hct leads to a low venous oxygen reserve, and the venous blood is the main oxygen supply for the heart (see discussion about arteriovenous pathways and possible effects of oxygen supply to the heart in the branchial circulation section below). So unless the polar fish hearts can function temporarily anaerobically, either burst activities must be minimized [because burst swimming results in precipitous decreases in venous oxygen tension in rainbow trout (Farrell and Clutterham, 2003)] or there must be a means of protecting the venous oxygen reserve.

All icefish species that lack Hb and RBCs also lack the oxygen-binding protein myoglobin in the oxidative muscles, which gives these muscles their distinctive white-yellow color. However, some of the icefish species do express myoglobin in the cardiac muscle (Acierno *et al.*, 1997; Sidell *et al.*, 1997). Diffusion of oxygen is temperature dependent and can be described by the formula $\delta O_2/\delta t = -K_{O2} \times A \times (\delta P_{O2}/x)$ ($-K_{O2}$ is the diffusion constant, which is the product of the diffusion coefficient [D_{O2}] and the solubility coefficient [a_{O2}], A is the surface area, δP_{O2} is the partial pressure difference and x is the diffusion distance). A decrease in temperature affects both the solubility coefficient (approximately $+1.4\%\,°C^{-1}$ decrease) and the diffusion coefficients (approximately $-3\%\,°C^{-1}$ decrease) (for further discussion, see Sidell, 1998). Myoglobin has a high oxygen affinity and acts as a store for oxygen and facilitates the diffusion of oxygen, but it has been debated whether this is the case at very low temperature. In a study of the cardiac performance in two icefish species (*Chionodraco rastrospinosus*, which has myoglobin in the myocardial muscle, and *C. aceratus*, which lacks cardiac myoglobin), it was shown that if the myoglobin in the heart of *C. rastrospinosus* was blocked, the mechanical performance of the heart was significantly impaired, indicating a role for myoglobin in the diffusion of oxygen into the myocytes in this species (Acierno et al., 1998).

In terms of intracardiac oxygen diffusion, two mechanisms are known to reduce the effect of the cold temperature and the lack of myoglobin. The mitochondrial density is higher in cold living species, thereby reducing the average diffusion distance for oxygen from the capillary wall to the mitochondria. Also, cold temperature in both temperate-zone and polar fishes is frequently correlated with a high content of lipids in oxidative muscles, enhancing diffusion of oxygen through the tissue. Even myoglobin structure has been modified for the polar environment because myoglobins from fish (in those species that express myoglobin) bind and release oxygen faster at cold temperatures than myoglobin from mammals. Together, these factors may compensate for the low level or absence of myoglobin and the cold temperature (Sidell, 1998; O'Brien and Sidell, 2000; O'Brien et al., 2000)

A. Cardiac Myocytes

Fish cardiac myocytes are much smaller (3–4 μm), compared with those in mammals (10–25 μm) (Santer, 1985). This leads to a much higher cell surface area to volume ratio in the fish myocytes, which has importance for gas and ion exchanges across the sarcolemma. In both red-blooded and Hb-free notothenioids, the diameter of the myocytes seems to be slightly larger, 4–6 μm, and due largely to an increased number of mitochondria per cell (Johnston et al., 1983; Tota et al., 1991; Zummo et al., 1995; O'Brien and Sidell, 2000; O'Brien et al., 2000). The high mitochondrial number is likely an adaptation to compensate for the lowered rates of ATP production at low temperature. Cardiac work, expressed as mW g^{-1} ventricular tissue in polar fishes does not differ from temperate species, so ATP demand should be similar (Table 6.6). What is particularly important in this comparison, however, is that the relatively large heart mass of icefish is not primarily achieved through cellular hypertrophy. This may mean that cold-induced/

anemia-induced cardiac remodeling in fish still requires the myocytes to maintain a high surface-to-volume ratio. The fish myocyte lacks T-tubules, and sarcolemmal Ca^{2+} exchange is of great importance for the excitation–contraction coupling (Shiels *et al.*, 2002). Thus, there may be a need to maintain the Ca^{2+} diffusion distance between the extracellular space and myofibril as small as possible. Indeed, the myofibrils are arranged as a ring close to the sarcolemmal membrane, whereas the mitochondria are more centrally located in the myocyte.

B. Cardiac Performance

When considering cardiac output in polar fish, a distinction must be made between the red-blooded group of fish and the icefish group. In the red-blooded group, cardiac output is similar or slightly higher compared to temperate species, whereas in the icefish group, cardiac output is very high (see Table 6.3). Because both routine and intrinsic heart rate is lower in polar species compared with temperate species, stroke volume must be elevated (see Table 6.5 and Figure 6.3). During periods of increased oxygen demand, both heart rate and stroke volume are increased in order to increase cardiac output, and the scope for this increase has been studied in three sculpins, one eurythermal, *M. scorpius*, and two stenothermal species (*Gymnocanthus tricuspis* and *M. scorpioides*). It was shown that an acute temperature change increased the scope for cardiac output in *M. scorpius*, whereas in the two more stenothermal species, the scope for cardiac output increase was abolished by this acute temperature increase (Axelsson *et al.*, unpublished data) (Figure 6.4).

C. Control of Heart Rate

Heart rate is a function, in the first instance, of the intrinsic rate of the cardiac pacemaker cells in the heart (located in the *sinus venous* or in the sinoatrial junction) and the modulatory effects of extrinsic factors such as humoral substances and/or a direct innervation via the autonomic nervous system (Morris and Nilsson, 1994). Temperature directly affects the rate of the firing of the pacemaker with a Q_{10} of about 2 (see Figure 6.5). Extrinsic modulation of heart rate is more complex to understand. The relative contributions of the modulatory systems are typically determined by injecting appropriate pharmacological antagonists (muscarinic and β-adrenergic) until the heart is operating at its intrinsic rate.

All fish species so far studied, with the exception of hagfishes and lampreys (*Cyclostomata*), have muscarinic cholinergic inhibitory innervation of the heart. This is revealed by injections of atropine. In most of these

Table 6.5

Heart Rate Data for Different Teleost Species with Calculated Cholinergic and Adrenergic Tone on the Heart[a]

Species	Temp (°C)	f_H (resting) (min^{-1})	f_H (intrinsic) (min^{-1})	Adrenergic tone (%)	Cholinergic tone (%)	Reference
Cold-Water Living Species						
Gymnodraco acuticeps	-1.0	17.4	18.4	20.2	23.7	Axelsson *et al.*, 2000
Pagothenia borchgrevinki	0.0	11.3	23.3	4.0	112.0	Axelsson *et al.*, 1992
Trematomus bernacchii	0.0	10.5	21.7	22.0	130.0	Axelsson *et al.*, 1992
Temperate Species						
Paranotothenia angustata	12		47.3	35	15	Egginton *et al.*, 2001
Oncorhynchus mykiss	7.0	42.2	43.5	22.7	22.0	Gamperl *et al.*, 1995
Gadus morhua	10.0	37.2	39.8	12.9	21.3	Altimiras *et al.*, 1997
Myoxocephalus scorpius	10.0	48.3	42.3	20.2	8.2	Axelsson *et al.*, 1987
Atlantic cod	10.5	30.5	36.6	17.0	39.0	Axelsson, 1988
Gadus morhua						
Hemitripterus americanus	11.0	37.6	33.0	29.0	23.0	Axelsson *et al.*, 1989
Polachius pollachius	11.5	46.0	40.0	25.0	13.0	Axelsson *et al.*, 1987
Labrus mixtus	11.5	52.0	50.0	14.0	12.0	Axelsson *et al.*, 1987
Labrus bergylta	11.5	41.0	49.0	14.0	36.0	Axelsson *et al.*, 1987
Ciliata mustela	11.5	67.0	58.0	23.0	9.0	Axelsson *et al.*, 1987
Raniceps raninus	11.5	31.0	28.0	22.0	8.0	Axelsson *et al.*, 1987
Zoarces viviparus	11.5	60.0	38.0	40.0	7.0	Axelsson *et al.*, 1987
Oncorhynchus kisutch	12.0	31.8	30.0	38.0	34.0	Axelsson and Farrell, 1993
Sparus aurata	16.0	63.4	73.8	38.8	16.8	Altimiras *et al.*, 1997
Labrus bergylta	20.0	84.5	88.3	20.8	30.4	Altimiras *et al.*, 1997
Carassius auratus	22.5	36.0	57.0	18.0	97.0	Cameron, 1979
Thunnus thunnus	25.0	75.5	116.0	4.1	58.1	Keen *et al.*, 1995
Katsuwonus pelamis	25.0	79.4	183.0	5.8	130.8	Keen *et al.*, 1995

[a]For the calculation of the cholinergic and adrenergic tone, the **R-R** interval method as suggested by Altimiras *et al.* (1997) was used. *Intrinsic heart rate* is defined as heart rate after a complete blockade of muscarine and β-adrenoceptors using atropine and sotalol or propranolol.

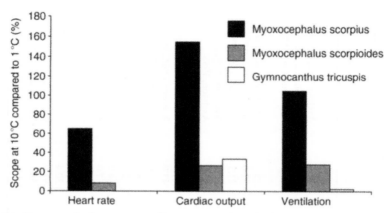

Fig. 6.4. The scope for heart rate, cardiac output, and ventilation in three species of sculpins from Arctic waters around Greenland (*Myoxocephalus scorpius*, *Myoxocephalus scorpioides*, and *Gymnocanthus tricuspis*) was tested at 1, 4, 7, and 10 °C by calculating the difference between resting values and values obtained after 10 minutes of chasing. The graph shows the change in scope between the control (1 °C) and 10 °C (acute temperature change with a time course of ~4 hours from 1 to 10 °C). Note that the scope for the eurythermal *M. scorpius* showed a small decrease for heart rate, whereas the scope for cardiac output and ventilation increased. The scope for all variables decreased markedly for the two other more stenothermal polar living species.

species, cholinergic innervation is tonic. The cholinergic tone in two Antarctic species, *P. borchgrevinki* and *T. bernacchii*, is high (130 and 112%, respectively) compared with most other fish species with the exception of goldfish, *Carassius auratus*, 97% and skipjack tuna, *Katsuwonus pelamis*, 131% (see Table 6.5). On the other hand, cholinergic tone is low and comparable with other studied teleosts in another Antarctic species *Gymnodraco acuticeps*. In fact, when all data are compared (Figure 6.5), it is evident that no clear trend exists for the species studied so far (Axelsson *et al.*, 1992, 2000a). Changing the cardiac cholinergic tone is one of the fastest ways to change heart rate. Cholinergic tone decreases during stress, and this fact is important to keep in mind when discussing the role of cholinergic tone because if the way we acquire our data stresses the fish, data interpretations may be affected (Altimiras *et al.*, 1997; Altimiras and Larsen, 2000). In fact, using heart rate variability, which is a function of active cholinergic and adrenergic influence on the heart, may be a better way of assessing stress levels compared with mean heart rate itself. Campbell *et al.* (2004) found that while resting mean heart rate stabilized during recovery in the shorthorn sculpin (*M. scorpius*), heart rate variability remained low, indicating that an even longer period of recovery was needed to reestablish the normal control and function of the heart.

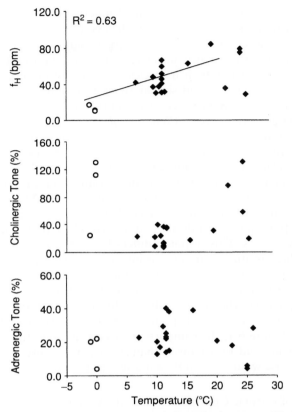

Fig. 6.5. The relationship between temperature and (A) resting heart rate, (B) cholinergic tone, and (C) adrenergic tone for selected resting temperature-acclimated polar (open circles) and temperate fishes (closed diamonds). Note the lack of any trend in the cholinergic and adrenergic tone on the heart versus acclimation temperature among fish species despite a clear relationship for resting heart rate. In the top panel, the regression line for the resting heart rate versus temperature is shown ($R^2 = 0.63$). See Table 6.5 for numerical values.

Many teleosts also have adrenergic innervation of the heart acting via β-adrenoceptors; this is complemented by circulating catecholamines released from the head kidney and, in some cases (cyclostomes, elasmobranchs, and dipnoans), from endogenous or other chromaffin stores (Morris and Nilsson, 1994). Adrenergic excitation also acts tonically to increase heart rate, because injection of β-adrenoceptor blockers such as sotalol or propranolol decreases heart rate. A small β-adrenergic cardiac tone was found in resting *P. borchgrevinki* (Axelsson *et al.*, 1994). What is less clear is whether this adrenergic stimulation is a result of direct innervation or circulating catecholamines (Franklin *et al.*, 2001). Nonetheless, the relatively small change in adrenergic tone compared with the larger change in cholinergic tone indicates that cholinergic mechanisms remain more important for the control of the heart in Antarctic fish species. This conclusion appears to be in accordance with an earlier and similar suggestion that cholinergic control was more pronounced than adrenergic control in cold-acclimated than in warm-acclimated rainbow trout (Wood *et al.*, 1979).

Cholinergic tone can also modulate the effect of an acute temperature change on intrinsic heart rate. In the two Antarctic species *P. borchgrevinki* and *T. bernacchii*, an acute temperature increase up to 2.5 °C did not alter heart rate in intact animals, due to a compensatory increase in cholinergic tone on the heart (Axelsson *et al.*, 1992; Franklin *et al.*, 2001) (Figure 6.6). Nevertheless, the intrinsic heart (recorded after total muscarinic and β-adrenoreceptor blockade) rate showed the typical temperature dependence in both these species. This response differs from that seen in three sculpin species in which routine heart rate in intact animals clearly increased with temperature. Even so, during acute temperature change, the scope for heart rate increase was attenuated in the two stenothermal species (*G. tricuspis* and *M. scorpioides*), whereas a smaller (-40%) reduction of the scope was found in the more eurythermal species *M. scorpius* (Axelsson *et al.*, unpublished data) (Figure 6.4). It is important to keep in mind that these results come from acute and rather rapid changes in temperature, and it is likely that the response would look different after temperature acclimation.

D. Control of Cardiac Contractility and Stroke Volume

To meet the variable metabolic demands of the organism, cardiac output must be constantly adjusted. This can be achieved by changes in stroke volume and heart rate. In Antarctic fish species both routine and maximum stroke volume tend to be higher than in temperate species, with extremely high values found in the icefish group (see Figure 6.3 and Table 6.3). The product of stroke volume and mean driving pressure (mean ventral aortic − venous pressure) is called *stroke work*, and this is an approximation of the work done by the heart, which would be more accurately calculated using pressure–volume loops (Farrell and Jones, 1992). Figure 6.3 shows the relationship between stroke volume and mean ventral aortic blood pressure (cardiac afterload). Most of the Antarctic species are found on the low-pressure/high-flow side of the graph. The most extreme examples are the two icefish species *Chaenocephalus hamatus* and *C. aceratus*, and it is likely that other members of the icefish group fall into this extreme group with a heart specialized for high-volume pumping (Tota *et al.*, 1991; Tota and Gattuso,

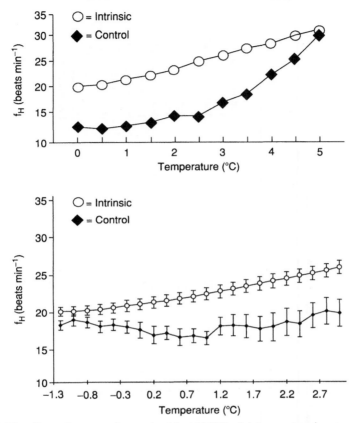

Fig. 6.6. The effects of an acute (approximately $+1°C/30$ min) temperature increase on heart rate in *Trematomus bernacchii* (upper panel) and *Pagothenia borchgrevinki* (lower panel). In both panels, intrinsic (open circles) and unblocked heart rate (closed diamonds) are shown. In *T. bernacchii*, no change in heart rate is seen until the temperature is increased above $2°C$, and at $5°C$, intrinsic heart rate and resting heart rate are the same. In *P. borchgrevinki* (lower panel), no change in heart rate in the unblocked animals is seen over the tested tempera- ture interval, whereas the intrinsic heart rate shows the expected increase with temperature. (Adopted from Axelsson *et al.*, 1992 [upper panel], and Franklin *et al.*, 2001 [lower panel].)

1996) (Table 6.6). There is another clear distinction between the fish on the right side of the graph compared with the left side; in the high-pressure/low-volume species, an increase in stroke work is mainly achieved by an increase in pressure, whereas in the other group, it is an increase in stroke volume with a relatively constant pressure generation that increases stroke work. This is most likely due to the difference in myoarchitecture (Agnisola and Tota, 1994; Tota and Gattuso, 1996).

Table 6.6
Cardiac Power Output and Heart Rate (f_H) for Polar, Cold-Water, and Temperate Fish Species

Species	Power output ($mW\ g^{-1}$)	($f_H\ min^{-1}$)	T (°C)	References
Cold-Water Living Species				
Eleginops maclovinus (Antarctic)	1.4	46	10	Agnisola et al., 1997
Paranotothenia magellanica (Antarctic)	0.8	61	10	Agnisola et al., 1997
Patagonotothen tessellata (Antarctic)	0.8	73	10	Agnisola et al., 1997
Pagothenia borchgrevinki (Antarctic)	2.3	37	0	Axelsson et al., UP
Trematomus bernacchii (Antarctic)	2.0	28	−0.5	Agnisola et al., UP
Chionodraco rastrospinosus (Antarctic)	1.1	28	0.5	Acierno et al., 1997
Chaenocephalus aceratus (Antarctic)	1.4	28	−0.5	Acierno et al., 1997
Chionodraco hamatus (Antarctic)	1.8–3.2[e]	26	0.6	Tota et al., 1991
Chaenocephalus aceratus (Antarctic)	0.98[d]	16		Hemmingsen and Douglas, 1977
Pseudochaenichthys georgianus (Antarctic)	0.72[d]	13		Hemmingsen and Douglas, 1977
Chaenocephalus aceratus (Antarctic)	1.80[d]			Hemmingsen et al., 1972
Notothenia corriiceps[c] (Antarctic)	0.35	21	0	Egginton, 1997
Gadus ogac (Arctic)	0.65	15	0	Axelsson et al., UP
M. scorpius (Arctic)	0.89	25	5	
	2.8[a]	28	1	Axelsson et al., UP
Temperate Species				
Oncorhynchus mykiss	5–6.5	45–55	10	Farrell et al., 1986
Thunnus albacares	7.6	123	25	Farrell et al., 1991

260

Species					Reference
Thunnus albacares	5.6			25	Bushnell and Brill, 1992
Katsuwonus pelamis	2.6	138		25	Farrell et al., 1991
Katsuwonus pelamis	6.3			25	Bushnell and Brill, 1992
Myxine glutinosa	0.2[b]		22	8–12	Axelsson et al., 1986
Myxine glutinosa	0.54		23	10–11	Johnsson and Axelsson, 1996
Eptatretus cirrhatus (portal heart)	0.20		40	10	Johnsson et al., 1996
Eptatretus cirrhatus	0.4				Forster et al., 1989
Oncorhynchus mykiss	2.7–3.9		30–58	10	Farrell et al., 1988
Oncorhynchus mykiss	4.1		58–60	15	Graham and Farrell, 1990
Squalus acanthias	2.3		41	16	Davie and Franklin, 1992
Anguilla dieffenbachii	3.0		33–41	15	Franklin and Davie, 1992
Hemitripterus americanus	1.2[b]		39	10	Farrell et al., 1985
	0.8[a]				
Gadus morhua	2.7[b]		41	10	Pettersson and Nilsson, 1980
Gadus morhua	1.4[b]		43 rest	10	Axelsson and Nilsson, 1986
			53 exercise		
Ophiodon elongatus	2.6		29	9–11	Farrell, 1981
Myoxocephalus scorpius	1.2		62	13	Axelsson et al., unpublished
	1.3[a]				

[a] *In situ* value.
[b] Calculated from *in vivo* data assuming zero venous pressure.
[c] *In vivo* data based on microspheres.
[d] Estimated from tabulated data.
[e] Maximum values taken from graph range representing different preloads.

The force generated by the myocytes depends on the duration and rate of contraction and relaxation. This in turn is affected by the calcium release and reuptake. Rates of contraction are tied to rates of calcium movements and myosin ATPase activity, among other factors. Temperature generally has a negative effect on the inotropic state of the heart through its effect on rates of reaction. However, the low absolute heart rates of cold-water species should be beneficial and improve inotropy through a positive staircase effect (see review by Shiels *et al.*, 2002). It has also been shown that cold acclimation, in addition to affecting the total mass and protein content of the fish heart, also increases rates of contraction (Driedzic and Gesser, 1994; Aho and Vornanen, 1999, 2001).

It has been postulated that at low temperatures, adrenergic stimulation may be more important for the inotropy of the cardiac muscle than for heart rate (Graham and Farrell, 1989). In fact, the current (I_{ca}) through L-type calcium channels in rainbow trout myocytes decreases dramatically at 5 °C, and this critical ion flux can be restored with adrenergic stimulation (Shiels *et al.*, 2003). Cardiac β-adrenoceptor density was higher in the cold-acclimated rainbow trout and this was thought to contribute to the increased sensitivity to adrenergic stimulation (Keen *et al.*, 1993). Few studies have examined the importance of adrenergic stimulation for the cardiac performance in polar fishes. In a multispecies study, Olsson *et al.* (2000) showed that the β-adrenoreceptor density was low in the Antarctic *T. bernacchii*, but they also concluded that there is a large interspecies variability that cannot fully be explained by temperature alone. In contrast to cold-acclimated rainbow trout hearts, adrenergic stimulation (10^{-7} M adrenaline) was without effect in *in situ* perfused *P. borchgrevinki* heart at 0 °C (Axelsson *et al.*, unpublished data). Similarly, adrenergic effects on heart rate, cardiac output, and power generation in the perfused heart from *M. scorpius* was to a much greater degree at 15 °C compared with 1 °C (Farrell *et al.*, unpublished data). Further studies of adrenergic control in polar fishes are needed to resolve these apparent differences between polar and temperate species.

III. THE BRANCHIAL AND SYSTEMIC VASCULATURE

A. Control of the Branchial Vasculature

The branchial circulation has not been studied to any great extent in polar fishes. Like all fish, deoxygenated blood leaving the heart enters the branchial circulation via the afferent branchial arteries. In the branchial system, there are three vascular pathways: the arterioarterial (or respiratory) pathway, the interlamellar pathway, and the nutrient pathway (see Olson,

2002, for further discussion). The arterioarterial pathway carries 90–92% of cardiac output, taking blood through the secondary lamellae, where gas and ion exchanges take place. The blood enters the systemic circulation via the efferent branchial vessels. The arteriovenous pathways usually arise from the efferent side of the gills and, therefore, deliver fully oxygenated blood to the body of the gill filaments, and then this blood drains back into the heart (Ishimatsu *et al.*, 1988; Sundin and Nilsson, 1992; Olson, 2002) (Figure 6.7). The extent to which the arteriovenous pathway may contribute to the oxygen supply of the heart is unknown. The only known striking difference in gill morphology between the Antarctic fish and other teleosts is the larger diameter branchial vessels found in the icefish group (Vogel and Koch, 1981; Davison *et al.*, 1997).

A number of different regulatory mechanisms are involved in the control of the blood flow through the different pathways in the gills. They include cholinergic, α- and β-adrenergic, and serotonergic (5-HT, 5-hydroxytryptamine) mechanisms, as well as different neuropeptides (Morris and Nilsson, 1994). Branchial vascular resistance is lowered in both Atlantic cod (*G. morhua*) and rainbow trout (*Oncorhynchus mykiss*) at sustained swimming speeds but increases in rainbow trout as U_{crit} is reached (Kiceniuk and Jones, 1977; Axelsson and Nilsson, 1986). In contrast, sustained swimming in the *P. borchgrevinki* is associated with a 30% increase in branchial vascular resistance (Figure 6.8). This increase could be blocked by atropine (a muscarine receptor antagonist), indicating a cholinergic mechanism (Axelsson *et al.*, 1994). Subsequent injection of a general β-adrenoceptor antagonist restored the increase in branchial resistance during exercise, indicating a β-adrenoceptor–mediated vasodilation of the branchial vasculature, but an α-adrenergic vasoconstriction of the branchial vasculature also dominated over the β-adrenoceptor–mediated vasodilation (Axelsson *et al.*, 1994). Serotonin is a very potent vasoconstrictor of the branchial vasculature in *P. borchgrevinki*, mimicking the effects of exercise on dorsal aortic pressure and lowering the oxygen partial pressure in the dorsal aortic blood (Sundin *et al.*, 1998) (Figure 6.9). The branchial sensitivity to serotonin in *P. borchgrevinki* was around 1000-fold higher compared to the rainbow trout (Sundin, 1995; Sundin *et al.*, 1995), leading to speculation that a very high serotonin sensitivity is a cold adaptation. This suggestion was subsequently tested by comparing notothenioid fish species from different habitats, *T. bernacchii* (Antarctic), *D. mawsoni* (Antarctic), *Paranotothenia angustata* (New Zealand waters 12–14 °C), *Bovichtus variegatus* (New Zealand waters 12–14°C) (Egginton *et al.*, 2001). They showed that *P. angustata*, a temperate/warm-water living species related to the Antarctic notothenioids, had the same serotonin sensitivity as that found in the *P. borchgrevinki* and *T. bernacchii*. Thus, the increased serotonin sensitivity in *Pagothenia*

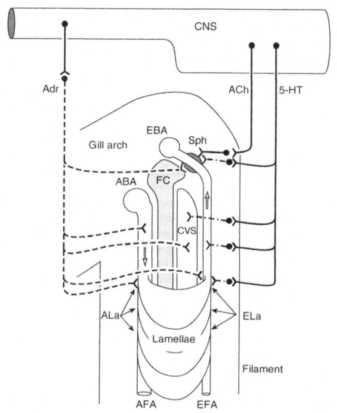

Fig. 6.7. A schematic diagram showing circulation pattern and the innervation of the branchial vasculature in teleost fish. Blood enters the gill arch via the afferent branchial artery (ABA) and reaches the lamellae where the gas exchange takes place via the afferent filamental artery (AFA) and the lamellar arterioles (ALa). The oxygenated blood then leaves the lamellae via the efferent lamellar arteriole (ELa) and efferent filamental artery (EFA) to the efferent branchial artery (EBA). Autonomic postganglionic adrenergic nerves (Adr: broken lines) innervate both the afferent and the efferent vessels and the central sinus (CVS). Autonomic cranial vagal postganglionic nerves including both cholinergic (Ach: solid lines) and serotonergic (5-HT, dash/dot lines) innervate the sphincter (Sph) at the base of the EFA; this sphincter also receives adrenergic and serotonergic innervation. Serotonergic fibers are also found in the CVS and in most vessels within the gills. The various receptors in the gill can also be affected by circulating substances such as adrenaline, noradrenaline, and angiotensin. FC, filamental cartilage. (Used with permission from Morris and Nilsson, 1992.)

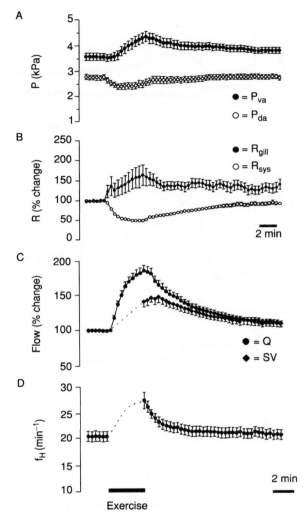

Fig. 6.8. The effects of exercise on (A) ventral (P_{VA}) and dorsal aortic (P_{DA}) blood pressure, (B) branchial (R_{gill}) and systemic (R_{sys}) vascular resistance, (C) cardiac output (Q) and stroke volume (SV), and (D) heart rate (f_H). Note the decrease in P_{DA} and R_{sys} during exercise and the increase in both heart rate and stroke volume. (Modified from Axelsson *et al.*, 1994.)

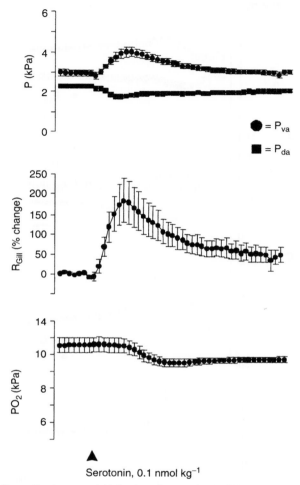

Fig. 6.9. The effects of an intraarterial injection of serotonin (5-HT) in *Pagothenia borchgrevinki* on ventral (P_{VA}) and dorsal aortic (P_{DA}) blood pressure (upper panel), branchial (R_{gill}) vascular resistance (middle panel), and dorsal aortic oxygen tension (lower panel). Note the large increase in R_{gill} and decrease in arterial oxygen tension. (Modified from Sundin *et al.*, 1998.)

borchgrevinki is most likely a consequence of evolutionary lineage instead of a response to cold adaptation (Egginton *et al.*, 2001).

Similar control mechanisms for the branchial vasculature have been described for the icefish *Chionodraco hamatus*, in which it was shown that acetylcholine and serotonin both induced a vasoconstriction, whereas nitric oxide induced a vasodilation. It was also found that nitric oxide exerted a tonic vasodilation in the isolated perfused head preparation (Pellegrino *et al.*, 2003). In conclusion, the control mechanisms described for the branchial vasculature in the Antarctic species do not differ from other teleosts, and variations in sensitivity to different antagonists are most likely due to evolutionary lineage than anything else even if ecological differences cannot be ruled out because of the low number of species tested so far. The large increase in branchial vascular resistance during exercise is unusual, and the functional significance of this is not fully understood.

B. Control of the Systemic Vasculature

Similar to the branchial circulation, information on systemic vascular control in polar fish species is limited and mostly is derived from a few Antarctic species. Systemic vascular resistance is calculated as a function of the pressure decrease across the systemic circulation. This calculation typically assumes that (a) dorsal aortic blood pressure represents the driving force, venous blood pressure is zero, and (b) all cardiac output passes through the major site of gill resistance (i.e., the lamellar arterioles) (Farrell, 1981; Olsson, 2002). With the exception of the icefish group, values for systemic resistance in polar fishes are similar to those reported for most temperate teleosts (Hemmingsen *et al.*, 1972; Rankin, 1989; Axelsson *et al.*, 1994; Franklin *et al.*, 2004) (see Table 6.3).

Although systemic vascular control mechanisms are poorly investigated in polar fishes, published data point to a fundamental difference to the commonly reported α-adrenergic control of systemic resistance in temperate fish species. Injection of adrenaline in *P. borchgrevinki* and *T. bernacchii* produced only a short lasting increase in the systemic vascular resistance mediated via α-adrenoceptors (Axelsson, *et al.*, 1994). This brief response contrasts with a longer lasting increase in the systemic vascular resistance seen after adrenaline injection in the rainbow trout and sea raven (*Hemitripterus americanus*) (Tuurola *et al.*, 1982; Axelsson *et al.*, 2000b). Injection of an α-receptor antagonist revealed a resting α-adrenergic tone in *P. borchgrevinki* (Axelsson *et al.*, 1994), but the pressor response to an adrenaline injection in *P. borchgrevinki* is due to an increase in cardiac output rather than an increase in systemic vascular resistance. Because there is an increase in stroke volume, the site of adrenergic action may be either the veins, increasing venous return to the heart, or directly on the heart, stimulating cardiac contractility.

Little is know about the innervation of the systemic vasculature in Antarctic fish, and only few studies in non-Antarctic species try to distinguish between adrenergic innervation and circulating plasma catecholamines in the control of the vascular tone (Smith, 1978; Axelsson, 1988). In a study of the systemic vascular control in *P. borchgrevinki*, indications of an α-adrenergic nervous tone on the systemic vasculature was found, but this could explain only part of the resting adrenergic vascular tone because the α-adrenoceptor antagonist phentolamine further lowered the systemic vascular resistance (Axelsson *et al.*, unpublished observations).

Another important control system for blood pressure control is the renin–angiotensin system (RAS). The RAS involves a chain of events starting with a release of the enzyme renin by the kidney, which acts on angiotensin I (Ang I) in the blood to generate the functional angiotensin II (Ang II). The RAS is present in the Antarctic species *P. borchgrevinki*, *T. bernacchii*, and *C. hamatus* (Uva *et al.*, 1991; Axelsson *et al.*, 1994). Injections of Ang II into *P. borchgrevinki* elicit marked increases in both ventral and dorsal aortic blood pressure, but similar to the response to adrenaline, a marked increase in stroke volume and gill vascular resistance rather than an increase in systemic vascular resistance is responsible for the pressure changes (Figure 6.10). Injection of Ang I also produces this effect, but with a slower time course, and pretreatment with the angiotensin-converting enzyme (ACE) inhibitor enalapril abolishes the effect, indicating a functional RAS in *P. borchgrevinki* (Axelsson *et al.*, 1994). Some studies suggest that part of the effects of Ang II are mediated via catecholamines released from adrenergic nerves and that this may be a more important system for the tonic control of systemic blood pressure in some fish species (Olson, 1992; Platzack *et al.*, 1993; Bernier *et al.*, 1999a,b).

If treated with an α-adrenoceptor antagonist before exercise, there is a massive fall in both prebranchial and postbranchial blood pressure in the Atlantic cod and rainbow trout, and this elicits an increase in the RAS that will at least partially compensate for the lack of α-adrenoceptor control (Platzack *et al.*, 1993; Bernier *et al.*, 1999a,b). However, similar compensatory mechanisms, as found in the Atlantic cod, were not found in *P. borchgrevinki* (Axelsson *et al.*, 1994).

IV. INTEGRATED CARDIOVASCULAR RESPONSES

A. Stress

During periods of stress (exhaustive exercise, hypoxia, thermal stress, etc.), a general increase in the adrenergic drive is a common feature in most studied fish species and results in a large increase in the concentration of circulating catecholamines (adrenaline and noradrenaline) (Mazeaud *et al.*, 1977; Wahlqvist and Nilsson, 1980; Van Dijk and Wood, 1988, Perry and

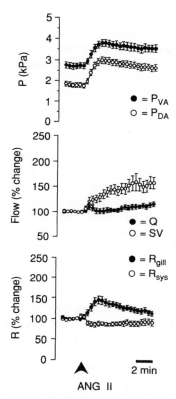

Fig. 6.10. The effects of an intraarterial injection of angiotensin II (ANG II) in *Pagothenia borchgrevinki* on ventral (P_{VA}) and dorsal aortic (P_{DA}) blood pressure (upper panel), cardiac output (Q) and stroke volume (SV) (middle panel), and systemic (R_{sys}) and branchial (R_{gill}) vascular resistance (lower panel). Note the large increase in both P_{DA} and P_{VA} and the increase in R_{gill}. (Modified from Axelsson *et al.*, 1994.)

Reid, 1992). Plasma levels of circulation catecholamines (adrenaline and noradrenaline) are a function of release from endocrine cells (chromaffin cells) and paracrine cells such as chromaffin cells in the heart of cyclostomes and lungfish that release catecholamines and spill over from adrenergic nerves (catecholamines leaking out of the synapses into the main circulation during nerve firing). In most temperate fish species studied, the resting catecholamine levels are in the range of 3–7 nM (Axelsson, 1988; Milligan *et al.*, 1989).

In the few studies that report plasma levels of catecholamines in Antarctic species (*N. coriiceps*, *C. aceratus*, and *T. bernacchii*), resting levels are similarly low. However, periods of stress, exercise, or hypoxia in the Antarctic fish species induce only small increases in circulating catecholamines (Egginton et al., 1991; Axelsson et al., 1994; Egginton, 1994, 1997a; Davison et al., 1995; Forster et al., 1998; Whiteley and Egginton, 1999). These species certainly have the capacity to increase plasma catecholamine levels further, but the stimuli needed (both intensity and time) to elicit such increases are higher compared to most studied temperate species (Davison et al., 1988; Egginton, 1991, 1994; Davison et al., 1995). In addition, the changes in ventral aortic blood pressure and heart rate that are seen during periods of stress cannot be correlated with circulating catecholamines because their levels do not change, indicating that unlike in temperate species, other control mechanisms are more important in these species for the adjustment of the cardiovascular system during stress (Davison et al., 1995). The blunted response to stress in these polar species may be explained by the low tyrosine hydroxylase activity revealed in a number of Antarctic species compared with temperate species (Whiteley and Egginton, 1999).

B. Exercise

It is important to make a distinction between exercise and stress (handling or chasing). Exercise in temperate fish species has been classified into three swimming speeds/modes/levels: sustained, prolonged, and burst swimming, depending on the length of time that the fish can maintain the swimming velocity (Beamish, 1978). This classification was created for temperate fish species using the subcarangiform swimming mode, and this may not apply to the labriform swimming, as found in many Antarctic fish species. Critical swimming speed (U_{crit}), based on various test protocols, estimates the prolonged swimming speed (Brett, 1973). During burst swimming and as fish approach U_{crit}, an increase in plasma catecholamines is usually found. At these swimming speeds, it is hard to make a clear distinction between stress and exercise in terms of cardiovascular responses. During nonexhaustive exercise (i.e., no increase in plasma catecholamines), the fish cardiovascular system, like all vertebrates studied so far, is adjusted to meet the increased demand in blood flow by changing heart rate, stroke volume, and arteriovenous oxygen content to varying degrees (Kiceniuk and Jones, 1977; Jones and Randall, 1978; Axelsson and Nilsson, 1986; Axelsson and Fritsche, 1991). A few studies have examined cardiovascular changes associated with nonexhaustive exercise in Antarctic fish species. In *P. borchgrevinki*, the increase seen in cardiac output during exercise is a result of increases in both stroke volume and heart rate, and at the same time, ventral aortic blood pressure increases while a large increase in the branchial vascular resistance causes a fall in the dorsal aortic blood pressure (Axelsson et al.,

1994) (Figure 6.8). Another Antarctic species *N. coriiceps* also exhibited decreased dorsal aortic pressure during exercise, whereas increases in both heart rate and stroke volume produced a 2.3-fold increase in cardiac output. Additionally, a small redistribution of blood flow away from the gastrointestinal circulation occurred (Egginton, 1997b). These results contrast with Atlantic cod and rainbow trout in which both the ventral and the dorsal aortic blood pressure increase during exercise, with an increase in systemic vascular resistance and an unchanged branchial vascular (Kicieniuk and Jones, 1977; Axelsson and Nilsson, 1986). This increased vascular tone in the Atlantic cod and rainbow trout serves two functions: (a) redistribute blood away from the gastrointestinal canal (Axelsson and Fritsche, 1991; Thorarensen *et al.*, 1993) and (b) counterbalance the metabolite-induced vasodilation in the active muscles. The decrease in systemic vascular resistance in *P. borchgrevinki* during exercise is unaffected by either muscarinic, α-adrenoceptor, or β-adrenoceptor antagonists (see Figure 6.8), something that contrasts with the α-adrenoceptor–controlled systemic vascular resistance in the Atlantic cod and rainbow trout.

C. Hypoxia

Fish living in temperate waters have to deal with changes in dissolved oxygen and do so via complex regulation of many physiological processes including adjustments of the cardiovascular system. In temperate fish species, variations in temperature and consequently water oxygen content are common (Henry's gas law in combination with Le Chatelier's principle). Perhaps the greatest variations in water oxygen content are seen in tropical climates. Conversely, polar and deep-sea species live in a stenothermal environment and see little natural change in water oxygen content. Even so, polar fishes show cardiovascular changes to environmental hypoxia.

Upon rapid acute exposure to hypoxia, a typical temperate fish responds with bradycardia and an increase in dorsal aortic blood pressure due to an increase in systemic vascular resistance (Wood and Shelton, 1980; Axelsson *et al.*, 1990; Fritsche, 1990; Fritsche and Nilsson, 1990; Farrell, 1991, 1992; Satchell, 1991). In contrast, heart rate in *T. bernacchii* increased by 2 beats min^{-1} during hypoxia (water PO_2 6.7 kPa), and both ventral aortic blood pressure and total systemic vascular resistance also increased. In some animals, the beat-to-beat interval became more variable during hypoxia, being prolonged for one to three beats, followed by a period of shorter intervals with little effect on the average heart rate. Similar to the vagal hypoxic bradycardia of temperate fish, the beat-to-beat variation of the heart rate in *T. bernacchii* was abolished by atropine pretreatment, indicating a vagal origin. Three individuals of the cryopelagic species *P. borchgrevinki* showed a variable response to hypoxia: One individual showed a clear bradycardia, and the two others showed a slight tachycardia at water PO_2 of 6 kPa; two of the fish also showed increases in blood pressure and systemic vascular resistance, similar to the response to hypoxia seen in temperate species (Axelsson *et al.*, 1992). In three separate studies of the icefish *C. aceratus*, two report a clear hypoxic bradycardia and increased systemic blood pressure (Hemmingsen and Douglas, 1972; Holeton, 1972), whereas the third study did not detect any change in heart rate but an increased dorsal aortic blood pressure (Hemmingsen *et al.*, 1972). In *Pseudochaenichthys georgianus*, cardiac output, heart rate, and ventral aortic blood pressure were all reduced during hypoxia (Hemmingsen and Douglas, 1977). Based on the results from these studies, it seems that some species, including both the red- and the white-blooded Antarctic fish, show a hypoxic response similar to the response seen in temperate species, whereas other species or individuals show no response or a reversed response compared with temperate fish species. The variable results among red-blooded species may be due to a variable and high cholinergic tone on the heart, so individuals with a high initial heart rate (low cholinergic tone) showed the largest changes in heart rate (*P. borchgrevinki*) or periods of vagal escape, causing phasic oscillations of heart rate during hypoxia (*T. bernacchii*). The conflicting results for icefish could also be due to differences in experimental protocol and/or differences in hypoxic thresholds.

V. SUMMARY

The physiological information that has been collected so far from fish living in polar environments is still rather fragmentary, with most of the information coming from relatively few species mostly from around the Antarctic continent. The waters around the Antarctic continent are unique in that the fish species today living inside the Antarctic convergence have been genetically isolated for many million years. This contrasts from the Arctic region in which fish species are freer to move over a larger latitudinal range. Some of the physiology of the polar fish species is unique and may be true adaptations to a life in the cold and stenothermal environment, but today we know that some aspects of the physiology of polar living fishes that were thought to be specific adaptations to the cold are instead ancestral traits of the fish group rather that adaptations to the cold environment. This is certainly the case with the high branchial vascular sensitivity to serotonin. In contrast, a high cardiac cholinergic tone, as reported for two Antarctic species, is not that different compared with that reported for tuna and goldfish. More studies on the sub-Antarctic and temperate relatives of the polar fishes are clearly needed to try to separate true adaptations to the cold environment from phylogenetic traits.

ACKNOWLEDGMENTS

I would like to thank A. P. Farrell and S. Nilsson for valuable help and discussion during the writing.

REFERENCES

Acierno, R., MacDonald, J. A., Agnisola, C., and Tota, B. (1995). Blood volume in the hemoglobinless Antarctic teleost *Chionodraco hamatus* (Lönnberg). *J. Exp. Zool.* **272,** 407–409.

Agnisola, C., Acierno, R., Calvo, J., Farina, F., and Tota, B. (1997). *In vitro* cardiac performance in the sub-Antarctic notothenioids *Eleginops maclovinus* (Subfamily Eleginopinae), *Paranotothenia magellanica*, and *Patagonotothen Tessellata* (Subfamily Nototheniinae). *Comp. Biochem. Physiol. A* **118,** 1437–1445.

Agnisola, C., and Tota, B. (1994). Structure and function of the fish cardiac ventricle: Flexibility and limitations. *Cardioscience* **5,** 145–153.

Aho, E., and Vornanen, M. (1999). Contractile properties of atrial and ventricular myocardium of the heart of rainbow trout *Oncorhynchus mykiss*: Effects of thermal acclimation. *J. Exp. Biol.* **202,** 2663–2677.

Aho, E., and Vornanen, M. (2001). Cold acclimation increases basal heart rate but decreases its thermal tolerance in rainbow trout (*Oncorhynchus mykiss*). *J. Comp. Physiol. B* **171,** 173–179.

Altimiras, J., Aissaoui, A., Tort, L., and Axelsson, M. (1997). Cholinergic and adrenergic tones in the control of the heart rate in teleosts. How should they be calculated? *Comp. Biochem. Physiol.* **118A,** 131–139.

Altimiras, J., and Axelsson, M. (2004). Intrinsic autoregulation of cardiac output in rainbow trout (*Oncorhynchus mykiss*) at different heart rates. *J. Exp. Biol.* **207,** 195–201.

Altimiras, J., and Larsen, E. (2000). Non-invasive recording of heart rate and ventilation rate in rainbow trout during rest and swimming. Fish go wireless. *J. Fish. Biol.* **57,** 197–209.

Axelsson, M. (1988). The importance of nervous and humoral mechanisms in the control of cardiac performance in the Atlantic cod, *Gadus morhua*, at rest and during non-exhaustive exercise. *J. Exp. Biol.* **137,** 287–303.

Axelsson, M., Abe, A. S., Bicudo, J. E., and Nilsson, S. (1989). On the cardiac control in the south American lungfish *Lepidosiren paradoxa*. *Com. Biochem. Physiol* **93A,** 561–565.

Axelsson, M., Agnisola, C., Nilsson, S., and Tota, B. (1998). Fish cardio-circulatory function in the cold. *In* "Cold Ocean Physiology" (Pörtner, H. O., and Playle, R. C., Eds.), pp. 327–364. Cambridge University Press, Cambridge.

Axelsson, M., Davison, B., Forster, M., and Nilsson, S. (1994). Blood pressure control in the Antarctic fish *Pagothenia borchgrevinki*. *J. Exp. Biol.* **190,** 265–279.

Axelsson, M., Davison, W., Forster, M. E., and Farrell, A. P. (1992). Cardiovascular responses of the red-blooded Antarctic fishes *Pagothenia bernacchii* and *P. borchgrevinki*. *J. Exp. Biol.* **167,** 179–201.

Axelsson, M., Davison, W., and Franklin, C. E. (2000a). Cholinergic and adrenergic tone on the heart of the Antarctic dragonfish, *Gymnodraco acuticeps*, living at sub-zero temperatures. *Exp. Biol. Online* **5,** 1–12.

Axelsson, M., Ehrenström, F., and Nilsson, S. (1987). Cholinergic and adrenergic influence on the teleost heart *in vivo. Exp. Biol.* **46,** 179–186.

Axelsson, M., and Farrell, A. P. (1993). Coronary blood flow *in vivo* in the coho salmon (*Oncorhynchus kisutch*). *Am. J. Physiol.* **264,** R963–R971.

Axelsson, M., and Fritsche, R. (1991). Effects of exercise, hypoxia and feeding on the gastrointestinal blood flow in the Atlantic cod, *Gadus morhua. J. Exp. Biol.* **158,** 181–198.

Axelsson, M., and Nilsson, S. (1986). Blood pressure regulation during exercise in the Atlantic cod, *Gadus morhua. J. Exp. Biol.* **126,** 225–236.

Axelsson, M., Thorarensen, H., Nilsson, S., and Farrell, A. P. (2000b). Gastrointestinal blood flow in the red Irish lord, *Hemilepidotus hemilepidotus*: Long-term effects of feeding and adrenergic control. *J. Comp. Physiol. B* **170,** 145–152.

Beamish, F. W. H. (1978). Swimming capacity. *In* "Fish Physiology" (W. S., Hoar, and D. J., Randall, Eds.), pp. 101–187. Academic Press Ltd, New York, San Fransisco, London.

Bernier, N. J., Gilmour, K. M., Takei, Y., and Perry, S. F. (1999a). Cardiovascular control via angiotensin II and circulating catecholamines in the spiny dogfish, *Squalus acanthias. J. Comp. Physiol. B* **169,** 237–248.

Bernier, N. J., Kaiya, H., Takei, Y., and Perry, S. F. (1999b). Mediation of humoral catecholamine secretion by the renin-angiotensin system in hypotensive rainbow trout (*Oncorhynchus mykiss*). *J. Endocrinol.* **160,** 351–363.

Brett, J. R. (1973). Metabolic rates and critical speed of sockeye salmon (*Oncorhyncus nerka*) in relation to size and temperature. *J. Fish. Res. Bd. Canada,* **30,** 379–387.

Bushnell, P. G., and Brill, R. W. (1992). Oxygen transport and cardiovascular responses in skipjack tuna (*Katsuwonus pelamis*) and yellowfin tuna (*Thunnus albacares*) exposed to acute hypoxia. *J. Comp. Physiol. B* **162,** 131–143.

Cameron, J. S. (1979). Autonomic nervous tone and regulation of heart rate in the goldfish *Carassius auratus. Comp. Biochem. Physiol.* **63C,** 341–349.

Campbell, H. A., Taylor, E. W., and Egginton, S. (2004). The use of power spectral analysis to determine cardiorespiratory control in the short-horned sculpin *Myoxocephalus scorpius. J. Exp. Biol.* **207,** 1969–1976.

Cocca, E., Ratnayake, L. M., Parker, S. K., Camardella, L., Ciarmella, M., Di, P. G., and Detrich, H. W., III (1995). Genomic remnants of alpha-globin genes in the hemoglobinless Antarctic icefishes. *Proc. Natl. Acad. Sci.* **92,** 1817–1821.

Davison, W. (2001). Respiration of the emerald rockcod, a notothenioid fish from Antarctica. *N Z Nat. Sci.* **26,** 13–20.

Davison, W., Axelsson, M., Forster, M., and Nilsson, S. (1995). Cardiovascular responses to acute handling stress in the Antarctic fish *Trematomus bernacchii* are not mediated by circulatory catecholamines. *Fish Physiol. Biochem.* **14,** 253–257.

Davison, W., Axelsson, M., Nilsson, S., and Forster, M. E. (1997). Cardiovascular function in Antarctic Notothenioid fishes. *Comp. Biochem. Physiol. A* **118,** 1001–1008.

Davison, W., Forster, M. E., Franklin, C. E., and Taylor, H. H. (1988). Recovery from exhausting exercise in an Antarctic fish *Pagothenia borchgrevinki. Polar Biol.* **8,** 167–172.

Davison, W., Franklin, C. E., and McKenzie, J. C. (1994). Haematological changes in an Antarctic teleost, *Trematomus bernacchii*, following stress. *Polar Biol.* **14,** 463–466.

DeVries, A. L. (1971). Glycoproteins as biological antifreeze agents in Antarctic fishes. *Science* **172**, 1152–1155.

Driedzic, W. R., and Gesser, H. (1994). Energy metabolism and contractility in ectothermic vertebrate hearts: Hypoxia, acidosis, and low temperature. *Physiol. Rev.* **74**, 221–258.

Eastman, J. T. (1993). "Antarctic Fish Biology." Academic Press, San Diego.

Egginton, S. (1994). Stress response in two Antarctic teleosts (*Notothenia coriiceps* Richardson and *Chaenocephalus aceratus* Lönnberg) following capture and surgery. *J. Comp. Physiol. B* **164**(6), 482–491.

Egginton, S. (1996). Blood rheology of Antarctic fishes: Viscosity adaptations at very low temperatures. *J. Fish Biol.* **48**(3), 513–521.

Egginton, S. (1997a). A comparison of the response to induced exercise in red- and white-blooded Antarctic fishes. *J. Comp. Physiol. B* **167**, 129–134.

Egginton, S. (1997b). Control of tissue blood flow at very low temperatures. *J. Thermal Biol.* **22** (6), 403–407.

Egginton, S., and Davison, B. (1998). Stress effects in Antarctic fish. *In* "Cold Ocean Physiology" (Pörtner, H. O., and Playle, R. C., Eds.), pp. 327–364. Cambridge University Press, Cambridge.

Egginton, S., Forster, M. E., and Davison, W. (2001). Control of vascular tone in notothenioid fishes is determined by phylogeny, not environmental temperature. *Am. J. Physiol.* **280**, R1197–R1205.

Egginton, S., and Rankin, J. C. (1998). Vascular adaptations for low pressure/high flow blood supply to locomotory muscles of icefish. *In* "Antarctic Fishes. A Biological Overview" (DiPrisco, G., Clarke, A., and Pisano, E., Eds.), pp. 185–195. Springer Verlag, Berlin.

Egginton, S., Skilbeck, C., Hoofd, L., Calvo, J., and Johnston, I. A. (2002). Peripheral oxygen transport in skeletal muscle of Antarctic and sub-Antarctic notothenioid fish. *J. Exp. Biol.* **205**, 769–779.

Egginton, S., Taylor, E. W., Wilson, R. W., Johnston, I. A., and Moon, T. W. (1991). Stress Response in the Antarctic Teleosts *Notothenia neglecta* Nybelin and *Notothenia rossii* Richardson. *J. Fish Biol.* **38**(2), 225–236.

Farrell, A. P. (1981). Cardiovascular changes in the unanaesthetized lingcod (*Ophiodon elongatus*) during short-term, progressive hypoxia and spontaneus activity. *Can. J. Zool.* **60**, 933–941.

Farrell, A. P. (1984). A review of cardiac performance in teleost heart: Intrinsic and humoral regulation. *Can. J. Zool.* **62**, 523–536.

Farrell, A. P. (1991). Circulation of body fluids. *In* "Environment Metabolic Animal Physiology" (Posser, C. L., Ed.), pp. 509–557. Wiley-Lise.

Farrell, A. P., and Clutterham, S. M. (2003). On-line venous oxygen tensions in rainbow trout during graded exercise at two acclimation temperatures. *J. Exp. Biol.* **206**, 487–496.

Farrell, A. P., Hammons, A. M., Graham, M. S., and Tibbits, G. F. (1988). Cardiac growth in rainbow trout, *Salmo gairdneri*. *Can. J. Zool.* **66**, 2368–2373.

Farrell, A. P., and Jones, D. R. (1992). The heart. *In* "Fish Physiology, The Cardiovascular System" (Hoar, W. S., Randal, D. J., and Farrell, A. P., Eds.), pp. 1–88. Academic Press Inc., San Diego.

Farrell, A. P., MacLeod, K., and Driedzic, W. R. (1982). The effects of preload, after load and epinephrine on cardiac performance in the sea raven, *Hemitripterus americanus*. *Can. J. Zool.* **60**, 3165–3171.

Farrell, A. P., Wood, S., Hart, T., and Driedzic, W. R. (1985). Myocardial oxygen consumption in the sea raven, *Hemitripterus americanus*: The effects of volume loading, pressure loading and progressive hypoxia. *J. Exp. Biol.* **117**, 237–250.

Feller, G., and Gerday, C. (1997). Adaptation of hemoglobinsless Antarctic Icefish (*Channichthyidae*) to hypoxia tolerance. *Comp. Biochem. Physiol.* **118A**, 981–987.

Forster, M. E., Davison, W., Axelsson, M., Sundin, L., Franklin, C. E., and Gieseg, S. (1998). Catecholamine release in heat-stressed Antarctic fish causes proton extrusion by the red cells. *J. Comp. Physiol. B* **168**, 345–352.

Franklin, C. E., Axelsson, M., and Davison, W. (2001). Constancy and control of heart rate during an increase in temperature in the Antarctic fish *Pagothenia borchgrevinki*. *Exp. Biol. Online* **6**, 1–8.

Franklin, C. E., Axelsson, M., Sundin, L., and Davison, W. (2004). Antarctic fish: Survival and performance at −1.86 °C. "Antarctic Challenges. Historical and Current Perspectives on Otto Nordenskjöld's Antarctic Expedition 1901–1903." Royal Society of Arts and Science; Acta Regiae Societal, Royal Society of Arts and Sciences, 2004. (*Acta Regiae Societatis Scientiarum et Litterarum Gothoburgensis. Interdisciplinaria.* **5**.) ISBN 91-85252-64-6.

Franklin, C. E., Davison, W., and Mckenzie, J. C. (1993). The role of the spleen during exercise in the Antarctic teleost, *Pagothenia borchgrevinki*. *J. Exp. Biol.* **174**, 381–386.

Fritsche, R. (1990). Effects of hypoxia on blood pressure and heart rate in three marine teleosts. *Fish Physiol. Biochem.* **8**, 85–92.

Fritsche, R., and Nilsson, S. (1990). Autonomic nervous control of blood pressure and heart rate during hypoxia in the cod, *Gadus morhua*. *J. Comp. Physiol. B* **160**, 287–292.

Gallaugher, P., Axelsson, M., and Farrell, A. P. (1992). Swimming performance and haematological variables in splenectomized rainbow trout, *Oncorhynchus mykiss*. *J. Exp. Biol.* **171**, 301–314.

Gallaugher, P., and Farrell, A. P. (1998). Hematocrit and blood oxygen carrying capacity. *In* "Fish Respiration" (Perry, S. F., and Tufts, B. L., Eds.), pp. 185–227. Academic Press, San Diego, London, Boston, New York, Sydney, Tokyo, Toronto.

Gallaugher, P., Thorarensen, H., and Farrell, A. P. (1995). Hematocrit in oxygen transport and swimming in rainbow trout (*Oncorhynchus mykiss*). *Resp. Physiol.* **102**, 279–292.

Gamperl, A. K., Axelsson, M., and Farrell, A. P. (1995). Effects of swimming and environmental hypoxia on coronary blood flow in rainbow trout. *Am. J. Physiol.* **269**, R1258–R1266.

Graham, M., and Farrell, A. P. (1989). The effect of temperature acclimation and adrenaline on the performance of a perfused trout heart. *Physiol. Żool.* **62**, 38–61.

Graham, M., and Farrell, A. P. (1990). Myocardial oxygen consumption in trout acclimated to 5 °C and 15 °C. *Physiol. Zool.* **63**, 536–554.

Graham, M. S., and Fletcher, G. L. (1983). Blood and plasma viscosity of winter flounder *Pseudopleuronectes americanus*: Influence of temperature red cell concentration and shear rate. *Can. J. Zool.* **61**, 2344–2350.

Graham, M. S., and Fletcher, G. L. (1985). On the low viscosity blood of two cold water, marine sculpins: A comparison with the winter flounder. *J. Comp. Physiol.* **155**, 455–459.

Graham, M. S., Fletcher, G. L., and Haedrich, R. L. (1985). Blood viscosity in Arctic fishes. *J. Exp. Zool.* **234**, 157–160.

Hemmingsen, E. A., and Douglas, E. L. (1972). Respiratory and circulatory responses in a hemoglobin-free fish. *Chaenocephalus aceratus*, to changes in temperature and oxygen tension. *Comp. Biochem. Physiol.* **43A**, 1031–1043.

Hemmingsen, E. A., and Douglas, E. L. (1977). Respiratory and circulatory adaptations to the absence of haemoglobin in Channichthyid fishes. *In* "Adaptations within Antarctic Ecosystems" (Llano, G. A., Ed.), pp. 479–487. Smithsonian Institution, Washington.

Hemmingsen, E. A., Douglas, E. L., Johansen, K., and Millard, R. W. (1972). Aortic blood flow and cardiac output in the hemoglobin-free fish. *Chaenocephalus aceratus*. *Comp. Biochem. Physiol.* **43A**, 1045–1051.

Hipkins, S. F. (1985). Adrenergic responses of the cardiovascular system of the eel *Anguilla australis, in vivo. J. Exp. Zool.* **235,** 7–20.

Holeton, G. F. (1972). Gas exchange in fish with and without hemoglobin. *Respir. Physiol.* **14,** 142–150.

Hughes, G. M., Peyraud, C., Peyraud-Waitzenegger, M., and Soulier, P. (1981). Proportion of cardiac output concerned with gas exchange in gills of the eel (*A. anguilla*). *J. Physiol.* **310,** 61P–62P.

Ishimatsu, A.,K., I., G., N., and H. (1988). *In vivo* analysis of partitioning of cardiac output between systemic and central venous sinus circuits in rainbow trout: a new approach using chronic cannulation of the branchial vein. *J. Exp. Biol.* **137,** 75–88.

Johnston, I. A., Fitch, N., Zummo, G., Wood, R. E., Harrison, P., and Tota, B. (1983). Morphometric and ultrastructural features of the ventricular myocardium of the haemoglobinless icefish *Chaenocephalus aceratus. Comp. Biochem. Physiol.* **76A,** 475–480.

Jones, D. R., and Randall, D. J. (1978). The respiratory and circulatory systems during exercise. *In* "Fish Physiology" (Randall, W. S. H., and a.D. J., Eds.), Vol. 7, pp. 425–501. Academic Press, New York, London.

Keen, J. E., Aota, S., Brill, R. W., Farrell, A. P., and Randall, D. J. (1995). Cholinergic and adrenergic regulation of heart rate and ventral aortic pressure in two species of tropical tunas, *Katsuwonus pelamis* and *Thunnus albacares. Can. J. Zool.* **73,** 1681–1688.

Keen, J. E., Vianzon, D.-M., Farrell, A. P., and Tibbits, G. F. (1993). Thermal acclimation alters both adrenergic sensitivity and adrenoceptor density in cardiac tissue of rainbow trout. *J. Exp. Biol.* **181,** 27–47.

Kent, J., Koban, M., and Prosser, C. L. (1988). Cold-acclimation-induced protein hypertrophy in channel catfish and green sunfish. *J. Comp. Physiol.* **B 58,** 185–198.

Kiceniuk, J. W., and Jones, D. R. (1977). The oxygen transport system in trout (*Salmo gairdneri*) during sustained exercise. *J. Exp. Biol.* **69,** 247–260.

Lecklin, T., Nash, G. B., and Egginton, S. (1995). Do fish acclimated to low temperature improve microcirculatory perfusion by adapting red cell rheology? *J. Exp. Biol.* **198,** 1801–1808.

Leonard, J. B. K., and McCormick, S. D. (1999). Changes in haematology during upstream migration in American shad. *J. Fish Biol.* **54,** 1218–1230.

Macdonald, J. A., Montgomery, J. C., and Wells, R. M. G. (1987). Comparative physiology of Antarctic species. *In* "Advances in Marine Biology" (di Prisco, G., Maresca, B., and Tota, B., Eds.), pp. 321–388. Academic Press, London.

Macdonald, J. A., and Wells, R. M. G. (1991). Viscosity of body fluids from Antarctic Notothenioid fish. *In* "Biology of Antarctic Fish" (di Prisco, G., Maresca, B., and B., Tota, Eds.), pp. 163–178. Springer Verlag, Berlin, Heidelberg, New York, London, Budapest, Paris, Tokyo, Hong Kong, Barcelona.

Mazeaud, M. M., Mazeaud, F., and Donaldson, E. M. (1977). Primary and secondary effects of stress in fish: some new data with general review. *Trans. Am. Fish. Soc.* **106,** 201–212.

Milligan, C. L., Graham, M. S., and Farrell, A. P. (1989). The response of trout red cells to adrenaline during seasonal acclimation and changes in temperature. *J. Fish. Biol.* **35,** 229–236.

Morris, J. L., and Nilsson, S. (1994). The circulatory system. *In* "Comparative Physiology and Evolution of the Autonomic Nervous System" (Nilsson, S., and Holmgren, S., Eds.), pp. 193–246. Harwood Academic Publisher, Chur.

Nilsson, S. (1994). The spleen. *In* "Comparative Physiology and Evolution of the Autonomic Nervous System" (Nilsson, S., and Holmgren, S., Eds.), pp. 247–256. Harwood Academic Publisher, Chur.

Nilsson, S., Forster, M. E., Davison, W., and Axelsson, M. (1996). Nervous control of the spleen in the red-blooded Antarctic fish, *Pagothenia borchgrevinki*. *Am. J. Physiol.* **270**, R599–R604.

Nilsson, S., and Grove, D. J. (1974). Adrenergic and cholinergic innervation of the spleen of the cod, *Gadus morhua*. *Eur. J. Pharmacol.* **28**, 138–143.

O'Brien, K. M., and Sidell, B. D. (2000). The interplay among cardiac ultrastructure, metabolism and the expression of oxygen-binding proteins in Antarctic fishes. *J. Exp. Biol.* **203**, 1287–1297.

O'Brien, K. M., Xue, H., and Sidell, B. D. (2000). Quantification of diffusion distance within the spongy myocardium of hearts from Antarctic fishes. *Respir. Physiol.* **122**, 71–80.

Olson, K. R. (1992). Blood and extracellular fluid volume regulation: Role of the renin angiotensin system, kallikrein-kinin system, and atrial natriuretic peptides. In "Fish Physiology" (Hoar, W. S., Randall, D. J., and Farrell, A. P., Eds.), Vol. XIIB, pp. 135–254. Academic Press, San Diego, New York London.

Olson, K. R. (2002). Vascular anatomy of the fish gill. *J. Exp. Zool.* **293**, 214–231.

Olsson, H. I., Yee, N., Shiels, H. A., Brauner, C., and Farrell, A. P. (2000). A comparison of myocardial beta-adrenoreceptor density and ligand binding affinity among selected teleost fishes. *J. Comp. Physiol.* **170B**, 545–550.

Pearson, M. P., and Stevens, E. D. (1991). Size and hematological impact of the splenic erythrocyte reservoir in rainbow trout, *Oncorhynchus mykiss*. *Fish. Physiol. Biochem.* **9**, 39–50.

Pellegrino, D., Acierno, R., and Tota, B. (2003). Control of cardiovascular function in the icefish *Chionodraco hamatus*: Involvement of serotonin and nitric oxide. *Comp. Biochem. Physiol.* **134A**, 471–480.

Perry, S. F., and Reid, S. D. (1992). The relationship between beta-adrenoceptors and adrenergic responsiveness in trout (*Oncorhynchus mykiss*) and eel (*Anguilla rostrata*) erythrocytes. *J. Exp. Biol.* **167**, 235–250.

Platzack, B., Axelsson, M., and Nilsson, S. (1993). The renin angiotensin system in blood pressure control during exercise in the cod, *Gadus morhua*. *J. Exp. Biol.* **180**, 253–262.

Santer, R. M. (1985). Morphology and innervation of the fish heart. *Adv. Anat. Embryol. Cell Biol.* **89**, 1–97.

Satchell, G. H. (1991). "Physiology and Form of Fish Circulation." Cambridge University Press, Cambridge.

Shiels, H. A., Vornanen, M., and Farrell, A. P. (2002). The force-frequency relationship in fish hearts—A review. *Comp. Biochem. Physiol.* **132A**, 811–826.

Shiels, H. A., Vornanen, M., and Farrell, A. P. (2003). Acute temperature change modulates the response of ICa to adrenergic stimulation in fish cardiomyocytes. *Physiol. Biochem. Zool.* **76**, 816–824.

Short, S., Butler, P. J., and Taylor, E. W. (1977). The relative importance of nervous, humoral and intrinsic mechanisms in the regulation of heart rate and stroke volume in the dogfish (*Scyliorhinus canicula*). *J. Exp. Biol.* **70**, 77–92.

Sidell, B. D. (1998). Intracellular oxygen diffusion: The roles of myoglobin and lipid at cold body temperature. *J. Exp. Biol.* **201**, 1118–1127.

Sidell, B. D., Vayda, M. E., Small, D. J., Moylan, T. J., Londraville, R. L., Yuan, M. L., Rodnick, K. J., Eppley, Z. A., and Costello, L. (1997). Variable expression of myoglobin among the hemoglobinless Antarctic icefishes. *Proc. Natl. Acad. Sci. USA* **94**, 3420–3424.

Smith, D. G. (1978). Neural regulation of blood pressure in rainbow trout (*Salmo gairdneri*). *Can. J. Zool.* **56**, 1678–1683.

Storelli, C., Acierno, R., and Maffia, M. (1998). Membrane lipid and protein adaptations in Antarctic fish. *In* "Cold Ocean Physiology" (Pörtner, H. O., and Playle, R. C., Eds.), pp. 327–364. Cambridge University Press, Cambridge.

Sundin, L. I. (1995). Responses of the branchial circulation to hypoxia in the Atlantic cod, *Gadus morhua*. *Am. J. Physiol.* **268**, R771–R778.

Sundin, L., Davison, W., Forster, M., and Axelsson, M. (1998). A role of 5-HT2 receptors in the gill vasculature of the Antarctic fish *Pagothenia borchgrevinki*. *J. Exp. Biol.* **201**, 2129–2138.

Sundin, L., and Nilsson, S. (1992). Arterio-venous branchial blood flow in the Atlantic cod *Gadus morhua*. *J. Exp. Biol.* **165**, 73–84.

Sundin, L., Nilsson, G. E., Block, M., and Lofman, C. O. (1995). Control of gill filament blood flow by serotonin in the rainbow trout, *Oncorhynchus mykiss*. *Am. J. Physiol.* **268**, R1224–R1229.

Thorarensen, H., Gallaugher, P. E., Kiessling, A. K., and Farrell, A. P. (1993). Intestinal blood flow in swimming Chinook salmon *Oncorhynchus tshawytscha* and the effects of haematocrit on blood flow distribution. *J. Exp. Biol.* **179**, 115–129.

Thorarensen, H., Knight, C., and Davie, P. S. (1996). Seasonal changes in 11-ketotestosterone and relative ventricle mass in wild rainbow trout, *Oncorhynchus mykiss*. *N Z J. Marine Freshwater Res.* **30**, 397–402.

Tiitu, V., and Vornanen, M. (2002). Regulation of cardiac contractility in a cold stenothermal fish, the burbot *Lota lota* L. *J. Exp. Biol.* **205**, 1597–1606.

Tota, B., Acierno, R., and Agnisola, C. (1991). Mechanical performance of the isolated and perfused heart of the hemoglobinless Antarctic icefish *Chionodraco hamatus* Lönnberg. Effects of loading conditions and temperature. *Phil. Trans. Royal Soc. Lond. B* **332**, 191–198.

Tota, B., Cimini, V., Salvatore, G., and Zummo, G. (1983). Comparative study of the arterial and lacunary systems of the ventricular myocardium of elasmobranch and teleost heart. *Am. J. Anat.* **167**, 15–32.

Tota, B., and Gattuso, A. (1996). Heart ventricle pumps in teleosts and elasmobranchs: A morphodynamic approach. *J. Exp. Zool.* **275**, 162–171.

Uva, B. M., Masini, M. A., Devecchi, M., and Napoli, L. (1991). Renin–angiotensin system in Antarctic fishes. *Com. Biochem. Physiol.* **100A**, 897–900.

Van Dijk, P. L. M., and Wood, C. M. (1988). The effect of β-adrenergic blockade on the recovery process after strenuous exercise in the rainbow trout, *Salmo gairdneri* Richardsson. *J. Fish Biol.* **32**, 557–570.

Vogel, W. O. P., and Koch, K.-H. (1981). Morphology of gill blood vessels in icefish. *Arch. Fischerei-Wissenschaft* **31**, 139–150.

Wahlqvist, I., and Nilsson, S. (1980). Adrenergic control of the cardio-vascular system of the Atlantic cod, *Gadus morhua*, during "stress." *J. Comp. Physiol.* **137**, 145–150.

Wells, R. M. G., Ashby, M. D., Duncan, S. J., and Macdonald, J. A. (1980). Comparative study of the erythrocytes and haemoglobins in nototheniid fishes from Antarctica. *J. Fish Biol.* **17**, 517–527.

Wells, R. M. G., Macdonald, J. A., and di Prisco, G. (1990). Thin blooded Antarctic fishes: A rheological comparison of the haemoglobin-free icefishes *Chionodraco kathleenae* and *Cryodraco antarcticus* with a red-blooded nototheniid, *Pagothenia bernacchii*. *J. Fish Biol.* **36**, 595–609.

Wells, R. M. G., Tetens, V., and Devries, A. L. (1984). Recovery from stress following capture and anesthesia of Antarctic fish hematology and blood chemistry. *J. Fish Biol.* **25**(5), 567–576.

Whiteley, N. M., and Egginton, S. (1999). Antarctic fishes have a limited capacity for catecholamine synthesis. *J. Exp. Biol.* **202,** 3623–3629.

Winberg, M., Holmgren, S., and Nilsson, S. (1981). Effects of denervation and 6-hydroxydopamine on the activity of choline acetyltransferase in the spleen of the cod, *Gadus morhua. Comp. Biochem. Physiol.* **69C,** 141–143.

Wood, C. h. M., Pieprazak, P., and Trott, J. N. (1979). The influence of temperature and anaemia on the adrenergic and cholinergic mechanisms controlling heart rate in the rainbow trout. *Can. J. Zool.* **57,** 2440–2447.

Wood, C. h. M., and Shelton, G. (1980). Cardiovascular dynamics and adrenergic responses of the rainbow trout *in vivo. J. Exp. Biol.* **87,** 247–270.

Yamamoto, K., Itazawa, Y., and Kobayashi, H. (1983). Erythrocyte supply from the spleen and hemo concentration in hypoxic yellowtail. *Seriola*-Quinqueradiata. Marine Biol. Berlin. **73,** 221–226.

Zummo, G., Acierno, R., Agnisola, C., and Tota, B. (1995). The heart of the icefish: Bioconstruction and adaptation. *Braz. J. Med. Biol. Res.* **28,** 1265–1276.

BLOOD-GAS TRANSPORT AND HEMOGLOBIN FUNCTION IN POLAR FISHES: DOES LOW TEMPERATURE EXPLAIN PHYSIOLOGICAL CHARACTERS?

R.M.G. WELLS

I. Introduction
II. Mechanism of Evolution in Physiological Systems
III. Expression and Significance of Multiple Hemoglobin Components
    A. Hemoglobinless Icefishes
    B. Oxygen Transport Capacity
    C. Viscosity
    D. Stress
IV. Functional Properties of Hemoglobins
    A. Temperature Effects
    B. Root Effect
V. Respiratory Functions of Blood: Role of Blood in Maintaining Homeostasis
    A. Metabolic Cold Adaptation
    B. Blood Buffering and Carbon Dioxide Transport
    C. Blood Oxygen Transport
    D. Allosteric Regulation of Hemoglobin–Oxygen Binding
    E. Role of Catecholamines
VI. Reflections and Perspectives
    A. Evolution of Oxygen Transport in Polar Fishes
    B. Challenges for the Future

I. INTRODUCTION

The Arctic and Antarctic regions differ fundamentally both in their tectonic histories and in the composition of their fish fauna. The isolation and greater age of the Antarctic ecosystem is reflected in the higher degree of endemism and lower species diversity compared to the Arctic region (Eastman, 1997a). At its highest latitudes, the Arctic Ocean is characterized

The Physiology of Polar Fishes: Volume 22
FISH PHYSIOLOGY

by a floating ice cap and the absence of a continent, but below 80°N, it meets substantial land masses that continue south to the tropics and beyond. By contrast, the continental shelf and inshore habitats surrounding Antarctica are isolated from other land masses at lower latitudes. These features have undoubtedly influenced the evolution of faunal diversity (Eastman and McCune, 2000).

The composition of the Antarctic fish fauna, especially at the high-latitude and ice-bound margins of the coast, is unique. The dominant and ecologically diverse fish are the highly endemic perciform notothenioids, and they represent a rare example of a monophyletic radiation derived from benthic ancestors without swim bladders (Eastman, 1993). The Notothenioidei are distributed right around the continental margins of Antarctica (i.e., presently sympatric), and though fairly uniform in terms of diet choice, they have radiated considerably in terms of habitat and physiology (Macdonald et al., 1987; Eastman, 1993). By contrast, the dominant fish fauna of the Arctic are polyphyletic and include functionally diverse gadids, cottids, salmonids, pleuronectids, and chondrichthyans (Eastman, 1997a)—all of which have representatives with wide latitudinal distribution. For example, the bullhead Cottus gobio is a small benthic sculpin found in the Arctic seas off Greenland and Scandinavia, around the British Isles to northern Italy, and in the Black Sea (Greenhalgh, 1999). There is also evidence for behavioral thermoregulation among Arctic gadinid fish (Schurmann and Christiansen, 1994). Accordingly, Arctic fishes are generally more eurythermal and euryhaline than the Antarctic fish fauna.

We should expect fish living in the cold Arctic and Antarctic marine environments to show adjustments to physiological and biochemical processes that depend on temperature. Thus, energy metabolism (Giardina et al., 1998; Somero et al., 1998; Kawall et al., 2002), ion and oxygen transport (Bargelloni et al., 1998; Maffia et al., 1998; di Prisco, 1998), and propagation of motor-neural information (Macdonald, 2000) appear sensibly cold adapted. Conversely, physiological features that are essentially independent of temperature either may reflect adaptation to specific habitats or are phylogenetically constrained, but are not the proximate result of selection for function at low temperature (Montgomery and Wells, 1993; Montgomery and Clements, 2000).

A significant problem arises in that there are few comparative studies that cover both Arctic and Antarctic fishes, particularly with respect to their blood-gas transport systems. In the absence of evidence for physiological convergence among phylogenetically distinct and secure species, how can we be sure that observed traits in one of the polar regions are adaptations to low temperature and have arisen through natural selection? Genetic drift and phylogenetic constraints are alternative evolutionary mechanisms to adaptation (Feder *et al.*, 2000). The key evolutionary innovation in both Arctic and Antarctic fishes, however, is the glycopeptide antifreeze, which provides essential protection from freezing in shallow ice-laden seas. Convergent evolution of antifreeze compounds is strongly supported by molecular evidence (Chen *et al.*, 1997) and enables all physiological attributes to be expressed in subzero water temperatures.

The aim of this chapter is to document advances describing the physiology of blood-gas transport in polar fishes and to make limited comparisons with information from less stenothermal species. In addition to providing a sinecure for literature assimilation, an explanation is offered for variation in respiratory characters of the blood, and the conclusions of some older work are reviewed within the context of evolutionary physiology. In other words, how strong is the evidence for evolutionary convergence in the oxygen transport systems of Arctic and Antarctic fishes, and is it explained by habitat similarities or from low temperature alone? It should be easier to recognize adaptation in the Arctic fauna precisely because it is polyphyletic (i.e., it has to have evolved multiple times). Thus, a central question is whether adaptation to low temperature has shaped the physiological traits of the blood-gas transport systems in polar fishes.

II. MECHANISM OF EVOLUTION IN PHYSIOLOGICAL SYSTEMS

The origin of species is the most fundamental of biological questions, and recent debate has focused on the importance of geographical isolation versus selection in the speciation process (Schilthuizen, 2000; Barton, 2001). Polar fishes provide a useful model to test whether selection of physiological characters in the absence of geographic barriers led to the evolution of unique radiations. Antarctica is isolated by geography, by currents, and by low temperature, with a continental shelf as a focus for biodiversity and evolution. The finding of an Antarctic nototheniid in the Arctic, possibly through trans-equatorial migration using cold isothermal corridors (Møller *et al.*, 2003), is remarkable in such a taxonomically diverse environment. Demonstration of sympatric speciation is difficult in the sea, where the combination of high larval dispersal and incomplete geographical barriers often leads to large ranges for many polar species. This does not, however, imply an absence of historical allopatry.

The dominant Antarctic notothenioid order most likely arose from a blennioid ancestor that was a physiological generalist. Thus, allopatric speciation occurred in a geographically separated population. Eastman (1993) uses the term "vicariance" to indicate passive allopatric speciation concomitant with the geological evolution of Antarctica. In the Arctic, speciation may have occurred in the absence of geographic barriers (sympatry) among species with sufficient phenotypic plasticity to migrate into colder waters. Subsequent evolution of these populations may be driven by natural selection.

The divergence in physiological traits of species that occupy different habitats in the Antarctic may well represent sympatric speciation. Fish from high Antarctic latitudes have more specialized biochemical adaptations than sub-Antarctic species (DeVries and Eastman, 1981). The origins of variation in metabolic and oxygen transport systems are difficult to explain, however, because of pleiotropy, where one gene may influence more than one aspect of the phenotype. The rapid progress in evolutionary genetics promises resolution of the fundamental problem of whether ecological divergence has been driven by selection of physiological traits that affect performance in different habitats. The tendency to view natural selection as operating with few constraints remains a problem in interpreting the physiological traits of polar fishes. Nonadaptive features such as those imposed by phyletic inertia and ontogenetic constraint (see Wells, 1990) appear to have played major roles in shaping the oxygen transport systems of Antarctic notothenioids.

III. EXPRESSION AND SIGNIFICANCE OF MULTIPLE HEMOGLOBIN COMPONENTS

Hemoglobin (Hb) is a remarkable protein found in all vertebrates, save the few Antarctic icefish in the family Channichthyidae. Accordingly, the protein is expected to meet the operational requirements for oxygen transport under a wide range of physicochemical conditions and for vertebrates displaying great diversity in activity levels. The structural and functional characteristics of the Hbs from Antarctic notothenioid fishes have received far more attention than those from fishes in any other ecosystem or phylogeny. Lack of comparable information, thus, complicates our attempts to evaluate the role of phylogeny or environment in shaping the observed traits.

In contrast with many teleost fishes in which multiple Hbs occur in individuals (isohemoglobins) or populations (polymorphisms), electrophoretic screening of species from the notothenioid suborder shows reduced multiplicity (Wells *et al.*, 1980; Kunzmann, 1991; di Prisco *et al.*, 1991, 1998). The reduced number of components has been linked to environmental constancy in the Antarctic and low notothenioid oxygen demand, although temperate versus tropical thermostable fishes differ little in Hb multiplicity (Jensen *et al.*, 1998). Nototheniids generally possess major and minor anodal Hb components designated *Hb1* and Hb2, respectively, in the ratio 95:5 (di Prisco, 1991; Kunzmann, 1991). These Hb pairs share a common

β-globin chain and are distinguished by their α-chains (di Prisco *et al.*, 1991; Fago *et al.*, 1992). Notable exceptions are the pelagic *Pleuragramma antarcticum* in which Hb2 comprises 25%, and a third cathodal component HbC is present (Kunzmann, 1991; Tamburrini *et al.*, 1997), and the semipelagic *T. newnesi* with Hb1 and Hb2 in the ratio 70:25 comprising 75% of the total components (D'Avino *et al.*, 1994). Unlike other notothenioids, *Gobionotothen gibberifrons* Hb1 and Hb2 do not have globin chains in common (Marinakis *et al.*, 2003). Up to five Hbs have been isolated from *Pagothenia borchgrevinki* (Riccio *et al.*, 2000). Interestingly, *Notothenia angustata*, the only temperate-water nototheniid found in southern New Zealand, also has a 95:5 ratio componentry (Fago *et al.*, 1992) and raises the question of whether the low multiplicity is indeed a cold adaptation. Moreover, *N. coriiceps* is characterized by the presence of electrophoretically distinct juvenile Hb components (Cocca *et al.*, 2000). Species of the more derived notothenioid families usually have a single Hb. For example, six of seven bathydraconid species examined have only one Hb (Kunzmann, 1991), and from seven species in the bathydraconid, harpagiferid, and artedidraconid families, only *Cygnodraco mawsoni* expressed a 5% minor component (di Prisco *et al.*, 1991).

Pseudaphritis urvillii is a primitive notothenioid in the Bovichtidae and, like most members of the family, are non-Antarctic, occurring in estuarine waters of southeastern Australia (Eastman, 1993). Interestingly, it also has the Hb1/Hb2 system yet shares high sequence identity with Hb1 and Hb2 from *N. angustata* (D'Avino and Di Prisco, 1997). For a more detailed analysis of the relationships between molecular structure, sequence data, and functional properties of Hbs in Antarctic fish, there are several excellent reviews (di Prisco *et al.*, 1991a,b, 1998; di Prisco and Tamburrini, 1992; Stam *et al.*, 1997; di Prisco, 1998, 2000).

Tropical and temperate teleosts typically have multiple Hbs that show diverse functional properties (di Prisco and Tamburrini, 1992; Wells, 1999a,b; Weber, 2000). The New Zealand gurnard *Chelidonichthys kuma* is exceptional in having a single Hb (Fago *et al.*, 1994). Arctic and sub-Arctic species, however, have received scant attention. The Arctic wolffish *Anarhichas minor* has multiple Hbs with heterogeneous oxygen-binding properties (Verde *et al.*, 2002). The eelpouts (Zoarcidae) are typically deep cold-water species with a cosmopolitan distribution and sedentary benthic habit. The northern *Zoarces viviparus* has at least six Hb components, some of which are polymorphic (Hjorth, 1974); Antarctic zoarcids are similarly characterized by multiple Hbs (di Prisco *et al.*, 1991a; Kunzmann, 1991). Salmonids are represented by a number of cold north temperate and Arctic freshwater species that are highly active and may undertake significant migrations. Arctic charr, *Salvelinus alpinus*, possess 10 isohemoglobins, of which several are polymorphic (Giles, 1991), whereas rainbow trout, *Oncorhynchus mykiss*, have at least nine hemoglobin components (Fago *et al.*, 2002). Atlantic cod (*Gadus morhua*) populations show distinctive intraspecific isohemoglobin patterns, and the Hb1 polymorphism provides a reliable marker for the identification of the Arctic populations (Fyhn *et al.*, 1994). A hemoglobin polymorphism also occurs in turbot (*Scophthalamos maximus*) and is strongly correlated with temperature (Imsland *et al.*, 2000).

The Antarctic fishes *Anotopterus pharao* and *Macrourus holotrachys* have at least four to five Hb components (Kunzmann, 1991), yet they are not endemic and they do not belong to the Notothenioidei. Thus, the Antarctic notothenioids have a low incidence of Hb multiplicity and, hence, form a major point of distinction between the northern polar and sub-Arctic fish. A search for polymorphic Hbs in the notothenioids has yet to be undertaken.

A note of caution concerns the interpretation of multiple Hbs. Despite more than 30 years of research on the four-component Hb system of the highly studied rainbow trout, *Oncorhynchus mykiss*, Fago *et al.* (2002) have now isolated nine components consisting of nine distinct globin chain types. Moreover, additional components have been described for latitudinally separated cod populations of *Gadus morhua* (Fyhn *et al.*, 1994). An increase in the proportion of cathodal Hbs is expressed in the cod at warmer temperatures (Brix *et al.*, unpublished results).

A. Hemoglobinless Icefishes

Some molecular studies suggest a single large-scale deletion of all icefish β-globin genes (Cocca *et al.*, 1997; di Prisco *et al.*, 2002), together with the 5′ end of the linked α-globin gene (Zhao *et al.*, 1998), occurred in the ancestral icefish. In all probability, an additional linkage of the larval Hb gene cluster present on a single chromosome pair in nototheniids (Pisano *et al.*, 2003) has been deleted in icefishes (Cocca *et al.*, 2000). The icefishes are, uniquely, the only adult vertebrates that lack Hb. The loss of Hb, however, does not necessarily entail loss of intracellular oxygen-binding hemoproteins, because some icefishes have retained myoglobin in the cardiac myotome (Moylan and Sidell, 2000; O'Brien and Sidell, 2000). The lack of oxygen-binding proteins in icefishes appears compensated by higher mitochondrial densities and intracellular lipid content that helps maintain oxygen flux because of the higher solubility of oxygen in lipid compared to cytoplasm (O'Brien and Sidell, 2000). Increasing the volume and surface density of mitochondrial clusters is the primary mechanism for enhancing the aerobic capacity of muscle in cold-water fish, but maximum rates of mitochondrial substrate oxidation have not been upregulated in Antarctic fishes to values comparable with perciform species living in warmer seas (Johnston *et al.*, 1998).

The loss of circulating red blood cells is expected to conserve energy (costs of maintaining flow and protein turnover) in a cold viscous medium when activity levels are extremely low (Wells *et al.*, 1990; Macdonald and Wells, 1991). However, substantial reengineering of the channichthyid cardiovascular and branchial systems (Feller and Gerday, 1997) and wide-bore capillaries with large blood volume are characteristics of the icefishes (Egginton *et al.*, 2002) and are required to maintain high volume outputs of the ventilatory apparatus and the cardiac pump. An unanswered question is whether the energy cost of increased cardiac pumping is offset by the reduced energy cost of circulating a erythrocyte-free blood. The icefishes may reflect altered selection pressure for a more active lifestyle than occurred in the low competitive environment of their ancestors at the base of the notothenioid radiation.

The loss of Hb in channichthyids may be viewed in the context of a general trend where haematocrit, erythrocyte Hb content, and Hb multiplicity all decrease with increasing phylogenetic divergence among the red-blooded Antarctic notothenioid fishes (di Prisco, 1998). Physiological studies employing carbon monoxide inactivation of Hb suggest that Hb is not essential for survival in *Trematomus bernacchii* (di Prisco *et al.*, 1992). Yet, in the evolutionary perspective, the loss of Hb appears to be detrimental. Molecular studies from *Gymnodraco acuticeps* (Bathydraconidae, sister taxon to Channichthyidae) provide clear evidence for positive selection of Hb through a higher rate of nonsynonymous substitutions than synonymous substitutions in the DNA sequences of the β-globin gene (Bargelloni *et al.*, 1998). It may be prudent to consider that in the course of evolution, detrimental traits may arise from pleiotropy or pedomorphosis (Schlichting and Pigliucci, 1998).

B. Oxygen Transport Capacity

Hb concentration provides a direct measure of the oxygen-carrying capacity of blood because 1 g binds 1.35 ml oxygen. Hematocrit is the fraction of blood volume taken up by the erythrocytes and, thus, provides an indirect measure of transport capacity. Gallaugher and Farrell (1998) summarize an extensive literature on fish hematocrits but also caution that measurement is subject to stress-induced erythrocyte swelling. Generally, aerobically active fish have higher Hb concentrations and hematocrits than less active species (Weber and Wells, 1989). Accordingly, the active cryopelagic *P. borchgrevinki* has a high Hb concentration compared to other notothenioids (Riccio *et al.*, 2000).

Arctic charr (*Salvelinus alpinus*) are unusual in that they are often anemic during the warm summer temperatures, indicating their inability to respond to increased temperature by an increase in erythropoiesis (Lecklin and Nikinmaa, 1998). Exposure temperature does not appreciably alter haematocrit in the trout, *O. mykiss* (Gallaugher and Farrell, 1998). Comparable studies are not available for Antarctic fishes, other than the observation by Davison *et al.* (1990) that *P. borchgrevinki* with X-cell gill disease show a compensatory increase in hematocrit. Lowering the hematocrit by venesection to 1–2% in *T. bernacchii* produced no apparent deleterious effect (Wells *et al.*, 1990), as was similarly observed after CO inactivation of Hb in *T. bernacchii* (di Prisco *et al.*, 1992). There is, however, a significant loss of mechanical performance of the channichthyid heart when myoglobin is inactivated (Acierno *et al.*, 1997a). Although the lowering of hematocrit does not have deleterious effects in resting animals, the effects on exercise or temperature tolerance have not been investigated. Adequate dissolved oxygen in the venous return and a significant cutaneous oxygen uptake appear adequate to sustain aerobic metabolism in notothenioids (Montgomery and Wells, 1993). The relationship between intraspecific variation in hematocrit and metabolic scope in nonpolar species is contradictory; swimming performance appears compromised in anemic salmonids, but not in marine teleosts (Gallaugher and Farrell, 1998). However, compensation for experimental anemia may occur down to a critical hematocrit, below which compensation via cardiac output and ventilation are unlikely to sustain a natural exercise regimen (Brauner and Randall, 1998) or adequate carbon dioxide excretion (Tufts and Perry, 1998).

Hematocrit values for red-blooded Antarctic notothenioids fall within the range of values observed for temperate and tropical obligate water breathers, but their Hb levels are slightly lower (Kooyman, 1963; Grigg, 1967; Wells *et al.*, 1980; Kunzmann, 1991; Kunzmann *et al.*, 1991). Thus, the mean cellular Hb concentration (MCHC) of Antarctic notothenioids appears reduced, as has also been observed in Arctic fishes (Kooyman, 1963). *Pleuragramma antarcticum* is a streamlined nototheniid with a high proportion of red muscle fibers and cruises using subcarangiform locomotion in preference to the labriform mode shown by other species in the family (Eastman, 1993). It is, therefore, surprising that this seasonally migrating species has a Hb concentration in the low range of values observed for nototheniids (Tamburrini *et al.*, 1997). However, Kunzmann (1991) surveyed a broad range of high-Antarctic notothenioid species and was unable to demonstrate a clear relationship between hematocrit and activity.

Kunzmann (1991) proposed that rather than simply reducing the number of circulating erythrocytes, the trend toward reduced Hb content in Antarctic fish is more efficiently achieved through reduction in MCHC, thereby providing a more even distribution of oxygen to tissues. This suggestion is consistent with the presence of small numbers of Hb-free erythrocytes in the icefishes, reflecting the progressive evolutionary decline in protein expression that ends with the complete elimination of Hb. The question that now arises is whether there is a trend toward reduced MCHC in cold-adapted fishes? Preliminary data from Wells *et al.* (1980), together with additional data from Antarctic fishes (Kunzmann, 1991), lend some support to this hypothesis. Accordingly, I have collated MCHC data on teleosts from a broad range of latitudes and constructed linear correlations against the environmental temperatures from which the fish were taken (Figure 7.1). MCHC was calculated in a number of instances from [Hb]/hematocrit or, where Hb concentration was not stated, from oxygen-carrying capacity data by assuming that 1 g of Hb maximally binds 1.35 ml of oxygen. Early investigators would not have been aware of the phenomenon of adrenergically induced erythrocyte swelling that arises from handling stresses (Perry and Reid,

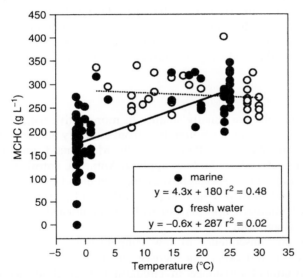

Fig. 7.1. Mean cellular hemoglobin concentrations (MCHCs) for marine and freshwater teleost fishes at their sampling temperature. Analysis is by linear regression. Data from Albers *et al.*, 1981, 1983; Bollard *et al.*, 1993; Boutilier *et al.*, 1988; Cameron, 1973; Cech *et al.*, 1984; Canfield *et al.*, 1994; Davison *et al.*, 1994, 1995; Dobson and Baldwin, 1982; Egginton *et al.*, 1991; Egginton, 1994; Forster *et al.*, 1998; Franklin *et al.*, 1993; Grigg, 1967, 1969; Herbert and Wells, 2001; Kunzmann and Zimmermann, 1992; Herbert *et al.*, 2002; Innes and Wells, 1985; Jensen and Weber, 1982; Kunzmann, 1989, 1991; Kunzmann *et al.*, 1991; Laursen *et al.*, 1985; Lecklin *et al.*, 1995; Lecklin and Nikinmaa, 1998; Ling and Wells, 1985; Lowe and Wells, 1996; Lowe *et al.*, 2000; Sadler *et al.*, 2000; Tamburrini *et al.*, 1997; Tetens and Lykkeboe, 1985; Tetens *et al.*, 1984; Wells and Baldwin, 1990; Wells and Weber, 1991; Wells *et al.*, 1980, 1984, 1986, 1989, 1990, 1997; Wood *et al.*, 1979.

1993; Davison *et al.*, 1994). Following extreme exercise, the nototheniid *P. borchgrevinki* exhibited substantial erythrocyte swelling (Franklin *et al.*, 1993), but two species of *Notothenia* did not (Egginton, 1997a). However, the extent of swelling from acutely sampled resting fish is not likely to be great in relation to the large range of observed MCHC values. I separated data from freshwater species on the basis that many of these fish inhabit highly unstable and often hypoxic environments so that any effect of temperature might be obscured. Indeed, the data show no correlation ($r^2 = 0.02$). Marine species from the Arctic and Antarctic, through to the tropics, do, however, show a correlation ($r^2 = 0.48$). Because the analysis includes species both active and sluggish in a particular thermal habitat, we may conclude that a significant component of the variation in MCHC is explained by temperature. However, it is surprising that there is so little hematological information from temperate marine teleosts.

An unusual role has been proposed for the high concentration of intracellular lipids in Antarctic fishes. In addition to their role in adding buoyancy and providing aerobic fuel reserves, Sidell (1991) noted their very high oxygen solubility and proposed that they may serve as an oxygen store during activity, thus reducing the burden on hemoproteins.

C. Viscosity

Red blood cells are usually smaller in more active fishes, enabling the circulation to overcome frictional drag from viscosity and enhanced cell-to-tissue diffusion (Wells and Baldwin, 1990; Macdonald and Wells, 1991). However, this correlation does not apply to Antarctic notothenioids because both active and sluggish species have similar red blood cell sizes (Kunzmann, 1991). The trend toward reduction in notothenioid erythrocyte Hb content rather than hematocrit appears not to be a rheological adaptation to low temperature (Egginton, 1996).

A universal consequence of low temperature is an increase in fluid viscosity. We should expect an increase in the energetic costs required to overcome frictional resistance in circulating body fluids as well as for swimming, and that these effects are proportional to body size. The viscosity of polar fluids is about twice that of fluids at 20 °C, and the effect of temperature on blood viscosity is similar in a range of temperate and Antarctic fish when hematocrit is accounted for (Wells *et al.*, 1990; Macdonald and Wells, 1991; Egginton, 1996). The temperature effect on viscosity is, therefore, partly offset by a reduction in hematocrit. Erythrocyte rigidity increases at low temperature, but there is no evidence for compensation through rheological adaptation in Antarctic fishes (Lecklin *et al.*, 1995).

D. Stress

Various stressors including severe exercise, hypoxia, and high temperature may trigger an increase in oxygen-carrying capacity of the blood to compensate increased respiratory demand. Antarctic nototheniids appear to show this response through stress-induced splenic contraction, and predictably, the effect is more marked in the cryopelagic species *P. borchgrevinki* than in the benthic *T. bernacchii* (Davison *et al.*, 1994) and recovery is slower (Davison *et al.*, 1988). The spleen in *P. borchgrevinki* provides a large reservoir of erythrocytes, allowing the hematocrit to more than double during exercise from its resting value of approximately 15%, an observation explained by their relatively low Hb content and the constraints of the nototheniid cardiovascular system (Franklin *et al.*, 1993). This response is similar to that shown by a number of active warm-water fishes (Gallaugher and Farrell, 1998). However, Egginton (1997a) was not able to demonstrate a rise in hematocrit for two species of *Notothenia* following a bout of forced exercise. The control of splenic contraction is either cholinergic or adrenergic in teleosts; Nilsson *et al.* (1996) proposed that cholinergic control predominates in *P. borchgrevinki* and other cold-adapted fish. Further, the cardiovascular system in Antarctic nototheniids appears regulated by the autonomic nervous system, rather than by circulating catecholamines as in most other teleosts (Davison *et al.*, 1995). Catecholamine effects are, however, manifested in nototheniids exposed to elevated temperatures (Franklin *et al.*, 1991; Forster *et al.*, 1998).

Stress responses in the icefishes are less well understood. Lacking red blood cells, the restoration of blood chemistry to resting values takes much longer after bouts of exercise-induced stress (Egginton, 1994).

IV. FUNCTIONAL PROPERTIES OF HEMOGLOBINS

The functional properties of temperate and tropical fish Hbs have been extensively reviewed, and in general, specific components have been linked with adaptive environmental or behavioral features, although some intraspecific isohemoglobins are functionally indistinguishable (Weber and Wells, 1989; di Prisco and Tamburrini, 1992; Wells, 1999a,b; Weber, 2000). The low level of Hb multiplicity in the notothenioids has frequently been interpreted as an adaptation to the relative constancy of physicochemical conditions of the Antarctic marine environment where constant low temperatures are coupled with high constant oxygen content (Caruso *et al.*, 1991; Jensen *et al.*, 1998; di Prisco *et al.*, 1998, 2002; Wells, 1999b). In this chapter, I challenge this interpretation. Phylogenetic studies have revealed appreciable functional differences in isolated Hb components despite high identity in primary structure (di Prisco *et al.*, 1991a). Typical of the most derived families of the Notothenioidei, the bathydraconid *Gymnodraco acuticeps* has a single Hb, but it is a most unusual teleost in that the Hb is not modulated by pH and allosteric factors, although the specific amino acid residues generally considered responsible for the Bohr and Root effects are conserved in the sequence (Tamburrini *et al.*, 1992). These functional studies confirm the earlier study by Wells and Jokumsen (1982) and correlate with the benthic sedentary habits of the species. The bathydraconid *Cygnodraco mawsoni* is also a highly derived notothenioid, and its major and minor Hbs display both Bohr and Root effects that are sensitive to affinity reduction via organic phosphates.

T. newnesi is the only Antarctic species in which neither the major nor the minor anodal Hb displays significant Bohr or Root effects, but a third cathodal Hb comprising 25% of the total components is sensitive to protons and other effectors and correlates with structural substitutions that stabilize the deoxy T-state (D'Avino *et al.*, 1994). These findings are at variance with the general teleost pattern of pH and phosphate-sensitive anodal Hbs, opposed by insensitive cathodal fractions (Weber, 2000), but may represent specialization for active swimming near the surface—a behavior not typical of the notothenioids (Eastman, 1993).

The triple Hb system of the nototheniid *P. antarcticum* is exceptional and is considered a specialized system rather than an evolutionary or larval remnant: All three Hbs show strong Bohr and Root effects, but Hb2 has a low temperature sensitivity (Tamburrini *et al.*, 1997). Hb 1 and Hb2 share common α-chains, and Hb2 and HbC share a common β-chain. The pH sensitivities of the Hbs accord with the relatively high aerobic behavior of the species (Tamburrini *et al.*, 1997). The closely related benthopelagic *Aethotaxis mitopterix* has a single Hb with a moderate Bohr effect, but the Root effect is absent (D'Avino *et al.*, 1992).

Hemolysate studies in the nototheniids, *Gobionotothen gibberifrons*, show that the Hb2 fraction is higher than that found in most notothenioids and has a lower oxygen affinity, indicating possible adaptation for hypoxic episodes (Marinakis *et al.*, 2003).

The major (Hb1) fraction from the cold-temperate nototheniid *N. angustata* displays both Bohr and Root effects and is sensitive to ATP binding (Fago *et al.*, 1992). The β-globin chain is shared in the major and minor nototheniid components, and so we would predict that both Hbs display similar allosteric effects. Moreover, the high sequence identity, especially in critical domains associated with stereochemical shifts from the R to T state transition, suggests common functional characteristics in nototheniids (di Prisco, 1998).

The Arctic populations of cod (*G. morhua*) are characterized by a higher affinity Hb polymorphism than that found in more southerly populations, and this variant is associated with higher growth rates (Brix *et al.*, 1998). Similar results have been found for turbot, *S. maximus*, (Samuelsen *et al.*, 1999). Further, Hb polymorphisms in the Arctic charr are associated with differences in the pH sensitivity of Hb–oxygen binding (magnitude of the Bohr effect) and correlate with differences in the thermal environments of the populations (Giles, 1991). Genetic population-level differences are thus established, but it has not been unequivocally established whether temperature causes changes in Hb polymorphisms at the level of the individual (Tun and Houston, 1986; Murad and Houston, 1991).

A. Temperature Effects

Values for the enthalpy of oxygenation reflect the temperature sensitivity of the Hb–oxygen binding equilibrium, and there are several reports of lower than expected heats of oxygenation ($\Delta H°$) in Antarctic fish Hbs. For example, low $\Delta H°$ characterizes the single Hb of *G. acuticeps* (Tamburrini *et al.*, 1992), the Hb2 component of *P. antarcticum* (Tamburrini *et al.*, 1997), and the isohemoglobins of *P. borchgrevinki* (Riccio *et al.*, 2000). These authors speculate that the reduced enthalpy is a cold-adaptation conserving energy. An alternative adaptive explanation is given for low $\Delta H°$ in sub-Arctic turbot Hbs where the effect is viewed as an adaptation for variable environmental temperatures (Samuelsen *et al.*, 1999). Conversely, a high value of $\Delta H°$ was hypothesized to compromise the warmer limit of the geographic range of the cool-temperate teleost *Odax pullus* (Brix *et al.*, 1998). But saving energy is not the same as saving heat, as may be appreciated from fish living in thermal equilibrium compared to endotherms that function under a substantial thermal disequilibrium, irrespective of their values for $\Delta H°$. We would, therefore, expect that energy is largely conserved in the oxygenation cycle of Antarctic fish. Brittain (1984) examined the thermodynamics of ligand binding to the T- and R-states of the isolated Hb1 Hb from *P. borchgrevinki* and concluded that thermal effects represented an intrinsic rather than functional property of Hb. Moreover, artifacts may arise from differences in the enthalpy of ionization, depending on the buffer systems used to measure the equilibrium constant (Fago *et al.*, 1997).

In a thermodynamic analysis of oxygen binding in Antarctic fish Hbs, Fago *et al.* (1997) observed that $\Delta H°$ was temperature dependent in *D. mawsoni*; that is, the van't Hoff plots were nonlinear, indicating that oxygen binding was more exothermic at low temperatures. The degree of temperature dependence of the $\Delta H°$ is a direct consequence of differences in allosteric effects rather than a specific molecular adaptation to low temperatures. Apparent inconsistencies in values for $\Delta H°$ may arise from assumptions of a linear van't Hoff relationship. Hbs with stronger Bohr and Root effects show greater departure from linearity (Fago *et al.*, 1997). Further, $\Delta H°$ appears pH dependent in both temperate species (Brix *et al.*, 1999) and the Antarctic *P. antarcticum* (di Prisco, 1998), in which the value becomes less negative at low pH. These observations merely reflect a gain in entropy of the Hb system in the transition from the deoxygenated to the oxygenated state. The magnitude of $\Delta H°$, thus, relates to the size of the Bohr and Root effects and should take into account the endothermic release of allosteric effectors on oxygenation. Reports of low enthalpy of oxygenation in Antarctic fish species such as *P. bernacchii* with Bohr and Root effects (di Prisco *et al.*, 1991b) may be because *in vitro* measurements of Hb oxygenation reactions at temperatures around 20 °C cannot *a priori* be extrapolated to subzero temperatures.

It is worth noting that values for $\Delta H°$ in native unfractionated hemolysates from three nototheniids were similar to values from temperate species (Wells and Jokumsen, 1982). Further, there is no evidence for temperature insensitivity of the oxygen dissociation or CO-association rate constants of Antarctic fish myoglobins (Cashon *et al.*, 1997).

B. Root Effect

Root effect Hbs are unique to fishes and operate within the countercurrent exchangers in the choroid body of the eye, or the swim bladder where high partial pressures of oxygen are generated for retinal metabolism or buoyancy control following acidification of blood pH. The swim bladder is absent in Antarctic notothenioids, but many species possess a choroid body (Eastman, 1993). The presence of Root effect Hbs in many Antarctic notothenioids (di Prisco *et al.*, 1988, 1998) is a puzzling aspect of the oxygen transport system. The residues from primary sequence analyses of the nototheniid Hbs do not conform to those responsible for expression of the Root effect in other fishes (di Prisco *et al.*, 1991a; D'Avino *et al.*, 1994). The cluster of seven functionally important β-chain residues in Root effect Hbs in both carp and trout HbIV differ from Hb1 in the Antarctic *P. borchgrevinki* (see Table 1 in Jensen *et al.*, 1998). New structural data comparing the high sequence identity of the principal Hb in *T. newnesi* (showing pH independent oxygen binding), and the Root effect component in *P. borchgrevinki* has shown that pH dependence cannot be related to the modulation of the R-state (Mazzarella *et al.*, 1999). Phylogenetic analysis of ancestral amino acid residues associated with the Root effect in Arctic and Antarctic fishes does not predict the expression of the physiological trait in extant fishes (Verde *et al.*, 2003). This emphasizes the caution needed in linking functional behavior to single or few amino acid residues. Perhaps the Root effect in Antarctic nototheniids is an example of convergent evolution.

The major Hb of the bathydraconid *Cygnodraco mawsoni* exhibits a strong Root effect, thereby preventing saturation with oxygen at low pH levels under normal oxygenating conditions (Caruso *et al.*, 1991). The Root effect is present in many notothenioid Hbs stripped of red cell organic phosphates and in the pH range 6.5–7.0 or in the presence of ATP levels less than pH 7.5 (di Prisco, 1998, 2000; di Prisco *et al.*, 1998). These results are not inconsistent with those of earlier workers who did not observe Root effects in either hemolysates or whole blood at higher pH values (Wells and Jokumsen, 1982; Tetens *et al.*, 1984). The Root effect of *T. newnesi* Hb is lost during the purification process and suggests a requirement of allosteric phosphate cofactors for expression of extreme pH sensitivity (D'Avino *et al.*, 1994; Mazzarella *et al.*, 1999).

We do not know what pH values apply within the erythrocyte microenvironment in Antarctic fish under different physiological states, so the role of the Root effect in them is unclear from structure-function studies. The structural basis of the Root effect Hbs in the nototheniids is not due to simple interactions between specific residues as in most fishes, but due to more complex associations in other domains (Fago *et al.*, 1992).

The benthic nototheniid *T. bernacchii* possesses both a rete and a significant Root effect, whereas the closely related *P. borchgrevinki* lacks a rete yet possesses a Root effect (Riccio *et al.*, 2000). Although both species possess Root effect Hbs, it comprises only a minor component in *P. borchgrevinki*, whereas it is the major component in *T. bernacchii* (di Prisco *et al.*, 1988). The physiological significance of this becomes apparent under the influence of red cell organic phosphates (Riccio *et al.*, 2000) because the magnitude of the whole blood Bohr effect, which includes the Root effect, is significantly greater for *T. bernacchii* than for *P. borchgrevinki* (Tetens *et al.*, 1984). Behavioral experiments with these closely related species show that the optomotor response (ability to visually resolve high-contrast objects on a moving background) of the former species is facilitated through the oxygen-concentrating apparatus in the eye coupled with the Root effect Hb enabling a higher ocular pO_2 (Herbert *et al.*, 2003). Antarctic nototheniids are visually unspecialized in that they lack high retinal cell densities and sensitive convergence ratios (Montgomery and Wells, 1993). The Root effect may provide visual augmentation, although with no rete, *P. borchgrevinki* may have an alternative role for Root effect Hb in buoyancy regulation (di Prisco *et al.*, 1988). It is more likely, however, that there has been no selection pressure for deletion of this character.

The larvae of notothenioids are generally pelagic, and retention of their low-multiplicity Hb characteristics might be a pedomorphic feature

(Eastman, 1997b). Because the timing of developmental processes will have an impact on a much larger set of characters, it is difficult to interpret the individual traits of Hb function as adaptive (Wells, 1990). It is not yet known whether the genes for the adult globins are contiguous with those expressed in early ontogeny. Tamburrini *et al.* (1997) do not consider the Hb system of *Pleuragramma* to be larval remnants, but this is a highly specialized and atypical Hb system.

V. RESPIRATORY FUNCTIONS OF BLOOD: ROLE OF BLOOD IN MAINTAINING HOMEOSTASIS

The blood-gas transport system must support the oxygen requirements for the scope of activity in the polar environment and should provide some buffering of carbon dioxide and lactic acid as the principal respiratory and metabolic end products. It might be expected that energy expenditure for ventilation and cardiac output is increased as a result of increased viscosity at low temperature. We should first consider, however, whether the overall energy costs of living in the polar environment are concordant with generalized rate–temperature relationships.

A. Metabolic Cold Adaptation

The oxygen transport system is critical in supporting aerobic energy expenditure. It is, therefore, relevant in this chapter to inquire into the metabolic requirements of polar fishes. The first experiments to support an idea that polar fishes might compensate for near-freezing conditions by elevating metabolic rate are now 50 years old (Scholander *et al.*, 1953). But a series of carefully conducted respirometry experiments determining the standard metabolic rate (SMR) and taking particular account of stressful states that might generate artifacts of elevated oxygen consumption have shown clearly that Arctic fish are not metabolically cold adapted (Steffensen *et al.*, 1994; Hop *et al.*, 1997; Jordan *et al.*, 2001). Differences in metabolic rates of benthic Arctic cottids are attributed to specific differences in their potential for activity (von Dorrien, 1993). The evidence from Antarctic fish is more equivocal. Older studies by Wells (1978) and Torres and Somero (1988) suggested that although metabolic rates of notothenioid fishes did not show the markedly elevated rates claimed in earlier studies, they still appeared higher than expected and higher than Antarctic fish that were not endemic to the region. These authors were well aware of metabolic enhancement arising from the handling stresses imposed on the experimental fish but may have recorded routine rather than basal oxygen uptake rates due to the longer integration times for data analysis. Steffensen (2002) refers to unpublished data from Antarctic notitheniid fishes that appear to confirm the absence of SMR cold adaptation, echoing earlier studies that attribute elevated rates to artifacts arising from the use of stressed fish.

In a study comparing standard metabolic rates of sluggish tropical scorpenid fishes, Zimmermann and Kunzmann (2001) found rates comparable to those from boreal and Arctic scorpenids, suggesting that there were no specific SMR adaptations for temperature in either tropical or polar environments. Alternatively, stenothermal tropical species may be showing compensatory metabolic depression.

Is this the final word on a vigorous debate that has continued for half a century? I think further discussion is warranted. First, whereas cold-adapted Atlantic cod (*G. morhua*) exhibit the general adrenergic response to circulating catecholamines manifested by other teleosts (Whiteley and Egginton, 1999), one of the most striking features of notitheniid physiology is the strongly attenuated or absent adrenergic stress response (Davison *et al.*, 1995; Nilsson *et al.*, 1996; Egginton, 1997a,b; Forster *et al.*, 1998). In an interesting series of experiments, Davison *et al.* (1995) examined catecholamine responses in a second set of rested control fish that had been maintained under controlled captive conditions for much longer. They found that several cardiovascular rate functions were related to not only time in captivity, but also according to the time of year when measurements were obtained. These downregulated adrenergic responses may, therefore, themselves be an artifact and result in pathologically low apparent rates of oxygen uptake. The curious endocrine system of Antarctic notitheniids may, however, derive from the phylogenetic constraints of a benthic ancestor.

Second, growth rates in notothenioid fishes are up to three times higher in the austral summer and are linked not to temperature, but to food availability (North, 1998). A higher metabolic rate might, therefore, be expected to support the additional energy costs, but oxygen uptake in *Notothenia neglecta* acclimated to summer and winter conditions was similar in both fasted and fed states (Johnston and Battram, 1993). Measurements during the Antarctic winter are not available.

Finally, Antarctic notothenioid fishes have an unusual body composition. In addition to a general lack of ossified skeletal material, high lipid reserves contribute to buoyancy compensation for loss of the swim bladder (Eastman, 1993). Comparative measurements of whole animal oxygen uptake are a relatively crude measure of metabolism because anatomical structures vary substantially in both proportion and composition. Increased mitochondrial abundance in the red muscle of Antarctic fish has been proposed as an adaptation to offset the effects of low temperature on aerobic capacity (Clarke and Johnston, 1996), but there are limits to packing density. The authors further note that the maximum power output of anaerobically contracting fast muscle fibers is only one third that of tropical fish and may limit the effectiveness of the escape response.

Comparisons of phylogenetically diverse groups are, thus, problematic. For example, oxygen uptake measurements in polar fishes have always been expressed in units of wet mass. Yet, the whole-body water content varies from 60% in the Antarctic notothenioid *D. mawsoni* (Eastman and DeVries, 1981) to more than 90% in other notothenioids and in cold deep-sea benthic species including polar liparids (Alexander, 1993; Eastman *et al.*, 1994). Moreover, the lipid content of *D. mawsoni* varies considerably over the seasons (Eastman, 1993), further confounding attempts to compare metabolic rates. Corrections for body water content based on species and seasonal differences will undoubtedly influence published values of oxygen uptake.

Comparisons of standard rates of metabolism may, therefore, be largely meaningless because they may be applied to any physiological state, including those approaching a zero metabolic rate (death). Perhaps as Clarke (1991) concurred, comparisons of metabolic rates are simply too complex to be meaningful.

B. Blood Buffering and Carbon Dioxide Transport

Low environmental temperatures in the polar seas accord with both low rates of metabolism and high solubility of respiratory gases. Just as increased oxygen solubility at low temperature provides high aquatic oxygen content, so too does respiratory carbon dioxide dissolve readily in solution, so CO_2 excretion should not require special adaptations for the cold environment. Red cell carbonic anhydrase from several fish groups plays a crucial role in the restoration of acid–base balance following bouts of exercise (Tufts and Perry, 1998). Nevertheless, following strenuous activity, Antarctic nototheniids seem to differ from many other fishes in that they manifest a significant respiratory increase, rather than metabolic acidosis (Egginton *et al.*, 1991). Thus, despite low production rates, the conversion of CO_2 to bicarbonate for transport to the gills for excretion may be limited by the paucity or even absence of erythrocytes. Consequently, the burden of hydrated CO_2 transfer may defer to the relatively large Bohr and Root effects characteristic of most nototheniid fishes (di Prisco, 1998). The presence of a Haldane effect in nototheniids (Powell *et al.*, 2000) supports a role for Hb in the oxygenation-linked transport of carbon dioxide. This mechanism of proton binding, as well as direct binding of CO_2 to Hb (carbamate effect), is obviously unavailable to the channichthyid icefishes, which lack erythrocytes. Maffia *et al.* (2001) have confirmed the absence of carbonic anhydrase activity in blood from the icefish *Chionodraco hamatus* but found exceptionally high activity in the gills relative to that in red-blooded nototheniids. Interestingly, the isoform of the branchial enzyme differed from that occurring in other teleosts and suggests both compensation for the lack of erythrocytes and a strong selection pressure for efficient discharge of CO_2 at the gill surfaces. Similar results have been demonstrated for another icefish *Chaenocephalus aceratus*, in which the enzyme is located in the cytoplasm (Tufts *et al.*, 2002). In the absence of Hb as one of the principal blood buffering mechanisms, the subcellular distribution of carbonic anhydrase is expected to play an important role in ionic regulation (Rankin and Tuurala, 1998). The effectiveness of blood in buffering excess protons in *Chionodraco* was similar in red-blooded fish (Acierno *et al.*, 1997b) and may result from a relatively greater blood volume in the icefish (Acierno *et al.*, 1995). But Wells *et al.* (1988) found that icefish blood was less effective in buffering fixed acid (arising from lactate acidosis following anaerobic burst activity) and residual buffering capacity could be attributed to relatively high plasma proteins with zwitterionic properties.

As with other teleosts, carbon dioxide excretion from nototheniid erythrocytes is inhibited by catecholamines through their action on the membrane Na^+/H^+ antiporter (Powell *et al.*, 2000).

C. Blood Oxygen Transport

The oxygen affinity coefficient, P_{50}, of blood is the partial pressure of oxygen at which half of the blood is oxygenated. P_{50} is generally intraspecifically temperature sensitive, so an increase in the body temperature of an ectotherm results in a decrease in Hb-oxygen affinity, thereby improving oxygen delivery to tissues. If this were also the case between species, oxygen would be bound to Hb so tightly in polar fishes that it would not be released to tissues until internal pO_2 became very low indeed. Hubold (1991) noted that Antarctic notothenioids are more active than their cottid Arctic counterparts, and we might predict that the Antarctic fish have comparatively lower blood oxygen affinities. Separating the effects of environmental temperature from activity on blood oxygen affinity is difficult. Analysis of data from both marine and freshwater fishes, and the effect of pH on affinity (the Bohr factor, $\Phi = \Delta\log P_{50}/\Delta pH$) as a function of environmental temperature is attempted in Figure 7.2. Temperature explains some of the variation in oxygen affinity data from fish in the marine, but not freshwater environment. This is to be expected in view of the relative stenothermy in the marine environment that contrasts with the wide range of seasonal and diurnal temperatures in freshwater habitats, including those in the Arctic. Moreover, ambient oxygen tensions are far more variable in freshwater than in marine

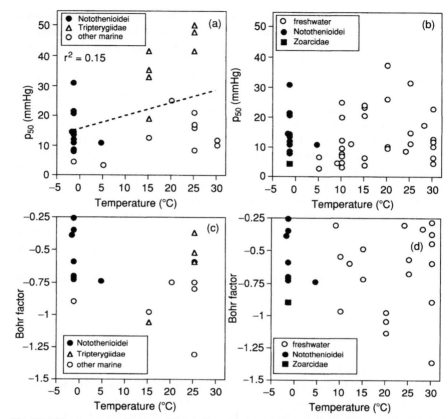

Fig. 7.2. Blood oxygen affinity and Bohr effect data for marine and freshwater fish as a function of environmental temperature. P_{50} values were selected from data assumed to reflect resting arterial pH values. In (a), analysis is by linear regression. (1 mmHg = 0.133 kPa.) Data from Albers *et al.*, 1981, 1983; Brix *et al.*, 1999; personal communication; Cameron, 1973; Cech *et al.*, 1984; Dobson and Baldwin, 1982; Giles, 1991; Grigg, 1967, 1969; Herbert and Wells, 2001; Innes and Wells, 1985; Jensen and Weber, 1982; Jones *et al.*, 1986; Laursen *et al.*, 1985; Lowe *et al.*, 2000; Qvist *et al.*, 1977; Takeda, 1991; Tetens and Lykkeboe, 1981; Tetens *et al.*, 1984; Val *et al.*, 1986; Verheyen *et al.*, 1985; Vorger, 1986; Wells and Davie, 1985; Wells *et al.*, 1984a,b, 1989; Wood *et al.*, 1979.

habitats and should be reflected in specific homeostatic adaptations that preserve oxygen uptake in the gills. There is no clear relationship between the Bohr factor and temperature for either marine or freshwater fishes. These observations suggest that adaptive adjustments to the oxygen transport system ensure transport from environment to tissues occurs effectively in all thermal habitats and are consistent with the expected range in activities between species. The relationship for essentially normoxic fishes in the marine environment, however, is not entirely robust, because not all data points share equivalent phylogenetic independence and they have been collated from diverse sources using different methodologies. Clearer thermal relationships may exist within phylogenetic boundaries, but such data sets are unavailable. Nonetheless, intact red blood cells, as well as purified Hbs, appear to have low affinities in the range similar to those from fish in temperate and tropical regions.

Blood pH in Antarctic fish is extremely high (>8.0), as one would expect from both the low rates of respiratory carbon dioxide production and the thermodynamic effects on hydrogen ion activity wherein *in vivo* pH varies in parallel with the neutral point of water (Qvist *et al.*, 1977; Tetens *et al.*, 1984). Magnetic resonance imaging measurements from *Harpagifer antarcticus* muscle, however, suggest that intracellular pH is relatively independent of temperature (Moerland and Egginton, 1998). Quite large transmembrane pH differences exist between cell types (Heisler, 1986) and between red blood cells from different species (Nikinmaa and Salama, 1998). Because the Hb is located intracellularly, the interpretation of whole-blood oxygen-affinity data in which Bohr and Root effects are expressed in terms of plasma pH becomes problematic.

D. Allosteric Regulation of Hemoglobin–Oxygen Binding

Concentrations of organic phosphates in the erythrocytes of Antarctic nototheniids are in the approximate range of $1-2 \, mol \, mol \, Hb^{-1}$ and similar to values for temperate species (Wells, 1978). The generalized vertebrate response to prolonged hypoxia includes a reduction in erythrocyte organic phosphates, thus increasing blood oxygen affinity and securing efficient loading of oxygen at the uptake sites (for recent reviews, see Jensen *et al.*, 1998; Nikinmaa and Salama, 1998; Wells, 1999a,b). The Antarctic nototheniid *P. borchgrevinki* shows a vigorous and graded response to hypoxia via ATP regulation, resulting in an increased blood oxygen affinity, in addition to increasing oxygen-carrying capacity (Wells *et al.*, 1989). There is only a modest accumulation of blood lactate following hypoxia (Wells *et al.*, 1989) or a temperature increase (Wells, 1978), so metabolic acidosis appears not to be a major problem for nototheniids (Lowe and Wells, 1997). Given the history of constant high levels of oxygen in the stenothermal Antarctic seas, the adaptive significance of the acclimatory responses must be called into question, and the observation may simply represent phenotypic plasticity, a feature of many eurythermal species (Wilson and Franklin, 2002).

A rise in temperature generally results in instability of the Hb–oxygen complex manifested as a reduction in blood oxygen affinity. Quantitatively, the thermal effect on whole blood of *P. borchgrevinki* is similar to that in temperate fishes (Tetens *et al.*, 1984), although the effect appears somewhat attenuated in *D. mawsoni* (Qvist *et al.*, 1977). The observation of low MCHC in polar fishes (Figure 7.1) is also significant for blood-oxygen transport. As MCHC decreases or erythrocytes swell, the equilibrium of the ATP–Hb complex shifts toward the dissociated state, thus increasing oxygen affinity. Partial compensation for the temperature effect through reduction in erythrocyte ATP is typical of most fishes and occurs in Antarctic nototheniids Qvist *et al.* (1977); Wells (1978) observed a decrease in ATP with an increase in temperature. Tetens *et al.* (1984), in contrast found increased ATP:Hb ratios in warm-acclimated *P. borchgrevinki*. These observations have no immediate significance for fishes which live at constant temperature but suggest a possible mechanism for adaptation under a scenario of warming of the Antarctic seas.

E. Role of Catecholamines

An unusual feature of erythrocytes in many fish species is the presence of surface receptors that are readily primed by circulating catecholamines to activate membrane sodium-proton pumps (Roig *et al.*, 1997; Wells, 1999b). Typically, when under stress, adrenaline stimulates fish erythrocytes to secrete protons, thus increasing internal pH and increasing Hb–oxygen affinity via the Bohr effect. An influx of water causes a further increase in oxygen affinity due to the dissociation of the Hb-organic phosphate complexes (Wells, 1999b). These actions are presumed to be homeostatic under acidosis in protecting oxygen uptake in the gills when oxygen demand is temporarily elevated. But not all fish show these responses, particularly sedentary benthic forms (Perry and Reid, 1993). Cold-acclimated trout (*O. mykiss*) lack the adrenergic red cell swelling response (Nikinmaa and Jensen, 1986). Curiously, Antarctic notothenioids show a strongly attenuated stress response, regardless of ecotype (Whiteley and Egginton, 1999), although there does appear to be some postexercise erythrocyte swelling in the more active species *P. borchgrevinki* (Lowe and Wells, 1997). Forster *et al.* (1998) have shown that the adrenergic effect on *P. borchgrevinki* erythrocytes is more strongly mediated when the fish are exposed to higher temperatures.

VI. REFLECTIONS AND PERSPECTIVES

The opportunity to study fish in the Antarctic seas has resulted in a disproportionate amount of information compared to fish from the Arctic and elsewhere. Several unusual physiological features of Antarctic fishes contrast with those of other teleosts, but we cannot always be certain whether these characteristics represent phylogenetic divergence within a single radiation or are adaptations to the polar environment. By contrast, fish of the Arctic Ocean show low endemism and may be found over a relatively broad latitudinal range. They appear adapted to low temperatures and demonstrate convergent evolution. In the Antarctic seas, the process of speciation and physiological adaptation differed because it occurred in the context of monophyletic radiation in a low competition environment (Eastman and Clarke, 1998). Accordingly, Antarctic fish show greater physiological diversity relating to niche shifts than we see in the Arctic. Although low temperatures close to freezing are the main characteristic of polar marine ecosystems, Eastman (1993) considers this unlikely to be a major factor in determining the composition of the present Antarctic ichthyofauna, given the relatively slow rate of cooling.

Ecological constraints reveal a distinction between Arctic and Antarctic ecosystems. The very much greater depth of the Antarctic continental shelf, lack of intertidal habitat, and extensive anchor ice formation in shallow habitats severely limit ecological opportunities for fishes (Eastman, 1993). The basal notothenioid characters for benthic existence and low activity were replaced, in several instances, by characters supporting pelagic modes of behavior. Secondary pelagicism has, thus, led to ecological and physiological diversity in the endemic Antarctic fishes. Accordingly, there is no *a priori* reason that their adaptations for energy expenditure should parallel those seen in the Arctic fishes. Several researchers are investigating Arctic and northern cold-temperate species and are likely to shed light on this problem.

A. Evolution of Oxygen Transport in Polar Fishes

The Hb molecule has been shown to be remarkably plastic during the course of vertebrate evolution (Wells, 1999a); however, speciation also leads to divergence in physiological mechanisms that are not easily interpreted in the context of respiratory adaptations (Wells, 1990, 1999b). Although an acclimatory response to hypoxia has been demonstrated in the allosteric control of oxygen delivery in *P. borchgrevinki* (Wells *et al.*, 1989), the presence of Hb in nototheniids cannot be assumed to provide greater protection against hypoxia than in icefishes, which lack Hb because the latter show higher resistance to low environmental oxygen (Montgomery and Wells, 1993).

It is generally believed that phenotypic plasticity should be greater in variable than in constant environments (Feder *et al.*, 2000). Acclimations and acclimatizations to temperature are examples of phylogenetic plasticity (Hochachka and Somero, 2002). Yet the magnitude of physiological plasticity in the case of hypoxic acclimation in Antarctic fishes, and temperature responses in Arctic fish are clearly not correlated with the magnitude of environmental variation in the polar regions. Phenotypic plasticity of the oxygen equilibrium enables *P. borchgrevinki* to survive improbable bouts of environmental hypoxia (Wells *et al.*, 1989). Hb concentration and MCHC are hematological parameters reflecting oxygen transport capacity and appear correlated with environmental temperatures. Features of the blood–oxygen equilibrium in notothenioids, including the Bohr and Root effects, and allosteric regulation cannot be explained by low temperature alone but appear to be linked to diversified metabolic demands. Blood oxygen transport has yet to be correlated with the strong seasonal variation in trophic and physical conditions of the Antarctic. Morphological examples of plasticity are also a feature of Arctic fishes (Eastman and DeVries, 1997).

The α- and β-chains of the major notothenioid Hb components form a monophyletic group (Stam *et al.*, 1997). The remarkable range of functional properties displayed among Hbs with high sequence identity is unlikely to have occurred in the blennioid ancestors to the group. Whether species in the radiation are gaining or losing isohemoglobin complexity is unclear, but it is reasonable to assume that the selection pressures now persistent are different to those operating in the early evolution of the group under a low competitive environment when other families of fishes disappeared from the cooling continent. It has proved difficult, however, to demonstrate that amino acid differences between Hbs isolated from Antarctic fishes have current adaptive significance. Although sequence convergence has not been established, there is strong evidence for neutral rate violation in the residues responsible for the Root effect phenomenon (di Prisco *et al.*, 1998), suggesting an unidentified selection pressure.

The burden of physiological control of oxygen transport appears to shift progressively from the level of proteins and cells in early development to organ systems in adulthood. A similar transition from lower to higher animals invokes the maxim "ontogeny recapitulates phylogeny" for the special circumstance of pedomorphosis (Wells, 1999a). Fish are generally characterized by a progressive shift in Hb forms during development, rather than an abrupt ontogenetic change such as occurs in most vertebrates (Wells, 1999a,b). Yet, bathydraconids possess a single Hb irrespective of developmental stage (di Prisco, work in progress). Only minor structural differences occur in the primary structures of nototheniid Hbs, and Hb multiplicity decreases from juvenile to adult in *T. bernacchii* and *P. antarcticum* (di Prisco *et al.*, 1998), suggesting selection, rather than pedomorphosis. In view of the high levels of multiplicity in Arctic marine fish, we cannot, therefore, claim reduced multiplicity as a cold adaptation. Curiously, during a recent examination of notothenioid specimens in the British Museum of

Natural History, I was struck by the much larger size of the holotypes than we have seen since the 1970s and suspect that most experimental work in the Ross Sea has since been carried out on immature fish. Low Hb multiplicity in notothenioids may have arisen through pedomorphic heterochrony, a change in the timing of developmental events that results in a novel adult phenotype (Schlichting and Pigliucci, 1998). Development is not fixed by a single timer, but by a set of them; thus, the persistence of the aquatic larval stage of some salamanders is nevertheless accompanied by the transition of Hbs into the adult form (Raff and Wray, 1989). The concept of ontogenetic arrest can account for several changes at once and could include the failure to ossify the notothenioid skeleton, thus increasing buoyancy (Eastman and DeVries, 1981).

Icefish and other derived forms such as the bathydraconids have extremely low blood oxygen transport capacities. However, it is not necessary to link the trend to low hematocrit with reduced Hb multiplicity. The presence of an unusually low number of isohemoglobins in the notothenioids may, therefore, be viewed as an evolutionary remnant rather than a specific adaptation to consistently low environmental temperature (Caruso et al., 1991). But di Prisco et al. consider the Hb systems of some secondarily pelagic notothenioids to be functionally specialized (D'Avino et al., 1994; di Prisco, 1998, 2000).

B. Challenges for the Future

The impact of environmental change on biodiversity through evolutionary process is not in question. The capacity for polar fishes to adapt to faster climate change resulting from human activity, however, is uncertain and is increasingly engaging public concern. Modeling studies based on the expression of temperature-sensitive Hb polymorphisms in two contrasting sub-Arctic species, a zoarcid and the cod, suggest a northern latitudinal shift in the distribution of both species (Pörtner et al., 2001). With little information on physiological or biochemical polymorphisms in the notothenioids, it is difficult to comment on the likely effect of global warming on Antarctic populations. Clearly though, with their low oxygen-carrying capacity and Hb multiplicity, there is limited capacity for these extreme stenotherms to acclimatize or adapt in the face of rapid environmental perturbation. Enzyme-ligand interactions are highly temperature sensitive in Antarctic fishes so that minor stereochemical shifts will have substantial effects on protein function (Somero and Hofmann, 1997). Thus, global warming is expected to affect Antarctic fish more than Arctic or eurythermal species because of their greater capacity for acclimatization aided by the expression of additional isoforms of proteins such as Hb.

A number of new species of snailfishes (Liparidae) have been described from the abyssal regions of Antarctica; the group has a worldwide distribution, including a boreal center of speciation (Eastman and Clarke, 1998). Physiological comparisons of polar representatives of this group should be undertaken to probe cold adaptive mechanisms of oxygen transport.

Much of the recent work on Hbs from polar fishes tends to be interpreted in the context of elucidating the mechanisms of the oxygen transport system at molecular and cellular levels, and their functional significance is explained in terms of what is now called the *beneficial acclimation hypothesis* (Leroi *et al.*, 1994; Wilson and Franklin, 2002). This approach falls short of the rigorous analysis required to interpret the origins of physiological diversity. The solution resides in a "post-genomic" approach to biology in which gene expression is linked to the study of how species respond to environmental constraints. Testing the benefits and costs of such responses will not be easy, and the broader challenge is to determine the limitations of adaptive processes in polar ectotherms so that we might understand the effects of a rapidly changing thermal environment. Recent work has shown that differential genome duplication has occurred in many fish lineages (Zhou *et al.*, 2002), but whether Hb chain variability has been influenced by this process is unknown. Notothenioids, however, are ideal subjects for testing hypotheses in the context of evolutionary physiology (Garland and Carter, 1994). It is already clear that factors other than temperature are driving speciation and physiological variation in the Antarctic notothenioids.

ACKNOWLEDGMENTS

I want to thank Thomas Brittain, Kendall Clements, and John Macdonald for helpful discussions.

REFERENCES

Acierno, R., Macdonald, J. A., Agnisola, C., and Tota, B. (1995). Blood volume in the haemoglobinless Antarctic teleost *Chionodraco hamatus* (Lonnberg). *J. Exp. Zool.* **272**, 403–409.
Acierno, R., Agnisola, C., Tota, B., and Sidell, B. D. (1997a). Myoglobin enhances cardiac performance in Antarctic icefish species that express the protein. *Am. J. Physiol.* **273**, R100–R106.
Acierno, R., Maffia, M., Rollo, M., and Storelli, C. (1997b). Buffer capacity in the blood of the hemoglobinless Antarctic fish *Chionodraco hamatus*. *Comp. Biochem. Physiol.* **118A**, 989–992.
Albers, C., Götz, K. H., and Welbers, P. (1981). Oxygen transport and acid–base balance in the blood of the sheatfish, *Silurus glanis*. *Resp. Physiol.* **46**, 223–236.

Albers, C., Manz, R., Muster, D., and Hughes, G. M. (1983). Effect of acclimation temperature on oxygen transport in the blood of the carp, *Cyprinus carpio*. *Resp. Physiol.* **52**, 165–179.

Alexander, R. McN. (1993). Buoyancy. *In* "The Physiology of Fishes" (Evans, D. H., Ed.), pp. 75–97. CRC Press, Boca Raton.

Bargelloni, L., Marcato, S., and Patarnello, T. (1998). Antarctic fish hemoglobins: Evidence for adaptive evolution at subzero temperature. *Proc. Natl. Acad. Sci. USA* **95**, 8670–8675.

Barton, N. H. (2001). Speciation. *Trends Ecol. Evol.* **16**, 325.

Bollard, B. A., Pankhurst, N. W., and Wells, R. M. G. (1993). Effects of artificially elevated plasma cortisol levels on blood parameters in the teleost fish *Pagrus auratus* (Sparidae). *Comp. Biochem. Physiol.* **106A**, 157–162.

Boutilier, R. G., Dobson, G., Hoeger, U., and Randall, D. J. (1988). Acute exposure to graded levels of hypoxia in rainbow trout (*Salmo gairdneri*): Metabolic and respiratory adaptations. *Resp. Physiol.* **71**, 69–82.

Brauner, C. J., and Randall, D. J. (1998). The linkage between oxygen and carbon dioxide transport. *In* "Fish Physiology" (Perry, S. F., and Tufts, B. L., Eds.), Vol. 17, "Fish Respiration", pp. 283–319. Academic Press, San Diego.

Brittain, T. (1984). A two-state thermodynamic and kinetic analysis of the allosteric functioning of the haemoglobin of an extreme poikilotherm. *Biochem. J.* **221**, 561–568.

Brix, O., Clements, K. D., and Wells, R. M. G. (1998). An ecophysiological interpretation of hemoglobin multiplicity in three herbivorous marine teleost species from New Zealand. *Comp. Biochem. Physiol.* **121A**, 189–195.

Brix, O., Clements, K. D., and Wells, R. M. G. (1999). Haemoglobin components and oxygen transport in relation to habitat distribution in triplefin fishes (Tripterygiidae). *J. Comp. Physiol. B* **169**, 329–334.

Brix, O., Forås, E., and Strand, I. (1998). Genetic variation and functional properties of Atlantic cod hemoglobins: Introducing a modified tonometric method for studying fragile hemoglobins. *Comp. Biochem. Physiol.* **119A**, 575–583.

Cameron, J. N. (1973). Oxygen dissociation and content of blood from Alaskan burbot (*Lota lota*), pike (*Esox lucius*) and grayling (*Thymallus arcticus*). *Comp. Biochem. Physiol.* **46A**, 491–496.

Canfield, P. J., Quartararo, N., Griffin, D. L., Tsoukalas, G. N., and Cocaro, S. E. (1994). Haematological and biochemical reference values for captive Australian snapper, *Pagrus auratus*. *J. Fish Biol.* **44**, 849–856.

Caruso, C., Rutigliano, B., Romano, M., and di Prisco, G. (1991). The hemoglobins of the cold-adapted Antarctic teleost *Cygnodraco mawsoni*. *Biochim. Biophys. Acta* **1078**, 273–282.

Cashon, R. E., Vayda, M. E., and Sidell, B. D. (1997). Kinetic characterization of myoglobins from vertebrates with vastly different body temperatures. *Comp. Biochem. Physiol.* **117B**, 613–620.

Cech, J. J., Laurs, R. M., and Graham, J. B. (1984). Temperature-induced changes in blood gas equilibria in the albacore, *Thunnus alalunga*, and a warm-bodied tuna. *J. Exp. Biol.* **109**, 21–34.

Chen, L., DeVries, A. L., and Cheng, C.-H. C. (1997). Convergent evolution of antifreeze glycoproteins in Antarctic notothenioid fish and Arctic cod. *Proc. Natl. Acad. Sci. USA* **94**, 3817–3822.

Clarke, A. (1991). What is cold adaptation and how should we measure it? *Am. Zool.* **31**, 81–92.

Clarke, A., and Johnston, I. A. (1996). Evolution and adaptive radiation of Antarctic fishes. *Trends Ecol. Evol.* **11**, 212–218.

Cocca, E., Ratnayake-Lecamwasam, M., Parker, S. K., Carmadella, L., Ciaramella, M., di Prisco, G., and Detrich, H. W. (1997). Do the hemoglobinless icefishes have globin genes? *Comp. Biochem. Physiol.* **118A**, 1027–1030.

Cocca, E., Detrich, H. W., Parker, S. K.. and di Prisco, G. (2000). A cluster of four globin genes from the Antarctic fish *Notothenia coriiceps. J. Fish Biol.* **57**(Suppl. A), 33–50.

D'Avino, R., and di Prisco, G. (1997). The hemoglobin system of Antarctic and non-Antarctic notothenioid fishes. *Comp. Biochem. Physiol.* **118A**, 1045–1049.

D'Avino, R., Caruso, C., Tamburrini, M., Romano, M., Rutigliano, B., de Laureto, P. P., Carmadella, L., Carratore, V., and di Prisco, G. (1994). Molecular characterization of the functionally distinct hemoglobins of the Antarctic fish *Trematomus newnesi. J. Biol. Chem.* **269**, 9675–9681.

D'Avino, R., Fago, A., Kunzmann, A., and di Prisco, G. (1992). The primary structure and oxygen-binding properties of the single haemoglobin of the high Antarctic fish *Aethotaxis mitopteryx* deWitt. *Polar Biol.* **12**, 135–140.

Davison, W., Axelsson, M., Forster, M., and Nilsson, S. (1995). Cardiovascular responses to acute handling stress in the Antarctic fish *Trematomus bernacchii* are not mediated by circulatory catecholamines. *Fish Physiol. Biochem.* **14**, 253–257.

Davison, W., Forster, M. E., Franklin, C. E., and Taylor, H. H. (1988). Recovery from exhausting exercise in an Antarctic fish, *Pagothenia borchgrevinki. Polar Biol.* **8**, 167–171.

Davison, W., Franklin, C. E., and Carey, P. W. (1990). Oxygen uptake in the Antarctic teleost *Pagothenia borchgrevinki.* Limitations imposed by X-cell gill disease. *Fish Physiol. Biochem.* **8**, 69–77.

Davison, W., Franklin, C. E., and McKenzie, J. C. (1994). Haematological changes in an Antarctic teleost, *Trematomus bernacchii,* following stress. *Polar Biol.* **14**, 463–466.

DeVries, A. L., and Eastman, J. T. (1981). Physiology and ecology of notothenioid fishes of the Ross Sea. *J. R. Soc. New Zeal.* **11**, 329–340.

di Prisco, G. (1998). Molecular adaptations of Antarctic fish hemoglobins. *In* "Fishes of Antarctica: A Biological Overview" (di Prisco, G., Pisano, E., and Clarke, A., Eds.), pp. 339–353. Springer-Verlag, Berlin.

di Prisco, G. (2000). Life style and biochemical adaptation in Antarctic fishes. *J. Mar. Systems* **27**, 253–265.

di Prisco, G., and Tamburrini, M. (1992). The hemoglobins of marine and freshwater fish: The search for correlations with physiological adaptation. *Comp. Biochem. Physiol.* **102B**, 661–671.

di Prisco, G., Giardina, B., D'Avino, R., Condó, S. G., Bellelli, A., and Brunori, M. (1988). Antarctic fish haemoglobin: An outline of the molecular structure and oxygen binding properties II. Oxygen binding properties. *Comp. Biochem. Physiol.* **90B**, 585–591.

di Prisco, G., Macdonald, J. A., and Brunori, M. (1992). Antarctic fishes survive exposure to carbon monoxide. *Experientia* **48**, 473–475.

di Prisco, G., D'Avino, R., Caruso, C., Tamburrini, M., Camardella, L., Rutigliano, B., Carratore, V., and Romano, M. (1991a). The biochemistry of oxygen transport in red-blooded Antarctic fish. *In* "Biology of Antarctic Fish" (di Prisco, G., Maresca, B., and Tota, B., Eds.), pp. 263–281. Springer-Verlag, Berlin.

di Prisco, G., Condó, S. G., Tamburrini, M., and Giardina, B. (1991b). Oxygen transport in extreme environments. *Trends Biochem. Sci.* **16**, 471–474.

di Prisco, G., Tamburrini, M., and D'Avino, R. (1998). Oxygen transport systems in extreme environments: Multiplicity and structure-function relationship in haemoglobins of Antarctic fish. *In* "Cold Ocean Physiology" (Pörtner, H. O., and Playle, R. C., Eds.), pp. 143–165. *Society for Experimental Biology.* Series 66, Cambridge University Press, United Kingdom.

di Prisco, G., Cocca, E., Parker, S. K., and Detrich, H. W. (2002). Tracking the evolutionary loss of hemoglobin expression by the white-blooded Antarctic icefishes. *Gene* **295**, 185–191.

Dobson, G. P., and Baldwin, J. (1982). Regulation of blood oxygen affinity in the Australian blackfish *Gadopsis marmoratus.* II. Thermal acclimation. *J. Exp. Biol.* **99**, 245–254.

Eastman, J. T. (1993). "Antarctic Fish Biology: Evolution in a Unique Environment." Academic Press, San Diego.

Eastman, J. T. (1997a). Comparison of the Antarctic and Arctic fish faunas. *Cybium* **21**, 335–352.

Eastman, J. T. (1997b). Phyletic divergence and specialization for pelagic life in the Antarctic notothenioid fish *Pleuragramma antarcticum*. *Comp. Biochem. Physiol.* **118A**, 1095–1101.

Eastman, J. T, and Clarke, A. (1998). Radiations of Antarctic and non-Antarctic fish. *In* "Fishes of Antarctica: A Biological Overview" (di Prisco, G., Pisano, E., and Clarke, A., Eds.), pp. 3–26. Springer-Verlag Italia, Milan.

Eastman, J. T., and DeVries, A. L. (1981). Buoyancy adaptations in a swim-bladderless Antarctic fish. *J. Morphol.* **167**, 91–102.

Eastman, J. T., and DeVries, A. L. (1997). Biology and phenotypic plasticity of the Antarctic nototheniid fish *Trematomus newnesi* in McMurdo Sound. *Antarctic Sci.* **9**, 27–35.

Eastman, J. T., and McCune, A. R. (2000). Fishes on the Antarctic continental shelf: Evolution of a marine species flock? *J. Fish Biol.* **57**(Suppl. A), 84–102.

Eastman, J. T., Hikida, R. S., and DeVries, A. L. (1994). Buoyancy studies and microscopy of skin and subdermal extracellular matrix of the Antarctic snailfish, *Paraliparis devriesi*. *J. Morphol.* **220**, 85–101.

Egginton, S. (1994). Stress response in two Antarctic teleosts (*Notothenia coriiceps* Richardson and *Chaenocephalus aceratus* Lönnberg) following capture and surgery. *J. Comp. Physiol. B.* **164**, 482–491.

Egginton, S. (1996). Blood rheology of Antarctic fishes: Viscosity adaptations at very low temperatures. *J. Fish Biol.* **48**, 513–521.

Egginton, S. (1997a). A comparison of the response to induced exercise in red-blooded and white-blooded Antarctic fishes. *J. Comp. Physiol. B* **167**, 129–134.

Egginton, S. (1997b). The physiological response of Antarctic fishes to environmental and experimental stresses. *Cybium* **21**, 415–421.

Egginton, S., Taylor, E. W., Wilson, R. W., Johnston, I. A., and Moon, T. W. (1991). Stress response in the Antarctic teleosts (*Notothenia neglecta* Nybelin and *N. rossi* Richardson). *J. Fish Biol.* **38**, 225–235.

Egginton, S., Skilbeck, C., Hoofd, L., Calvo, J., and Johnston, I. A. (2002). Peripheral oxygen transport in skeletal muscle of Antarctic and sub-Antarctic notothenioid fish. *J. Exp. Biol.* **205**, 769–779.

Fago, A., D'Avino, R., and di Prisco, G. (1992). The hemoglobins of *Notothenia angustata*, a temperate fish belonging to a family largely endemic to the Antarctic Ocean. *Eur. J. Biochem.* **210**, 963–970.

Fago, A., Forest, E., and Weber, R. E. (2002). Hemoglobin and subunit multiplicity in the rainbow trout (*Oncorhynchus mykiss*) hemoglobin system. *Fish Physiol. Biochem.* **24**, 335–342.

Fago, A., Romano, M., Tamburrini, M., Coletta, M., D'Avino, R., and di Prisco, G. (1994). A polymerising Root-effect fish hemoglobin with high subunit heterogeneity. Correlation with primary structure. *Eur. J. Biochem.* **218**, 829–835.

Fago, A., Wells, R. M. G., and Weber, R. E. (1997). Temperature-dependent enthalpy of oxygenation in Antarctic fish hemoglobins. *Comp. Biochem. Physiol.* **118B**, 319–326.

Feder, M. E., Bennett, A. F., and Huey, R. B. (2000). Evolutionary physiology. *Annu. Rev. Ecol. Syst.* **31**, 315–341.

Feller, G., and Gerday, C. (1997). Adaptations of the hemoglobinless Antarctic icefish (Channichthyidae) to hypoxia tolerance. *Comp. Biochem. Physiol.* **118A**, 981–987.

Forster, M. E., Davison, W., Axelsson, M., Sundin, L., Franklin, C. E., and Giesen, S. (1998). Catecholamine release in heat-stressed Antarctic fish causes proton extrusion by the red cells. *J. Comp. Physiol. B* **168**, 345–352.

Franklin, C. E., Davison, W., and Carey, P. W. (1991). The stress response of an Antarctic teleost to an acute increase in temperature. *J. Therm. Biol.* **16,** 173–177.

Franklin, C. E., Davison, W., and McKenzie, J. C. (1993). The role of the spleen during exercise in the Antarctic teleost, *Pagothenia borchgrevinki. J. Exp. Biol.* **174,** 381–386.

Fyhn, U. E. H., Brix, O., Nævdal, G., and Johansen, T. (1994). New variants of the haemoglobins of Atlantic cod: A tool for discriminating between coastal and Arctic cod populations. *ICES Mar. Sci. Symp.* **198,** 666–670.

Gallaugher, P., and Farrell, A. P. (1998). Hematocrit and blood oxygen-carrying capacity. *In* "Fish Physiology" (Perry, S. F., and Tufts, B. L., Eds.), pp. 185–227. Vol. 17, Fish Respiration. Academic Press, San Diego.

Garland, T., and Carter, P. A. (1994). Evolutionary physiology. *Annu. Rev. Physiol.* **56,** 579–621.

Giardina, B., Mordente, A., Zappacosta, B., Calla, C., Colacicco, L., Gozzo, M. L., and Lippa, S. (1998). The oxidative metabolism of Antarctic fish: some peculiar aspects of cold adaptation. *In* "Fishes of Antarctica: A Biological Overview" (di Prisco, G., Pisano, E., and Clarke, A., Eds.), pp. 129–138. Springer-Verlag Italia, Milan.

Giles, M. A. (1991). Strain differences in hemoglobin polymorphism, oxygen consumption, and blood oxygen equilibria in three hatchery broodstocks of Arctic charr, *Salvelinus alpinus. Fish Physiol. Biochem.* **9,** 291–301.

Greenhalgh, M. (1999). "Freshwater Fish." Mitchell Beazley, London.

Grigg, G. C. (1967). Some respiratory properties of the blood of four species of Antarctic fishes. *Comp. Biochem. Physiol.* **23,** 139–148.

Grigg, G. C. (1969). Temperature-induced changes in the oxygen equilibrium curve of the blood of the brown bullhead, *Ictalurus nebulosus. Comp. Biochem. Physiol.* **28,** 1203–1223.

Heisler, N. (1986). Acid–base regulation in fishes. *In* "Acid–Base Regulation in Animals" (Heisler, N., Ed.), pp. 309–356. Elsevier, Amsterdam.

Herbert, N. A., and Wells, R. M. G. (2001). The aerobic physiology of the air-breathing blue gourami, *Trichogaster trichopterus*, necessitates behavioural regulation of breath-hold limits during hypoxic stress and predatory challenge. *J. Comp. Physiol.* **171,** 603–612.

Herbert, N. A., Wells, R. M. G., and Baldwin, J. (2002). Correlates of choroid rete development with the metabolic potential of various tropical reef fish and the effect of strenuous exercise on visual performance. *J. Exp. Mar. Biol. Ecol.* **275,** 31–46.

Herbert, N. A., Macdonald, J. A., Wells, R. M. G., and Davison, W. (2003). A difference in the optomotor behaviour of two Antarctic nototheniid fishes is correlated with the presence of the choroid *rete mirabile* and Root effect. *Polar Biol.* **26,** 411–415.

Hjorth, J. P. (1974). Molecular and genetic structure of multiple hemoglobins in the eelpout, *Zoarces viviparus* L. *Biochem. Genet.* **13,** 379–391.

Hochachka, P. W., and Somero, G. N. (2002). "Biochemical Adaptation. Mechanism and Process in Physiological Evolution." Oxford University Press, Oxford.

Hop, H., Tonn, W. M., and Welch, H. E. (1997). Bioenergetics of Arctic cod (*Boreogadus saida*) at low temperatures. *Can. J. Fish. Aquatic Sci.* **54,** 1772–1784.

Hubold, G. (1991). Ecology of notothenioid fish in the Weddell Sea. *In* "Biology of Antarctic Fish" (di Prisco, G., Maresca, B., and Tota, B., Eds.), pp. 3–22. Springer-Verlag, Berlin.

Imsland, A. K., Foss, A., Stefansson, S. O., and Nævdal, G. (2000). Hemoglobin genotypes of turbot (*Scophthalamus maximus*): Consequences for growth and variations in optimal temperature for growth. *Fish Physiol. Biochem.* **23,** 75–81.

Innes, A. J., and Wells, R. M. G. (1985). Respiration and oxygen transport functions of the blood from and intertidal fish, *Helcogramma medium* (Trypterygiidae). *Envir. Biol. Fishes* **14,** 213–226.

Jensen, F. B., and Weber, R. E. (1982). Respiratory properties of tench blood and hemoglobin adaptation to hypoxic-hypercapnic water. *Mol. Physiol.* **2**, 235–250.

Jensen, F. B., Fago, A., and Weber, R. E. (1998). Hemoglobin structure and function. *In* "Fish Physiology" (Perry, S. F., and Tufts, B. L., Eds.), Vol. 17, Fish Respiration, pp. 1–40. Academic Press, San Diego.

Johnston, I. A., and Battram, J. (1993). Feeding energetics and metabolism in demersal fish species from Antarctic, temperate and tropical environments. *Marine Biol.* **115**, 7–14.

Johnston, I. A., Calvo, J., Guderley, H., Fernandez, D., and Palmer, L. (1998). Latitudinal variation in the abundance and oxidative capacities of muscle mitochondria in perciform fishes. *J. Exp. Biol.* **201**, 1–12.

Jones, D. R., Brill, R. W., and Mense, D. C. (1986). The influence of blood gas properties on gas tensions and pH of ventral and dorsal aortic blood in free-swimming tuna, *Euthynnus affinis*. *J. Exp. Biol.* **120**, 201–213.

Jordan, A. D., Jungersen, M., and Steffensen, J. F. (2001). Oxygen consumption of East Siberian cod: No support for the metabolic cold adaptation theory. *J. Fish Biol.* **59**, 818–823.

Kawall, H. G., Torres, J. J., Sidell, B. D., and Somero, G. N. (2002). Metabolic cold adaptation in Antarctic fishes: Evidence from enzymatic activities of brain. *Mar. Biology* **140**, 279–286.

Kooyman, G. L. (1963). Erythrocyte analysis of some Antarctic fishes. *Copeia* **1963**, 457–458.

Kunzmann, A. (1989). Blutphysiologie hochantarktischer Fische. *Polarforschung* **59**, 129–139.

Kunzmann, A. (1991). Blood physiology and ecological consequences in Weddell Sea fishes (Antarctica). *Berichte zur Polarforschung* **91**, 1–79.

Kunzmann, A., and Zimmermann, C. (1992). *Aethotaxis mitopteryx*, a high-Antarctic fish with benthopelagic mode of life. *Mar. Ecol. Prog. Ser.* **88**, 33–40.

Kunzmann, A., Caruso, C., and di Prisco, G. (1991). Haematological studies on a high-Antarctic fish: *Bathydraco marri* Norman. *J. Exp. Mar. Biol. Ecol.* **152**, 243–255.

Laursen, J. S., Andersen, N. A., and Lykkeboe, G. (1985). Temperature acclimation and oxygen binding properties of blood of the European eel, *Anguilla anguilla*. *Comp. Biochem. Physiol.* **81A**, 79–86.

Lecklin, T., Nash, G., and Egginton, S. (1995). Do fish acclimated to low temperature improve microcirculatory perfusion by adapting red cell rheology? *J. Exp. Biol.* **198**, 1801–1808.

Lecklin, T., and Nikinmaa, M. (1998). Erthropoiesis in Arctic charr is not stimulated by anaemia. *J. Fish Biol.* **53**, 1169–1177.

Leroi, A. M., Bennett, A. F., and Lenski, R. E. (1994). Temperature acclimation and competitive fitness: An experimental test of the beneficial acclimation assumption. *Proc. Natl. Acad. Sci. USA* **91**, 1917–1921.

Ling, N., and Wells, R. M. G. (1985). Plasma catecholamines and erythrocyte swelling following capture stress in a marine teleost fish. *Comp. Biochem. Physiol.* **82C**, 231–234.

Lowe, T. E., and Wells, R. M. G. (1996). Primary and secondary responses to line capture in the blue mao mao. *J. Fish Biol.* **49**, 287–300.

Lowe, T. E., and Wells, R. M. G. (1997). Exercise challenge in Antarctic fish: Do haematology and muscle metabolite levels limit swimming performance? *Polar Biol.* **17**, 211–218.

Lowe, T. E., Brill, R. W., and Cousins, K. L. (2000). Blood oxygen-binding characteristics of bigeye tuna (*Thunnus obesus*), a high-energy-demand teleost that is tolerant of low ambient oxygen. *Marine Biol.* **136**, 1087–1098.

Macdonald, J. A. (2000). Neuromuscular temperature compensation in fish: A comparative study. *In* "Antarctic Ecosystems: Models for Wider Ecological Understanding" (Davison, W., Howard-Williams, C., and Broady, P., Eds.), pp. 134–141. New Zealand Natural Sciences, Christchurch.

Macdonald, J. A., Montgomery, J. C., and Wells, R. M. G. (1987). Comparative physiology of Antarctic fishes. *In* "Advances in Marine Biology" (Blaxter, J. H. S., and Southward, A. J., Eds.), Vol. 24, pp. 321–388. Academic Press, London.

Macdonald, J. A., and Wells, R. M. G. (1991). Viscosity of body fluids from Antarctic notothenioid fish. *In* "Biology of Antarctic Fish" (di Prisco, G., Maresca, B., and Tota, B., Eds.), pp. 163–178. Springer-Verlag, Berlin.

Maffia, M., Acierno, R., Rollo, M., and Storelli, C. (1998). Ion and metabolite transport through the intestinal luminal membranes of the Antarctic fish *Trematomus bernacchii*. *In* "Fishes of Antarctica: A Biological Overview" (di Prisco, G., Pisano, E., and Clarke, A., Eds.), pp. 237–246. Springer-Verlag Italia, Milan.

Maffia, M., Rizzello, A., Acierno, R., Rollo, M., Chiloiro, R., and Storelli, C. (2001). Carbonic anhydrase activity in tissues of the icefish *Chionodraco hamatus* and of the red-blooded teleosts *Trematomus bernacchii* and *Anguilla anguilla*. *J. Exp. Biol.* **204,** 3893–3992.

Marinakis, P., Tamburrini, M., Carratore, V., and di Prisco, G. (2003). Unique features of the haemoglobin system of the Antarctic notothenioid fish *Gobionotothen gibberifrons*. *Eur. J. Biochem.* **270,** 3981–3987.

Mazzarella, L., D'Avino, R., di Prisco, G., Savino, C., Vitagliano, L., Moody, P. C. E., and Zagari, A. (1999). Crystal structure of *Trematomus newnesi* haemoglobin re-opens the Root effect question. *J. Mol. Biol.* **287,** 897–906.

McCune, A. R., and Lovejoy, N. R. (1998). The relative rate of sympatric and allopatric speciation in fishes. *In* "Endless Forms. Species and Speciation" (Howard, D. J., and Berlocher, S. H., Eds.), pp. 172–185. Oxford University Press, New York.

Moerland, T. S., and Egginton, S. (1998). Intracellular pH of muscle and temperature: Insight from *in vivo* ^{31}P NMR measurements in a stenothermal Antarctic teleost (*Harpagifer antarcticus*). *J. Therm. Biol.* **23,** 275–282.

Montgomery, J., and Clements, K. (2000). Disaptation and recovery in the evolution of Antarctic fishes. *Trends Ecol. Evol.* **15,** 267–271.

Montgomery, J. C., and Wells, R. M. G. (1993). Recent advances in the ecophysiology of Antarctic notothenioid fishes: Metabolic capacity and sensory performance. *In* "Fish Ecophysiology" (Rankin, J. C., and Jensen, F. B., Eds.), pp. 341–374. Chapman and Hall, London.

Moylan, T. J., and Sidell, B. D. (2000). Concentrations of myoglobin and myoglobin mRNA in heart ventricles from Antarctic fishes. *J. Exp. Biol.* **203,** 1277–1286.

Murad, A., and Houston, A. (1991). Haemoglobin isomorph abundances in splenectomized rainbow trout, *Oncorhynchus mykiss* (Walbaum). *J. Fish Biol.* **38,** 641–651.

Møller, P. R., Nielsen, J. C., and Fossen, I. (2003). Patagonian toothfish found off Greenland. *Nature* **421,** 599.

Nikinmaa, M., and Jensen, F. B. (1986). Blood oxygen transport and acid–base status of stressed trout (*Salmo gairdnerii*): Pre- and postbranchial values in winter fish. *Comp. Biochem. Physiol.* **84A,** 391–396.

Nikinmaa, M., and Salama, A. (1998). Oxygen transport in fish. *In* "Fish Physiology" (Perry, S. F., and Tufts, B. L., Eds.), Vol. 17, Fish Respiration, pp. 141–184. Academic Press, San Diego.

Nilsson, S., Forster, M. E., Davison, W., and Axelsson, M. (1996). Nervous control of the spleen in the red-blooded Antarctic fish, *Pagothenia borchgrevinki*. *Am. J. Physiol.* **270,** R599–R604.

North, A. W. (1998). Growth of young fish during winter and summer at South Georgia, Antarctica. *Polar Biol.* **19,** 198–205.

O'Brien, K. M., and Sidell, B. D. (2000). The interplay among cardiac ultrastructure, metabolism and the expression of oxygen-binding proteins in Antarctic fishes. *J. Exp. Biol.* **203,** 1287–1297.

Perry, S. F., and Reid, S. D. (1993). β-Adrenergic signal transduction in fish: Interactive effects of catecholamines and cortisol. *Fish Physiol. Biochem.* **11**, 195–203.

Pisano, E., Cocca, E., Mazzei, F., Ghigliotti, L., di Prisco, G., and Detrich, H. W. (2003). Mapping of the α and β-globin genes on Antarctic fish chromosomes by fluorescence *in situ* hybridization. *Chromosome Res.* **11**, 633–640.

Pörtner, H. O., Berdal, B., Blust, R., Brix, O., Colosimo, A., De Wachter, B., Giuliani, A., Johansen, T., Fischer, T., Knust, R., Lannig, G., Nævdal, G., Nedenes, A., Nyhammer, G., Sartoris, F. J., Serendero, I., Sirabella, P., Thorkildsen, S., and Zahartsev, M. (2001). Climate induced temperature effects on growth performance, fecundity and recruitment in marine fish: Developing a hypothesis for cause and effect relationships in Atlantic cod (*Gadus morhua*) and common eelpout (*Zoarces viviparus*). *Continental Shelf Res.* **21**, 1975–1997.

Powell, M. D., Forster, M. E., and Davison, W. (2000). *In vitro* carbon dioxide excretion from erythrocytes of two species of Antarctic fishes and its inhibition by catecholamines. *J. Fish Biol.* **57**(Suppl. A), 112–120.

Qvist, J., Weber, R. E., DeVries, A. L., and Zapol, W. M. (1977). pH and haemoglobin oxygen affinity in blood from the Antarctic cod *Dissostichus mawsoni*. *J. Exp. Biol.* **67**, 77–88.

Raff, R. A., and Wray, G. A. (1989). Heterochrony: Developmental mechanisms and evolutionary results. *J. Evol. Biol.* **2**, 409–434.

Rankin, J. C., and Tuurala, H. (1998). Gills of Antarctic fish. *Comp. Biochem. Physiol.* **119A**, 149–163.

Riccio, A., Tamburrini, M., Carratore, V., and di Prisco, G. (2000). Functionally distinct haemoglobins of the cryopelagic Antarctic teleost *Pagothenia borchgrevinki*. *J. Fish Biol.* **57**(Suppl. A), 20–32.

Roig, T., Sánchez, J., Tort, L., Altimiras, J., and Bermúdez, J. (1997). Adrenergic stimulation of sea bream (*Sparus auratus*) and blood cells in normoxia and anoxia: Effects on metabolism and on the oxygen affinity of haemoglobin. *J. Exp. Biol.* **200**, 953–961.

Sadler, J., Wells, R. M. G., Pankhurst, P. M., and Pankhurst, N. W. (2000). Blood oxygen transport, rheology and haematological responses to confinement stress in diploid and triploid Atlantic salmon, *Salmo salar. Aquaculture* **184**, 349–361.

Samuelsen, E. N., Imsland, A. K., and Brix, O. (1999). Oxygen binding properties of three different hemoglobin genotypes in turbot (*Scophthalmus maximus* Rafinesque): Effect of temperature and pH. *Fish Physiol. Biochem.* **20**, 135–141.

Schilthuizen, M. (2000). Dualism and conflicts in understanding speciation. *BioEssays* **22**, 1134–1141.

Schlichting, C. D., and Pigliucci, M. (1998). "Phenotypic Evolution: A Reaction Norm Perspective." Sinauer Associates, Massachusetts.

Scholander, P. F., Flagg, W., Walters, V., and Irving, L. (1953). Climatic adaptations in arctic and tropical poikilotherms. *Physiol. Zool.* **26**, 67–92.

Schurmann, H., and Christiansen, J. S. (1994). Behavioural thermoregulation and swimming activity of two arctic teleosts (subfamily Gadinae)—The Polar cod (*Boreogadus saida*) and the Navaga (*Eleginus navaga*). *J. Thermal Biol.* **19**, 207–212.

Sidell, B. D. (1991). Physiological roles of high lipid content in tissues of Antarctic fish species. *In* "Biology of Antarctic Fish" (di Prisco, G., Maresca, B., and Tota, B., Eds.), pp. 220–231. Springer-Verlag, Berlin.

Somero, G. N., and Hofmann, G. E. (1997). Temperature thresholds for protein adaptation: When does temperature start to "hurt"? *In* "Global Warming: Implications for Freshwater and Marine Fish" (Wood, C. M., and McDonald, D. G., Eds.), pp. 1–24. Cambridge University Press, Cambridge, UK.

Somero, G. N., Fields, P. A., Hofmann, G. E., Weinstein, R. B., and Kawall, H. (1998). Cold adaptation and stenothermy in Antarctic notothenioid fishes: What has been gained and what has been lost. *In* "Fishes of Antarctica: A Biological Overview" (di Prisco, G., Pisano, E., and Clarke, A., Eds.), pp. 97–109. Springer-Verlag Italia, Milan.

Stam, W. W., Beintema, J. J., D'Avino, R., Tamburrini, M., and di Prisco, G. (1997). Molecular evolution of hemoglobins of Antarctic fishes (Notothenioidei). *J. Mol. Evol.* **45**, 437–445.

Steffensen, J. F. (2002).Metabolic cold adaptation of polar fish based on measurements of aerobic oxygen consumption: Fact or artefact? Artefact! Comp. Biochem. Physiol. A**132**, 789–795.

Steffensen, J. F., Bushnell, P. G., and Schurmann, H. (1994). Oxygen consumption in four species of teleosts from Greenland: No evidence of metabolic cold adaptation. *Polar Biol.* **14**, 49–54.

Takeda, T. (1991). Regulation of blood oxygenation during short-term hypercapnia in the carp, *Cyprinus carpio. Comp. Biochem. Physiol.* **98A**, 517–521.

Tamburrini, M., Brancaccio, A., Ippoliti, R., and di Prisco, G. (1992). The amino acid sequence and oxygen-binding properties of the single hemoglobin of the cold-adapted Antarctic teleost *Gymnodraco acuticeps. Arch. Biochem. Biophys.* **292**, 295–302.

Tamburrini, M., Davino, R., Carratore, V., Kunzmann, A., and Di Prisco, G. (1997). The hemoglobin system of *Pleuragramma antarcticum*—Correlation of hematological and biochemical adaptations with life style. *Comp. Biochem. Physiol.* **118A**, 1037–1044.

Tetens, V., and Lykkeboe, G. (1981). Blood respiratory properties of rainbow trout, *Salmo gairdneri*: Responses to hypoxic acclimation and anoxic incubation of blood *in vitro. J. Comp. Physiol.* **145**, 117–125.

Tetens, V., and Lykkeboe, G. (1985). Acute exposure of rainbow trout to mild and deep hypoxia: O2 affinity and O2 capacitance of arterial blood. *Resp. Physiol.* **61**, 221–235.

Tetens, V., Wells, R. M. G., and DeVries, A. L. (1984). Antarctic fish blood: Respiratory properties and the effects of thermal acclimation. *J. Exp. Biol.* **109**, 265–279.

Torres, J. J., and Somero, G. N. (1988). Metabolism, enzyme activities and cold adaptation in Antarctic mesopelagic fishes. *Mar. Biol.* **98**, 169–180.

Tufts, B., and Perry, S. F. (1998). Carbon dioxide transport and excretion. *In* "Fish Physiology" (Perry, S. F., and Tufts, B. L., Eds.), Vol. 17, Fish Respiration, pp. 229–281. Academic Press, San Diego.

Tufts, B. L., Gervais, M. R., Staebler, M., and Weaver, J. (2002). Subcellular distribution and characterization of gill carbonic anhydrase and evidence for a plasma carbonic anhydrase inhibitor in Antarctic fish. *J. Comp. Physiol. B* **172**, 287–295.

Tun, N., and Houston, A. H. (1986). Temperature, oxygen, photoperiod, and the haemoglobin system of the rainbow trout, *Salmo gairdneri. Can. J. Zool.* **64**, 1883–1888.

Val, A. L., Schwantes, A. R., and de Almeida-Val, V. M. F. (1986). Biological aspects of Amazonian fishes-VI. Hemoglobins and whole blood properties of *Semaprochilodus* species (Prochilodontidae) at two phases of migration. *Comp. Biochem. Physiol.* **83B**, 659–667.

Verdi, C., Carratore, V., Riccio, A., Tamburrini, M., Parisi, E., and di Prisco, G. (2002). The functionally distinct hemoglobins of the Arctic spotted wolffish Anarhichas *minor. J. Biol. Chem.* **277**, 36312–36320.

Verdi, C., Parisi, E., and di Prisco, G. (2003). The evolution of polar fish haemoglobin: A phylogenetic analysis of the ancestral amino acid residues linked to the Root effect. *J. Mol. Evol.* **57**(Suppl. 1), S258–S267.

Verheyen, E., Blust, R., and Doumen, C. (1985). The oxygen uptake of *Sarotherodon niloticus* L., and the oxygen binding properties of its blood and hemolysate (Pisces: Cichlidae). *Comp. Biochem. Physiol.* **81A**, 423–426.

von Dorrien, C. F. (1993). Ökologie und Respiration ausgewählter arktischer Bodenfischarten. *Ber. Polarforsch* **125**, 1–99.

Vorger, P. (1986). The effect of temperature on oxygen equilibrium curves of trout blood (*Salmo gairdnerii*). *Comp. Biochem. Physiol.* **83A**, 525–531.

Weber, R. E. (2000). Adaptations for oxygen transport: Lessons from fish hemoglobins. *In* "Hemoglobin Function in Vertebrates: Molecular Adaptation in Extreme Environments." pp. 23–37. Springer-Verlag, Italia.

Weber, R. E., and Wells, R. M. G. (1989). Haemoglobin structure and function. *In* "Comparative Pulmonary Physiology" (Wood, S. C., Ed.), pp. 279–310. Marcell Dekker, New York.

Wells, R. M. G. (1990). Hemoglobin physiology in vertebrate animals: A cautionary approach to adaptationist thinking. *In* "Advances in Comparative and Environmental Physiology" (Boutilier, R. G., Ed.), Vol. 6, pp. 143–161. Springer-Verlag, Berlin.

Wells, R. M. (1978). Respiratory adaptation and energy metabolism in Antarctic nototheniid fishes. *New Zeal. J. Zool.* **5**, 813–815.

Wells, R. M. G. (1999a). Evolution of haemoglobin function: Molecular adaptations to environment. *Clin. Exp. Pharm. Physiol.* **26**, 591–595.

Wells, R. M. G. (1999b). Haemoglobin function in aquatic animals: Molecular adaptations to environmental challenge. *Mar. Freshwater Res.* **50**, 933–939.

Wells, R. M. G., and Baldwin, J. (1990). Oxygen transport potential in tropical reef fish with special reference to blood viscosity and haematocrit. *J. Exp. Mar. Biol. Ecol.* **141**, 131–143.

Wells, R. M. G., and Davie, P. S. (1985). Oxygen binding by the blood and hematological effects of capture stress in two big gamefish: Mako shark and striped marlin. *Comp. Biochem. Physiol.* **81A**, 643–646.

Wells, R. M. G., and Jokumsen, A. (1982). Oxygen binding properties of hemoglobins from Antarctic fishes. *Comp. Biochem. Physiol.* **71B**, 469–473.

Wells, R. M. G., and Weber, R. E. (1991). Is there an optimal haematocrit for rainbow trout, Oncorhynchus mykiss (Walbaum)? An interpretation of recent data based on blood viscosity measurements. *J. Fish Biol.* **38**, 53–65.

Wells, R. M. G., Ashby, M. D., Duncan, S. J., and Macdonald, J. A. (1980). Comparative study of the erythrocytes and haemoglobins in nototheniid fishes from Antarctica. *J. Fish Biol.* **17**, 517–527.

Wells, R. M. G., Baldwin, J., Seymour, R. S., and Weber, R. E. (1997). Blood oxygen transport and hemoglobin function in three tropical fish species from northern Australian freshwater billabongs. *Fish Physiol. Biochem.* **16**, 247–258.

Wells, R. M. G., Grigg, G. C., Beard, L. A., and Summers, G. (1989). Hypoxic responses in a fish from a stable environment: Blood oxygen transport in the Antarctic fish *Pagothenia borchgrevinki*. *J. Exp. Biol.* **141**, 97–111.

Wells, R. M. G., Macdonald, J. A., and di Prisco, G. (1990). Thin-blooded Antarctic fishes: A rheological comparison of the haemoglobin-free icefishes, *Chionodraco kathleenae* and *Cryodraco antarcticus*, with a red-blooded nototheniid, *Pagothenia bernacchii*. *J. Fish Biol.* **36**, 595–609.

Wells, R. M. G., McIntyre, R. H., Morgan, A. K., and Davie, P. S. (1986). Physiological stress responses in big gamefish after capture: Observations on plasma chemistry and blood factors. *Comp. Biochem. Physiol.* **84A**, 565–571.

Wells, R. M. G., Summers, G., Beard, L. A., and Grigg, G. C. (1988). Ecological and behavioural correlates of intracellular buffering capacity in the muscles of Antarctic fishes. *Polar Biol.* **8**, 321–325.

Wells, R. M. G., Forster, M. E., and Meredith, A. S. (1984a). Blood oxygen affinity in the amphibious fish *Neochanna burrowsius* (Galaxiidae: Salmoniformes). *Physiol. Zool.* **57**, 261–265.

Wells, R. M. G., Tetens, V., and DeVries, A. L. (1984b). Recovery from stress following capture and anaesthesia of Antarctic fish: Haematology and blood chemistry. *J. Fish Biol.* **25,** 567–576.

Whiteley, N. M., and Egginton, S. (1999). Antarctic fishes have a limited capacity for catecholamine synthesis. *J. Exp. Biol.* **202,** 3623–3629.

Wilson, R. S., and Franklin, C. E. (2002). Testing the beneficial acclimation hypothesis. *Trends Ecol. Evol.* **17,** 66–70.

Wood, S. C., Weber, R. E., and Powers, D. A. (1979). Respiratory properties blood and hemoglobin solutions from the piranha. *Comp. Biochem. Physiol.* **62A,** 163–167.

Zhao, Y., Ratnayake-Lecamwasam, M., Parker, S. K., Cocca, E., Carmadella, L., di Prisco, G., and Detrich, H. W. (1998). The major adult α-globin gene of Antarctic teleosts and its remnants in the hemoglobinless icefishes: Calibration of the mutational clock for nuclear genes. *J. Biol. Chem.* **272,** 14745–14752.

Zhou, R., Cheng, H., and Tiersch, T. R. (2002). Differential genome duplication and fish diversity. *Rev. Fish Biol. Fisheries* **11,** 331–337.

Zimmermann, C., and Kunzmann, A. (2001). Baseline respiration and spontaneous activity of sluggish marine tropical fish of the family Scorpaenidae. *Mar. Ecol. Prog. Ser.* **219,** 229–239.

8

ANTARCTIC FISH SKELETAL MUSCLE AND LOCOMOTION

WILLIAM DAVISON

I. Introduction
II. Basic Fish Muscle Anatomy
III. Mechanical and Physiological Properties of Isolated Muscle
IV. Protecting the Cell
V. Exercise and Energy Supply
VI. Swimming
VII. Exercise and Temperature

I. INTRODUCTION

The extant fish fauna of the Antarctic is relatively recent. An earlier temperate fauna, represented by fossil remains on Seymour Island, died out in association with the opening of the Drake Passage and the cooling of the Southern Ocean (Eastman, 1993, 2000). This led to the opening of shallow-water niches around Antarctica and rapid radiations to fill these. Nevertheless, although the Southern Ocean contains about one-tenth of the world's seawater, with a continental shelf of $2 \times 10^6 \, km^2$ (Pakhomov, 1997), the Antarctic Zoogeographic Region has a depauperate ichthyofauna, containing only 313 species, in 50 families, representing only 1.2% of the world's fish fauna (Eastman, 2000). The fauna is dominated by a single suborder, the Notothenioidei. There are about 96 species of notothenioid, although this relatively low number of species is misleading, as the notothenioids make up most of the biomass of the Southern Ocean. Non-notothenioid fish make up two-thirds of the species list, but they are deep-water specialists and stragglers into northern regions and, other than the myctophids, are numerically small. Endemism is a major feature of the Antarctic fauna, with 88% of

The Physiology of Polar Fishes: Volume 22
FISH PHYSIOLOGY

the total fish fauna endemic, rising to 97% if only the notothenioids are considered (Eastman, 1993, 2000).

The notothenioids are a unique group in that they are Antarctic specialists. Of 122 species in eight families, 96 species from five families are endemic to the Southern Ocean, whereas the rest are found in southern hemisphere sub-Antarctic waters (Eastman and Clarke, 1998; Eastman, 2000). Shallow continental regions are populated almost exclusively by notothenioids, and as a consequence, the bulk of fish research in Antarctica has concentrated on this group. Rapid radiation from a common benthic ancestor has led to a range of ecotypes. In addition, although most of the fish have remained benthic, there are demersal and pelagic species, making it possible to conduct comparative studies within the group. However, the bulk of the work has concentrated on comparisons between Antarctic and temperate, and even tropical, fish. There has been much debate about their phylogeny, hampered by the presence of different morphs and cryptic species, perhaps indicative of the continuing rapid radiation of the group (Bernardi and Goswami, 1997; Eastman and DeVries, 1997; Ritchie *et al.*, 1997; see Chapter 2).

The success of the notothenioids clearly has been their ability to survive in waters that are almost permanently below 0°C. Production of antifreeze proteins has been crucial to this (Cheng, 2000), as well as the ability of the fish to carry out all of its physiological and biochemical processes at these low temperatures. One extreme modification has been the loss of the respiratory pigment hemoglobin in the blood of the icefish (Channichthyidae), thought to be possible because of high levels of dissolved oxygen in water at low temperatures and the low metabolic rates of these animals (Egginton *et al.*, 2002), although it should be noted that there are several sub-Antarctic icefish living at relatively high temperatures. Living in permanently cold waters is not, of course, exclusive to Antarctic organisms, as cold waters also occur in the Arctic and the deep oceans. In the latter case, however, physiological and biochemical adaptations are geared toward coping with high pressures and lack of food, as well as minimizing energy expenditure rather than upregulating systems to cope with Q_{10} effects (Torres and Somero, 1988; Somero, 1998, 2003). There has been much debate about mechanisms of cold adaptation in polar fish, particularly as it relates to resting metabolism (Holeton, 1974; Clarke, 1991; Drud Jordan *et al.*, 2001) and the need to save energy. The debate continues and is beyond the scope of this chapter, other than where it relates directly to muscle structure and function (Egginton *et al.*, 2002). In Antarctic waters, with most research focused on a single group of fish, there has been debate over whether any findings relate to adaptation to the cold waters of the Southern Ocean or represent a phylogenetic anomaly, the so-called adaptations being simply a feature of the group (Fernandez *et al.*, 2000; Egginton *et al.*, 2001; Tuckey and Davison, 2004).

Antarctic fish are regarded as specialists. Adaptation to extreme temperatures (hot or cold) is thought to be associated with a move away from a generalized eurytherm animal toward specialization and stenothermy (Huey and Berrigan, 1996; Pörtner *et al.*, 2000; Pörtner, 2002). Thus, animals living close to extreme thermal limits, such as the freezing point of seawater, have achieved this by becoming restricted to a very narrow thermal range (Peck, 2002; Pörtner, 2002) and lack the phenotypic plasticity to cope with changing environments (Johnston and Ball, 1997). Following this reasoning, it would be expected that Antarctic fish muscle has become adapted to function at cold temperatures, but that it is now locked into a specialized niche, unable to respond to rising temperatures (Pörtner, 2002). Some evidence, however, discussed at the end of this chapter, may force us to reassess this.

This chapter examines the physiology of skeletal muscle of polar fish, concentrating on those animals found in Antarctic waters. Cardiac muscle is not covered here because much of this information is covered in Chapter 6. In addition, smooth muscle is not covered, as there is very little information relating to this tissue in Antarctic fish.

II. BASIC FISH MUSCLE ANATOMY

Fish musculature follows a standard pattern, from the most primitive hagfish (Korneliussen and Nicolaysen, 1973) to the most modern of actinopterygians, irrespective of temperature (Hoar and Randall, 1978; Johnston, 2001). In post-larval fish, four fiber types are recognized, classified initially according to the color of the fresh tissue. White muscle fibers constitute the bulk ($\geq$90%) of the trunk (myotome) of a fish. They are large, have a poor blood supply, and possess few mitochondria, and the cells are packed with myofibrillar material with a fast-type myosin light-chain molecule (Johnston *et al.*, 1977; Feller and Bionet, 1984). They are fast-contracting muscle fibers used for sprinting and high-speed sustained swimming, corresponding to type IIb fibers of mammals. Because of the poor blood supply and limited mitochondrial capacity, this muscle tends to use glycogen-fueled anaerobic metabolism for generation of ATP, with the concomitant buildup of lactic acid. In fish, including Antarctic eelpouts, the lactate ions are not released readily from the muscle during recovery, with *in situ* regeneration of glucose taking many hours (Milligan and Wood, 1987; Hardewig *et al.*, 1998).

By contrast, red muscle fibers have smaller diameters and contain many mitochondria, with their cytochrome pigments contributing to the color of the muscle. Located at the outer edge of the myotome, there is a good blood supply and high levels of intracellular myoglobin, ensuring that oxygen reaches the innermost areas of each fiber. The myofibrillar ATPase has a slow cycle time (Johnston *et al.*, 1977), so the muscle contracts relatively slowly. It is resistant to fatigue and is the cruising, sustained swimming muscle of the fish, equivalent to type I fibers of mammals. Used for most of the animal's swimming activity (Johnston *et al.*, 1977), it is aerobic, using lipid, and often protein, as preferred fuel sources.

Pink muscle, often termed *intermediate*, has not been particularly well studied. It lies somewhere between red and white in both its structure and its actions (and its location), perhaps corresponding to type IIa fibers of mammals. It certainly has a fast type myosin ATPase (Johnston *et al.*, 1977), and electromyographic (EMG) studies have shown that it is active at speeds intermediate between red and white (Johnston *et al.*, 1977). However, its appearance is very variable between species and possibly even has different functions; Davison and Goldspink (1984) suggested that in goldfish, pink muscle was acting as a "fast red muscle," whereas in brown trout, it had the characteristics of an "intermediate white muscle."

A fourth muscle type is found in many, if not all, teleost fish. Called *small-diameter fibers* (SDFs), they are small with a poor blood supply and few mitochondria and typically fail to stain with most histochemical stains. Direct physiological experimentation describing their function is lacking, but it is assumed that they have a similar function to the tonic fibers of amphibians and are used to hold the body in a particular attitude, particularly during fin-based swimming (Davison and Macdonald, 1985).

The method of swimming used by a fish is a major determinant of the distribution of muscle fibers. The default swimming mode is carangiform or subcarangiform, with the fish using its myotomal muscles of the trunk for propulsion. A cross-section of the myotome shows that the bulk of the musculature is composed of white fibers. Red muscle forms a thin band of muscle around the periphery of the animal, with a wedge of tissue at the level of the median transverse septum. SDFs are typically also found here. Pink muscle forms a transition zone a few fibers thick between the red and white fibers. In animals that are essentially sit-and-wait predators, the red muscle is not well developed, comprising perhaps 2% of the muscle bulk. In active swimmers, there is a great deal more red muscle, with suggestions that it might reach as much as 30% or more of the myotome (Dunn *et al.*, 1989), although this is highly unusual, and 10% is regarded as high. In carangiform swimming fish, it has been shown using EMG techniques that the different fiber types become active at increasing speeds in the order red, pink, white (Johnston *et al.*, 1977; Beddow and McKinley, 1999).

Many teleost fish use their fins for low-speed propulsion, and there has been an associated shift in the distribution of muscle fibers. For example, balistiform swimming fish have the bulk of the red fibers associated with the median fins (Davison, 1987), whereas in labriform swimming fish, the bulk of the red muscle is in the pectoral fins (Davison and Macdonald, 1985; Davison, 1988). In these animals, the shift of red muscle bulk to the fins is associated with a reduction or even an absence of red muscle fibers in the myotome (Davison and Macdonald, 1985).

With a single exception (Eastman, 1997), adult Antarctic notothenioid fish use drag-based labriform swimming (Montgomery and Macdonald, 1984; Forster et al., 1987; Walker and Westneat, 2000). Low-speed swimming is achieved almost exclusively by rhythmical contractions of the pectoral fins. Higher sustainable speeds are achieved using the pectoral fins and myotome in a flick-and-guide action (Wilson et al., 2001), and top speeds are achieved using carangiform or subcarangiform swimming utilizing the white muscle of the myotome (Forster et al., 1987). Larval Antarctic notothenioids are typically pelagic and swim using subcarangiform locomotion, changing to labriform during transformation to the adult form, with corresponding changes to muscle fiber distribution (Johnston and Camm, 1987). Although many adult notothenioids have become secondarily pelagic, they have retained the labriform swimming mode. *Pleuragramma antarcticum* is the exception. As an adult, it is a small mid-water schooling fish that has retained subcarangiform locomotion as the major mode of swimming, a trait described as *pedomorphic* (Johnston et al., 1988; Eastman, 1997).

Early microscopic detail of Antarctic fish muscle comes from a number of authors (Walesby and Johnston, 1979; Smialowska and Kilarski, 1981; Kilarski et al., 1982; Walesby et al., 1982; Davison and Macdonald, 1985). In adult notothenioids, the myotome is dominated by white fibers in both Antarctic (Davison and Macdonald, 1985; Johnston and Harrison, 1985) and sub-Antarctic species (Fernandez et al., 1999, 2000), although Dunn et al. (1989) suggested that many fish have considerable amounts of red muscle in the myotome, up to 31% of the myotome in posterior regions of *Notothenia gibberifrons*. This latter work contradicts the other studies and is difficult to reconcile. Oxidative fibers comprise a small amount of tissue around the periphery of the myotome, one or two cells thick, plus a small wedge of tissue at the region of the major horizontal septum, making up at most 2% of the muscle mass (Johnston and Harrison, 1985). Several authors (Walesby et al., 1982; Davison and Macdonald, 1985) have suggested that red muscle has been lost from the myotome during the ontogenetic switch from carangiform to labriform locomotion and that these oxidative fibers are pink fibers rather than red, based on the low numbers of mitochondria in the cells. However, pink muscle that is distinct from red has been shown at least in the myotome of sub-Antarctic notothenioids (Fernandez et al., 2000), and Walesby and Johnston (1979) indicate that enzyme profiles showed the presence of red muscle. Detecting the presence of, and differentiating between, red and pink fibers is hampered by the odd histochemical staining characteristics of Antarctic fish muscle for myofibrillar ATPase. This is a standard technique used for typing muscle fibers and relies on differential resistance to denaturation of muscle proteins at extreme pH levels (both acid and alkali). Antarctic fish muscle, irrespective of fiber type or pH used, does not stain well with this technique and cannot be used to identify the major muscle types (Davison and Macdonald, 1985; Harrison *et al.*, 1987; Fernandez *et al.*, 2000).

Small-diameter fibers are a distinct feature of the myotome in all Antarctic species studied (Davison and Macdonald, 1985; Johnston and Camm, 1987; Fernandez *et al.*, 2000), suggesting that in these labriform swimmers, tonic positioning of the bulk of the myotome is important, as might be expected in fish that use their fins for swimming, holding the trunk stationary. SDFs are undoubtedly used to keep the trunk rigid, but also to position the trunk of the fish so that it acts as a rudder (Johnston and Camm, 1987).

Larval fish use subcarangiform swimming. In the few studies that have looked at muscle type distribution in the myotomes of these juvenile fish, red muscle is well developed, comprising 30% of the myotome in larval *Notothenia coriiceps* (Camm and Johnston, 1985; Johnston and Camm, 1987). *Pleuragramma* adults continue to use subcarangiform locomotion, and in these fish, the red muscle is retained and well developed in the myotome (Johnston *et al.*, 1988; Eastman, 1997). A similar trait of retaining red muscle in the trunk has been seen with a move to pelagism in the typically benthic temperate-water tripterygiids (Hickey and Clements, 2003).

The pectoral fins of notothenioids are large and well developed. The gross anatomy of the pectoral fins has been described in some detail and need not be elaborated here (Johnston and Harrison, 1985; Harrison *et al.*, 1987; Fernandez *et al.*, 2000). Several muscle blocks are present, controlling abduction and adduction of the whole fin, as well as individual fin rays, although most work has concentrated on the major (profundus) pectoral adductor muscle. This is a well-defined muscle with parallel fibers running the length of the muscle block (Davison and Macdonald, 1985; Harrison *et al.*, 1987). The muscle is dominated by red oxidative fibers, as might be expected of the major cruising muscle of the animal. However, there are differences between species, and Davison and Macdonald (1985) have suggested that lifestyle is important. For example, the pelagic *Pagothenia borchgrevinki* has its adductor muscle almost uniformly red, whereas the benthic *Trematomus bernacchii* has a mixture of red, pink, and white fibers. The benthic fish requires the fast contracting pink and white fibers to initiate swimming, overcoming the inertia of moving the animal from the sea floor, with red fibers taking over once the animal is moving. Benthic fish tend not

Table 8.1
Mean Fiber Diameters of Several Antarctic Fish and Two Temperate-Water Fish[a]

	Myotome			Pectoral Fin	
	Red	White	SDF	Red	White
Antarctic					
Pagothenia borchgrevinki	65 (125)	230 (515)	29 (62)	49 (109)	101 (190)
Trematomus bernacchii	56 (130)	222 (515)	23 (39)	52 (182)	94 (182)
Lycodichthys dearborni	54 (104)	169 (309)	21 (43)		
Dissostichus mawsoni		157 (520)			
Temperate					
Salmo trutta	29 (62)	49 (104)			
Notolabrus fucicola	30 (62)	61 (108)		44 (86)	67 (102)

[a]Maximum recorded diameter in parentheses (measured in μm).

Source: Data from Davison and Goldspink, 1977; Davison and Macdonald, 1985; Davison, 1994.

to move far during swimming, moving mainly to chase prey or avoid predators, so it might be expected that the pectoral muscle would contain fibers such as pink and white capable of moving the animal quickly. By contrast, the constantly swimming pelagic fish has a much reduced need to overcome inertia but has an enhanced need for long periods of swimming at a constant speed, requiring large numbers of red fibers. Even within the red fibers themselves, the spread of fiber size is much greater in the benthic fish, suggesting a greater diversity of roles within this single fiber type (Davison and Macdonald, 1985).

A great number of studies has commented on the large size of all muscle fibers within the notothenioids. Many studies indicate that in temperate and tropical fish, red fibers reach about $30 \mu m$ and white fibers less than $100 \mu m$. For example, brown trout have myotomal red fiber diameters of $30 \mu m$, pink $45 \mu m$, and white $50 \mu m$, whereas in a comparison of swimming capacities of temperate wrasses and Antarctic fish, the wrasses had large pectoral fin red fibers at $39 \mu m$ and white myotomal fibers at $58 \mu m$ (Davison and Goldspink, 1977; Tuckey and Davison, 2004) (Table 8.1). Red fibers in Antarctic fish are typically greater than $50 \mu m$ in diameter, whereas white fibers are enormous, reaching in excess of $500 \mu m$ diameter (Fitch *et al.*, 1984; Davison and Macdonald, 1985; Harrison *et al.*, 1987; Londraville and Sidell, 1990; Archer and Johnston, 1991; Egginton *et al.*, 2002; Johnston 2003; Sänger *et al.*, 2005). Icefish seem to have particularly large fibers (Egginton *et al.*, 2002, O'Brien *et al.*, 2003), although Dunn *et al.* (1989) suggested that they were not. Even in sub-Antarctic notothenioids, muscle fibers are big (Fernandez *et al.*, 1999, 2000). This tendency toward large fibers is seen even at the larval stage (Dunn *et al.*, 1989).

Teleost fish grow throughout their lives, with muscle growth dominated by increasing numbers of fibers (hyperplasia) during juvenile phases and, by hypertrophy of existing fibers during later phases. In their classic work on fish muscle growth, Weatherley *et al.* (1987) indicated that in temperate-water fish, hyperplasia is positively correlated with the final size of a fish, large numbers of fibers in juveniles, indicating a large final body size. Hyperplasia stops at about 44% of final size, and growth from that point relies on hypertrophy. Growth in Antarctic fish muscle does not follow this general rule. Growth in Antarctic fish is assumed to be slow, but of note are the toothfish that can reach in excess of 25 cm in their first year (Ashford *et al.*, 2002). Muscle growth is primarily by hypertrophy of existing fibers rather than creation of new fibers from satellite cells, even in young animals. Indeed Battram and Johnston (1991) showed that myotomal fiber number does not change in *N. coriiceps* in fish spanning a mass range of 12–889 g, with all fish having about 15,000 red fibers and 60,000 white. White fibers with a diameter less than 10 μm were seen only in fish less than 20 g, emphasizing the dominance of hypertrophy in growth of notothenioids. The net result of this is that Antarctic fish have fewer but much bigger muscle cells (Johnston, 2003; Johnston *et al.*, 2003). Despite this, Antarctic fish are not particularly small, with many attaining 1 kg or more (Gon and Heemstra 1990), icefish all attain a large size, and the toothfish (*Dissostichus eleginoides* and *Dissostichus mawsoni*) reach over 100 kg. Thus, Antarctic fish clearly do not follow the Weatherley principle.

Feeding stimulates muscle growth, and this has been investigated in terms of fiber recruitment in both Antarctic and sub-Antarctic notothenioids. Feeding stimulates satellite cell activity (also known as *myogenic progenitor cells*), those cells responsible for producing new muscle fibers in temperate-water fish. After feeding, these cells become more active in notothenioids and increase in number (Brodeur *et al.*, 2002, 2003). The activity of the cells is cold adapted, with cell cycle times in Antarctic fish much shorter than times in sub-Antarctic counterparts (Brodeur *et al.*, 2003). However, in Antarctic fish, stimulation of the myogenic progenitor cells does not lead to production of new muscle cells but provides nuclei for the ever enlarging existing cells (Brodeur *et al.*, 2002, 2003). Johnston (2003) has suggested that large fiber size is a particular trait of the notothenioids, appearing early in the move to cold waters. However, there is a paucity of data relating to non-Antarctic notothenioids, and an Antarctic eelpout (*Rhigophila dearborni*) has been shown to have myotomal red fibers in excess of 50 μm (range 26–124) and white fibers averaging 185 μm (range 31–309) (Davison and Macdonald, 1985). Clearly, there is a need for more information from both notothenioid and non-notothenioid cold-water fish.

Capillary supply to the muscles is poor, especially to white myotomal muscle, although as white muscle constitutes such a large part of the fish, much of the systemic supply actually goes to this tissue (Egginton, 1997a). There is an auxiliary supply to the pectoral fin muscles in notothenioid fish, the hypobranchial arteries that run from the gills straight to the fins, with an estimated 30% of cardiac output flowing in this system (Egginton and Rankin, 1998). Fanta *et al.* (1989) implanted O_2 electrodes into white muscle and showed that under normoxic conditions, PO_2 was only 5–20 mmHg. The poor blood supply appears to be a consequence of lack of capillary proliferation during growth, coupled with hypertrophy of muscle fibers, so that any individual capillary has to supply a large volume of fiber. As the fish grows, capillarity (measured as capillaries per volume of muscle) decreases (Johnston and Camm, 1987; Dunn, 1988b; Egginton *et al.*, 2002). This is apparent in icefish muscle in which the bigger fibers have a markedly reduced capillary density (Egginton *et al.*, 2002; O'Brien *et al.*, 2003), although in icefish, this is somewhat offset by large-bore highly tortuous capillaries (Hemmingsen and Douglas, 1977; Fitch *et al.*, 1984; O'Brien *et al.*, 2003).

Antarctic fish skeletal muscle generally lacks myoglobin (but see Walesby *et al.*, 1982; Morla *et al.*, 2003) irrespective of whether the fish is red blooded or devoid of hemoglobin, although cardiac muscle does seem to require it, perhaps associated with the much lower PO_2 of blood bathing the spongy myocardium (Egginton and Sidell, 1989; Archer and Johnston, 1991; Hoofd and Egginton, 1997; Moylan and Sidell, 2000; O'Brien and Sidell, 2000; O'Brien *et al.*, 2003; Small *et al.*, 2003). Any myoglobin present is cold adapted, working as well at 0 °C as mammalian myoglobins at 37 °C (Cashon *et al.*, 1997). Lack of or low concentrations of myoglobin are possibly a general feature of cold-water fishes, as Grove and Sidell (2002) examined a number of cold-water northern hemisphere fish and noted a lack of myoglobin in the muscle and hearts of several species.

Antarctic fish muscle cells, especially the red fibers, contain much lipid, as extracellular adipocytes, as intracellular lipid droplets, and the lipids associated with mitochondrial and other membranes (Lin *et al.*, 1974; Eastman and DeVries, 1982; Eastman, 1988; Londraville and Sidell, 1990; Pörtner, 2002). The extracellular lipid not only is used as an energy store but also has a very important function of providing buoyancy in pelagic species (Eastman and DeVries, 1982). The intracellular lipid has been implicated in transport of oxygen through the cell. Antarctic fish red muscle has an extremely high mitochondrial volume, especially in pectoral red muscle where mitochondria can make up more than 50% of the volume of the fiber, much higher than found in temperate-water fish (Johnston, 1987; Londraville and Sidell, 1990; O'Brien *et al.*, 2003). In many fish, these mitochondria run throughout the fiber, so there are as many mitochondria at the center of the cell as there are at the periphery (O'Brien *et al.*, 2003). There is evidence that mitochondrial density is associated with activity levels, with the benthic *N. gibberifrons* and *N. coriiceps* having lower volume densities (25% and 27% of the cell, respectively) than the active pelagic *Trematomus newnesi* (35%) and *P. antarcticum* (56%) in similar sized cells (Johnston *et al.*, 1988; Londraville and Sidell, 1990; O'Brien *et al.*, 2003). The pelagic *P. borchgrevinki* has a red muscle cell mitochondrial volume of 38% (Sänger *et al.*, 2005). Changes to mitochondrial volume can be correlated with changes to lifestyle. Pelagic juveniles of *N. coriiceps* swimming continuously using subcarangiform locomotion have about 35% mitochondrial volume in both pectoral and myotomal red muscle. In the benthic adults, using labriform swimming, the oxidative fibers of the myotome and pectoral fin have only 13% and 27% mitochondria by volume, respectively (Camm and Johnston, 1985; Johnston and Camm, 1987).

There appears to be a paradox in that red muscle fibers need to be able to contract aerobically for long periods, they are very big (compared with temperate water fish red fibers), they are stacked full of oxygen demanding mitochondria, yet blood supply is poor and there is a lack of myoglobin, the haem pigment normally associated with transport of oxygen through the cell. This is particularly so for icefish that lack hemoglobin and thus have a limitation of oxygen-carrying capacity in the blood, estimated at about 10% of the oxygen content of red-blooded notothenioids (Fitch *et al.*, 1984; Egginton, 1994). This is nicely debated by Egginton *et al.* (2002) and O'Brien *et al.* (2003). Lipid is regarded as the cold-water equivalent of myoglobin because oxygen is highly soluble in lipids (Hoofd and Egginton, 1997). In particular, mitochondrial membrane lipids are seen as presenting a pathway for diffusion of oxygen. The large volume density and spread of mito-chondria through the cell results in a small separation distance between individual mitochondria, and this enhances diffusion (Archer and Johnston, 1991). Egginton *et al.* (2002) modeled oxygen transport through the cell, taking into account such variables as cell size, capillarity, PO_2 at the cell surface, and temperature. They showed that for Antarctic red-blooded notothenioids, fiber size was the important variable. In small fibers from *T. newnesi,* there was no oxygen limitation right to the center of the cell, whereas in the larger red fibers from *N. coriiceps*, PO_2 at the center of the cell was about half of that at the periphery, a significant drop, but still presenting sufficient O_2 for aerobic production of ATP. In the icefish *Chaenocephalus aceratus*, the model predicted that much of the interior of the cell would be severely hypoxic. The low capillary supply in Antarctic fish is clearly a major limitation, particularly at higher temperatures, as sub-Antarctic notothe-nioids have a much higher capillarity. This allowed Egginton's model to predict that red-blooded sub-Antarctic notothenioids at 4°C would show similar O_2 gradients across the cell to that seen in Antarctic fish at subzero temperatures. At higher temperatures (10°C), fiber size becomes important, with the centers of larger fibers becoming hypoxic. The sub-Antarctic icefish *Champsocephalus esox* suggested severe intracellular hypoxia at 4°C, and even much anoxia at 10°C. The model presented by Egginton *et al.* (2002) goes a long way to explaining O_2 diffusion through muscle cells of red-blooded fish, although how icefish manage, particularly at higher temperatures, remains a mystery.

III. MECHANICAL AND PHYSIOLOGICAL PROPERTIES OF ISOLATED MUSCLE

Fish skeletal muscle has a typical structure of thick (myosin) and thin (actin) filaments that interact to produce crossbridges that in turn generate the tension for contraction. A great deal of work has investigated the physiology and mechanics of contraction using single isolated muscle fibers. Much of the work has concentrated on mammalian tissues, although fish muscle has received a good deal of attention. This has extended to work on Antarctic fish, although virtually all of the work has come from Ian Johnston's lab. Early work was carried out on chemically skinned fibers, ensuring that the contractile apparatus was being investigated without any potential contamination from cell membranes, including mitochondria, potentially important in fibers having mitochondrial volumes of 50% or more (Johnston and Brill, 1984; Johnston and Altringham, 1985; Johnston and Harrison, 1985). Later work has used intact living fibers, achieving similar results (Johnson and Johnston, 1991; Franklin and Johnston, 1997). McVean and Montgomery (1987) used isolated strips of muscle without stating how many fibers might be involved.

Isolated fibers can be stimulated tetanically at a fixed length to produce maximum isometric tension (force). Different fiber types produce different tensions, with white muscle greater than red, which in turn is greater than SDF (Altringham and Johnston, 1986a). Antarctic fish require a lower stimulus frequency to induce tetanus (~50 Hz) compared with temperate (80 Hz) or tropical water fish (400 Hz) (Johnson and Johnston, 1991). Irrespective of the fish or habitat temperature, isometric tension is temperature sensitive, with maximum tension produced at the habitat temperature of the species (Altringham and Johnston, 1986b; Johnson and Johnston, 1991). Johnson and Johnston (1991) showed nicely that in species living at each end of the temperature spectrum, maximum tension was developed over a narrow range of temperatures, illustrating the stenothermal nature of muscle mechanics in both tropical and Antarctic fish. Temperate-water fish, however, showed relatively broad peaks, allowing maximum tension to be developed over at least a 10°C range. Maximum isometric tension is similar in all fish at their habitat temperature at around 200 kN m^{-2} (Johnston and Brill, 1984; Altringham and Johnston, 1986a; Johnson and Johnston, 1991; Franklin, 1998), although a few studies have suggested that Antarctic fish might produce higher tensions (Johnston and Altringham, 1985, 1988). Interestingly, although maximum tension appears to be temperature independent, the fuel used during maintenance of this shows a typical Q_{10} response, indicating little thermal compensation. Myofibrillar ATPase activities of stimulated fibers plotted against temperature all fall on the same line irrespective of species or habitat temperature, with a Q_{10} of about 1.6 (Johnston and Altringham, 1988). Thus, to achieve the same tension, warmer fish must spend more energy. It is cheaper for Antarctic fish to maintain tension, with a ratio of 1:2:4 Antarctic:temperate:tropical. It is worth noting that fibers from Antarctic fish fail to relax fully at temperatures above 5°C because of the formation of abnormal crossbridges. Myofibrillar ATPases *in vitro* from these fish are thermally unstable (Johnston *et al.*, 1975). Fish muscle myosin is capable of change to suit thermal conditions. For example, Goldspink (1995) isolated 28 clones for the myosin heavy chain of carp, several of which were associated with acclimation temperature. In a similar study, Gauvry *et al.* (2000) isolated heavy chains from Antarctic and tropical fish and noted distinct differences. In particular, they noted similarities between the tropical white muscle and the Antarctic red muscle chains and suggested that this went some way to explaining the different ATPase characteristics of the muscles.

A very different picture emerges with unloaded isotonic contraction. Maximum contraction velocity (V_{max}) is correlated with temperature, irrespective of species (Q_{10} 1.8–2.0), thus showing no thermal compensation (Johnston and Brill, 1984; McVean and Montgomery, 1987; Johnston and Altringham, 1988; Franklin, 1998). This means that isolated Antarctic fish muscle contracts much more slowly at its habitat temperature than muscle from temperate or tropical fish at their habitat temperatures, about 1 muscle fiber length s^{-1} for Antarctic fish compared with 16 muscle fiber lengths s^{-1} for tropical fish (Johnson and Johnston, 1991). This affects force–velocity curves such that power generation (W kg^{-1}) of Antarctic fish muscle is only 60% of that of temperate fish (Johnston and Brill, 1984).

It has been recognized that experiments looking at isometric and isotonic contraction in isolated muscles say little about how the muscles might function in the living animal, where a combination of the two are seen in any cycle of contraction, especially in swimming muscle of fish. Methods have been devised for recording muscle contraction in free swimming fish, then using this to stimulate isolated muscle fibers, through simulations of cycles of contraction and relaxation (Rome and Swank, 1992; Swank and Rome, 1999). This technique has been applied to Antarctic fish white muscle (Franklin and Johnston, 1997). The authors noted a maximum speed of shortening of 1.7 fiber lengths^{-1} in *N. coriiceps*. Stretching of muscle fibers before activation caused significant force enhancement during contraction, giving greater power outputs than seen in traditionally stimulated muscles. Interestingly, a comparison with an earlier study on the temperate sculpin (*Myoxocephalus scorpius*) showed that despite a 15°C difference and the sculpin having a V_{max} of 3.8 fiber lengths s^{-1}, both species showed maximum instantaneous power outputs of 250–300 W kg^{-1} and mean power outputs of 15–25 W kg^{-1} (Johnston *et al.*, 1995).

IV. PROTECTING THE CELL

There is an inverse relationship between dissolved oxygen and temperature in seawater. Thus, the cold waters inhabited by Antarctic fish contain high levels of O_2. Oxygen at high concentrations poses problems in terms of generation of oxygen-free radicals and it has long been speculated that Antarctic fish must possess high levels of antioxidants to protect their tissues. Life at cold temperatures requires increased levels of low-molecular-weight polyunsaturated lipids in membranes, making them more vulnerable to free radical attack (Hazel and Williams, 1990). In addition, because the aerobic generation of high-energy phosphates involves shuttling of electrons, the very high numbers of mitochondria in many tissues, especially red muscle, increase the likelihood of free radical damage (Gieseg *et al.*, 2000). Antarctic fish have levels of vitamins C and E up to six times higher than those in temperate-water fish (Giardina *et al.*, 1997; Ansaldo *et al.*, 2000; Gieseg *et al.*, 2000). Ansaldo *et al.* (2000) also measured high levels of the antioxidant enzymes superoxide dismutase, catalase, and glutathione peroxidase in several tissues, including muscle. Vitamin E is present in Antarctic fish tissues in much greater quantities than in marine mammals, 52 pmol mg tissue^{-1} in pectoral muscle of the icefish *Champsocephalus gunnari* compared with 20 pmol mg tissue^{-1} in the minke whale *Balaenoptera acutorostrata* (Dunlap *et al.*, 2002), with the highest levels (417 pmol mg tissue^{-1}) recorded in the liver of *Gobionotothen gibberifrons*. The predominant form of vitamin E is α-tocopherol, although in Antarctic fish, up to 20% is present in a form known as *marine-derived tocopherol* (MDT), thought to have greater activity at low temperatures (Dunlap *et al.*, 2002). Red muscle vitamin E content is four to five times higher than in white muscle, reflecting the differences in mitochondrial volume. The high levels of vitamin E are acquired via the diet, with Antarctic plankton, especially krill (*Euphausia superba*) having high levels of both α- and MDT-tocopherol (Dunlap *et al.*, 2002). Of interest is that although the whale sampled in this study consumes large quantities of Antarctic plankton, MDT-tocopherol is not a major constituent of its muscle vitamin E content, emphasizing its role in both Antarctic invertebrates and fish as a cold-water antioxidant.

The buffering capacity of Antarctic fish muscle is not particularly different from that seen in temperate-water fish. White muscle has a buffering capacity of 50–60 slykes, with red muscle up to half this value, putting the buffering capacity of Antarctic fish on a par with benthic and deep water temperate fish (Wells *et al.*, 1988). Intracellular pH of white muscle of the spiny plunderfish, *Harpagifer antarcticus*, was determined noninvasively using nuclear magnetic resonance spectroscopy (NMRS) to be 7.36 at 1°C (Moerland and Egginton, 1998). Although the data appear to indicate that pHi of Antarctic fish muscle fits the alphastat model, the authors used a review of measured values of glycolytic muscle from a number of non-Antarctic species to show that muscle does not follow the model. However, using NMRS techniques on both Antarctic and temperate-water zoarcids (eelpouts), Bock *et al.* (2001) showed that the Antarctic fish had a resting pHi of 7.4 at habitat temperature, decreasing by 0.015 pH units per degree increase in temperature, thus clearly following the alphastat model. They warn against using interspecific comparisons to determine temperature-dependent pH shifts. Similar values were obtained using invasive techniques (homogenates) with the Antarctic eelpout by Van Dijk *et al.* (1999). Following their work, Mark *et al.* (2002), also using Antarctic eelpouts, using a range of techniques including NMRS, showed a pHi at 0°C of 7.41 and a change of -0.012 pH units per degree increase, following alphastat rules. This only held up to 6°C. The authors also used their noninvasive techniques to measure relative blood flow in the dorsal aorta and discovered that the failure of pH regulation at high temperature was linked to maximum blood flow, indicating a failure of the cardiovascular system to supply sufficient oxygen to the tissues and a move to anaerobic metabolism. Further details relating to pHi in Antarctic fish muscle can be found in Chapter 3.

V. EXERCISE AND ENERGY SUPPLY

Notothenioid fish use labriform locomotion for low-speed swimming and subcarangiform for high-speed sprinting (Montgomery and Macdonald, 1984; Forster *et al.*, 1987), and there is a spatial separation of the major muscle groups involved, with red muscle in the pectoral fins and white muscle in the myotome, making analysis of their functions straightforward. Antarctic fish are not noted for their swimming ability with critical swimming speeds (U_{crit}) up to 2.5 body lengths per second (bl s^{-1}) and maximum speeds using pectoral fins alone of 0.8–5 bl s^{-1} (Archer and Johnston, 1989). However, it has been suggested that this indifferent performance is associated with the labriform mode of locomotion, rather than specifically to the fish or to temperature (Webb, 1973; Davison, 1988; Tuckey and Davison, 2004). Certainly high-speed exercise is limited and is associated with a relative inability to produce lactic acid in the white muscle fibers (Davison et al., 1988; Egginton, 1994, 1997a; Lowe and Wells, 1997; Tuckey and Davison, 2004) and a reliance on existing levels of phosphocreatine (PCr) and ATP for sprinting (Dunn and Johnston, 1986; Archer and Johnston, 1991; Lowe and Wells, 1997). There is a good deal of information in the literature relating to the activity of enzymes of energy production in these cold-water fish, often associated with speculation about adaptation during movement of the ancestral stock from a higher to a lower temperature. Guderley (1998) indicated that there are at least two strategies relating to cold acclimation in fish. Trout and whitefish, for example, show increased levels of glycolytic enzymes with cold acclimation, whereas goldfish and striped bass show decreases. This is inversely related to mitochondrial volume changes during acclimation and is associated with the activity of the animals at their new (lower) habitat temperature. The very large mitochondrial volumes in notothenioid muscle suggest that reduced glycolytic activity would be expected.

There is compelling evidence that cold adaptation has maintained aerobic capacity in the muscles of Antarctic fish. Activities of mitochondrial enzymes are certainly much higher than would be expected by simply extrapolating activities of temperate-water fish enzymes down to subzero temperatures using a Q_{10} of about 2 (Johnston and Harrison, 1985; Dunn and Johnston, 1986; Dunn, 1988a; Crockett and Sidell, 1990; Kawall et al., 2002; Pörtner, 2002). Activities are as high at 0 °C in Antarctic fish as can be found in tench, trout, and even tuna at their respective operating temperatures (Johnston and Harrison, 1985). Walesby and Johnston (1979) noted high activities of hexokinase in red fibers and suggested that this was indicative of a preference for glucose as a substrate, although this does not tie in with an apparent lack of adaptation of glycolytic enzymes. An early study on respiration of homogenized tissues indicated that lipid was a preferred fuel for red muscle (Lin et al., 1974). Enzymes involved with lipid metabolism are high in red muscle, in particular the use of monoenoic fatty acids (Crockett and Sidell, 1990; Sidell et al., 1995; Sidell and Hazel, 2002). Monounsaturated and polyunsaturated fatty acids make up more than 80% of the lipids present in red muscle (Kamler et al., 2001), whereas red muscle fatty acid binding proteins are highly conserved in icefish (74% identity with mammalian FABPs) (Londraville et al., 1995). However, it appears that rather than

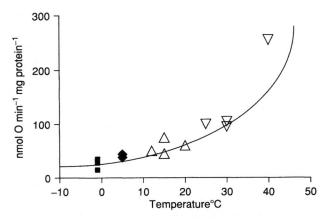

Fig. 8.1. Maximum state 3 respiration of red muscle mitochondria using pyruvate as a substrate measured at the habitat temperature of the animal. ■ Antarctic notothenioids, ◆ sub-Antarctic notothenioids, △ temperate water fish, ▽ tropical fish. (Data from Guderley, 1998; and Johnston *et al.*, 1998.)

cold adaptation influencing the structure of the major mitochondrial enzymes, notothenioids have maintained aerobic capacity by increasing the numbers of enzymes, thus increased expression or differences in turnover of the same isoenzymes (Johnston *et al.*, 1994, 1998). This seems to be a feature of all fish mitochondrial enzymes, irrespective of species or habitat temperature, as all isolated mitochondrial preparations oxidizing substrates at a maximal rate fit on a single line when rate is plotted against temperature (Johnston *et al.*, 1994; Guderley, 1998, 2004) (Figure 8.1). However, there have been some modifications to Antarctic fish mitochondria, as Weinstein and Somero (1998) showed that *Arrhenius* break temperatures for oxygen consumption of isolated mitochondria are the lowest reported for any animal. Assuming that cold adaptation has involved increases in numbers of mitochondrial enzymes, then this requires changes to the mitochondria themselves. The strategy employed by notothenioids has been to increase numbers of mitochondria (seen as an increase in mitochondrial volume density) rather than packing more enzymes into individual mitochondria (which would be seen as an increase in mitochondrial cristae density), hence the very high mitochondrial volume seen in Antarctic fish muscle (Johnston *et al.*, 1994, 1998; Pörtner, 2002). Cristae density is actually lower in icefish than in most other fish muscles (O'Brien *et al.*, 2003).

There is general agreement that anaerobic metabolism is not cold adapted in Antarctic fish, although the data are somewhat equivocal. Early work (Dunn and Johnston, 1986; Dunn *et al.*, 1989; Crocket and Sidell,

1990) has indicated that activity levels of pyruvate kinase (PK) and lactate dehydrogenase (LDH) are low compared with those of temperate-water fish. However, Bacila *et al.* (1989) measured extremely high activities of LDH in muscle, brain, and heart of icefish, and Kawall *et al.* (2002) showed high levels in brain and red and white muscle. In this latter study, brain activities of LDH were higher in eight species of Antarctic fish compared with a similar number of tropical species, although all assays were carried out at 10°C. Nevertheless, extrapolation to their habitat temperatures indicated that cold compensation had occurred. LDH levels were low in Antarctic fish red muscle but equivalent to tropical fish in white muscle, particularly in icefish (Kawall *et al.*, 2002). The authors concluded that there had been cold adaptation in LDH, and it showed up well in brain because that organ is not subject to the vagaries of locomotion style and ecotype. The mode of life of notothenioids meant that it was not so obvious in muscle. At habitat temperatures, the Km (pyruvate) for LDH-A orthologues is similar in Antarctic fish and a range of temperate- and tropical-water fish (Hochachka and Somero, 2002), but see Fitch (1988, 1989) for different values. However, for the Antarctic fish, Km rises very rapidly as temperature rises, especially more than 5°C, suggesting that the enzyme is unstable above this temperature (Somero, 1998; Sharpe *et al.*, 2001). These changes to LDH activity in Antarctic fish are related to changes in flexibility of mobile regions of the molecule associated with the overall structure of the molecule (Fields and Somero, 1998; Fields, 2001; Hochachka and Somero, 2002). The actual pyruvate binding regions of the enzyme appear to be highly conserved (Hochachka and Somero, 2002; Somero, 2003). Marshall *et al.* (2000) determined that there were essentially no sequence differences between Antarctic and temperate notothenioids and that the small changes that were noted were not associated with the active site. They suggest that performance changes to the enzyme are a consequence of subtle changes to the enzyme's environment, and Somero (2003) has indicated that ionic composition of the cytoplasm is crucial for activity.

A study by Tuckey and Davison (2004) has highlighted the inability of Antarctic fish muscle to produce lactic acid. Neither exercise to U_{crit}, nor direct electrical stimulation of isolated muscle blocks elicited changes in lactate levels in red or white muscles in *P. borchgrevinki*. Two temperate-water wrasses, both labriform swimmers, produced significant rises in lactate in white myotomal muscle following exercise, as well as increases in both red (pectoral) and white muscle following electrical stimulation. This was perhaps expected in red muscle where LDH activities were low in all three fish. It was particularly low in *P. borchgrevinki* measured at 2°C. However, in white muscle, LDH activities were up to three times higher in *P. borchgrevinki* at 2°C than seen in the wrasse muscles measured at 15°C

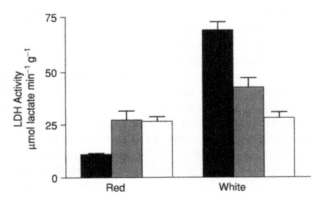

Fig. 8.2. Lactate dehydrogenase (LDH) activity in red and white muscle from an Antarctic fish (black bars) and two temperate-water wrasses (gray and white bars) measured at their habitat temperatures. (Data from Tuckey and Davison, 2004.)

(Figure 8.2). This study, and others, has demonstrated that *in vitro*, Antarctic fish white muscle is capable of converting pyruvate to lactate and can do so as well as temperate-water fish muscle. In the intact muscle fiber, however, this does not happen to any great extent. Presumably there is some other constraint other than the final step in glycolysis, although what this is has not been elucidated. It has been suggested that some aspects of notothenioid physiology are phylogenetically determined, rather than a consequence of living at subzero temperatures (Fernandez *et al.*, 2000; Egginton *et al.*, 2001), and the restricted ability to produce lactic acid may be an example of this. Certainly, Antarctic eelpouts produce significant amounts of lactate during exhaustive exercise, equivalent to temperate-water relatives, and enough to influence intracellular pH (Hardewig *et al.*, 1998). Recovery was much faster in the Antarctic eelpouts, suggesting compensation of LDH in this species.

VI. SWIMMING

Muscle makes up a significant proportion of the body of a fish and is used for propelling the animal through its dense environment, so ultimately the performance of the fish must relate to how this tissue functions in the living animal. However, although there is a great deal of information on structural and biochemical aspects of muscles in Antarctic fish, there is surprisingly little information on the ability of these cold-water fish to swim. Much of the problem lies with the fish themselves. The ancestral notothenioid stock was benthic and heavy and lacked a swim bladder. Most of the

Antarctic notothenioids have retained these characteristics and have remained benthic fish. They are sit-and-wait predators, show little spontaneous activity in aquarium tanks, and do not swim in exercise tunnels. They do respond to chasing, and hence, there are data relating to anaerobic exercise (Dunn and Johnston, 1986; Egginton, 1994, 1997a; Lowe and Wells, 1997). Some notothenioids have become pelagic (Eastman, 1993, 1997), and there is the potential for using these animals to investigate aspects of sustained swimming. Almost all of the work in this area, however, has focused on a single species, the cryopelagic *P. borchgrevinki*.

With the exception of *P. antarcticum*, Antarctic fish move using labriform locomotion. Sustained swimming in Antarctic fish can be maintained using the large pectoral fins up to about $1.5 \, \mathrm{bl} \, \mathrm{s}^{-1}$ (Montgomery and Macdonald, 1984; Forster *et al.*, 1987; Davison *et al.*, 1988, 1990; Archer and Johnston, 1989) after which the fish is forced to switch to carangiform locomotion, leading to rapid fatigue, without much increase in swimming speed. Pectoral fin beat is positively correlated with swimming speed up to about $2 \, \mathrm{Hz}$, though typically, maximum fin beat is reached before U_{crit} (Archer and Johnston, 1989; Forster *et al.*, 1987). In a study of free living fish attracted to a bait station, Yau *et al.* (2002) noted that the demersal toothfish *D. eleginoides* (mean length 700 mm) swam at a mere $0.22 \, \mathrm{bl} \, \mathrm{s}^{-1}$ but were capable of rapid subcarangiform locomotion of $3 \, \mathrm{bl} \, \mathrm{s}^{-1}$ when startled. Labriform locomotion in Antarctic fish is cost effective, averaging $0.095 \, \mathrm{ml} \, O_2 \, \mathrm{kg}^{-1} \, \mathrm{m}^{-1}$, compared with values of 0.07–0.26 measured for a range of temperate-water teleosts (Jones and Randall, 1978; Forster *et al.*, 1987). Antarctic fish, even the pelagic species are not noted athletes. U_{crit} values are low and metabolic scope for activity is only 3.9–5.7 (Forster *et al.*, 1987; Dunn *et al.*, 1989). Sustained swimming is largely aerobic with lactic acid not featuring as an end product, even at U_{crit} velocities (Dunn and Johnston, 1986; Forster *et al.*, 1987; Egginton, 1997a; Tuckey and Davison, 2004), and as a consequence, recovery is relatively rapid, at least in red-blooded notothenioids (Davison *et al.*, 1988; Lowe and Wells, 1997). Though regarded as stressful, swimming to U_{crit} does not elicit catecholamine release, not surprising, as a lack of a catecholamine response is a general feature of notothenioids to any stressor other than high temperatures (Egginton and Davison, 1998; Forster *et al.*, 1998).

Sustained swimming is aerobic, which ultimately is determined by the ability of the gills and cardiovascular system to supply oxygen to the tissues (Axelsson, 1992, 1994; Egginton, 1997b). This was highlighted in a study of the swimming ability of *P. borchgrevinki* affected by gill disease. Maximum oxygen uptake (VO_2 max) was directly related to the extent of gill damage, oxygen uptake being severely compromised at high levels of the disease. This in turn determined sustained swimming ability (Davison *et al.*,

1990). Sustained swimming certainly increases blood flow to the working muscles (Egginton, 1997b), though interestingly, in this study, the increased flow was accomplished entirely by increased cardiac output, rather than a redistribution from other tissues, such as the viscera, as seen during exercise in trout (Thorarensen *et al.*, 1993).

There have been a number of studies on exhaustive exercise in Antarctic fish, mainly showing little change to the muscles or transient changes with rapid recovery. Muscle tissue at exhaustion shows major drops in high-energy phosphates (ATP, PCr) with corresponding increases in Pi and IMP (Dunn and Johnston, 1986; Lowe and Wells 1997). These high-energy phosphates are clearly the source of energy for much of the work done, as there is little change to muscle levels of lactate or glycolytic intermediates (Dunn and Johnston, 1986; Davison *et al.*, 1988; Egginton *et al.*, 1994, 1997; Lowe and Wells, 1997). The pH of blood can be affected, although this has been shown to be a short-lived respiratory acidosis, rather than metabolic (Egginton, 1997a).

Antarctic fish are able to accelerate rapidly from a standing start, a process often termed the *startle response*, or *C*-start. This involves maximum myotomal white muscle contraction for several tail beats, lasts for less than a second, and is used for predator avoidance and prey capture. It is certainly not exhaustive exercise, as just described, relying only on PCr and on-board ATP for fuel. The startle response is well developed in Antarctic fish (Archer and Johnston, 1989; Franklin and Johnston, 1997; Franklin *et al.*, 2003), reaching maximum velocities (U_{max}) of 1–$2\,m\,s^{-1}$, within 60–$200\,ms$ (Montgomery and Macdonald, 1984; Wilson *et al.*, 2001). Acceleration is comparable to that seen in temperate-water fish (Beddow *et al.*, 1995; Temple and Johnston, 1998; Franklin *et al.*, 2003), whereas U_{max} is similar to velocities seen in cold-water species, but less than seen in tropical fish (Wilson *et al.*, 2001). The limited available data suggest a correlation between both acceleration and U_{max}, and ecotype, with pelagic fish having the better fast starts (Wilson *et al.*, 2001). Franklin *et al.* (2003) expanded the work of Wilson *et al.* (2001), examining the startle response of five Antarctic fish. They noted that habitat was important for fast starts, with pelagic fish having greater acceleration and maximum velocity compared with benthic forms. Within the Nototheniidae, maximum velocity was correlated with the shape of the animal expressed as a ratio of body mass to total length (Figure 8.3). Thus, the streamlined *P. borchgrevinki* had the best startle response, whereas the benthic *T. bernacchii* showed a relatively poor response. The benthic fish *Gymnodraco acuticeps*, a notothenioid within the family Bathydraconidae, did not fit the regression because of its elongated body form, especially the large triangular head. Interestingly though, when maximum velocity was plotted against buoyancy, using values taken from

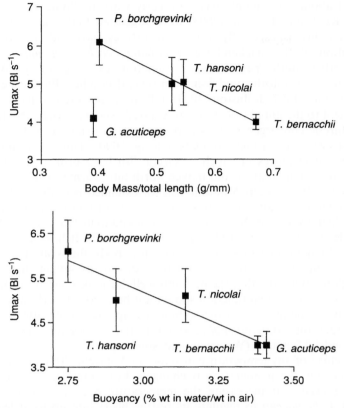

Fig. 8.3. Relationship between (a) body mass/length ratio and (b) buoyancy and maximum swimming velocity (U_{max}) for a number of nototheniid fish and a dragonfish (*Gymnodraco acuticeps*). (Data from Franklin *et al.*, 2002.)

Eastman (1993), all five species fitted the regression well. Eastman (1993) had shown that buoyancy was related to the degree of pelagism shown by the fish (Figure 8.3).

VII. EXERCISE AND TEMPERATURE

There has been an increase in interest in how Antarctic fish (and other organisms) might fare in an environment of increasing water temperatures (Peck, 2002; Pörtner, 2002; Davison and Franklin, 2003). There is a general consensus that Antarctic fish are extreme stenotherms with upper lethal limits of about 5°C, mainly based on early work by Somero and DeVries (1967). Kock and Everson (2003) have suggested that changing environmental conditions, especially rising temperatures, have had an effect on the distribution of the mackerel icefish (*Champsocephalus gunnari*), pushing them further south. Yet, there are numerous examples of sub-Antarctic distributions of notothenioid fish from several families. Phylogenetic trees (Ritchie *et al.*, 1997; Johnston, 2003) indicate that these fish have emerged subsequent to the initial radiations in Antarctic waters, moving north across the Antarctic convergence. This is backed up by evidence of the presence of antifreeze genes in these animals (Cheng, 2000). Thus, there must have been several instances in which Antarctic fish have broken free from the constraints of temperature and moved north into warmer water.

It is possible to keep Antarctic fish for extended periods at 4 °C. These high temperatures cause decreases in osmolarity of up to 150 mOsm. This is a controlled decrease in Na^+ and Cl^- ions linked to increases in gill ATPases (Gonzalez-Cabrera *et al.*, 1995; Guynn *et al.*, 2002). It has been suggested that the high osmolarities of Antarctic fish body fluids at their subzero habitat temperatures are associated with energy-saving mechanisms, although the potential that the high intracellular ionic concentrations could be necessary for protein function should not be overlooked (Somero, 2003), especially as changes to plasma sodium are apparent even after a week at high temperatures (Guynn *et al.*, 2002). Life at 4 °C does not apparently alter mitochondrial function (Weinstein and Somero, 1998).

The effect of any temperature increase on Antarctic fish can depend very much on how the temperature is changed. For example, Axelsson *et al.*, (1992) showed that increasing the temperature over a 40-minute period allowed *T. bernacchii* to regulate its resting heart rate up to at least 3 °C, after which heart rate climbed rapidly. Being able to control heart rate allows maintenance of cardiac scope and, thus, scope for activity. On the other hand, acute transfer of fish into 3 °C water produced a rapid increase in heart rate, narrowing the gap between resting and maximum heart rate and having detrimental effects on swimming performance (Wilson *et al.*, 2002; Lowe *et al.*, 2005).

For many of the benthic ambush predators that make up the bulk of Antarctic notothenioids, sustained swimming is not critical, but fast starts and high acceleration are very important in catching prey and escaping from predators. The fast-start response seen in two species of benthic nototheniid from McMurdo Sound was shown to be independent of temperature over at least a 9°C temperature range (Wilson *et al.*, 2001). In the same study, the response was independent of temperature over a 12 °C range in the pelagic *P. borchgrevinki*, and there was no difference between these fish acclimated for several weeks to their habitat temperature (−1 °C) or to 4 °C

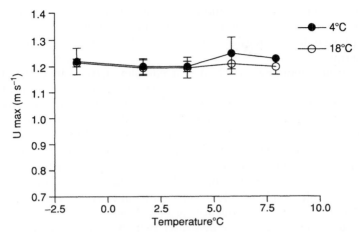

Fig. 8.4. Maximum velocity (U_{max}) during fast start for *Pagothenia borchgrevinki* in animals acclimated to $-1\,°C$ and $4\,°C$ (data from Wilson *et al.*, 2001).

(Figure 8.4). Acute changes to temperature have little effect on intracellular levels of PCr in Antarctic eelpouts (Van Dijk *et al.*, 1999). Assuming that this also applies to Antarctic notothenioids, it indicates that there is a constant fuel supply to the white muscle irrespective of the temperature of the water into which fish are placed. The lack of a Q_{10} response in the startle response, and the lack of any differences between the two acclimation groups indicate that there is some restraining mechanism, fixing the rate of energy consumption and muscle contraction, although what this might be is unknown. In the two benthic fish *T. bernacchii* and *T. pennellii*, there was some indication of impaired acceleration and lower U_{max} at the upper end of the temperature range, similar to that seen in cold-temperate cottids (Temple and Johnston, 1998; Wilson *et al.*, 2001). This may be associated with a failure of the cardiovascular system to supply sufficient oxygen to the rapidly metabolizing tissues, leading to a fall in PCr levels in white muscle, before the induced C-start, although with the limited information available, other factors such as failure of nervous transmission at high temperature, impaired excitation coupling within the muscle fibers, or depletion of fuel stores such as glycogen cannot be discounted. Fernandez *et al.* (2002) looked at fast-start performance in *Eleginops maclovinus*, a South American non-Antarctic notothenioid, and showed a distinct effect of temperature. However, their fish were acclimated to a relatively high temperature (8.5°C), most of the recordings were taken at temperatures below this, and much of the Q_{10} effect seemed to be a failure of the animals to perform at 2 and 4°C. Clearly more work is needed in this area, though the available evidence suggests that C-starts, involving contraction of white muscle utilizing available high-energy phosphates, do not fit the assumption that Antarctic fish are extreme stenotherms.

Sustained swimming (to U_{crit}) is a complex activity involving mainly red muscle working aerobically, with energy derived primarily from mito-chondria. Far more steps are involved than in sprint swimming, involving supply of oxygen to the working muscles (involving both the gills and cardiovascular system), transport of oxygen from the blood to the working muscles, making oxygen available to the mitochondria, and the many enzy-matic steps involved in taking a molecule of lipid or carbohydrate and converting it to ATP, water and carbon dioxide. By contrast, sprinting involves a simple supply of ATP to myosin from existing cytoplasmic stores of ATP and PCr. As a consequence, sustained swimming appears to be very temperature sensitive. In *P. borchgrevinki*, swimming to U_{crit} is severely com-promised at temperatures above 2°C, and the animals are essentially unable to swim at temperatures at 8°C or higher, apparently confirming the ste-nothermal nature of the fish (Figure 8.5). A very different picture emerges if the fish are acclimated for a month at 4°C (Seebacher *et al.*, 2005). There are signs that swimming ability may actually be impaired at −1°C, the fish swim well up to at least 8°C. and the animals are even able to swim, albeit poorly, at 10°C. A remarkable aspect of this work is that fish acclimated to −1°C lose equilibrium after 10 minutes at 10°C and release large quantities of catechola-mines, the only stimulus guaranteed to elicit catecholamine release in Antarctic

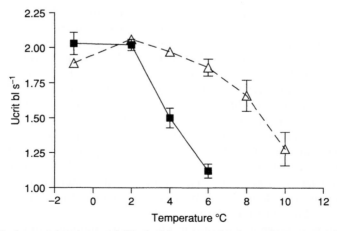

Fig. 8.5. Maximum swimming speed (U_{crit}) of *Pagothenia borchgrevinki* at a range of tempera-tures. ■ Fish acclimated to −1°C, △ fish acclimated to 4°C (data from Seebacher *et al.*, 2005).

fish (Egginton and Davison, 1998; Forster *et al.*, 1998). The fish kept at 4°C, however, were able to tolerate a 30-minute settling-in period at 10°C and then were able to swim for at least another 30 minutes. Indeed, several fish were even able to survive in 12°C water for up to 45 minutes but were unwilling to swim. Indications are that the acclimated fish have greater aerobic enzyme activities, though as discussed earlier, whether this is due to increased numbers of enzyme units or production of isoenzymes is not known. Preliminary information indicates that changes to the cardiovascular system are crucial to the success of this enhanced performance, in particular a resetting of resting heart rate and cardiac output, thus maintaining cardiac scope at elevated temperatures. Fish maintained at −1°C have a much reduced scope for activity at temperatures above 4°C and, thus, must rely increasingly on anaerobic metabolism during exercise. As has been discussed, this is not an available option for notothenioids, hence the impaired performance. Following acclimation to higher temperatures, resting heart rate and cardiac output are reset to levels expected of fish living at −1°C, presumably maintaining scope for activity at these new elevated temperatures.

These preliminary data are exciting in that they show that given time to acclimate, Antarctic notothenioids are able to adjust their physiology to suit their new environment. In other words, they acclimate rather than just tolerate elevated temperatures, and this has significance in terms of global warming. What this means in terms of changes to muscles or even other systems awaits discovery. In addition, the enhanced swimming abilities of these fish at temperatures well above their acclimation temperature suggests that 5°C may not be their upper lethal limit. This brings into question the concept that Antarctic notothenioids are extreme stenotherms. Given enough time at higher temperatures, we may find that these fish have a temperature range equivalent to that of many eurythermal animals.

REFERENCES

Altringham, J. D., and Johnston, I. A. (1986a). Evolutionary adaptation to temperature in fish muscle cross bridge mechanisms: Tension and ATP turnover. *J. Comp. Physiol. B* **156**, 819–821.

Altringham, J. D., and Johnston, I. A. (1986b). Energy cost of contraction in fast and slow muscle fibres isolated from an elasmobranch and an Antarctic teleost fish. *J. Exp. Biol.* **121**, 239–250.

Ansaldo, M., Luquet, C. M., Evelson, P. A., Polo, J. M., and Llesuy, S. (2000). Antioxidant levels from different Antarctic fish caught around South Georgia Island and Shag Rocks. *Polar Biol.* **23**, 160–165.

Archer, S. D., and Johnston, I. A. (1989). Kinematics of labriform and subcarangiform swimming in the Antarctic fish *Notothenia neglecta*. *J. Exp. Biol.* **143**, 195–210.

Archer, S. D., and Johnston, I. A. (1991). Density of cristae and distribution of mitochondria in the slow muscle fibers of an Antarctic fish. *Physiol. Zool.* **64,** 242–258.

Ashford, J., Jones, C., Bobko, S., and Everson, I. (2002). Length-at-age in juvenile Patagonian toothfish (*Dissostichus eleginoides*). *CCAMLR Sci.* **9,** 1–10.

Axelsson, M., Davison, W., Forster, M. E., and Farrell, A. P. (1992). Cardiovascular responses of the red-blooded Antarctic fishes *Pagothenia bernacchii* and *P. borchgrevinki*. *J. Exp. Biol.* **167,** 179–201.

Axelsson, M., Davison, W., Forster, M. E., and Nilsson, S. (1994). Blood pressure control in the Antarctic fish *Pagothenia borchgrevinki*. *J. Exp. Biol.* **190,** 265–279.

Bacila, M., Rosa, R., Rodrigues, E., Lucchiari, P. H., and Rosa, C. D. (1989). Tissue metabolism of the ice-fish *Chaenocephalus aceratus* Loenburg. *Comp. Biochem. Physiol.* **92B,** 313–318.

Battram, J. C., and Johnston, I. A. (1991). Muscle growth in the Antarctic teleost, *Notothenia neglecta* (Nybelin). *Antarctic Sci.* **3,** 29–33.

Beddow, T. A., and McKinley, R. S. (1999). Importance of electrode positioning in biotelemetry studies estimating muscle activity in fish. *J. Fish Biol.* **54,** 819–831.

Beddow, T. A., Van Leeuwen, J. L., and Johnston, I. A. (1995). Swimming kinematics of fast starts are altered by temperature acclimation in the marine fish *Myoxocephalus scorpius*. *J. Exp. Biol.* **198,** 203–208.

Bernardi, G., and Goswami, U. (1997). Molecular evidence for cryptic species among the Antarctic fish *Trematomus bernacchii* and *Trematomus hansoni*. *Antarctic Sci.* **9,** 381–385.

Bock, C., Sartoris, F. J., Wittig, R. M., and Pörtner, H. O. (2001). Temperature-dependent pH regulation in stenothermal Antarctic and eurythermal temperate eelpout (Zoarcidae): An *in vivo* NMR study. *Polar Biol.* **24,** 869–874.

Brodeur, J. C., Peck, L. S., and Johnston, I. A. (2002). Feeding increases MyoD and PCNA expression in myogenic progenitor cells of *Notothenia coriiceps*. *J. Fish Biol.* **60,** 1475–1485.

Brodeur, J. C., Calvo, J., Clarke, A., and Johnston, I. A. (2003). Myogenic cell cycle duration in Harpagifer species with sub-Antarctic and Antarctic distributions: Evidence for cold compensation. *J. Exp. Biol.* **206,** 1011–1016.

Camm, J.-P., and Johnston, I. A. (1985). Developmental changes in the differentiation of muscle fibres in the Antarctic teleost *Notothenia neglecta*. *J. Physiol.* **367,** 98P.

Cashon, R. E., Vayda, M. E., and Sidell, B. D. (1997). Kinetic characterization of myoglobins from vertebrates with vastly different body temperatures. *Comp. Biochem. Physiol.* **117B,** 613–620.

Cheng, C.-H. C. (2000). Antifreeze glycoprotein genes in temperate water notothenioid fish infer an Antarctic origin of speciation. *Am. Zool.* **40,** 972–973.

Clarke, A. (1991). What is cold adaptation and how should we measure it? *Am. Zool.* **31,** 81–92.

Crockett, E. L., and Sidell, B. D. (1990). Some pathways of energy metabolism are cold adapted in Antarctic fishes. *Physiol. Zool.* **63,** 472–488.

Davison, W. (1987). The median fin muscles of the leatherjacket, *Parika scaber* (Pisces: Balistidae). *Cell Tissue Res.* **248,** 131–135.

Davison, W. (1988). The myotomal muscle of labriform swimming fish is not designed for high speed swimming. *New Zeal. Nat. Sci.* **15,** 37–42.

Davison, W. (1994). Exercise training in the banded wrasse *Notolabrus fucicola* affects muscle fibre diameter, but not muscle mitochondrial morphology. *New Zeal. Nat. Sci.* **21,** 11–16.

Davison, W., and Goldspink, G. (1977). The effect of prolonged exercise on the lateral musculature of the brown trout (*Salmo trutta*). *J. Exp. Biol.* **70,** 1–12.

Davison, W., and Goldspink, G. (1984). The cost of swimming for two teleost fish. *New Zeal. J. Zool.* **11,** 225–232.

Davison, W., and Macdonald, J. A. (1985). A histochemical study of the swimming musculature of Antarctic fish. *New Zeal. J. Zool.* **12**, 473–483.

Davison, W., and Franklin, C. E. (2003). The Antarctic nemertean *Parborlasia corrugatus*: An example of an extreme oxyconformer. *Comp. Biochem. Physiol.* **134A,** S90.

Davison, W., Franklin, C. E., and Carey, P. W. (1990). Oxygen uptake in the Antarctic teleost *Pagothenia borchgrevinki.* Limitations imposed by X-cell gill disease. *Fish Physiol. Biochem.* **8,** 69–78.

Davison, W., Forster, M. E., Franklin, C. E., and Taylor, H. H. (1988). Recovery from exhausting exercise in an Antarctic fish, *Pagothenia borchgrevinki. Polar Biol.* **8**, 167–171.

Drud Jordan, A. D., Jungersen, M., and Steffensen, J. F. (2001). Oxygen consumption of East Siberian cod: No support for the metabolic cold adaptation theory. *J. Fish Biol.* **59**, 818–823.

Dunlap, W. C., Fujisawa, A., Yamamoto, Y., Moylan, T. J., and Sidell, B. D. (2002). Notothenioid fish, krill and phytoplankton from Antarctica contain a vitamin E constituent (alpha-tocomonoenol) functionally associated with cold-water adaptation. *Comp. Biochem. Physiol. B* **133**, 299–305.

Dunn, J. F. (1988a). Low-temperature adaptation of oxidative energy production in cold water fishes. *Can. J. Zool.* **66**, 1098–1104.

Dunn, J. F. (1988b). Muscle metabolism in Antarctic fish. *Comp. Biochem. Physiol.* **90B,** 539–545.

Dunn, J. F., and Johnston, I. A. (1986). Metabolic constraints on burst-swimming in the Antarctic teleost *Notothenia neglecta. Mar. Biol.* **91**, 433–440.

Dunn, J. F., Archer, S. D., and Johnston, I. A. (1989). Muscle fibre types and metabolism in post-larval and adult stages of notothenioid fish. *Polar Biol.* **9**, 213–223.

Eastman, J. T. (1988). Lipid storage systems and the biology of two neutrally buoyant Antarctic notothenioid fishes. *Comp. Biochem. Physiol.* **90B**, 529–537.

Eastman, J. T. (1993). "Antarctic Fish Biology. Evolution in a Unique Environment." Academic Press, New York.

Eastman, J. T. (1997). Phyletic divergence and specialization for pelagic life in the Antarctic nototheniid fish *Pleuragramma antarcticum. Comp. Biochem. Physiol.* **118A**, 1095–1101.

Eastman, J. T. (2000). Antarctic notothenioid fishes as subjects for research in evolutionary biology. *Antarctic Sci.* **12**, 276–287.

Eastman, J. T., and DeVries, A. L. (1982). Buoyancy studies of notothenioid fishes in McMurdo Sound, Antarctica. *Copeia* **1982**, 385–393.

Eastman, J. T., and DeVries, A. L. (1997). Biology and phenotypic plasticity of the Antarctic nototheniid fish *Trematomus newnesi* in McMurdo Sound. *Antarctic Sci.* **9**, 27–35.

Eastman, J. T., and Clarke, A. (1998). A comparison of adaptive radiations of Antarctic fish with those of non-Antarctic fish. *In* "Fishes of Antarctica, A Biological Overview" (di Prisco, G., Pisano, E., and Clarke, A., Eds.), pp. 3–26. Springer-Verlag, Milan.

Egginton, S. (1994). Stress response in two Antarctic teleosts (*Notothenia coriiceps* Richardson and *Chaenocephalus aceratus* Lonnberg) following capture and surgery. *J. Comp. Physiol. B* **164**, 482–491.

Egginton, S. (1997a). A comparison of the response to induced exercise in red- and white-blooded Antarctic fishes. *J. Comp. Physiol. B* **167**, 129–134.

Egginton, S. (1997b). Control of tissue blood flow at very low temperatures. *J. Therm. Biol.* **22**, 403–407.

Egginton, S., and Sidell, B. D. (1989). Thermal acclimation induces adaptive changes in subcellular structure of fish skeletal muscle. *Am. J. Physiol.* **256**, R1–R9.

Egginton, S., and Rankin, J. C. (1998). Vascular adaptations for a low pressure/high flow blood supply to locomotory muscles of Antarctic icefish. *In* "Fishes of Antarctica:

A Biological Overview" (di Prisco, G., Clarke, A., and Pisano, E., Eds.), pp. 185–195. Springer-Verlag, Milan.

Egginton, S., and Davison, W. (1998). Effects of environmental and experimental stress on Antarctic fish. In "Cold Ocean Physiology" (Pörtner, H. O., and Playle, R. C., Eds.), pp. 299–326. Cambridge University Press, Cambridge.

Egginton, S., Forster, M. E., and Davison, W. (2001). Control of vascular tone in notothenioid fishes is determined by phylogeny, not environmental temperature. Am. J. Physiol. 280, R1197–1205.

Egginton, S., Skilbeck, C., Hoofd, L., Calvo, J., and Johnston, I. A. (2002). Peripheral oxygen transport in skeletal muscle of Antarctic and sub-Antarctic notothenioid fish. J. Exp. Biol. 205, 769–779.

Fanta, E., Lucchiari, P. H., and Bacila, M. (1989). The effect of environmental oxygen and carbon dioxide levels on the tissue oxygenation and the behavior of Antarctic fish. Comp. Biochem. Physiol. 93A, 819–831.

Feller, G., and Bionet, P. (1984). Myofibrillar composition of some red- and white-blooded Antarctic teleosts. Mol. Physiol. 5, 179–188.

Fernandez, D. A., Calvo, J., and Johnston, I. A. (1999). Characterisation of the swimming muscles of two sub-Antarctic notothenioids. Sci. Mar. 63, 477–484.

Fernandez, D. A., Calvo, J., Franklin, C. E., and Johnston, I. A. (2000). Muscle fibre types and size distribution in sub-Antarctic notothenioid fishes. J. Fish Biol. 56, 1295–1311.

Fernandez, D. A., Calvo, J., Wakeling, J. M., Vanella, F. A., and Johnston, I. A. (2002). Escape performance in the sub-Antarctic notothenioid fish Eleginops maclovinus. Polar Biol. 25, 914–920.

Fields, P. A. (2001). Protein function at thermal extremes: Balancing stability and flexibility. Comp. Biochem. Physiol. 129A, 417–431.

Fields, P. A., and Somero, G. N. (1998). Hot spots in cold adaptation: Localized increases in conformational flexibility in lactate dehydrogenase A(4) orthologs of Antarctic notothenioid fishes. Proc. Natl. Acad. Sci. USA 95, 11476–11481.

Fitch, N. A. (1988). Lactate dehydrogenases in Antarctic and temperate fish species. Comp. Biochem. Physiol. 91B, 671–676.

Fitch, N. A. (1989). Lactate dehydrogenase isozymes in the trunk and cardiac muscles of an Antarctic teleost fish, Notothenia neglecta Nybelin. Fish Physiol. Biochem. 6, 187–195.

Fitch, N. A., Johnston, I. A., and Wood, R. E. (1984). Skeletal muscle capillary supply in a fish that lacks respiratory pigments. Resp. Physiol. 57, 201–211.

Forster, M. E., Franklin, C. E., Taylor, H. H., and Davison, W. (1987). The aerobic scope of an Antarctic fish, Pagothenia borchgrevinki and its significance for metabolic cold adaptation. Polar Biol. 8, 155–159.

Forster, M. E., Davison, W., Axelsson, M., Sundin, L., Franklin, C. E., and Gieseg, S. (1998). Catecholamine release in heat-stressed Antarctic fish causes proton extrusion by the red cells. J. Comp. Physiol. B 168, 345–352.

Franklin, C. E. (1998). Studies of evolutionary temperature adaptation: Muscle function and locomotor performance in Antarctic fish. Clin. Exp. Pharmacol. Physiol. 25, 753–756.

Franklin, C. E., and Johnston, I. A. (1997). Muscle power output during escape responses in an Antarctic fish. J. Exp. Biol. 200, 703–712.

Franklin, C. E., Wilson, R. S., and Davison, W. (2003). Locomotion at −1.0°C: Burst swimming performance of five species of Antarctic fish. J. Therm. Biol. 28, 59–65.

Gauvry, L., Ennion, S., Ettelaie, C., and Goldspink, G. (2000). Characterisation of red and white muscle myosin heavy chain gene coding sequences from Antarctic and tropical fish. Comp. Biochem. Physiol. 127B, 575–588.

Giardina, B., Gozzo, M. L., Zappacosta, B., Colacicco, L., Calla, C., Mordente, A., and Lippa, S. (1997). Coenzyme a homologs and trace elements content of Antarctic fishes *Chionodraco hamatus* and *Pagothenia bernacchii* compared with the Mediterranean fish *Mugil cephalus*. *Comp. Biochem. Physiol.* **118A**, 977–980.

Gieseg, S., Cuddihy, S., Hill, J. V., and Davison, W. (2000). A comparison of plasma vitamin C and E levels in two Antarctic and two temperate water fish species. *Comp. Biochem. Physiol.* **125B**, 371–378.

Goldspink, G. (1995). Adaptation of fish to different environmental temperature by qualitative and quantitative changes in gene expression. *J. Therm. Biol.* **20**, 167–174.

Gon, O., and Heemstra, P. C. (1990). "Fishes of the Southern Ocean." J. L. B Smith Institute of Ichthyology, Grahamstown, South Africa.

Gonzalez-Cabrera, P. J., Dowd, F., Pedibhotla, V. K., Rosario, R., Stanley-Samuelson, D., and Petzel, D. (1995). Enhanced hypo-osmoregulation induced by warm-acclimation in Antarctic fish is mediated by increased gill and kidney Na^+/K^+- ATPpase activities. *J. Exp. Biol.* **198**, 2279–2291.

Grove, T. J., and Sidell, B. D. (2002). Myoglobin deficiency in the hearts of phylogenetically diverse temperate-zone fish species. *Can. J. Zool.* **80**, 893–901.

Guderley, H. (1998). Temperature and growth rates as modulators of the metabolic capacities of fish muscle. *In* "Cold Ocean Physiology" (Pörtner, H. O., and Playle, R. C., Eds.), pp. 58–87. Cambridge University Press, Cambridge.

Guderley, H. (2004). Metabolic responses to low temperature in fish muscle. *Biol. Rev.* **79**, 409–427.

Guynn, S., Dowd, F., and Petzel, D. (2002). Characterization of gill Na/K-ATPase activity and ouabain binding in Antarctic and New Zealand notothenioid fishes. *Comp. Biochem. Physiol.* **131A**, 363–374.

Hardewig, I., Van Dijk, P. L. M., and Pörtner, H. O. (1998). High-energy turnover at low temperatures: Recovery from exhaustive exercise in Antarctic and temperate eelpouts. *Am. J. Physiol.* **43**, R1789–R1796.

Harrison, P., Nicol, C. J. M., and Johnston, I. A. (1987). Gross morphology, fibre composition and mechanical properties of pectoral fin muscles in the Antarctic teleost *Notothenia neglecta* Nybelin. *In* Proceedings of the fifth Congress of European Ichthyologists, Stockholm pp. 459–465.

Hazel, J. R., and Williams, E. E. (1990). The role of alterations in membrane lipid composition in enabling physiological adaptations of organisms to their physical environment. *Prog. Lipid Res.* **29**, 167–227.

Hemmingsen, E. A., and Douglas, E. L. (1977). Respiratory and circulatory adaptations to the absence of hemoglobin in Channichthyid fishes. *In* "Adaptations within Antarctic Ecosystems" (Llano, A., Ed.), Proceedings of the 3rd SCAR symposium on Antarctic Biology. pp. 479–487. Smithsonian Institute, Washington.

Hickey, A. J. R., and Clements, K. D. (2003). Key metabolic enzymes and muscle structure in triplefin fishes (Tripterygiidae): A phylogenetic comparison. *J. Comp. Physiol. B* **173**, 113–123.

Hoar, W. S., and Randall, D. J. (1978). "Locomotion, Fish Physiology," Vol VII. Academic Press, New York.

Hochachka, P. W., and Somero, G. N. (2002). "Biochemical Adaptation. Mechanism and Process in Physiological Evolution." Oxford University Press, Oxford.

Holeton, G. F. (1974). Metabolic adaptation of polar fish: Fact or artefact? *Physiol. Zool.* **47**, 137–152.

Hoofd, L., and Egginton, S. (1997). The possible role of intracellular lipid in determining oxygen delivery to fish skeletal muscle. *Resp. Physiol.* **107**, 191–202.

Huey, R. B., and Berrigan, B. (1996). Testing evolutionary hypotheses of acclimation. *In* "Animals and Temperature. Phenotypic and Evolutionary Adaptation" (Johnston, I. A., and Bennett, A. F., Eds.), pp. 205–237. Cambridge University Press, Cambridge.

Johnson, T. P., and Johnston, I. A. (1991). Temperature adaptation and the contractile properties of live muscle from teleost fish. *J. Comp. Physiol. B* **161**, 27–36.

Johnston, I. A. (1987). Respiratory characteristics of muscle fibres in a fish (*Chaenocephalus aceratus*) that lacks haem pigments. *J. Exp. Biol.* **133**, 415–428.

Johnston, I. A. (2001). "Muscle Development and Growth." Academic Press, San Diego.

Johnston, I. A. (2003). Muscle metabolism and growth in Antarctic fishes (suborder Notothenioidei): Evolution in a cold environment. *Comp. Biochem. Physiol.* **136B**, 701–713.

Johnston, I. A., and Brill, R. W. (1984). Thermal dependence of contractile properties of single skinned muscle fibres from Antarctic and various warm water marine fishes including skipjack tuna (Katsuwonus pelamis) and kawakawa (Euthynnus affinis). *J. Comp. Physiol. B* **155**, 63–70.

Johnston, I. A., and Harrison, P. (1985). Contractile and metabolic characteristics of muscle fibres from Antarctic fish. *J. Exp. Biol.* **116**, 223–236.

Johnston, I. A., and Altringham, J. D. (1985). Evolutionary adaptation of muscle power output to environmental temperature: Force-velocity characteristics of skinned fibres isolated from Antarctic, temperate and tropical marine fish. *Pflugers Arch.* **405**, 136–140.

Johnston, I. A., and Camm, J.-P. (1987). Muscle structure and differentiation in pelagic and demersal stages of the Antarctic teleost *Notothenia neglecta*. *Mar. Biol.* **94**, 183–190.

Johnston, I. A., and Altringham, J. D. (1988). Muscle contraction in polar fishes: Experiments with demembranated muscle fibres. *Comp. Biochem. Physiol.* **90B**, 547–555.

Johnston, I. A., and Ball, D. (1997). Thermal stress and muscle functioning fish. *In* "Global Warming: Implications for Freshwater and Marine Fish" (Wood, C. M., and McDonald, D. G., Eds.), pp. 79–104. Cambridge University Press, Cambridge.

Johnston, I. A., Davison, W., and Goldspink, G. (1977). Energy metabolism of carp swimming muscles. *J. Comp. Physiol. B* **114**, 203–216.

Johnston, I. A., Camm, J.-P., and White, M. (1988). Specialisations of swimming muscles in the pelagic Antarctic fish *Pleuragramma antarcticum*. *Mar. Biol.* **100**, 3–12.

Johnston, I. A., Walesby, N. J., Davison, W., and Goldspink, G. (1975). Temperature adaptation in myosin of Antarctic fish. *Nature* **254**, 74–75.

Johnston, I. A., Van Leeuwen, J. L., Davies, M. L. F., and Beddow, T. A. (1995). How fish power predation fast starts. *J. Exp. Biol.* **198**, 1851–1861.

Johnston, I. A., Guderley, H., Franklin, C. E., Crockford, T., and Kamunde, C. (1994). Are mitochondria subject to evolutionary temperature adaptation? *J. Exp. Biol.* **195**, 293–306.

Johnston, I. A., Calvo, J., Guderley, H., Fernandez, D., and Palmer, L. (1998). Latitudinal variation in the abundance and oxidative capacities of muscle mitochondria in perciform fishes. *J. Exp. Biol.* **201**, 1–12.

Johnston, I. A., Fernandez, D. A., Calvo, J., Vieira, V. L. A., North, A. W., Abercromby, M., and Garland, T. (2003). Reduction in muscle fibre number during the adaptive radiation of notothenioid fishes: A phylogenetic perspective. *J. Exp. Biol.* **206**, 2595–2609.

Jones, D. R., and Randall, D. J. (1978). The respiratory and circulatory systems during exercise. *In* "Fish Physiology" (Hoar, W. S., and Randall, D. J., Eds.). Academic Press, New York.

Kamler, E., Krasicka, B., and Rakusa-Suszczewski, S. (2001). Comparison of lipid content and fatty acid composition in muscle and liver of two notothenioid fishes from Admiralty Bay (Antarctica): an eco-physiological perspective. *Polar Biol.* **24**, 735–743.

Kawall, H. G., Torres, J. J., Sidell, B. D., and Somero, G. N. (2002). Metabolic cold adaptation in Antarctic fishes: evidence from enzymatic activities of brain. *Mar. Biol.* **140**, 279–286.

Kilarski, W., Smialowska, E., and Friedhuber, A. (1982). Histological analysis of fibres in myotomes of Antarctic fish. II. Morphometry of muscle fibres and capillaries. *Z. Mikrosk. Anat. Forsch* **96**, 791–801.

Kock, K. H., and Everson, I. (2003). Shedding new light on the life cycle of mackerel icefish in the Southern Ocean. *J. Fish Biol.* **63**, 1–21.

Korneliussen, H., and Nicolaysen, K. (1973). Ultrastructure of four types of striated muscle fibers in the Atlantic hagfish (*Myxine glutinosa*, L.). *Z. Zellforsch.* **143**, 273–290.

Lin, Y., Dobbs, G. H., and Devries, A. L. (1974). Oxygen consumption and lipid content in red and white muscles of Antarctic fishes. *J. Exp. Zool.* **189**, 379–386.

Londraville, R. L., and Sidell, B. D. (1990). Ultrastructure of aerobic muscle in Antarctic fishes may contribute to maintenance of diffusive fluxes. *J. Exp. Biol.* **150**, 205–220.

Londraville, R. L., and Sidell, B. D. (1995). Purification and characterization of fatty-acid binding-protein from aerobic muscle of the Antarctic icefish *Chaenocephalus aceratus*. *J. Exp. Zool.* **273**, 190–203.

Lowe, C. J., Seebacher, F., and Davison, W. (2005). Thermal sensitivity of heart rate and insensitivity of blood pressure in the Antarctic nototheniid *Pagothenia borchgrevinki*. *J. Comp. Physiol.* **175B**, 97–105.

Lowe, T. E., and Wells, R. M. G. (1997). Exercise challenge in Antarctic fishes: Do haematology and muscle metabolite levels limit swimming performance? *Polar Biol.* **17**, 211–218.

Mark, F. C., Bock, C., and Pörtner, H. O. (2002). Oxygen-limited thermal tolerance in Antarctic fish investigated by MRI and P-31-MRS. *Am. J. Physiol.* **283**, R1254–R1262.

Marshall, C., Crossman, D., Love, C., McInnes, S., and Fleming, R. (2000). Structural studies of cold-adaptation: Lactate dehydrogenases from notothenioid fish. *In* "Antarctic Ecosystems: Models for Wider Ecological Understanding" (Davison, W., Broady, P., and Howard-Williams, C., Eds.), pp. 123–133. New Zeal. Nat. Sci., Christchurch.

McVean, A. R., and Montgomery, J. C. (1987). Temperature compensation in myotomal muscle: Antarctic versus temperate fish. *Env. Biol. Fishes* **19**, 27–33.

Milligan, C. L., and Wood, C. M. (1987). Muscle and liver intracellular acid–base and metabolite status after strenuous activity in the inactive, benthic starry flounder *Platichthys stellatus*. *Physiol. Zool.* **60**, 54–68.

Moerland, T. S., and Egginton, S. (1998). Intracellular pH of muscle and temperature: Insight from *in vivo* ^{31}P NMR measurements in a stenothermal Antarctic teleost (*Harpagifer antarcticus*). *J. Therm. Biol.* **23**, 275–282.

Montgomery, J. C., and Macdonald, J. A. (1984). Performance of motor systems in Antarctic fishes. *J. Comp. Physiol. A* **154**, 241–248.

Morla, M., Agusti, A. G. N., Rahman, I., Motterlini, R., Saus, C., Morales-Nin, B., Company, J. B., and Busquets, X. (2003). Nitric oxide synthase type I (nNOS), vascular endothelial growth factor (VEGF) and myoglobin-like expression in skeletal muscle of Antarctic icefishes (Notothenioidei: Channichthyidae). *Polar Biol.* **26**, 458–462.

Moylan, T. J., and Sidell, B. D. (2000). Concentrations of myoglobin and myoglobin mRNA in heart ventricles from Antarctic fishes. *J. Exp. Biol.* **203**, 1277–1286.

O'Brien, K. M., and Sidell, B. D. (2000). The interplay among cardiac ultrastructure, metabolism and the expression of oxygen-binding proteins in Antarctic fishes. *J. Exp. Biol.* **203**, 1287–1297.

O'Brien, K. M., Skilbeck, C., Sidell, B. D., and Egginton, S. (2003). Muscle fine structure may maintain the function of oxidative fibres in haemoglobinless Antarctic fishes. *J. Exp. Biol.* **206**, 411–421.

Pakhomov, E. A. (1997). Feeding and exploitation of the food supply by demersal fishes in the Antarctic part of the Indian Ocean. *J. Ichthyology* **37**, 360–380.

Peck, L. S. (2002). Ecophysiology of Antarctic marine ectotherms: Limits to life. *Polar Biol.* **25**, 31–40.

Pörtner, H. O. (2002). Physiological basis of temperature-dependent biogeography: Trade-offs in muscle design and performance in polar ectotherms. *J. Exp. Biol.* **205**, 2217–2230.

Pörtner, H. O., Van Dijk, P. L. M., Hardewig, I., and Sommer, A. (2000). Levels of metabolic cold adaptation: Tradeoffs in eurythermal and stenothermal ectotherms. *In* "Antarctic Ecosystems: Models for Wider Ecological Understanding" (Davison, W., Howard-Williams, C., and Broady, P., Eds.), pp. 109–122. New Zeal. Nat. Sci., Christchurch.

Ritchie, P. A., Lavoue, S., and Lecointre, G. (1997). Molecular phylogenetics and the evolution of Antarctic Notothenioid fishes. *Comp. Biochem. Physiol.* **118A**, 1009–1025.

Rome, L., and Swank, D. M. (1992). The influence of temperature on power output of scup red muscle during cyclical length changes. *J. Exp. Biol.* **171**, 261–281.

Sänger, A. M., Davison, W., and Egginton, S. (2005). Muscle fine structure reflects ecotype in Antarctic fish. *J. Fish Biol.* **66**, 1371–1386.

Seebacher, F., Davison, W., Lowe, C. J., and Franklin, C. E. (2005). A falsification of the thermal specialization paradigm: Compensation for elevated temperatures in Antarctic fishes. *Biol. Lett.* **2**, 151–154.

Sharpe, M., Love, C., and Marshall, C. (2001). Lactate dehydrogenase from Antarctic eelpout, *Lycodichthys dearborni*. *Polar Biol.* **24**, 258–269.

Sidell, B. D., and Hazel, J. R. (2002). Triacylglycerol lipase activities in tissues of Antarctic fishes. *Polar Biol.* **25**, 517–522.

Sidell, B. D., Crockett, E. L., and Driedzic, W. R. (1995). Antarctic fish-tissues preferentially catabolize monoenoic fatty-acids. *J. Exp. Zool.* **271**, 73–81.

Small, D. J., Moylan, T., Vayda, M. E., and Sidell, B. D. (2003). The myoglobin gene of the Antarctic icefish, *Chaenocephalus aceratus*, contains a duplicated TATAAAA sequence that interferes with transcription. *J. Exp. Biol.* **206**, 131–139.

Smialowska, E., and Kilarski, W. (1981). Histological analysis of fibres in myotomes of Antarctic fish (Admiralty Bay, King George Island, South Shetland Islands) I. Comparative analysis of muscle fibres size. *Pol. Polar Res.* **2**, 109–129.

Somero, G. N. (1998). Adaptation to cold and depth: contrasts between polar and deep-sea animals. *In* "Cold Ocean Physiology" (Pörtner, H. O., and Playle, R. C., Eds.), pp. 33–57. Cambridge University Press, Cambridge.

Somero, G. N. (2003). Protein adaptations to temperature and pressure: complementary roles of adaptive changes in amino acid sequence and internal milieu. *Comp. Biochem. Physiol.* **136B**, 577–591.

Somero, G. N., and DeVries, A. L. (1967). Temperature tolerance of some Antarctic fishes. *Science* **156**, 257–258.

Swank, D. M., and Rome, L. (1999). The influence of temperature on power production during swimming. I. *In vivo* length change and stimulation pattern. *J. Exp. Biol.* **202**, 321–331.

Temple, G. K., and Johnston, I. A. (1998). Testing hypotheses concerning the phenotypic plasticity of escape performance in fish of the family Cottidae. *J. Exp. Biol.* **201**, 317–331.

Thorarensen, H., Gallaugher, P. E., Kiessling, A. K., and Farrell, A. P. (1993). Intestinal blood flow in swimming Chinook salmon Oncorhynchus tshawytscha and the effects of hematocrit on blood flow distribution. *J. Exp. Biol.* **179**, 115–129.

Torres, J. J., and Somero, G. N. (1988). Vertical distribution and metabolism in Antarctic mesopelagic fishes. *Comp. Biochem. Physiol.* **90B**, 521–528.

Tuckey, N., and Davison, W. (2004). Mode of locomotion places selective pressures on Antarctic and temperate labriform swimming fishes. *Comp. Biochem. Physiol.* **138A**, 391–398.

Van Dijk, P. L. M., Tesch, C., Hardewig, I., and Pörtner, H. O. (1999). Physiological disturbances at critically high temperatures: A comparison between stenothermal Antarctic and eurythermal temperate eelpouts (Zoarcidae). *J. Exp. Biol.* **202**, 3611–3621.

Walesby, N. J., and Johnston, I. A. (1979). Activities of some enzymes of energy metabolism in the fast and slow muscles of an Antarctic teleost fish (*Notothenia rossii*). *J. Physiol.* **7**, 659–661.

Walesby, N. J., Nicol, C. J. M., and Johnston, I. A. (1982). Metabolic differentiation of muscle fibres from a haemoglobinless (*Champsocephalus gunnari* Lonnberg) and a red-blooded (*Notothenia rossii* Fischer) Antarctic fish. *Br. Antarctic Surv. Bull.* **51**, 201–214.

Walker, J. A., and Westneat, M. W. (2000). Mechanical performance of aquatic rowing and flying. *Proc. R. Soc. Lond. Series B Biol. Sci.* **267**, 1875–1881.

Weatherley, A. H., Gill, H. S., and Casselman, J. M. (1987). "The Biology of Fish Growth." Academic Press, London.

Webb, P. W. (1973). Kinematics of pectoral fin propulsion in *Cymatogaster aggregata*. *J. Exp. Biol.* **59**, 697–710.

Weinstein, R. B., and Somero, G. N. (1998). Effects of temperature on mitochondrial function in the Antarctic fish *Trematomus bernacchii*. *J. Comp. Physiol. B* **168**, 190–196.

Wells, R. M. G., Summers, G., Beard, L. A., and Grigg, G. C. (1988). Ecological and behavioural correlates of intracellular buffering capacity in the muscles of Antarctic fishes. *Polar Biol.* **8**, 321–325.

Wilson, R. S., Franklin, C. E., Davison, W., and Kraft, P. (2001). Stenotherms at sub-zero temperatures: Thermal dependence of swimming performance in Antarctic fish. *J. Comp. Physiol. B* **171**, 263–269.

Wilson, R. S., Kuchel, L. J., Franklin, C. E., and Davison, W. (2002). Turning up the heat on sub-zero fish: Thermal dependence of sustained swimming in an Antarctic fish. *J. Therm. Biol.* **27**, 381–386.

Yau, C., Collins, M. A., Bagley, P. M., Everson, I., and Priede, I. G. (2002). Scavenging by megabenthos and demersal fish on the South Georgia slope. *Antarctic Sci.* **14**, 16–24.

9

THE NERVOUS SYSTEM

JOHN MACDONALD
JOHN MONTGOMERY

I. Introduction
II. Nervous System: Structure and Function
   A. Cellular Function
   B. Temperature Compensation in the Nervous System
   C. The Brains of Polar Fishes
III. Sensory Physiology of Polar Fishes
   A. Visual Function
   B. Chemical Senses
   C. Mechanosenses
IV. Concluding Remarks

I. INTRODUCTION

Polar fishes are defined by their environment. In addition to the obvious semantic sense of this statement, there is also a biological sense in which the polar environment has shaped the evolution of these fishes. The most distinctive features of the polar marine environment are cold temperature and extreme seasonality. However, it is important to realize that this seasonality relates to light and energy flow but does not extend to temperature. Polar seas are buffered by the presence of sea ice such that though extremely cold, they remain at nearly the same temperature year round. Polar fishes feed, grow, and reproduce at this constant low temperature, making them an important model for understanding temperature adaptation. Other fish may survive cold winters, but polar fish conduct their entire life cycle in frigid water. This chapter deals with the physiology, or function, of the nervous system and the sensory organs of polar fishes. It addresses the issue of temperature adaptation in the nervous system and discusses what is special about the neurophysiology of fish from stenothermal arctic and Antarctic

The Physiology of Polar Fishes: Volume 22
FISH PHYSIOLOGY

environments, what themes are common to both groups, and how these polar species differ from fishes that are more eurythermal yet can acclimatize to severe cold. The physiological function of sensory systems is considered principally from the point of view of the extreme seasonality and periods of low light in the polar environment.

The majority of the work on nervous systems and sensory systems of polar fishes has been concentrated on Antarctic fishes, in particular the notothenioids. This is probably for two reasons. First, the Antarctic Treaty System, with its emphasis on the scientific study of Antarctica, has created a well-funded and productive environment for research and international collaboration. Second, and equally important, the seas surrounding Antarctica have been isolated from other oceans by the Circum-Antarctic Current and a sharp temperature gradient for several tens of millions of years, permitting a unique ichthyofauna to develop. The notothenioids, phylogenetically distinct from fishes of the northern seas, dominate the Antarctic fish fauna, with a high level of endemicity. Many species are confined to the waters of the Antarctic continental shelf and never experience temperate conditions. The notothenioid radiation in particular has attracted scientific attention across a range of disciplines. Thus, the emphasis in this chapter is on Antarctic rather than arctic species, simply because we know a great deal more about them.

The principles of temperature adaptation, however, should be common to both faunas, and we would predict that most of the temperature-related adaptations reported in Antarctic species will also be found in fishes from the Arctic. The sharing of analogous adaptive features would be evidence of convergence in unrelated groups, but precisely because we are dealing with unrelated groups, it should be no surprise if some adaptations are not shared or if common functional principles have been achieved in different ways.

In considering life at low temperature, we should make a functional distinction between resistance adaptation and capacity adaptation (Precht, 1958). Resistance adaptations manifest themselves as changes in tolerance limits, permitting physiological processes, and life itself, to continue outside "normal" or "usual" environmental limits. On the other hand, capacity (or capacitative) adaptations result in compensatory changes in rate functions; in the case of cold adaptation, these would enable physiological processes to at least partially reverse the inhibiting effects of low temperature. The fact that polar fishes survive year round and even thrive in ice-covered polar seas where water temperature hovers in the vicinity of $-1.9\,^{\circ}\mathrm{C}$ makes these fishes important models of both resistance and capacitative adaptations to low temperature.

The extent to which Antarctic fish have optimized their critical cellular and neural systems to operate at subzero temperatures is reflected in their extreme stenothermy. For high-latitude species such as the inhabitants of

McMurdo Sound, heat death occurs at only a few degrees above 0°C, only 4–5°C above the normal ambient temperature of −1.9°C (Somero and DeVries, 1967). One can make the comparison with homeotherms such as ourselves, where our systems are optimized at around 37°C, with about a 5° margin to our upper lethal temperature. If cold adaptation and reduced thermal tolerance are coupled processes, then the relatively small margin between ambient and upper lethal temperatures for polar fishes argues against the idea that there is substantial scope for further temperature adaptation in these fishes. However, this does not imply that capacitative temperature adaptations have compensated fully for the loss in functionality due to low temperature or that temperature adaptation has reached a final plateau in all subsystems. There are still significant challenges in gauging the degree of temperature compensation, in elucidating its underlying mechanisms, and determining the remaining potential for and ultimate limits to such compensation. For many physiological attributes, the function–temperature relationships themselves remain to be quantified. Likewise, the relative contributions of mechanisms with different time scales (e.g., acute [acclimation], seasonal [acclimatization], or multigenerational [evolutionary adaptation]) are not clear. Documenting the degree of temperature compensation still leaves unanswered the questions of the degree to which the capacity for temperature adaptation has been met and the degree to which it might be constrained by the direct effects of temperature on physical and chemical processes such as diffusion, fluid viscosity, and the rate of chemical reactions. In no other tissue are these effects more pronounced than in the nervous system, so this chapter will explore these issues with respect to neural function.

The sensory biology of polar fishes is discussed briefly from an evolutionary perspective, considering not only what is distinctive about the sensory milieu and capabilities of polar fishes, but also how those capabilities have been shaped by both environment and phylogeny. The emphasis is on sensory physiology rather than a full review of the anatomy of sense organs, and we concentrate on the progress made since the reviews of the physiology of Antarctic fish (Macdonald *et al.*, 1987) and the sensory biology of notothenioid fish (Macdonald and Montgomery, 1991).

II. NERVOUS SYSTEM: STRUCTURE AND FUNCTION

A. Cellular Function

Neural function depends on cellular and molecular competency. Comparative physiologists and biochemists have detailed the effects of acute temperature change on many aspects of cellular and molecular function,

Table 9.1
Low-Temperature Adaptations of Antarctic Fishes

Adaptations	References
Molecular adaptations	
Tubulin polymerization at $-2\,^\circ$C	Detrich (1991, 1997)
Heat shock gene expression at $5\,^\circ$C	Maresca *et al.* (1988)
Membrane homeoviscous adaptation	Cossins (1994)
Brain gangliosides	Becker *et al.* (1995)
Enzyme functional properties	Somero (1991); Genicot *et al.* (1996)
	Crockett and Sidell (1990)
Systems Adaptations	
Mitochondrial respiration	Johnston *et al.* (1994)
Muscle contraction	
Maximal stress production	Johnston and Johnston (1991)
Deactivation rate	Johnston and Johnston (1991)
Mitochondrial density	Clarke and Johnston (1996)
Neural conduction	Macdonald (1981)
Neurotransmitter release	Macdonald and Montgomery (1982)
Post-synaptic rate constants	Macdonald and Montgomery (1986)

and there is a growing understanding of mechanisms of temperature adaptation (Table 9.1). Although there has been little work on critical neural elements such as receptors and ion channels, studies of other macromolecules do provide us with insight into the likely nature of molecular temperature adaptation. The summary result in neural tissues, as elsewhere, is that decreasing temperature adversely affects most processes; for example, there is a reduction in the catalytic performance of enzymes as temperature decreases. One facet of this failure is the loss of molecular flexibility at low temperatures, reducing the conformational changes required for the proper function of enzymes and structural proteins. How can this functionality be restored? Once again using enzymes as our example, reaction rates can be restored by changing electrostatic potential and hydrogen bonding to alter substrate affinity (Leiros, 2000), and primary protein structure may be changed in ways that are predicted to restore structural flexibility at low temperature (Genicot *et al.*, 1996; Feller and Gerday, 1997; Feller *et al.*, 1997; Ciardiello *et al.*, 2000; Chessa *et al.*, 2000; Leiros, 2000; Zecchinon, 2001). Loosening up the structure by reducing the intramolecular bonding can restore the flexibility necessary for proper function. It is also likely that this increased structural flexibility contributes to the upper temperature limit at which cold-adapted enzymes can still operate (Leiros, 2000). So there is a sense in which the trade-off between structural flexibility and stability contributes to temperature adaptation in physiological function and to the reduction in upper temperature limits that accompanies reduced temperature optima. However, the upper temperature limit for many enzymes is well above the upper lethal limit, meaning that the underlying causes of heat death are to be found elsewhere.

As one example of enzyme function, acetylcholine esterase is an important constituent of the neuromuscular junction and other cholinergic synapses. The Michaelis-Menten constant K_m provides an index of the capacity of the enzyme to bind its substrate, and a rapid increase of K_m with changing temperature indicates a loss of that capacity (Somero, 1995). A steep rise in the K_m of acetylcholine esterase for acetylcholine at temperatures only slightly above 0°C has been observed in *Pagothenia borchgrevinki* (Baldwin, 1971), the functional implications of which are discussed later in our consideration of the neuromuscular junction.

In addition to single protein function, cellular competency depends on complex assemblages of macromolecules such as microtubule assembly and microtubule motors. Extensive studies of these complex assemblages show clear evidence of cold adaptation in Antarctic fishes and close to full temperature compensation when their functional properties are compared with those of animals adapted to warmer temperatures (Detrich, 1991, 1997, 1998; Detrich *et al.*, 2000). As with enzymatic function, macromolecular flexibility and changes in electrostatic properties contribute to cold adaptation.

Membrane physiology reflects a similar story to protein structure; again, there is a trade-off between mobility and stability. Low temperature decreases membrane fluidity. Cold-adapted animals have membranes with higher amounts of unsaturated phospholipid classes, which increase fluidity, counteracting the effects of low temperature. The systematic changes in membrane fluidity of living organisms inhabiting different thermal environments provide near-perfect compensation in membrane viscosity and have been termed "homeoviscous adaptation" (Nozawa *et al.*, 1974; Sinensky, 1974; Cossins, 1977; Cossins and Prosser, 1978; Behan-Martin *et al.*, 1993; Cossins, 1994). In addition to the compensatory changes in phospholipids, proper function of the nervous systems of polar fishes may be further promoted by the presence of gangliosides that are more highly sialylated (more polar groups) than those of warm-climate fishes (Rahmann and Hilbig, 1980; Rahmann *et al.*, 1984; Cossins, 1994; Becker *et al.*, 1995). One consequence of the enhanced membrane fluidity is that the cell membranes of Antarctic fish would be predicted to be less stable at high temperatures, and this may be a factor contributing to their stenothermal range. There is some experimental evidence for this in that both nerve conduction and neuromuscular transmission have upper temperature limits that are lower than those reported for temperate fishes (Macdonald and Ensor, 1975; Macdonald, 1981; Macdonald and Montgomery, 1982; Pockett and Macdonald, 1986).

However, the upper neurophysiological limits are considerably higher than survival limits for the intact fish.

B. Temperature Compensation in the Nervous System

1. EXCITABILITY AND CONDUCTION OF THE NERVOUS IMPULSE

From the earliest days of comparative physiology, neurophysiologists have recognized the inhibitory effects of low temperature. Although membrane excitability and conduction of the nervous impulse are depressed by low temperatures, the nervous system is capable of adapting to compensate for the effects of low temperature, thus maintaining the system's competency.

Compensatory differences in nerve function correlated with the temperature ranges experienced by particular tissues have been reported in both heterothermic endotherms (Chatfield et al., 1948, 1953; Kehl and Morrison, 1960; Miller and Irving, 1967; Miller, 1970) and in differently acclimated poikilotherms (McDonald et al., 1963; Lagerspetz and Talo, 1967; Bass, 1971; Dierolf and Brink, 1973).

Compensation for low temperature in the peripheral nerves of polar fishes has been described in a series of papers dealing with teleosts of the family Nototheniidae from McMurdo Sound, Antarctica (latitude ~78°S) (Macdonald and Ensor, 1975; Macdonald and Wells, 1978; Macdonald, 1981; Macdonald et al., 1987). These fish live year round at a nearly constant −1.9 °C, varying only by a few tenths of a degree (Littlepage, 1965). The salient points of this neural compensation include the following observations (Figure 9.1): (a) Excitability and impulse conduction are maintained in Antarctic fish nerves down to temperatures well below 0°C; (b) at low temperatures, conduction velocities in the Antarctic fish nerves are higher than velocities observed in temperate fish nerves at the same temperature; (c) impaired function is seen at high temperatures; for example, conduction velocities in the Antarctic fish are lower than expected at high temperatures— that is, the velocity–temperature relation is reduced in comparison to that of warm-climate fish nerves; and (d) reduced resistance to high temperatures is seen; on the upper end of the temperature range, propagation of the nervous impulse fails at lower temperatures than seen in temperate fishes.

As one example of neural competency at subzero temperatures, the posterior lateral line nerve (a branch of cranial nerve X) of the Antarctic toothfish *Dissostichus mawsoni* conducts compound action potentials at well over 10 m s^{-1} at −1.9°C, more than twice the velocity reported at the same temperature for the New Zealand snapper *Pagrus auratus* (Macdonald, 1981). This represents a significant compensation but is only partial, as the snapper nerve has a higher conduction velocity of 20–23 m s^{-1} at its normal environmental temperature of 12–15°C.

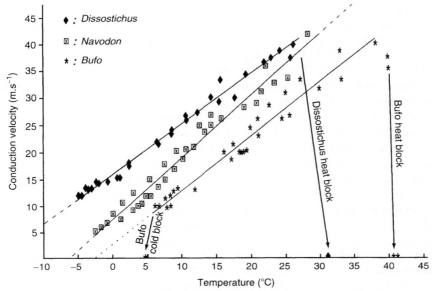

Figure 9.1. Temperature compensation in peripheral nerves. Data from individual spinal nerves (*Dissostichus, Navodon*) and sciatic nerve (*Bufo*). The speed at which nerve impulses travel (conduction velocity) is proportional to temperature. Nerves of temperate New Zealand fishes (e.g., *Navodon*) and tropical toad (*Bufo*) conduct slowly at low temperature and are inactivated near 0 °C. Inactivation by low temperature ("cold block") is shown for *Bufo*. Antarctic fish nerves (*Dissostichus*) continue to conduct at high velocities well below 0 °C. The slope of the velocity/temperature line is reduced for Antarctic fishes, showing a clockwise rotation relative to the temperature relationship for *Navodon*, and the upper temperature limit ("heat block") is lower than that of warm-climate fishes.

In common with conduction in many other nerves, impulse velocity in Antarctic fish nerves is a linear function of temperature over most of its range, permitting easy extrapolation to an x-intercept (zero velocity). The predicted temperature of zero velocity (T_0) is outside the normal temperature range of the nerve but differs systematically between nerves acclimated or otherwise adapted to different temperature regimes. Within any one animal, temperature appears to affect propagation in both large and small fibers in the same proportion to the normal velocity of the fibre. Thus, the velocity–temperature plots of high-velocity fibers and low-velocity fibers converge on the same value of T_0. For the Antarctic nototheniids *D. mawsoni*, *Trematomus bernacchii*, and *P. borchgrevinki*, T_0 is in the vicinity of $-18°C$, in contrast to values of -3 to $-8°C$ for a variety of temperate teleosts (Macdonald, 1981). It should be emphasized that T_0 is never actually attained, as impulse conduction fails suddenly at a temperature well above T_0, a phenomenon referred to as *cold block*. Cold block is easily demonstrated in the nerves of homeotherms and in tropical poikilotherms, where nerve conduction fails at temperatures above 0°C (e.g., *Bufo marinus* [Macdonald, 1981]). However, in temperate and Antarctic fish nerves, where impulse propagation persists below 0°C, no measurements of cold block were reported, except when the nerve tissue froze, which is not equivalent to cold block above freezing. It would be interesting to add antifreeze to the saline, but apparently this has never been done.

There is a clockwise rotation for the velocity–temperature plot for Antarctic fish nerves in comparison to similar plots from temperate poikilotherms. This clockwise rotation of a function-versus-temperature plot reflects a decrease in activation energy for the processes underlying membrane excitability and is seen in a wide variety of other studies on adaptive responses to temperature (Precht, 1958; Precht *et al.*, 1973). The enhancement of function at low temperatures is balanced by a relative loss of function at higher temperatures. This could be due, at least in part, to the loss of high-temperature stability often associated with increased molecular flexibility.

Moran and Melani (2001) report similar evidence of partial temperature compensation for nerve conduction in four species of arctic fishes: two eelpouts (*Lycodes* spp.), an eel blenny (*Lumpenus lampretaeformis*), and the arctic cod (*Boreogadus saida*). At their normal physiological temperature of 4°C, these species had conduction velocities approximately half that reported for the Mediterranean *Trachurus trachurus* at its physiological temperature of 20°C (Melani and Moran, 1998), whereas at 12°C, both fast and slow components of the compound action potential in arctic fish had conduction velocities exceeding those of *T. trachurus* measured at the same temperature. The upper temperature limit for conduction in the polar species was between 23 and 27°C, whereas it was about 30°C for the Mediterranean species.

One apparent difference between the effects of temperature on impulse conduction reported for arctic versus Antarctic fishes is the shape of the velocity–temperature curve. In the arctic species, the response is reported to be nonlinear and appears to follow a biphasic Arrhenius relationship, with breakpoints in the vicinity of 17°C. On the other hand, the response appears to be linear, with no breakpoints, in the Antarctic species (Macdonald, 1981). Subtle differences in experimental techniques may be the cause of this discrepancy. The two studies differed in the length of peripheral nerve obtainable and the number of recording sites used in the calculations made to determine conduction velocity. Regardless of these details, the two papers still agree on the important point that both groups of polar fishes show clear signs of partial compensation for low temperatures.

2. Neuromuscular Transmission

As the mechanism for distance transmission in the nervous system, conduction has a high "safety factor," continuing to operate well outside the animal's normal physiological limits. So in general terms, conduction of nerve action potentials is more robust than synaptic transmission, which is a more complex process, requiring a number of cellular processes, including anabolic and catabolic enzyme activity, exocytosis, receptor binding and ion channel activation/deactivation, and impulse conduction. The synapse is one of the major sites of information processing in the nervous system and the expectation is that synaptic function might be particularly susceptible to temperature induced disruption and an important site for temperature adaptation. Because of its relative accessibility, the neuromuscular junction has long acted as a model for synaptic transmission, and the information we have for temperature adaptation of synaptic function in polar fishes comes from studies of neuromuscular transmission.

As might be expected, all aspects of neuromuscular function persist to low temperatures in Antarctic fish (Macdonald and Montgomery, 1982; McVean and Montgomery, 1987). One way of measuring synaptic function is to study the miniature endplate currents (MEPCs) in the muscle caused by spontaneous release of packets of neurotransmitter from the motor nerve ending (Figure 9.2). The frequency of MEPCs is strongly temperature dependent and decreases with decreasing temperature. However, in Antarctic fish, this process shows strong temperature compensation (Table 9.2). Spontaneous synaptic release continues at subzero temperatures (Macdonald and Montgomery, 1982), as does release of transmitter evoked by a nerve impulse, retaining a relatively high quantal number (Pockett and Macdonald, 1986).

Low temperature also retards the closing of postsynaptic ion channels, causing the MEPC to return more slowly to baseline. In Antarctic fishes, this process of MEPC decay shows a substantial degree of temperature compensation. At the ambient McMurdo Sound water temperature of $-1.9°C$, the decay rate of individual MEPCs in Antarctic fish is much higher than would be expected for temperate or tropical fish measured at the same temperature. (Macdonald, 1983, 2000; Macdonald and Balnave, 1984; Macdonald and Montgomery, 1986; Cox and Macdonald, 2001). At $0°C$, MEPCs recorded from *P. borchgrevinki* extraocular muscle had a mean time constant of exponential decay equal to 2243 μs, which is not significantly different from that reported for two tropical species, *Scleropages jardini* (Osteoglossidae, 1805 μs) and *Megalops cyprinoides* (Megalopidae, 1649 μs), recorded at $20°C$. However, the ambient temperature for these tropical species was about $25°C$, so the Antarctic fish has not achieved a complete compensation.

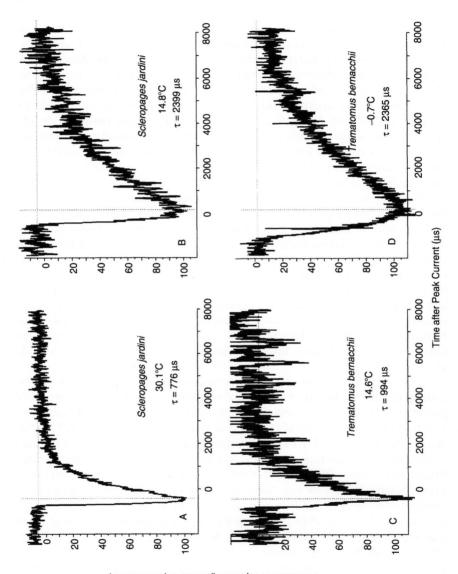

There is, however, an anomaly with the Australian tupong (*Pseudaphritis urvillii*, family Pseudaphritidae). Despite its being a temperate notothenioid, MEPC decay in this species seems to fit in better with its Antarctic relatives than with other temperate fishes (Macdonald, 2000). At 10°C, close to its normal ambient temperature, *Pseudaphritis* MEPCs had a time constant of 1126 μs, very similar to *P. borchgrevinki* (1006 μs), but in marked contrast to another non-notothenioid temperate fish *Trachurus novaezelandiae* (1666 μs [Macdonald and Balnave, 1984]). This may mean that fast MEPC decay at low temperature has a phylogenetic component in addition to adaptation to environmental conditions. However, part of the puzzle is that *Pseudaphritis* does not seem to have any evolutionary history in the Antarctic or any close living relatives there.

Table 9.2
Comparison of Thermal Properties of MEPCs in a Range of Fishes

Species	Habitat	T_r (°C)	T_a (°C)	T_{RL} (°C)	T_{RU} (°C)	τ_a μs	τ_{10} μs
S. jardini	Tropical	23–30	24	14	35	1600	3300
M. cypinoides	Tropical	23–30	24	13	32	1100	3000
P. urvillii	Temperate	5–15	8	8	30	1100	1000
P. borchgrevinki	Antarctic	−2	−2	−3	20	1900	1000

T_r is the ambient temperature range.
T_a temperature selected to represent normal ambient.
T_{RL} is the lower recording limit for MEPCs.
T_{RU} is the upper recording limit for MEPCs.
τ_a is the time constant interpolated for ambient temperature.
τ_{10} is the time constant extrapolated for a temperature of 10°C.
Source: Macdonald, 2000.

Figure 9.2. Miniature endplate currents (MEPCs) from a tropical teleost (*Scleropages jardini* [family Osteoglossidae]) and an Antarctic notothenioid fish *Trematomus bernacchii* (family Nototheniidae). The vertical axes give the normalized amplitude of the trans-membrane current as a percentage of the peak inward current. For each MEPC, the decay phase is approximately exponential and the time constant of decay (t) has been estimated by linear regression of the logarithm of amplitude versus time. The rate of decay is temperature sensitive and shows partial temperature compensation in the Antarctic fish. (A) MEPC from *Scleropages* extraocular muscle recorded at 30.1 °C; peak amplitude 431 μV. (B) *Scleropages* MEPC recorded at 14.8°C; peak amplitude 203 μV. The temperature drop of about 15° C has approximately trebled the MEPC duration. (C, D) MEPCs from *T. bernacchii* muscle recorded at 14.6°C and −0.7°C; peak amplitudes 226 and 130 μV, respectively. These MEPCs are recorded over the same temperature interval as for the tropical fish. The duration of *Trematomus* MEPCs is very similar to that of *Scleropages* at temperatures 15° higher, reflecting a high degree of temperature compensation. See also Table 9.2.

As previously mentioned, Baldwin (1971) observed a steep rise in the K_m of acetylcholine esterase for acetylcholine at temperatures only slightly above $0^{\circ}C$ in *P. borchgrevinki*. The rise in K_m denotes a fall in enzyme–substrate binding capacity and a concomitant reduction in the ability to break down acetylcholine. Coupled with the rapid rise in quantal release of acetylcholine from the motor endplate above $7^{\circ}C$ (Macdonald and Montgomery, 1982), this would lead to an accumulation of acetylcholine. The resulting neurochemical autointoxication may well be one of the causal factors underlying heat death in stenothermic polar fishes (Somero, 1991, 1995; Somero *et al.*, 1998).

3. TEMPERATURE COMPENSATION OF REFLEX AND FAST MOTOR BEHAVIOR

One way of evaluating the net effect of temperature compensation across the full range of neural subsystems and processes is to examine the function of integrated reflexes and motor behavior. The vestibuloocular reflex (VOR) is a particularly appropriate subject, as it is universal across vertebrates and has a clearly defined functional purpose: to stabilize the visual world on the retina. To achieve this purpose, the eye must rotate at the same rate but in the opposite direction of head movements. In engineering terms, the ratio of output to input of a system is called the "gain." So the desired gain of the VOR is a negative one: that is where eye movements are equal but opposite to head movements. The VOR of Antarctic fish has the appropriate gain to stabilize the visual field on the retina (Montgomery and Macdonald, 1985; Montgomery and McVean, 1987). In examining the temperature sensitivity of the constituent elements of the VOR, the counterintuitive finding is that the gain of motor systems actually increases with decreasing temperature. In the oculomotor system, the "gain" is defined as the degrees of eye rotation per impulse per second on the motor nerve. At lower temperatures, the rate of muscle action, particularly relaxation, is slowed, with the result that temporal summation (that is, the additive effect of overlapping muscle twitches) occurs at lower frequencies of nerve stimulation. This increase in neuromuscular gain may provide a degree of automatic temperature compensation, counteracting decreases in gain and maximum firing rates in sensory or central pathways.

In addition to slow compensatory eye movements, there is another class of eye movements called *saccades*. These provide for rapid repositioning of the eye. Vision is disrupted during saccades, hence the rapid nature of saccadic movements. Like many other performance measures of cold adaptation, saccadic movements in Antarctic fish are slower than those in temperate fish but faster than would be predicted for a temperate fish cooled to subzero temperatures (Montgomery and Macdonald, 1984). The rate of saccadic eye movements is a function of the viscoelastic properties of the eye movement system, the power capabilities of the muscle, and the drive to the muscle provided by neural input and neuromuscular transmission. The precise contribution of each of these elements to cold adaptation has not been determined, but it is likely that compensation of the neural drive to the muscle plays an important role.

Where muscle power is limiting, such as in fast-start behavior, there appears to be less capacity for temperature adaptation. Muscle power is the product of force and shortening velocity. At low temperature, muscle contracts more slowly, so power output decreases, which in turn limits the fast-start performance of cold-water species (Johnston and Johnston, 1991; Wakeling and Johnston, 1998). Burst-swimming performance is particularly poor in larval and juvenile Antarctic fish and shows little evidence for temperature compensation even in adults (Montgomery and Macdonald, 1984; Archer and Johnston, 1989).

C. The Brains of Polar Fishes

Over the past decade, J. T. Eastman and M. J. Lannoo have published a detailed series of reports on the central nervous systems (CNSs) of notothenioid and other Antarctic species. Although this chapter is oriented toward physiology rather than anatomy, the Eastman and Lannoo papers are briefly covered because they provide useful insights into CNS and sensory function across a wide range of Antarctic species.

Most notothenioids have bodies that are rounded in cross section, with a low aspect ratio. In a flattened fish species with a high aspect ratio, the head and brain cavity are also flattened and shortened. In Antarctic fish, the brain cavity remains fairly long and wide, and the brain itself is not compressed or folded anteroposteriorly. Thus, the constituent parts are easily visible externally in a more or less "primitive" configuration. In general, notothenioid brains are quite similar to the brains of temperate perciform fishes. Variation in brain morphology is evident in the family Nototheniidae, particularly in the sensory regions of the olfactory bulb, the eminentia granularis and the crista cerebellaris. The last two regions are related to the CNS processing of lateral line input. The cerebellum and telencephalon exhibit moderate variation in size, shape, and lobation patterns. Regulatory areas of the brain including the saccus vasculosus and the subependyma of the third ventricle are also variable. The degree of expansion of these regions is correlated inversely with environmental temperature, being greater in the high-latitude species living at lowest temperatures (Eastman and Lannoo, 1995; Lannoo and Eastman, 1995) (Figure 9.3).

Quantitative comparisons of the volume of specific brain areas reveal consistent differences related to ecological factors between six species of the genus *Trematomus*, three benthic species (*T. bernacchii, Trematomus pennellii,*

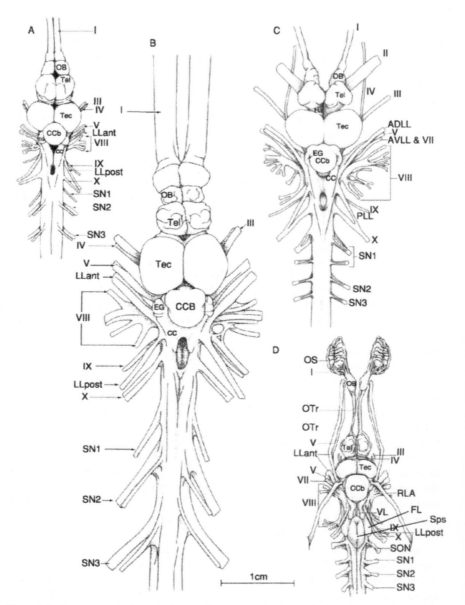

Figure 9.3. Brain and anterior spinal cord of representative Antarctic fish, drawn to the same scale. (A) *Trematomus bernacchii* (family Nototheniidae), a small (TL = 174 mm) benthic generalized predator. (B) *Dissostichus mawsoni* (family Nototheniidae), a large midwater predator, TL ~1000 mm. (C) *Chionodraco hamatus* (family Channichthyidae), a hemoglobin-free icefish (SL = 378 mm). (A), (B), and (C) are all members of the suborder Notothenioidei. (D) *Muraenolepis microps* (family

and *Trematomus scotti*) and three epibenthic species (*Trematomus eulepidotus, Trematomus lepidorhinus,* and *Trematomus loennbergii*) (Lannoo and Eastman, 2000). Epibenthic species tended to have larger olfactory lobes, cerebella and lateral line regions, whereas the benthic species have larger inferior lobes of the diencephalon and show a slight expansion of the telencephalon. Some other brain features are not related to position in the water column, but to other aspects of habitat. For example, the volume of optic tectum could be correlated with preferred depth, with the greatest volume in a shallow-water epibenthic species, *T. eulepidotus* (20% of total brain volume), and the smallest in a deep-water epibenthic species, *T. loennbergii* (13% of brain volume).

By way of comparison with Notothenioid fishes, Eastman and Lannoo (1998, 2001) have examined the morphology of the brains and the sense organs in the Antarctic snailfish (*Paraliparis devriesi,* Liparidae) and Antarctic eel cod (*Muraenolepis microps,* Muraenolepididae). Snailfish have well-developed olfactory lamellae, thick epithelia, large olfactory nerves and bulbs, and large telencephalic lobes. The retina contains only rods and has a high convergence ratio between receptors and ganglion cells, and the optic nerve is small. The corpus cerebellum is large, whereas the valvula is vestigial. The hindbrain includes expanded vagal and spinal sensory lobes. The cephalic lateral line is well developed, but the trunk line is reduced. The suggestion is that the expanded somatosensory input to the pectoral fin compensates for the reduction in the trunk lateral line. The brain of the eel cod is typical of the gadid group, with the only distinctive feature the diencephalic subependymal expansion previously found in the Liparid and Notothenioid species. An expanded cephalic lateral line system and associated expansion of brain regions associated with the lateral line is also characteristic of both shallow- and deep-dwelling members of the Bathydraconid subfamily Bathydraconini (Eastman and Lannoo, 2003).

The Antarctic dragonfishes (Notothenioidei: Bathydraconidae) occupy a wide depth range from surface waters to 3000 m. Bathydraconid brain

---

Muraenolepididae) a non-notothenioid eel-cod (TL = 236 mm). Antarctic fish brains. CC: crista cerebellaris of the rhombencephalon, CCb: corpus portion of the cerebellum, EG: eminentia granularis portion of the cerebellum, FL: facial lobe, LLant: anterior lateral line nerve, LLpost: posterior lateral line nerve, OB: olfactory bulb, OS: olfactory sac containing rosette, OTr: olfactory tract, RLA: ramus lateralis accessorius, SN1, SN2, SN3: 1st, 2nd, 3rd spinal nerves, SON: spino-occipital nerve, Sps: spinal sensory nucleus, Tec: tectum of the mesencephalon, Tel: telencephalon, VL: vagal lobe; Cranial nerves- I: olfactory nerve, II: optic nerve, III: oculomotor nerve, IV: trochlear nerve, V: trigeminal nerve, VII: facial nerve, VIII: auditory/vestibular nerve, IX: glossopharyngeal nerve, X: vagus nerve. (A, Modified from Eastman and Lannoo, 1995. B, Modified from Eastman, 1993. C, Modified from Eastman and Lannoo, 2004. D, Modified from Eastman and Lannoo, 2001.)

morphology of deeper living species shows a reduced telencephalon and tectum (Eastman and Lannoo, 2003). All species show duplex retinas, containing both rods and cones, but with considerable variation across species. All deep-dwelling members of the family show an expanded cephalic lateral line system, including large pores, wide membranous canals, hypertrophied canal neuromasts, and large anterodorsal lateral line nerves, eminentia granularis, and crista cerebellaris.

A detailed study of another Notothenioid family, the Channichthyidae, concluded that the most interesting features of brains are not in the nervous tissue but in support structures: The vasculature and the subependymal expansions show considerable elaboration (Eastman and Lannoo, 2004). Channichthyids have large accessory nasal sacs, and olfactory lamellae are more numerous than in other notothenioids (Figure 9.7). The eyes are relatively large and laterally oriented with similar duplex retinae in all eight species, and possess the most extensive system of hyaloid arteries known in teleosts (Figure 9.6). Cephalic lateral line canals are membranous and some exhibit extensions (canaliculi), but canals are more ossified than those of deeper living bathydraconids.

III. SENSORY PHYSIOLOGY OF POLAR FISHES

The sense organs are part of the nervous system and as such must be party to the same temperature considerations and adaptations as the peripheral nerves and the brain. However, little direct work has been done on the direct effects of temperature on sensory organs or on the presence or the nature of mechanisms to compensate for low temperatures. Most of our consideration of sensory systems revolves around the sensory environment of polar seas. Low light levels during winter, and even during summer under ice cover, mean that polar fishes share some of the sensory features of fish from other low-light environments, namely eyes geared to increased sensitivity and enhanced nonvisual sensory systems. Both the gross anatomy of the eye and the fine structure and biochemistry of the retina may be altered to enhance visual sensitivity. Nonvisual systems such as the mechanosensory lateral line and tactile and chemosensory systems show some evidence of enhanced function. In general terms, the sensory adaptations of polar fishes are not extreme (Figure 9.4) and reflect the phylogenetic origins of the fauna in addition to adaptive responses to the sensory environment.

A. Visual Function

Within the polar fish faunas, we can recognize an endemic Antarctic radiation of notothenioid fishes and the intrusion of deep-water–related groups with worldwide distribution such as zoarcids, liparids, and gadids

Figure 9.4. Lateral view of the Antarctic fish *Trematomus hansoni.* The eyes are well developed, but not outside typical values for diurnal coastal fishes in temperate waters. The most noticeable feature is the cephalic lateral line, with its enlarged pores (arrowheads). tll: trunk lateral line, na: nares. (Photo: Courtesy I. R. MacDonald, University of Auckland.)

(Eastman, 1993; Eastman and Clarke, 1998). There has been considerable work on the visual systems of fish within the notothenioid radiation, and we can recognize a number of sensory features that can be related to low-light environments (in addition to those mentioned earlier). Relative eye size (eye diameter as a proportion of head length) of notothenioids is roughly equivalent to that found in shallow-water teleosts, although some of the deeper water notothenioids have relatively larger eyes. Most have lateral directed vision, but a few species show a slight shift in visual axis that would enable them to use the brighter down-welling light for vision. This shift in the visual axis is reported for *Trematomus hansoni* (Pankhurst and Montgomery, 1989) and is apparent in *T. nicolai*. In his comprehensive study of the retinae of notothenioids, Eastman (1988) showed that notothenioids generally had duplex retinas with a low ratio of rods to cones (e.g., *T. bernacchii* 2:1, *T. hansoni* 7:1); only *Dissostichus mawsoni* had a rod dominated retina (rods to cone ratio of 57:1). Evidence of an adaptive response to low light levels was also found in the convergence ratio of receptor cells to ganglion cells, which correlated with light sensitivity. Of the notothenioids, *D. mawsoni* again shows the highest convergence ratio (58:1).

Functional studies of vision have depended on behavioral and electrophysiological estimates of visual threshold and spectral response (Pankhurst and Montgomery, 1989; Morita *et al.*, 1997). Both studies found that the spectral sensitivity of the eye of *P. borchgrevinki* has a single maximum at wavelengths of 480–490 nm corresponding to blue-green light. This sensitivity peak matches the spectral quality of down-welling light, at least in the early part of the summer.

Visual thresholds are limited theoretically by retinal "noise" due to random isomerization of visual pigments, which is a function of temperature. The prediction that thermal noise and visual thresholds should be strongly correlated and should both decrease at low temperatures has been supported by behavioral and electrophysiological studies of frogs, toads, and humans over a temperature range of 10–37°C (Aho *et al.*, 1988). This finding is of particular interest in considering the capabilities of vision in Antarctic fish; absolute thresholds in terms of photoisomerization events should be more than two orders of magnitude lower than for humans.

Pankhurst and Montgomery (1989) determined the visual threshold for feeding using behavioral methods and determined scotopic (dark-adapted) and photopic (light-adapted) thresholds electrophysiologically. The scotopic threshold was estimated to be between 0.001 and 0.005 μE m^{-2} s^{-1} These thresholds were then compared with the light profiles under the ice, and the conclusion drawn is that visual feeding should be possible down to about 40–60 m at midday under 1–2 m of fast ice. Morita *et al.* (1997) found visual thresholds 2 log units more sensitive than those reported earlier and argue that vision could be used successfully at correspondingly greater depths. However, visual feeding is critically dependent on target size and contrast and cannot easily be equated with scotopic vision. Moreover, differences in methodology, in particular the use of a retinal cup preparation and longer light flashes by Morita *et al.*, reduce the apparent discrepancy between these two studies to about an order of magnitude. Behavioral studies should provide a better guide to threshold light levels for feeding, but the caveat here is that it is impossible to recreate the normal environmental light conditions in experimental tanks. *P. borchgrevinki* has a lateral visual axis and must detect small, low-contrast prey against a low level horizontal background space light. Quite what this means for visual threshold is unclear, but the visual threshold for feeding under natural light conditions will likely lie somewhere between the best behavioral performance in the tank experiments and the absolute threshold of light detection for the dark-adapted eye given by Morita *et al.* (1997). However, even the lower threshold light value for feeding does not alter the conclusion that at some depth within the normal range of *P. borchgrevinki*, and at times of low sun angle or nighttime/winter darkness, the use of vision in feeding will be problematic and other senses may provide better prey detection and feeding opportunities.

One other line of evidence for the increased importance of the nonvisual sensory systems in Antarctic fish comes from the Montgomery and Sutherland (1997) study of the development of relative eye size and the timing of development of the parts of the brain associated with vision and lateral line input in *Pleuragramma antarcticum*. They found evidence for rapid somatic growth (decreasing relative eye size) during the first few summers of life, with little or no somatic growth during winter. Following the full development of lateral line structures, the inferred strong seasonal growth pattern evened out. These findings argue for an ontogenetic shift in feeding from vision-dominated and strongly seasonal to vision supplemented by mechanosense, with greater opportunity for winter feeding. The hypertrophy of the lateral line regions of the hindbrain in *P. antarcticum* provides additional evidence for the importance of mechanosensory systems in this species (Eastman, 1997).

Retinal function is highly dependent on oxygenation, and some fishes employ a combination of biochemistry and anatomy to ensure that the retina is adequately supplied with oxygen. In combination with a Root effect hemoglobin, and local acidification of the blood by secretion of lactic acid, the *rete mirabile* of the choroid body acts as a countercurrent multiplier to elevate the partial pressure of oxygen in retinal arteries (Pelster and Randall, 1998). Some of the Antarctic notothenioids possess choroid bodies (Eastman, 1993), and these are usually accompanied by Root effect hemoglobins (di Prisco, 1988). One functional consequence, a heightened sensitivity to movement in a species with a choroid body, has been demonstrated by behavioral studies (Herbert *et al.*, 2003).

Whereas a vestibuloocular response can be demonstrated easily in Antarctic fishes (Montgomery and McVean, 1987), it has been notoriously difficult to observe an optomotor response (OMR) mediated solely by movement of the visual field. Several investigators have made repeated attempts to elicit an OMR in *P. borchgrevinki*, but without success. This is one of the more active high Antarctic nototheniids, living near the top of the water column, swimming nearly continuously, and feeding mainly on free-swimming planktonic species.

Unexpectedly, the first unambiguous OMR came from *T. bernacchii*, a relatively sedentary demersal relative of *P. borchgrevinki*. The *T. bernacchii* OMR consisted of a directional swimming bias, with a threshold response to moving vertical bars subtending visual angles of 5 degrees (Figure 9.5). This is less sensitive than temperate and tropical fishes, which show angular thresholds of 2 and 3 degrees in *Oncorhynchus mykiss* and *Lutjanus carponotatus*, respectively (Herbert and Wells, unpublished data; Herbert *et al.*, 2002).

The critical difference between the two species seems to lie in their respective abilities to oxygenate the retina. *T. bernacchii* is one of the species

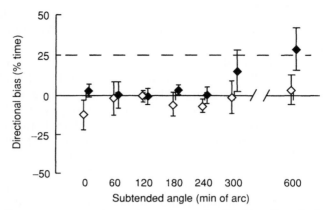

Figure 9.5. Optomotor response in two Antarctic nototheniids. The effect of subtended angle (0–600 minutes of arc) on the directional bias of *Pagothenia borchgrevinki* (◊) and *Trematomus bernacchii* (◆). Dashed line at 25% indicates the initiation of positive optomotor reactions (Harden-Jones, 1963). Error bars represent standard error. Two-way analysis of variance (ANOVA) highlighted a significant effect of subtended angle for *T. bernacchii* only. No specific significant differences were detected at any subtended angle. (From Herbert *et al.*, 2003.)

with both a choroid body (Eastman, 1988) and a Root effect hemoglobin (di Prisco *et al.*, 1988), whereas *P. borchgrevinki* lacks a choroid rete entirely, has only a minor Root effect component in its hemoglobin, and shows far less tendency to release O_2 from whole blood at low pH (Bohr effect) than *T. bernacchii* (Tetens *et al.*, 1984). The choroid rete is known to elevate ocular pO_2 in other teleosts (Wittenberg and Wittenberg, 1974), and in turn higher pO_2 has been shown to enhance retinal function (Fonner, 1973). Predictably the intraocular pO_2 is significantly higher in *T. bernacchii* than in *P. borchgrevinki* (Herbert *et al.*, 2003). Thus, the ability to maintain a high ocular pO_2 is a critical factor favoring an OMR in *T. bernacchii*, a conclusion supported also by the decrement in optomotor performance seen in tropical fish subjected to strenuous exercise and low arterial pO_2 (Herbert *et al.*, 2002).

The retinal dependence on oxygenation is apparent also in the organization of ocular blood vessels. In the absence of erythrocytes, hemoglobin, and choroid body, icefishes (family Channichthyidae) maintain an adequate supply of oxygen to the retina by means of a uniquely dense network of anastomosing hyaloid arteries over the vitreal (inner) surface of the retina (Eastman and Lannoo, 2004) (Figure 9.6B). In *P. borchgrevinki*, possessing hemoglobin but lacking a choroid rete, the hyaloid arteries are thinner, less branched, and more sparsely distributed (Figure 9.6A). In those red-blooded nototheniids, such as *T. bernacchii*, with a choroid body capable of boosting the intraocular pO_2, the pattern of hyaloid arteries is even less dense (Eastman, 1988).

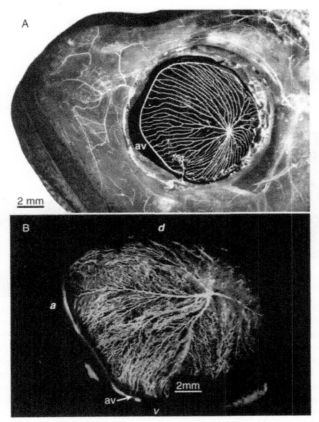

Figure 9.6. Ocular vasculature in two Antarctic notothenioid fishes. Both photos have the same orientation and scale, and show hyaline arteries perfused with microfil, oriented as for the left eye ([A] has been flipped horizontally). (A) *Pagothenia borchgrevinki*: The hyaloid arteries branch dichotomously before joining the annular vein (av). (B) *Chionodraco myersi*, an icefish (family Channichthyidae): The hyaloid arteries arise from five main branches and repeatedly anastomose to form a network that is much denser than that of the red-blooded species. a, anterior; av, annular vein; d, dorsal; p, posterior; v, ventral. (A, Modified from Eastman, 1988. B, Modified from Eastman and Lannoo, 2004.)

B. Chemical Senses

Olfaction and gustation are both chemosensory systems. There have been morphological and behavioral studies of these senses in polar fish, but there are no functional or physiological studies of either sense. All five of the notothenioid families are characterized by a single external nostril on

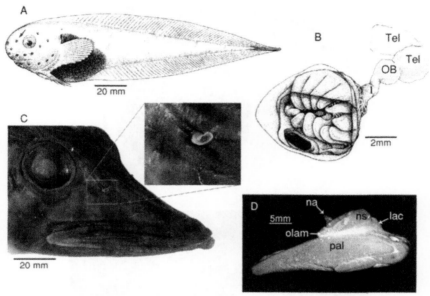

Figure 9.7. Olfactory organs of antarctic fish. (A) The liparid (snailfish) *Paraliparis devriesi*, showing location of the left naris (arrow) and of lateral line pores. (B) Left olfactory organ and neural connections of *Paraliparis devriesi*. The opening of the accessory olfactory sac shows as the dark oval at the bottom of the figure. (C) Head of *Chionodraco hamatus* (family Channichthyidae), showing the right naris; the area immediately surrounding the opening of the olfactory sac is shown in an enlargement. (D) Internal cast of the right olfactory organ and accessory nasal sac of *Chionodraco hamatus*. lac, lacrimal portion of accessory nasal sac; na, nasal aperture; ns, nasal sac; olam, olfactory lamellae; pal, palatal portion of accessory nasal sac. (A, From Eastman and Lannoo, 1998. B, Modified from Eastman and Lannoo, 1998. C, Photo courtesy I. R. MacDonald, University of Auckland. D, Modified from Eastman and Lannoo, 2004.)

either side. The single naris is associated with an accessory nasal sac (Iwami, 1986; Eastman and Lannoo, 2004), which contributes to water exchange over the lamellae of the olfactory epithelium, driven by pressure fluctuations from the buccal cavity via the soft palatal tissues (J. Janssen, personal communication) (Figure 9.7). Meyer-Rochow (1981) examined the distribution and morphology of taste buds on the tongues of three Antarctic nototheniid species with the scanning electron microscope. A relatively low average density of taste receptors was found in the toothfish *D. mawsoni* (mean interreceptor distance = 1000 μm), whereas receptor density was five to seven times greater in *P. borchgrevinki* and *T. bernacchii* (mean interreceptor distance = 200 and 150 μm, respectively), about the same as that found in the rainbow trout (*O. mykiss*: 150 μm). The tentative conclusion was that the abundance of taste receptors is higher in fish with a greater variety of prey species from which to choose. Fanta *et al.* (2001) studied the morphology of chemosensory systems and feeding behavior of the Antarctic eelpout *Ophthalmolycus amberensis*. These fish possess a large olfactory epithelium and a high density of taste buds on the snout and close to the nostrils. Behavioral studies show that feeding responses are mainly induced by chemical stimuli. Likewise, the Antarctic snailfish, *Paraliparis devriesi*, and *Muraenolepis microps*, an eel cod, both show well-developed olfactory systems, and cutaneous taste buds (Eastman and Lannoo, 1998, 2001).

C. Mechanosenses

As conditions for vision deteriorate, nonvisual sensory systems should become relatively more important. Mechanosensory systems form an important class of nonvisual senses and include both the tactile senses and the octavolateralis senses of the inner ear and lateral line.

1. TACTILE SENSES, BARBELS, AND FIN RAYS

Tactile senses are most obvious in the Antarctic plunderfish (Artedidraconidae), where each fish bears a single highly innervated barbel on the chin. These mental barbels are reminiscent of the tactile organs borne by deep-sea fishes, and physical contact with the barbel has been shown to evoke a strong feeding response (Janssen *et al.*, 1993). Tactile receptors on the pectoral fin rays of benthic species appear to play an important role in alerting fish to the presence of prey. In *T. bernacchii* and *T. pennellii*, light touch of the pectoral fin generated a repositioning response to bring the head and its associated lateral line organs close to the source of the contact (Janssen, 1996). The Antarctic snailfish (*Paraliparis devriesi*) is reported to have "expanded somatosensory innervation" in the lower rays of its pectoral fins, in addition to the taste buds mentioned earlier (Eastman and Lannoo, 1998).

2. INNER EAR ORGANS OF BALANCE AND HEARING

The inner ear, or vestibular system, has both otolithic organs and semicircular canals. Otolithic organs have a dense calcareous otolith nested on a bed of mechanosensory hair cells, in effect a differential density accelerometer. Otolithic input plays an important role in balance and posture and provides the principle sense of hearing in teleost fishes. Little is known of the importance of hearing in polar fishes, and there have been no physiological studies of otolithic function.

The semicircular canals are angular accelerometers and provide information on rotational movement of the head. Input from the semicircular canals drives the VOR (see Section II.B, p. 362) and is important in the control of movement. Electrophysiological recordings from secondary vestibular neurons in the hindbrain of *P. borchgrevinki* show firing rate responses that are virtually identical to those found in temperate fish (Montgomery and McVean, 1987). The only features that differed were a limited upper firing frequency of these neurons and a higher gain (i.e., an elevated change in firing frequency for a given head rotation) in comparison with temperate fish. The elevated gain can be related to the size of the labyrinth in this species (Macdonald and Montgomery, 1991). The larger labyrinth and the increased rotational sensitivity may be temperature adaptations in their own right or may be (perhaps fortuitous) consequences of the relatively large head and rounded profile of notothenioid fish.

3. MECHANOSENSORY LATERAL LINE

The mechanosensory lateral line of some species shows features that are correlated with increased sensitivity and match lateral line features found in fish from other low-light habitats. For example, some have head canal systems with large pores and widened canals, and the family Channichthyidae have semi-membranous canals. Behavioral studies also show the ability of Antarctic fish to capture prey solely on the basis of hydrodynamic cues (Janssen, 1996). The form and physiology of the lateral line system in notothenioid fish has been examined in a series of papers (Montgomery and Macdonald, 1987; Montgomery *et al.*, 1988; Montgomery and Coombs, 1992; Coombs and Montgomery, 1994a,b; Montgomery *et al.*, 1994, 1997; Carton and Montgomery, 2002).

The sensory elements of the mechanosensory lateral line consist of "neuromasts," patches of hair cells coupled to the immediate fluid environment with a gelatinous cupula. Flow over the cupula activates the hair cells. Neuromasts are found on the surface of the skin as "superficial neuromasts" or are embedded in subdermal canals known as "canal neuromasts." An extensive body of work shows that the canals operate as mechanical filters, reducing the response of the canal systems to the flow of water over the skin of the fish, but preserving their response to higher frequency oscillating water movements. So in effect, the superficial subsystem is mainly responsible for the detection of bulk flow over the surface of the skin and, in Antarctic fish, as in other species, is involved in rheotactic behavior (Montgomery *et al.*, 1997; Baker and Montgomery, 1999; Carton and Montgomery, 2002). The canal system is protected from the direct effects of flow but is highly sensitive to the oscillatory water movements produced in the water by the movement of other animals. Rephrasing this in more formal terms, with respect to the velocity of water movements, canal neuromasts have a "bandpass" filter response. They are not highly tuned but do show a maximal response in the region of 40 Hz. Recordings of the vibrations produced by swimming plankton show a significant amount of vibration energy at this frequency, and simultaneous recording of plankton movement and lateral line inputs shows that plankton swimming is a potent stimulus to the lateral line (Montgomery and Macdonald, 1987; Montgomery *et al.*, 1988).

Detailed electrophysiological and hydrodynamical studies of lateral line function over a range of notothenioid fish show that the functional characteristics of lateral line end organs are remarkably consistent within and across species despite a considerable range of morphological variation (Montgomery and Coombs, 1992; Coombs and Montgomery, 1994a, b; Montgomery *et al.*, 1994) (Figure 9.8). The upper roll-off frequency (>40 Hz) appears to be determined by the effects of low temperature on the hair cell/afferent fiber system. The low-frequency characteristics of canal organs are remarkably similar despite the semi-membranous canals found in channichthyids, as well as the large wide canals of *D. mawsoni*. This latter species is much larger than the other species studied, and the lateral line canals are correspondingly wide. Modeling studies predict that canals this wide would not provide an effective low-frequency filter. However, close inspection shows that there are narrow sections of the *D. mawsoni* canal in the region of the neuromast, and modeling studies confirm that this would be effective in restoring the filter function of the canals (Figure 9.8A). So the range of morphological variation observed may not arise from adaptive evolution to differing lateral line demands, so much as maintaining the effective function of the lateral line system as other factors (such as body size) change.

In a similar vein, there is evidence for pedomorphic change in notothenioid species that reinvaded the pelagic habitat. From a lateral line point of view, pedomorphic change can result in incomplete formation of the lateral line canal. For example, this is evident on the trunk canal of *P. antarcticum*. Loss of the canal mechanical filter could be detrimental to effective lateral line function; however, sensory competency can be restored by behavior means (remaining motionless in the water column) or perhaps by additional central processing of sensory information (Montgomery and Clements, 2000).

IV. CONCLUDING REMARKS

Considerable progress has been made in understanding neural and sensory function in polar fishes. A broad outline can be painted of the nature of temperature adaptation in proteins and membranes through to cellular and neural systems (Table 9.1). We have a general sketch of their sensory world, as well as their evolutionary responses to the cold, dark, and relatively quiet world they inhabit. It is not surprising that these outlines are still sketchy. Animals are complex systems, and even for the best-studied

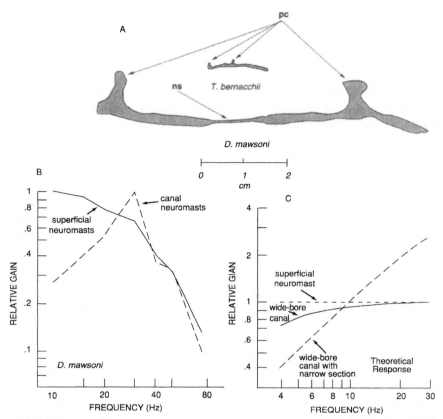

Figure 9.8. Lateral line form and function. (A) Morphology of the lateral line canals from the small *Trematomus bernacchii* (TL = 200 mm) and the much larger *Dissostichus mawsoni* (TL = 1000 mm). The profiles were determined by injection of Bateson's casting compound into the canals. The upwards projections on each canal (pc) represent the tubules connecting the canals to pore openings. Note the narrow section (ns) in the canal of *D. mawsoni*, where the neuromast is located. (B) Frequency response curves (gain vs. frequency, constant velocity stimulus) for superficial neuromasts (solid line) and canal neuromasts (dashed line) for *D. mawsoni*. The form of the curve for the superficial neuromasts is that of a low-pass filter, whereas low frequencies are attenuated by the thin section of the canal for the canal neuromasts. (C) Expected response of a superficial neuromast (horizontal line), wide-bore canal (solid line), and wide-bore canal with narrow section (dashed line) to a constant maximum velocity stimulus. A wide-bore canal only attenuates very low frequencies, whereas the effect of the narrow section of the canal is to increase the attenuation of low frequencies and amplify higher frequencies. (Modified from Montgomery *et al.*, 1994.)

laboratory models, we still lack complete descriptions of the complex interacting dynamics of cellular and neural function and find the complex
interplay between vision and other nonvisual sensory systems frustratingly
elusive. The techniques we have to observe, manipulate, and model neural
processes, brain function, and sensory interactions have their limitations. So,
what are the practical prospects for advancing our understanding of
the neural/sensory physiology of Antarctic fish; what is it we would like
to know?

For neural systems, it would be nice to be able to better document the
potentials and limitations for temperature adaptation. We have a range from
full compensation (as in membrane homeoviscosity) through to no compensation (as in muscle power output). There is still not a satisfying synthesis
across this range of the underlying mechanisms and constraints. Where is
compensation constrained by the direct effects of temperature on physical
and chemical processes such as diffusion, fluid viscosity, and the rate of
chemical reactions? What is the coupling between low-temperature adaptation and stenothermy? Which processes are critical for systems function
and which are not? Are the more critical processes those pushed closer to
their physical constraints? Antarctic fish are the best vertebrate model
for stenothermic temperature adaptation. The answers to these questions
will be of relevance not only to Antarctic fishes, but also to fundamental
understanding of temperature and the nervous system.

With respect to sensory systems, the notothenioid radiation also provides
a model system, not only in terms of the known evolutionary history and
phylogeny of the group, but also in terms of the distinctive features of the
Antarctic environment. Attention has already been drawn to the parallels
between the high-latitude Antarctic and the deep sea. Notothenioid species
do show some modest convergence with true deep-sea fishes in some aspects
of sensory biology, including aspects of their nonvisual sensory systems
(Montgomery and Macdonald, 1998). The lack of extreme specialization
can be interpreted in terms of the much more recent evolutionary history
of Antarctic fish in comparison with the deep-sea fauna and in terms of
the strong seasonality of light levels in the Antarctic which does not match
the year-round darkness of the deep sea. In addition to the distinctive
physical features of the Antarctic environment, Antarctic ecosystems are
not as diverse as those in temperate or tropical systems. Thus, we have a
combination of simplifying factors: an extreme but uniform environment,
with minimal temperature fluctuations in spite of major seasonal changes
in illumination, low faunal diversity, and a relatively recent monophyletic
radiation. The combination of these factors provides unique opportunities for matching sensory physiology to functional requirements. Initial
attempts to define light levels for visual feeding and correlate them with

environmental light levels might well be open to criticism, but the interaction between *P. borchgrevinki* and its planktonic prey in the simplified visual environment of the subice pelagic zone does at least offer the prospect of measuring sensory function in a biologically meaningful way. The general point is that the evolutionary and environmental context of the Antarctic fishes does provide an exciting model system for the study of sensory physiology, with significance well beyond the Antarctic perspective.

Finally, as mentioned at the beginning of this chapter, the bulk of our knowledge of the neuro/sensory physiology of polar fishes comes from Antarctic species. This may have resulted from historical accident, but as indicated earlier, there are also good biological reasons why the notothenioid radiation in particular should be the subject of special attention. However, a broad sweep of comparative studies on other Antarctic and Arctic species is also warranted to fill out a more compete picture of the neuro/sensory physiology of polar fishes.

ACKNOWLEDGMENTS

Figure 9.3 B: Copyright (1993, Academic Press.

Figure 9.3 A, C, D; Figure 9.6 A, B; Figure 9.7 A, B, D; Copyright (1988, 1995, 1998, 2004. John wiley & Sons, Inc. Reprinted with permission of Wiley-Liss, Inc., a subsidiary of John Wiley & Sons, Inc.

Figure 9.5, Herbert *et al.*, 2003, *Polar Biology* **26**, p. 412, Fig. 1: Copyright (2003, Springer-Verlag.

Figure 9.8: Copyright (1994. S. Karger AG, Basel.

REFERENCES

Aho, A. C., Donner, K. O., Hyden, C., Larsen, L. O., and Reuter, T. (1988). Low retinal noise in animals with low body temperature allows high visual sensitivity. *Nature* **334**, 348–350.

Archer, S. D., and Johnston, I. A. (1989). Kinematics of labriform and subcarangiform swimming in the Antarctic fish *Notothenia coriiceps. J. Exp. Biol.* **143**, 195–210.

Baldwin, J. (1971). Adaptation of enzymes to temperature: acetylcholinesterases in the central nervous system of fishes. *Comp. Biochem. Physiol.* **40B**, 181–187.

Baker, C. F., and Montgomery, J. C. (1999). Lateral line mediated rheotaxis in the Antarctic fish *Pagothenia borchgrevinki. Polar Biol.* **21**, 305–309.

Bass, E. L. (1971). Temperature acclimation in the nervous system of the brown bullhead (*Ictalurus nebulosus*). *Comp. Biochem. Physiol.* **40A**, 833–849.

Behan-Martin, M. K., Jones, G. R., Bowler, K., and Cossins, A. R. (1993). A near perfect temperature adaptation of bilayer order in vertebrate brain membranes. *Biochim. Biophys. Acta* **1151**, 216–222.

Becker, K., Wohrmann, P. A., and Rahmann, H. (1995). Brain gangliosides and cold-adaptation in high-Antarctic fish. *Biochem. Syst. Ecol.* **23**, 695–707.

Carton, A. G., and Montgomery, J. C. (2002). Responses of lateral line receptors to water flow in the Antarctic notothenioid, *Trematomus bernacchii*. *Polar Biol.* **25**, 789–793.

Chatfield, P. O., Battista, A. F., Lyman, C. P., and Garcia, J. P. (1948). Effects of cooling on nerve conduction in a hibernator (golden hamster) and nonhibernator (albino rat). *Am. J. Physiol.* **155**, 179–185.

Chatfield, P. O., Lyman, C. P., and Irving, L. (1953). Physiological adaptation to cold of peripheral nerve in the leg of the herring gull (*Larus argentatus*). *Am. J. Physiol.* **172**, 639–644.

Chessa, J. P., Petrescu, I., Bentahir, M., Van Beeumen, J., and Gerday, C. (2000). Purification, physico-chemical characterization and sequence of a heat labile alkaline metalloprotease isolated from a psychrophilic *Pseudomonas* species. *Biochim. Biophys. Acta Protein Structure Mol. Enzymol.* **1479**, 265–274.

Ciardiello, M. A., Camardella, L., Carratore, V., and di Prisco, G. (2000). L-glutamate dehydrogenase from the Antarctic fish *Chaenocephalus aceratus*—Primary structure, function and thermodynamic characterisation: relationship with cold adaptation. *Biochim Biophys Acta Protein Structure Mol. Enzymol.* **1543**, 11–23.

Clarke, A., and Johnston, I. A. (1996). Evolution and adaptive radiation of Antarctic fishes. *Tree* **11**, 212–218.

Coombs, S., and Montgomery, J. C. (1994a). Function and evolution of superficial neuromasts in an Antarctic notothenioid fish. *Brain Behav. Evol.* **44**, 287–298.

Coombs, S., and Montgomery, J. C. (1994b). Fibers innervating different parts of the lateral line system of an Antarctic notothenioid, *Trematomus bernacchii*, have similar frequency responses, despite large variation in the peripheral morphology. *Brain Behav. Evol.* **40**, 217–233.

Cossins, A. R. (1977). Adaptation of biological membranes to temperature. The effect of temperature acclimation of goldfish upon the viscosity of synaptosomal membranes. *Biochim. Biophys. Acta* **470**, 395–411.

Cossins, A. R. (1994). Homeoviscous adaptation of biological membranes and its functional significance. *In* "Temperature Adaptation of Biological Membranes" (Cossins, A. R., Ed.), pp. 63–75. Portland Press, Portland.

Cossins, A. R., and Prosser, C. L. (1978). Evolutionary adaptation of membranes to temperature. *Proc. Natl. Acad. Sci. USA* **75**, 2040–2043.

Cox, S. L., and Macdonald, J. A. (2001). Temperature compensation of postsynaptic currents from the extraocular muscle of temperate teleost fishes. *J. Comp. Physiol. A* **187**, 197–203.

Crockett, E. L., and Sidell, B. D. (1990). Some pathways of energy metabolism are cold adapted in Antarctic fishes. *Physiol. Zool.* **63**, 472–488.

Detrich, H. W. (1991). Polymerization of microtubule proteins from Antarctic fish. *In* "Biology of Antarctic Fish" (Di Prisco, G., Maresca, B., and Tota, B., Eds.), pp. 248–262. Springer-Verlag, Berlin.

Detrich, H. W. (1997). Microtubule assembly in cold-adapted organisms: functional properties and structural adaptations of tubulins from Antarctic fishes. *Comp. Biochem. Physiol.* **118A**, 501–513.

Detrich, H. W. (1998). Molecular adaptations of microtubules and microtubule motors from Antarctic fish. *In* "Fishes of Antarctica: A Biological Overview" (di Prisco, G., Pisano, E., and Clarke, A., Eds.), pp. 139–149. Springer-Verlag, Italia.

Detrich, H. W., Parker, S. K., Williams, R. C., Jr., Nogales, E., and Downing, K. H. (2000). Cold adaptation of microtubule assembly and dynamics: Structural interpretation of primary sequence changes present in the alpha- and beta-tubulins of Antarctic fishes. *J. Biol. Chem.* **275**, 37038–37047.

Dierolf, B. M., and Brink, P. M. (1973). Effects of thermal acclimation on cable constants of the earthworm median giant axon. *Comp. Biochem. Physiol.* **44A,** 401–406.

di Prisco, G., Giardina, B., D'Avino, R., Condò, S. G., Bellelli, A., and Brunori, M. (1988). Antarctic fish hemoglobin: an outline of the molecular structure and oxygen binding properties. II. Oxygen binding properties. *Comp. Biochem. Physiol.* **90B,** 585–591.

Eastman, J. T. (1988). Ocular morphology in Antarctic notothenioid fishes. *J. Morphol.* **196,** 283–306.

Eastman, J. T. (1993). "Antarctic Fish Biology: Evolution in a Unique Environment." Academic Press, San Diego.

Eastman, J. T. (1997). Phyletic divergence and specialization for pelagic life in the Antarctic Notothenioid fish *Pleuragramma antarcticum* . *Comp. Biochem. Physiol.* **118A,** 1095–1101.

Eastman, J. T., and Clarke, A. (1998). A comparison of adaptive radiations of Antarctic fish with those of non-Antarctic fish. *In* "Fishes of Antarctica: A Biological Overview" (Di Prisco, G., Pisano, E., and Clarke, A., Eds.), pp. 3–28. Springer-Verlag, Berlin.

Eastman, J. T., and Lannoo, M. J. (1995). Diversification of brain morphology in Antarctic notothenioid fishes: basic descriptions and ecological considerations. *J. Morphol.* **223,** 47–83.

Eastman, J. T., and Lannoo, M. J. (1998). Morphology of the brain and sense organs in the snailfish *Paraliparis devriesi*: Neural convergence and sensory compensation on the Antarctic shelf. *J. Morphol.* **237,** 213–236.

Eastman, J. T., and Lannoo, M. J. (2001). Anatomy and histology of the brain and sense organs of the Antarctic eel cod *Muraenolepis microps* (Gadiformes; Muraenolepididae). *J. Morphol.* **250,** 34–50.

Eastman, J. T., and Lannoo, M. J. (2003). Diversification of brain and sense organ morphology in Antarctic dragonfishes (Perciformes: Notothenioidei: Bathydraconidae). *J. Morphol.* **258,** 130–150.

Eastman, J. T., and Lannoo, M. J. (2004). Brain and sense organ anatomy and histology in hemoglobinless Antarctic icefishes (Perciformes: Notothenioidei: Channichthyidae). *J. Morphol.* **260,** 117–140.

Fanta, E., Sant' Anna Rios, F., Meyer, A. A., Grotzner, S. R., and Zaleski, T. (2001). Chemical and visual sensory systems in feeding behaviour of the Antarctic fish *Ophthalmolycus amberensis* (Zoarcidae). *Antarctic Record* (Tokyo) **45,** 27–42.

Feller, G., Arpigny, J. L., Narinx, E., and Gerday, C. H. (1997). Molecular adaptations of enzymes from psychrophilic organisms. *Comp. Biochem. Physiol.* **118A,** 495–499.

Feller, G., and Gerday, C. (1997). Psychrophilic enzymes: molecular basis of cold adaptation. *Cell. Mol. Life Sci.* **53,** 830–841.

Fonner, D. B., Hoffert, J. R., and Fromm, P. O. (1973). The importance of the counter current oxygen multiplier mechanism in maintaining retinal function in the teleost. *Comp. Biochem. Physiol.* **46A,** 559–567.

Genicot, S., Rentier-Delrue, F., Edwards, D., VanBeeumen, J., and Gerday, C. (1996). Trypsin and trypsinogen from an Antarctic fish: Molecular basis of cold adaptation. *Biochim. Biophys. Acta* **1298,** 45–57.

Harden-Jones, F. R. (1963). The reaction of fish to moving backgrounds. *J. Exp. Biol.* **40,** 437–446.

Herbert, N. A., Wells, R. M. G., and Baldwin, J. (2002). Correlates of choroid rete development with the metabolic potential of various tropical reef fish and the effect of strenuous exercise on visual performance. *J. Exp. Mar. Biol. Ecol.* **275,** 31–46.

Herbert, N. A., Macdonald, J. A., Wells, R. M. G., and Davison, W. (2003). A difference in optomotor behaviour of two Antarctic nototheniid fishes is correlated with the presence of a choroid rete mirabile and Root effect. *Polar Biol.* **26,** 411–415.

Iwami, T. (1986). A note on the nasal structures of fishes of the suborder Notothenioidei (Pisces, Perciformes). *Mem. Natl. Inst. Polar Res. Spec. Issue* **44**, 151–152.

Janssen, J. (1996). Use of the lateral line and tactile senses in feeding in four Antarctic nototheniid fishes. *Envir. Biol. Fish.* **47**, 51–64.

Janssen, J., Slattery, M., and Jones, W. (1993). Feeding responses to mechanical stimulation of the barbel in *Histiodraco velifer* (Artedidraconidae). *Copeia* **1993**, 885–889.

Johnston, T. P., and Johnston, I. A. (1991). Temperature adaptation and the contractile properties of live muscle fibres from teleost fish. *J. Comp. Physiol.* **16B**, 27–36.

Johnston, I. A., Guderley, H. E., Franklin, C. E., Crockford, T., and Kamunde, C. (1994). Are mitochondria subject to evolutionary temperature adaptation? *J. Exp. Biol.* **195**, 293–306.

Kehl, T. H., and Morrison, P. R. (1960). Peripheral nerve function and hibernation in the thirteen-lined ground squirrel, *Spermophilus tridecemlineatus*. *Bull. Mus. Comp. Zool. Harvard Univ.* **124**, 387–402.

Lagerspetz, K. Y. H., and Talo, A. (1967). Temperature acclimation of the functional parameters of the giant nerve fibres in *Lumbricus terrestris* L. I. Conduction velocity and the duration of the rising and falling phase of the action potential. *J. Exp. Biol.* **47**, 471–480.

Lannoo, M. J., and Eastman, J. T. (1995). Periventricular morphology in the diencephalons of Antarctic notothenioid teleosts. *J. Comp. Neurol.* **361**, 95–107.

Lannoo, M. J., and Eastman, J. T. (2000). Nervous and sensory system correlates of an epibenthic evolutionary radiation in Antarctic notothenioid fishes, genus *Trematomus* (Perciformes; Nototheniidae). *J. Morphol.* **245**, 67–79.

Leiros, H. K. S., Willassen, N. P., and Smalas, A. O. (2000). Structural comparison of psychrophilic and mesophilic trypsins—Elucidating the molecular basis of cold-adaptation. *Eur. J. Biochem.* **267**, 1039–1049.

Littlepage, J. L. (1965). Oceanographic investigations in McMurdo Sound, Antarctica. *In* "Biology of the Antarctic Seas" (Llano, G. A., Ed.), Vol. 2, pp. 1–37. American Geophysical Union, Washington, DC..

Macdonald, J. A. (1981). Temperature compensation in the peripheral nervous system: Antarctic vs. temperate poikilotherms. *J. Comp. Physiol.* **142**, 411–418.

Macdonald, J. A. (1983). Fast decay of fish synaptic currents. *Experientia* **39**, 230–231.

Macdonald, J. A. (2000). Neuromuscular temperature compensation in fish: a comparative study. *In* "Antarctic Ecosystems: Models for Wider Ecological Understanding" (Davison, W., Howard-Williams, C., and Broady, P., Eds.), pp. 134–141. New Zealand Natural Sciences, Christchurch.

Macdonald, J. A., and Balnave, R. J. (1984). Miniature end plate currents from teleost extraocular muscle. *J. Comp. Physiol.* **155**, 649–659.

Macdonald, J. A., and Ensor, D. R. (1975). Low temperature nerve function in Antarctic marine poikilotherms. *Antarct. J. US* **10**, 130–132.

Macdonald, J. A., and Montgomery, J. C. (1982). Thermal limits of neuromuscular function in Antarctic fish. *J. Comp. Physiol.* **147**, 237–250.

Macdonald, J. A., and Montgomery, J. C. (1986). Rate compensated synaptic events in Antarctic fish: Consequences of homeoviscous cold adaptation. *Experientia* **42**, 806–808.

Macdonald, J. A., and Montgomery, J. C. (1991). The sensory biology of notothenioid fishes. *In* "Biology of Antarctic Fishes" (di Prisco, G., Maresca, B., and Tota, B., Eds.), pp. 145–162. Springer-Verlag, New York.

Macdonald, J. A., Montgomery, J. C., and Wells, R. M. G. (1987). Comparative physiology of Antarctic fishes. *Adv. Marine Biol.* **24**, 321–388.

Maresca, B., Patriarca, E., Goldenberg, C., and Sacco, M. (1988). Heat shock and cold adaptation in Antarctic fishes: a molecular approach. *Comp. Biochem. Physiol.* **90B**, 623–629.

McDonald, H. S., Chiaramonte, G., and Tarnowski, L. E. (1963). Influence of temperature acclimation on conduction velocity in sciatic nerves of *Rana pipiens*. *Am. Zool.* **3,** 547–548.

McVean, A. R., and Montgomery, J. C. (1987). Temperature compensation in myotomal muscle: Antarctic vs. temperate fish. *Env. Biol. Fishes* **19,** 27–33.

Melani, R., and Moran, O. (1998). Characterization of the conduction properties in Mediterranean fish peripheral nerves. *Ital. J. Zool.* **65,** 155–158.

Meyer-Rochow, V. B. (1981). Fish tongues—surface fine structures and ecological considerations. *Zool. J. Linnean Soc.* **71,** 413–426.

Miller, L. K. (1970). Temperature-dependent characteristics of peripheral nerves exposed to different thermal conditions in the same animal. *Can. J. Zool.* **48,** 75–81.

Miller, L. K., and Irving, L. (1967). Temperature-related nerve function in warm- and cold-climate muskrats. *Am. J. Physiol.* **213,** 1295–1298.

Montgomery, J. C., and Clements, K. (2000). Disaptation and recovery in the evolution of Antarctic fish. *Trends Ecol. Evol.* **15,** 267–271.

Montgomery, J. C., and Macdonald, J. A. (1984). Performance of motor systems in Antarctic fishes. *J. Comp. Physiol.* **154A,** 241–248.

Montgomery, J. C., and Macdonald, J. A. (1985). Oculomotor function at low temperature: Antarctic vs. temperate fish. *J. Exp. Biol.* **117,** 181–192.

Montgomery, J. C., and Macdonald, J. A. (1987). Sensory turning of lateral line receptors in Antarctic fish to the movements of planktonic prey. *Science* **235,** 195–196.

Montgomery, J. C., and Macdonald, J. A. (1998). Evolution of sensory systems: A comparison of Antarctic and deep-sea ichthyofauna. *In* "Fishes of Antarctica: A Biological Overview" (Di Prisco, G., Pisano, E., and Clarke, A., Eds.), pp. 329–338. Springer-Verlag, Berlin.

Montgomery, J. C., Macdonald, J. A., and Housley, G. D. (1988). lateral line function in an Antarctic fish related to the signals produced by planktonic prey. *J. Comp. Physiol.* **163A,** 827–834.

Montgomery, J. C., and McVean, A. R. (1987). Brain function in Antarctic fish: activity of central vestibular neurons in relation to head rotation and eye movement. *J. Comp. Physiol.* **160A,** 289–293.

Montgomery, J. C., and Coombs, S. (1992). Physiological characterization of lateral line function in the Antarctic fish *Trematomus bernacchii*. *Brain Behav. Evol.* **40,** 209–216.

Montgomery, J. C., Coombs, S., and Janssen, J. (1994). Aspects of structure and function in the anterior lateral line of six species of Antarctic fish of the suborder Notothenioidei. *Brain Behav. Evol.* **44,** 299–306.

Montgomery, J. C., and Sutherland, K. B. W. (1997). Sensory development of the Antarctic silverfish *Pleurogramma antarcticum*: A test for the ontogenetic shift hypothesis. *Polar Biol.* **18,** 112–115.

Moran, O., and Melani, R. (2001). Temperature-dependent conduction properties in Arctic fish peripheral nerves. *Polar Biol.* **24,** 9–15.

Morita, Y., Meyer-Rochow, V. B., and Uchida, K. (1997). Absolute and spectral sensitivities in dark- and light-adapted *Pagothenia borchgrevinki*, an Antarctic nototheniid fish. *Physiol. Behav.* **61,** 159–163.

Nozawa, Y., Iida, H., Fukushima, H., Ohki, K., and Ohnishi, S. (1974). Studies on *Tetrahymena* membranes: temperature-induced alterations in fatty acid composition of various membrane fractions in *Tetrahymena pyriformis* and its effect on membrane fluidity as inferred by spin-label study. *Biochim. Biophys. Acta* **367,** 134–147.

Pankhurst, N. W., and Montgomery, J. C. (1989). Visual function in four Antarctic nototheniid fishes. *J. Exp. Biol.* **142,** 311–324.

Pelster, B., and Randall, D. (1998). The physiology of the Root effect. *In* "Fish Respiration. Fish Physiology" (Perry, S. F., and Tufts, B. L., Eds.), pp. 113–139. Vol. 17. Academic Press, London..

Pockett, S., and Macdonald, J. A. (1986). Temperature dependence of neurotransmitter release in the Antarctic fish *Pagothenia borchgrevinki*. *Experientia* **42**, 414–415.

Precht, H. (1958). Concepts of the temperature adaptation of unchanging reaction systems of cold-blooded animals. *In* "Physiological Adaptation" (Prosser, C. L., Ed.), pp. 50–78. American Association for the Advancement of Science. Washington, DC.

Precht, H., Christophersen, J., Hensel, H., and Larcher, W. (1973). "Temperature and Life." Springer-Verlag, Berlin.

Rahmann, H., and Hilbig, R. (1980). Brain gangliosides are involved in the adaptation of Antarctic fish to extreme low temperatures. *Naturwissenschaften* **67**, 259–260.

Rahmann, H., Schneppenheim, R., Hilbig, R., and Lauke, G. (1984). Variability in brain ganglioside composition: a further molecular mechanism beside serum antifreeze-glycoproteins for adaptation to cold in Antarctic and arctic-boreal fishes. *Polar Biol.* **3**, 119–125.

Sinensky, M. (1974). Homeoviscous adaptation—a homeostatic process that regulates the viscosity of membrane lipids in *Escherichia coli*. *Proc. Natl. Acad. Sci. USA* **71**, 522–525.

Somero, G. N. (1991). Biochemical mechanisms of cold adaptation and stenothermality in Antarctic fish. *In* "Biology of Antarctic Fish" (Di Prisco, G., Maresca, B., and Tota, B., Eds.), pp. 232–247. Springer-Verlag, Berlin.

Somero, G. N. (1995). Proteins and temperature. *Ann. Rev. Physiol.* **57**, 43–68.

Somero, G. N., Fields, P. A., Hofmann, G. E., Weinstein, R. B., and Kawall, H. (1998). Cold adaptation and stenothermy in Antarctic fishes: what has been gained and what has been lost? *In* "Fishes of Antarctica: A Biological Overview" (Di Prisco, G., Pisano, E., and Clarke, A., Eds.), pp. 97–109. Springer-Verlag, Berlin.

Somero, G. N., and DeVries, A. L. (1967). Temperature tolerance of some Antarctic fishes. *Science* **156**, 257–258.

Tetens, V., Wells, R. M. G., and DeVries, A. L. (1984). Antarctic fish blood: respiratory properties and the effects of thermal acclimation. *J. Exp. Biol.* **109**, 265–279.

Wakeling, J. M., and Johnston, I. A. (1998). Muscle power output limits fast-start performance in fish. *J. Exp. Biol.* **201**, 1505–1526.

Wittenberg, J. B., and Wittenberg, B. A. (1974). The choroid *rete mirabile* of the fish eye. I. Oxygen secretion and structure: comparison with the swim bladder *rete mirabile*. *Biol. Bull.* **146**, 116–136.

Zecchinon, L., Claverie, P., Collins, T., D'Amico, S., Delille, D., Feller, G., Georlette, D., Gratia, E., Hoyoux, A., Meuwis, M. A., Sonan, G., and Gerday, C. (2001). Did psychrophilic enzymes really win the challenge? *Extremophiles* **5**, 313–321.

INDEX

A

ABW. *See* Antarctic bottom water
ACC. *See* Acetyl-CoA carboxylase
ACC. *See* Antarctic Circumpolar Current
Acetyl-CoA carboxylase (ACC), 108
Achiropsettidae (armless or southern
    flounders), 52, 53
Acid-base regulation, 119
Acipenseridae (sturgeons), 35, 53
Adenosine diphosphate (ADP), 99, 105
Adenosine triphosphate (ATP) production,
    86, 88, 93, 214
ADP. *See* Adenosine diphosphate
Adsorption-inhibition mechanism, of
    noncolligative antifreeze activity,
    171–4
Aerobic/anaerobic capacities, schematic
    of, 103
AFGPs. *See* Antifreeze glycoproteins
AFP. *See* Antifreeze peptide
Agonidae (poachers), 46, 53
Alaskan Arctic sculpin (Myoxocephalus
    verrucosus), 189
Allies (Salmonidae), 38, 66
Allosteric regulation, of hemoglobin-oxygen
    binding, 301–2
Alphastat hypothesis, of acid–base
    regulation, 119
Ammodytidae (sand lances), 51, 54
Anaerobic glycolytic capacity, 99
Anaerobic metabolism, in fish muscle, 98–9
Anarhichadidae (wolffishes), 49, 54
Anchor ice, 18, 19
Anguilla australis schmidtii (short-finned
    eels), 105
Anguilla dieffenbachii (long-finned eels), 105

Anguillidae (freshwater eels), 36, 54
Anotopteridae (daggertooth), 40, 54
Antarctic bottom water (ABW), 14
Antarctic Circumpolar Current (ACC), 13, 14
Antarctic Coastal Current, 14
Antarctic cods (Nototheniidae), 49–50, 63–4
Antarctic Continental Shelf, 17–18
Antarctic Convergence, 15
Antarctic dragonfishes (Bathydraconidae), 51,
    52, 55
Antarctic fishes, endogenous ice in, 175–9
Antarctic ice cap, weight of, 17
Antarctic intermediate water, 15
Antarctic marine region, 12–7, 28. *See also*
    Arctic marine region; Polar regions
    extreme environment of, 187–9
    ice cover of, 9
    sea ice cover of, 13
Antarctic marine system, 3–4
Antarctic Ocean, 6, 12
    surface water temperatures of, 2
Antarctic polar front, 15
Antarctic surface waters, temperature of,
    15–16
Antarctic toothfish (Dissostichus mawsoni),
    105, 129
Antarctic zoarcids, 106
Antarctica, 2, 4
    thermal isolation of, 13
Antifreeze glycoproteins (AFGPs), 163–7
    molecular weights of, 165
Antifreeze peptide (AFP), 167
Antifreeze protein solutions, freezing
    behavior of, 168–71
Antifreeze proteins (APs), 26, 155–93

Antifreeze proteins (APs) (*continued*)
  body fluid synthesis/distribution of, 181–4
  noncolligative freezing point lowering by, 168–74
  structures of, 166
  types of, 163–8
  undercooled fish fluids without, 185–6
Antioxidants, 86
  in polar fishes, 329–30
Anvers Island, 16
APs. *See* Antifreeze proteins
Arctic char (Salvelinus alpinus), blood viscosity of, 241
Arctic islands, 2, 3
Arctic marine region, 8, 28. *See also* Antarctic marine region; Polar regions
  environmental thermal variability, 189–91
  ice cover of, 9
  sea ice cover of, 13
Arctic marine system, 2–3
Arctic Ocean, 2, 3
  gyre of, 10
  shelf waters of, 11
  surface water temperatures of, 2
  thermal characteristics of, 8
  winter sea ice of, 8
Arctogadus glacialis (East Siberian cod), 189, 191, 221
Arenicola marina, 126
Armless or southern flounders (Achiropsettidae), 52, 53
Arrhenius activation energy (Ea), 126
Arrhenius equation, 126
Artedidraconidae (barbeled plunderfishes), 50, 51, 54
Atlantic cod (Gadus morhua), 98, 129, 190, 191, 248
Atlantic Ocean, 8
ATP production. *See* Adenosine triphosphate production

B

Baffin Bay, 11
Balance/hearing, inner ear organs for, 373–4
Balitoridae (River loaches), 37, 55
Banzare Bank, 29
Barbeled dragonfish (Stomiidae), 39, 67

Barbeled plunderfishes (Artedidraconidae), 50, 51, 54
Barbels, 373
Barracudinas (Paralepididae), 39, 64
Basking sharks (Cetorhinidae), 34, 56
Bathydraconidae (Antarctic dragonfishes), 51, 52, 55
Bathylagidae (deepsea smelts), 37–8, 55
Bathylutichthyidae, 45, 55
Bathymasteridae (ronquils), 47, 55
Bellingshausen Sea, 14
Benthic fishes, 8
Benthic North Sea zoarcids, 106
Bering Sea, 8
Bidirectional mechanistic links, 86
Bigscale fishes (Melamphaidae), 44, 62
Blackdevils (Melanocetidae), 43, 62
Blennioid ancestor, 20
Blood, respiratory functions of, 296–302
Blood buffering, carbon dioxide transport and, 298–9
Blood-gas transport, 296
  hemoglobin function and, 281–306
Blood oxygen affinity, Bohr effect data and, 300
Blood oxygen transport, 299–301
Blood viscosity
  cardiac work and, 240–5
  Hct and, 241
  of Myoxocephalus octodecemspinosus (longhorn sculpin), 241
  of Myoxocephalus scorpius (shorthorn sculpin), 241
  RBCs and, 241
  of Salvelinus alpinus (Arctic char), 241
  at two shear rates, 242
Body fluid viscosity, hemoglobin and, 290
Bohr effect data, blood oxygen affinity and, 300
Boreogadus saida (polar cod), 7, 11, 88, 189, 190, 191, 218
Bottom water, 6
Bovichtidae (thornfishes), 49, 55
Brains, of fish, 363–6
Branchial vasculature
  control of, 262–7
  exercise and, 265
  intraarterial serotonin injection and, 266
  schematic of, 264
Bristlemouths (Gonostomatidae), 39, 59

Burbots (Lotidae), 42–3, 61
Burst exercise capacity, 99. *See also*
  Exercise

C

Canadian Archipelago, 8, 10
Canadian Arctic Strait, 11
Canadian continental shelf, 30
Capelin (Mallotus mallotus), 225
Carapidae (pearlfishes), 43, 56
Carassius auratus (goldfish), 203
Cardiac afterload, 250
Cardiac contracitility, stroke volume and,
  258–62
Cardiac myocytes, 253–4
Cardiac output
  in icefish, 249
  scope of, 256
  vascular resistance values and, 246
Cardiac performance, in polar fish, 254
Cardiac power output, heart rate and, 260–1
Cardiac work, blood viscosity and, 240–5
Cardiovascular responses
  to exercise, 270–1
  to hypoxia, 271–2
  to stress, 268–70
Cardiovascular system, in the
  cold, 239–51
Carnitine palmitoyl-transferase (CPT),
  104, 105
Cataecholamines, role of, 302
Catalytic efficiency, 125
Catostomidae (suckers), 37, 56
CDW. *See* Circumpolar Deep Water
Cell protection, fish muscle and, 329–30
Cellular oxygen supply, biochemistry/
  physiology of, 109–13
Centrolophidae (medusafishes), 52, 56
Ceratiidae (seadevils), 43, 44, 56
Cetomimidae (flabby whalefishes), 44, 56
Cetorhinidae (basking sharks), 34, 56
Chaenocephalus aceratus (icefish), 99, 124,
  189, 204
  oxygen consumption in, 225
Champsocephalus esox, 204
Champsocephalus gunnari, 209
Channichthyidae (icefishes), 20, 51, 52, 56

Chars (Salmonidae), 38, 66
Chemical senses, of polar fishes, 371–3
Circulatory system
  control of, 239–73
  secondary, 211
Circumpolar Deep Water (CDW), 14, 16
Citrate synthase (CS), 94, 95
Citric synthase, 104
Clupeidae (herrings), 36–7, 56
Cobitidae (loaches), 37, 56
Cod, polar, 6, 11
Cods/haddocks (Gadidae), 42, 58, 83
Cold adaptation, of polar fishes, 88–113
Cold compensation, 97
Cold eurytherms, v. cold stenotherms, 90
Cold stenotherms, v. cold eurytherms, 90
Cold water acclimation/adaptation, 96
Conductivity temperature depth (CTD)
  logging instrument, 5
Congelation ice, 18
Copepods, 8
Cottidae (sculpins), 45–6, 57
COX. *See* Cytochrome c oxidase activity
CPT. *See* Carnitine palmitoyl-transferase
Creatine kinase, 99
Critical threshold temperatures (Tc), 85
CS. *See* Citrate synthase
CTD logging instrument. *See* Conductivity
  temperature depth logging
  instrument
Cusk-eels/brotulas (Ophidiidae), 43, 64
Cutaneous oxygen uptake, 212–14
Cyclopteridae (lumpsuckers), 46, 47, 58
Cyprinidae (minnows and carps), 37, 58
Cytochrome c oxidase (COX) activity, 94, 95
Cytosolic malate dehydrogenase activities, 108

D

Daggertooth (Anotopteridae), 40, 54
Dalatiidae (sleeper sharks), 35, 58
Davis Strait, 8, 10, 11
Deep-sea cods (Moridae), 40–1, 62
Deepsea smelts (Bathylagidae), 37–8, 55
Deepwater migration, 32
Dicentrarchus labrax (sea bass), 94
Dissostichus mawsoni (Antarctic toothfish),
  105, 129

DNA nucleotide sequences, 25
Dogfish sharks (Squalidae), 35, 67
Drake Passage, 20
Dreamers (Oneirodidae), 43, 64

E

Ea. *See* Arrhenius activation energy
East Greenland Current, 10, 11
East Siberian cod (Arctogadus glacialis), 189,
    191, 221
East-Wind Drift, 14
Eel cods (Muraenolepididae), 40, 41, 62
Eelpouts (Zoarces viviparus), 94, 100, 106,
    118, 119, 120, 129, 190
Ekman flow, 14
Eleginus gracilis (saffron cod), 166, 190
Ellesmere Island, 10, 11
Emperor penguin, 18
Endogenous ice
    in Antarctic fishes, 175–9
    fate of, 177–9
Energy budgets, functional capacities and,
    128–38
Energy supply, exercise and, 330–4
Enhanced fatty acid synthesis, 107
Environmental ice, in polar fish, 174–9
Environmental severity, serum hysteresis
    levels and, 187–91
Eocene fish, 20
EPOC. *See* Postexercise oxygen consumption
Esocidae (pikes and mudminnows), 37, 58
Euphausia crystallorophias, 17
Euphausia superba, 17, 106
Euphausiids, 8
Euphotic zone, 6
Euryhaline species, 6, 21
Eurythermal cold adaptation, 92
Evolution mechanisms, in physiological
    systems, 283–4
Exercise. *See also* Burst exercise capacity;
    Postexercise oxygen consumption
    (EPOC)
    branchial vasculature and, 265
    cardiovascular responses to, 270–1
    energy supply and, 330–4
    temperature and, 337–41
Exogenous ice, in polar fish, 174–5

F

Fast ice, 12, 13
Fathead sculpins (Psychrolutidae), 46, 65
Filchner Ice Shelf, 17, 160
Fin rays, 373
Fish families, cosmopolitan, 81
Flabby whalefishes (Cetomimidae), 44, 56
Fram Strait, 2, 3, 10, 11
Freezing avoidance strategies, 158–61
Freezing point, 156
    of seawater, 6
Freshwater eels (Anguillidae), 36, 54
Functional protein levels, gene expression
    regulation and, 120–3
Fundulus grandis (Gulf killifish), 106
Fundulus heteroclitus, 120

G

Gadidae (cods and haddocks), 42, 58, 83
Gadus-Arctogadus clade, 26
Gadus-Microgadus clade, 26
Gadus morhua (Atlantic cod), 98, 129, 190,
    191, 248
Gadus ogac (Greenland cod), 98
Gadus ogac (Labrador Fjord cod), 190
Gadus uvak (Greenland cod), 221
    oxygen comsumption in, 222
GAPDH. *See* Glyceraldehyde-
    phosphatedehydrogenase
Gasterosteidae (sticklebacks), 44–5, 59
Gempylidae (snake mackerels), 52, 59
Gene expression regulation, functional
    protein levels and, 120–3
Gills
    acrifix casts of, 210
    filaments of, 212
    lamellae of, 206–7
    structure/morphometrics of, 204–12
Global warming, 10, 62, 80
Glyceraldehyde-phosphatedehydrogenase
    (GAPDH), 127
Gobionotothen gibberifrons, 94, 99, 100
Gobionotothen nudifrons, 106
Goldfish (Carassius auratus), 203
Gondwanaland, 33
    breakup of, 20

Gonostomatidae (bristlemouths), 39, 59
Goosefish (Lophiodes), 204
Graylings (Salmonidae), 38, 66
Graylings (Thymallinae), 38
Greenland, 10, 11
Greenland cod (Gadus ogac), 98
Greenland cod (Gadus uvak), 221
Greenland Sea, 2, 3, 8, 10
Greenlings (Hexagrammidae), 45, 59
Grenadiers/rattails (Macrouridae), 40, 41,
   61–2
Growth performance, in notetheioid fish, 130
Gulf killifish (Fundulus grandis), 106
Gulf Stream, 10
Gunnels (Pholidae), 48, 49, 64
Gyre, of Arctic Ocean, 10

H

Hagfishes (Myxinidae), 34, 63
Harpagiferidae (spiny plunderfishes), 51, 59
Hct. See Hematocrit
Heart comparison, 249
Heart, fish
   anatomy of, 251–3
   control of, 253–62
   heart rate control in, 254–8
Heart rate
   acute temperature and, 259
   cardiac power output and, 260–1
   temperature and, 257
Heart rate control, in fish heart, 254–8
Heart rate data, on polar fishes, 255
Heat shock proteins, 86
Hematocrit (Hct)
   blood viscosity and, 241
   splenic regulation of, 245–51
   stressors effects on, 247
Hemitripteridae (sailfin sculpins), 46, 59
Hemoglobin
   body fluid viscosity and, 290
   functional properties of, 291–6
   root effect, 294–6
   stress and, 291
   temperature effects on, 293–4
Hemoglobin capacity, for oxygen transport,
   287–90

Hemoglobin components, expression/
   significance of, 284–91
Hemoglobin concentrations, normal values
   for, 243
Hemoglobin-oxygen binding, of allosteric
   regulation, 301–2
Herrings (Clupeidae), 6, 36–7, 56
Hexagrammidae (greenlings), 45, 59
Hexose monophosphate shunt (HMPS)
   pathway, 108
HIF-1. See Hypoxia-inducible factor 1
HMPS pathway. See Hexose monophosphate
   shunt pathway
Homeostasis, blood's role in, 296–302
Homogenous nucleation, temperature of, 159
Hydroxy-acylcoenzyme A, 104
Hyposmotic teleost fish, freezing challenge to,
   157–8
Hypoxia
   cardiovascular responses to, 271–2
   events of, 87
Hypoxia-inducible factor 1 (HIF-1), 87

I

Ice algae, 16
Ice cover
   in Antarctic marine region, 9
   in Arctic marine region, 9
Ice formation, 160
Ice-free habitat migration, as freezing
   avoidance strategy, 158–9
Ice propagation barrier, integument as,
   179–81
Ice shelves, 160
Icefish (Chaenocephalus aceratus), 99, 124,
   189, 204
Icefish gill arch, diagram of, 209
Icefishes (Channichthyidae), 20, 51, 52, 56
   without hemoglobin, 286–7
IDH. See Isocitrate dehydrogenase
Illex illecebrosus, 101
Inner ear organs, of balance/hearing, 373–4
Integument, as ice propagation barrier,
   179–81
Ion/acid-base regulation, constraints in,
   113–20
Isocitrate dehydrogenase (IDH), 127

K

Katabatic winds, 12
Katsuwonas pelamis (tuna), 92
Krill, 17, 329
  Antarctic, 17
Krogh diffusion constant, 109

L

Labrador Current, 11
Labrador Fjord cod (Gadus ogac), 190
Lactate accumulation, in zoarces viviparus
    (eelpouts), 97
Lactate clearance, 98
Lactate dehydrogenase (LDH), 99, 100, 101,
    103, 104, 127
Lactate dehydrogenase-B isoform, 120
Lamellar perfusion, 209
Lamnidae (mackerel sharks), 34, 59
Lampreys (Petromyzontidae), 34, 64
Lampridae (opahs), 40, 59
Lanternfishes (Myctophidae), 40, 62–3
Larval fish, 102
LDH. See Lactate dehydrogenase
Liparidae (snailfishes), 47, 59–61
Lipid metabolism, substrates in, 104–9
Lipoprotein lipase, 105
Loaches (Cobitidae), 37, 56
Loligo pealei, 101
Long-finned eels (Anguilla dieffenbachii), 105
Longhorn sculpin (Myoxocephalus
    octodecemspinosus), 167
Lophiodes (goosefish), 204
Lotidae (burbots), 42–3, 61
Lumpsuckers (Cyclopteridae), 46, 47, 58
Lycodichthys dearborni, 167, 185, 188
  AP in, 161

M

Mackerel (Scomber), 205
Mackerel sharks (Lamnidae), 34, 59
Macrouridae (grenadiers or rattails), 40, 41,
    61–2

Malic enzyme, 108
Mallotus mallotus (capelin), 225
Mallotus villosus (capelin), 106
Marine-derived tocopherol (MDT), 329–30
Mb. See Myoglobin
MCA. See Metabolic cold adaptation
MCHC. See Mean cell hemoglobin
    concentration
McMurdo Sound, 7, 13, 160
  environment of, 18–9
  water temperatures of, 16
MDT. See Marine-derived tocopherol
Mean cell hemoglobin concentration
    (MCHC), 241
  temperature and, 289
Mechanosenses, of polar fishes, 372, 373–5
Mechanosensory lateral line, 374–5, 376
Medusafishes (Centrolophidae), 52, 56
Melamphaidae (bigscale fishes), 44, 62
Melanocetidae (blackdevils), 43, 62
Melanonidae (pelagic cods), 42, 62
Melt water, from pack ice, 16
Membrane functions/capacities, 113
MEPCs. See Miniature endplate currents
Metabolic cold adaptation (MCA), 204,
    296–8
  concept of, 217–30
  traditional interpretation of, 230
Metabolic functions, molecular physiology of,
    120–8
Metabolic gene expression/design, 80–139
Metabolic rates, 214–33
  divisions of, 214–15
  factors in, 215–16
  of polar fishes, 226–9
  temperature and, 216–17
Metabolism, in cold environments, 80
Microgadus tomcod, 190, 191
Microstomatidae, 37, 62
Miniature endplate currents (MEPCs),
    359–62
  thermal properties of, 361
Minnows/ carps (Cyprinidae), 37, 58
Mitochondria, 86
Mitochondrial ATP synthesis, 89
Mitochondrial cristae, 93
Mitochondrial densities, 92, 93, 96
Mitochondrial uncoupling protein
    (UCP2), 126

Molecular physiology, of metabolic functions, 120–8
Moridae (deep-sea cods), 40–1, 62
Morphometrics, gill structure and, 204–12
Muraenolepididae (eel cods), 40, 41, 62
Muscle fiber diameters, 323
Muscle, fish
  anatomy of, 319–27
  buffering effect of, 330
  cell protection and, 329–30
  fibers of, 319–322
  mechanical/physiological properties of, 327–9
  swimming and, 334–7
  types of, 319–20
Muscle growth, feeding and, 324
Myctophidae (lanternfishes), 40, 62–3
Myofibrillar ATPase, 100
Myoglobin (Mb), 94
Myoxocephalus, 21
Myoxocephalus octodecemspinosus
    (longhorn sculpin), 167
  blood viscosity of, 241
Myoxocephalus scorpius (shorthorn sculpin), 125
  blood viscosity of, 241
Myoxocephalus verrucosus (Alaskan Arctic
    sculpin), 189
Myxinidae (hagfishes), 34, 63

N

Nares Strait, 11
Nervous impulse, excitability/conduction of, 356–8
Nervous system
  cellular function of, 353–6
  of polar fishes, 351–78
  temperature compensation in, 353–63
Neuromuscular transmission, 359–62
Newnesi, 94
Noncolligative antifreeze activity,
    adsorption-inhibition
    mechanism of, 171–4
Norwegian Seas, 8
Notacanthidae (spiny eels), 35, 63
Notosudidae (waryfishes), 39, 63

Nototheioid fish, growth performance
  in, 130
Notothenia angustata, 118
Notothenia coriiceps, 99, 189
Notothenia gibbifrons, 204
Notothenia neglecta, 231
Nototheniidae (Antarctic cods), 49–50, 63–4
Notothenioidae, 12, 81
  monophyletic perciform suborder
    of, 19–20

O

Ocean currents, in Arctic marine region, 10
Okhotsh Seas, 8
Olfactory organs, of polar fishes, 371–3
Oncorhynchus mykiss (steelhead trout), 124
Oneirodidae (dreamers), 43, 64
Opahs (Lampridae), 40, 59
Ophidiidae (cusk-eels and brotulas), 43, 64
Oreosomatidae (oreos), 44, 64
Organismal freeze avoidance, mechanism of, 191–3
Organismal freezing points, 161–3
Osmeridae (smelts), 38, 64
β–oxidation, 104–9
Oxygen-carrying capacity, 205, 210
Oxygen consumption, 205
  in Chaenocephalus aceratus, 225
  of exercising fish, 230–231
  in Gadus uvak (Greenland cod), 222
  of rainbow trout, 220
Oxygen deficiency, 83, 85
Oxygen solubility, temperature and, 7
Oxygen transport, 211
  evolution of, 303–5
  hemoglobin capacity for, 287–90

P

Pachycara antarcticum, 106, 129
Pachycara brachycephalum, 94, 97, 100, 118,
    119, 126, 185, 188, 191
  AFPs in, 168

Pachycara brachycephalum (*continued*)
   AP in, 161
   CS activity in, 121
   protein synthesis capacities in, 135
   thermal tolerance of, 84
Pack ice, 7, 12, 13
   melt water from, 16
Pagothenia borchgrevinki, 18, 97, 99, 106,
      175, 176, 181, 230, 231
Paralepididae (barracudinas), 39, 64
Paraliparis devriesi (snail fish), 160
Pearleyes (Scopelarchidae), 39, 66
Pearlfishes (Carapidae), 43, 56
Pejus temperatures, 83, 85
Pelagic cods (Melanonidae), 42, 62
Perches (Percidae), 47, 64
Percidae (perches), 47, 64
Percopsidae (trout-perches), 40, 64
Petromyzontidae (lampreys), 34, 64
PFK. *See* Phosphofructokinase
Pholidae (gunnels), 48, 49, 64
Phosphofructokinase (PFK), 127
Phytoplankton, 6, 8, 11, 16
Phytoplankton grazers, 7
Pikes/ mudminnows (Esocidae), 37, 58
Platelet ice, 18
   formation of, 18
Pleuragramma antarcticum, 92, 204
Pleuronectidae (righteye flounders),
      52–3, 65
Poachers (Agonidae), 46
Polar cod (Boreogadus saida), 7, 11, 88, 189,
      190, 191, 218
Polar fishes
   biogeography of, 30–3
   brains of, 363–6
   cardiac performance in, 254
   chemical senses of, 371–3
   cold adaptation of, 88–113
   distribution patterns of, 30–2
   environmental ice in, 174–9
   exogenous ice in, 174–5
   family classification of, 33–53
   heart rate data on, 255
   historical description of, 29–30
   low-temperature adaptations of, 354
   mechanosenses of, 372, 373–5
   metabolic design of, 81–8
   metabolic mode of, 88–113
   metabolic rates of, 226–9

   muscular performance levels of, 88–113
   olfactory organs of, 371–3
   origin of, 32–3
   oxygen supply enhancement in, 90
   phylogenetic analyses of, 25
   physiological definitions of, 27–9
   sensory physiology of, 366–75
   species diversity limitations for, 33
   species number of, 29
   systematics of, 25–53
   thermal tolerance of, 81–8
   visual function of, 366–71
Polar Front, 12
Polar marine system, 2–4
Polar regions. *See also* Antarctic marine
      region; Arctic marine region
   glacial/geological history of, 19–22
Polar seawater. *See* Seawater
Polynyas, 13
Postexercise oxygen consumption (EPOC),
      98. *See also* Exercise
Practical salilnity units (PSUs), 6
Pricklebacks (Stichaeidae), 48, 49, 67
Primary production, 11, 16
   seasonal light availability and, 7–8
Protein structure/function, temperature effects
      on, 123–8
Proton leakage, 92
Prowfish (Zaproridae), 49, 67
Pseudochaenichthys georgianus, 204
Pseudopleuronectes americanus (winter
      flounder), AP in, 163
PSUs. *See* Practical salilnity units
Psychrolutidae (fathead sculpins), 46, 65
Pycnocline, 6

                              R

Rajidae (skates), 35, 36, 65–6
RBCs. *See* Red blood cells
Red blood cells (RBCs), blood viscosity
      and, 241
Reflex/fast motor behavior, temperature
      compensation of, 362–3
Respiratory system, of polar fish, 203–14
Respirometer, intermitten-flow-through, 224
Righteye flounders (Pleuronectidae),
      52–3, 65

River loaches (Balitoridae), 37, 55
Riverine input, 6
Rockfishes (Sebastidae), 45, 66–7
Ronne Ice Shelf, 17
Ronquils (Bathymasteridae), 47, 55
Root effect, hemoglobin, 294–6
Ross Ice Shelf, 17, 160
Ross Sea, 11, 14, 17
    water temperatures of, 16
Rutilus rutilus, 96, 102

S

Saffron cod (Eleginus gracilis), 166, 190
Sailfin sculpins (Hemitripteridae), 46, 59
Salinity/freezing, of seawater, 5–6
Salmonidae (chars, graylings, salmons, trouts,
        whitefishes, and allies), 38, 66
Salmons (Salmonidae), 38, 66
Salvelinus alpinus (Arctic char), blood
        viscosity of, 241
Sand lances (Ammodytidae), 51, 54
Scomber (mackerel), 205
Scopelarchidae (pearleyes), 39, 66
Sculpins (Cottidae), 45–6, 57
SDH. See Succinate dehydrogenase
Sea bass (Dicentrarchus labrax), 94
Sea ice cover
    of Antarctic marine region, 13
    of Arctic marine region, 13
Seadevils (Ceratiidae), 43, 44, 56
Seasonal light availability, primary
        production and, 7–8
Seawater
    density/viscosity of, 5
    properties of, 4
    salinity/freezing of, 5–6
    surface layer of, 6
    temperature effects on, 5
    temperature of, 178
Sebastidae (rockfishes), 45, 66–7
Sensory physiology, of polar fishes, 366–75
Serum hysteresis levels, environmental
        severity and, 187–91
Shear rate, temperature and, 244
Short-finned eels (Anguilla australis
        schmidtii), 105
Shorthorn sculpin (Myoxocephalus
        scorpius), 125

Siberian continental shelf, 30
Signaling mechanisms, 87
Signy Island, 16
Skates (Rajidae), 35, 36, 65–6
Sleeper sharks (Dalatiidae), 35, 58
Smelts (Osmeridae), 6, 38, 64
SMRs. See Standard metabolic rates
Snail fish (Paraliparis devriesi), 160
Snailfishes (Liparidae), 47, 59–61
Snake mackerels (Gempylidae), 52, 59
South Georgia Island, 16
South Georgia waters, 27
South Polar waters, 27
Southern Ocean, 2, 4, 12
    cross section of, 15
    temperature of, 16
Spiny eels (Notacanthidae), 35, 63
Spiny plunderfishes (Harpagiferidae), 51, 59
Splenic contraction, 246–7
Squalidae (dogfish sharks), 35, 67
Standard metabolic rates (SMRs), 88, 90, 92,
        101, 103, 107
Steelhead trout (Oncorhynchus mykiss), 124
Stenohaline, 20
Stenothremal, 20
Stichaeidae (pricklebacks), 48, 49, 67
Sticklebacks (Gasterosteidae), 44–5, 59
Stomiidae (barbeled dragonfish), 39, 67
Stress
    cardiovascular responses to, 268–70
    hemoglobin and, 291
Stroke volume, cardiac contracitility and,
        258–62
Sturgeons (Acipenseridae), 35, 53
Sub-ice platelet layer, 18, 19
Subtropical Convergence, 12
Succinate dehydrogenase (SDH), 94
Suckers (Catostomidae), 37, 56
Surface layer, of seawater, 6
Svalbard, 8
Swimming, fish muscle and, 334–7
Systematics, of polar fishes, 25–53
Systemic vasculature, control of, 267–8

T

Tactile senses, of polar fishes, 373
Teleost, 107

Temperature
  exercise and, 337–41, 339, 340
  heart rate and, 257
  hemoglobin and, 293–4
  of homogenous nucleation, 159
  MCHC and, 289
  metabolic rates and, 216–17
  oxygen solubility and, 7
  protein structure/function and, 123–8
  shear rate and, 244
Temperature compensation
  in nervous system, 356–63
  of reflex/fast motor behavior, 362–3
Thermal specialization, in marine
    environments, 80–1
Thermal tolerance
  of pachycara brachycephalum, 84
  of polar fishes, 81–8
Thornfishes (Bovichtidae), 49, 55
Threefin blennies (Tripterygiidae), 52, 67
Thysanoessa macrura, 106
Trematomus bernacchii, 118, 188, 222, 240
  oxygen comsumption in, 223
Trematomus fishes, 18
Trematomus hansoni, 223
Trematomus loennbergii, 161, 188
Trematomus scotti, 129
Tripterygiidae (threefin blennies), 52, 67
Troponin C, 124
Trout-perches (Percopsidae), 40, 64
Tuna (Katsuwonas pelamis), 92

U

UCP2. See Mitochondrial uncoupling protein
Undercooled fish fluids, without APs, 185–6
Undercooling, 159–61

V

Vascular geometry, vascular resistance and,
    251
Vascular resistance
  cardiac output and, 246
  vascular geometry and, 251
Visual function, of polar fishes, 366–71
Vitamin C, 329
Vitamin E, 329

W

Waryfishes (Notosudidae), 39, 63
Weddell Sea, 11, 14, 17
West Greenland Curent, 10, 11
Whitefishes (Coregonidae), 38
Whitefishes (Salmonidae), 38, 66
Winter flounder (Pseudopleuronectes
    americanus), AP in, 163
Winter sea ice, of Arctic Ocean, 8
Wolffishes (Anarhichadidae), 49, 54

Z

Zaproridae (prowfish), 49, 67
Zoarces viviparus (eelpouts), 94, 100, 106,
    118, 119, 120, 129, 190
  COX activity in, 121
  CS activity in, 121
  lactate accumulation in, 97
  temperature acclimation in, 122
Zoarcidae, 47–8, 67–9
Zooplankton, 6

OTHER VOLUMES IN THE
FISH PHYSIOLOGY SERIES

VOLUME 1   Excretion, Ionic Regulation, and Metabolism
            Edited by W. S. Hoar and D. J. Randall

VOLUME 2   The Endocrine System
            Edited by W. S. Hoar and D. J. Randall

VOLUME 3   Reproduction and Growth: Bioluminescence,
            Pigments, and Poisons
            Edited by W. S. Hoar and D. J. Randall

VOLUME 4   The Nervous System, Circulation, and Respiration
            Edited by W. S. Hoar and D. J. Randall

VOLUME 5   Sensory Systems and Electric Organs
            Edited by W. S. Hoar and D. J. Randall

VOLUME 6   Environmental Relations and Behavior
            Edited by W. S. Hoar and D. J. Randall

VOLUME 7   Locomotion
            Edited by W. S. Hoar and D. J. Randall

VOLUME 8   Bioenergetics and Growth
            Edited by W. S. Hoar, D. J. Randall, and J. R. Brett

VOLUME 9A   Reproduction: Endocrine Tissues and Hormones
            Edited by W. S. Hoar, D. J. Randall, and E. M. Donaldson

VOLUME 9B   Reproduction: Behavior and Fertility Control
            Edited by W. S. Hoar, D. J. Randall, and E. M. Donaldson

VOLUME 10A Gills: Anatomy, Gas Transfer, and Acid-Base Regulation
            Edited by W. S. Hoar and D. J. Randall

VOLUME 10B Gills: Ion and Water Transfer
            Edited by W. S. Hoar and D. J. Randall

VOLUME 11A The Physiology of Developing Fish: Eggs and Larvae
Edited by W. S. Hoar and D. J. Randall

VOLUME 11B The Physiology of Developing Fish: Viviparity and
Posthatching Juveniles
Edited by W. S. Hoar and D. J. Randall

VOLUME 12A The Cardiovascular System
Edited by W. S. Hoar, D. J. Randall, and A. P. Farrell

VOLUME 12B The Cardiovascular System
Edited by W. S. Hoar, D. J. Randall, and A. P. Farrell

VOLUME 13 Molecular Endocrinology of Fish
Edited by N. M. Sherwood arid C. L. Hew

VOLUME 14 Cellular and Molecular Approaches to Fish
Ionic Regulation
Edited by Chris M. Wood and Trevor J. Shuttleworth

VOLUME 15 The Fish Immune System: Organism, Pathogen,
and Environment
Edited by George Iwama and Teruyuki Nakanishi

VOLUME 16 Deep Sea Fishes
Edited by D. J. Randall and A. P. Farrell

VOLUME 17 Fish Respiration
Edited by Steve F. Perry and Bruce Tufts

VOLUME 18 Muscle Growth and Development
Edited by Ian A. Johnson

VOLUME 19 Tuna: Physiology, Ecology, and Evolution
Edited by Barbara A. Block and E. Donald Stevens

VOLUME 20 Nitrogen Excretion
Edited by Patricia A. Wright and Paul M. Anderson

VOLUME 21 Tropical Fishes
*Edited by Adalberto L. Val, Vera Maria F. De Almeida-Val, and
David J. Randall*

VOLUME 22 Polar Fishes
Edited by Anthony P. Farrell and John F. Steffensen

Printed and bound by CPI Group (UK) Ltd, Croydon, CR0 4YY

08/05/2025

01864951-0002